Integrating Face and Voice in Person Perception

Pascal Belin · Salvatore Campanella
Thomas Ethofer
Editors

Integrating Face and Voice in Person Perception

Springer

Editors
Pascal Belin
Voice Neurocognition Laboratory
Institute of Neuroscience and Psychology
College of Medical, Veterinary
and Life Sciences
University of Glascow
Glascow, UK

International Laboratories for Brain
Music and Sound (BRAMS)
Université de Montréal &
McGill University
Montreal, Quebec, Canada

Thomas Ethofer
Department of General Psychiatry
University of Tübingen
Tübingen, Germany

Salvatore Campanella
Laboratory of Psychological Medicine
Free University of Brussels
Brussels, Belgium
and
Psychiatry Department (EEG)
CHU Brugmann
The Belgian Fund for Scientific
Research (FNRS)
Brussels, Belgium

ISBN 978-1-4614-3584-6 ISBN 978-1-4614-3585-3 (eBook)
DOI 10.1007/978-1-4614-3585-3
Springer New York Heidelberg Dordrecht London

Library of Congress Control Number: 2012942182

Printed on acid-free paper

Springer is part of Springer Science+Business Media (www.springer.com)

Foreword

Most of our social interactions involve combining information from both the face and voice of other persons: speech information, and also crucial nonverbal information on a person's identity and affective state. The cerebral bases of the multimodal integration of speech have been intensively investigated; by contrast, only a few studies have focused on *nonverbal aspects of face–voice integration.*

Until recently, the quite different approaches used by investigators in auditory and visual perception have hindered efforts at bringing these two fields together: auditory perception largely concentrated on speech processing, while visual perception essentially investigated object and face recognition. Such emphasis on different types of information in the two modalities has not facilitated the understanding of how social signals from the face and voice are combined and integrated in everyday behaviour.

In an effort towards a broader perspective on these two research fields, we noted, as several other authors before us did, that information carried by faces and voices—speech, affect, identity—is largely similar in kind (if not in the underlying physical signals) and proposed that this similarity could extend to the underlying cerebral processing functional architecture. We suggested that Bruce and Young's (1986) influential model of face processing could be meaningfully extended to the processing of voice information.

In the "auditory face" model of voice processing (Belin, Fecteau, & Bedard, 2004), we proposed that the three main types of vocal information—speech, affect, identity—are processed, as for faces, in three partially independent functional pathways that interact during normal behaviour but can be impaired selectively from one another. In a subsequent refining of this model (Campanella & Belin, 2007) we proposed that multimodal face–voice integration occurs between corresponding pathways across the visual and auditory modalities (Fig. 1). This model, as most models, is probably wrong, but it was proposed in the hope that it could provide a useful heuristic to guide further research on the way our brain combines signals from the face and voice of other persons.

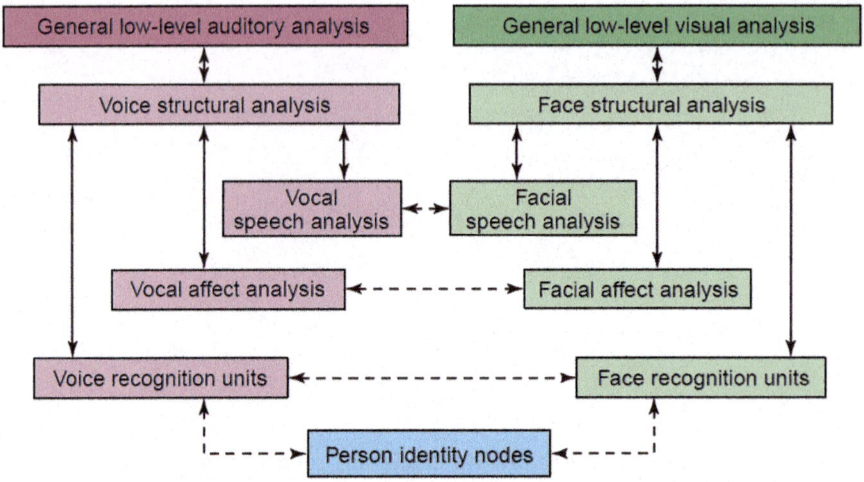

Fig. 1 The "auditory face" model of voice perception. The *right-hand part* of the figure is adapted from Bruce and Young's (1986) model of face perception. The *left-hand part* proposes a similar functional organization for voice processing. *Dashed arrows* indicate multimodal interactions. Reprinted (permission pending) from Belin, Fecteau, and Bedard (2004)

The present book aims to highlight recent exciting advances in the investigation of the behavioural and cerebral bases of face–voice integration in the context of person perception, focusing on the integration of *affective* and *identity* information. Several research domains are brought together: behavioural and neuroimaging work in healthy adult humans, and also evidence from several other relevant research fields to provide complementary insights.

Part I: Evolution and development provides both evolutionary and developmental perspectives on face–voice integration. Do other animals show evidence of face–voice integration? And how early do these abilities develop in human infants? *Ipek G. Kulahci* and *Asif A. Ghazanfar* review evidence in primates showing that multisensory processes enhance multiple types of behaviour and that cortical processes are largely multisensory by default. *Akihiro Izumi* reviews research paradigms employed in probing auditory–visual conceptual representations and highlights evidence for clear multimodal integration processes in non-human primates. *Maria M. Diehl* and *Liz M. Romanski* examine integrative properties of neurons in the macaque ventro-lateral prefrontal cortex and show that these cells respond optimally to face and vocalization stimuli, exhibiting multisensory enhancement or suppression when face and vocalization stimuli are combined. *Ross Flom* reviews evidence on the development face and voice perception and integration and interprets this evidence in the context of the "intersensory redundancy hypothesis". *Tobias Grossman* presents evidence on the development of face–voice integration with a focus on affective information and shows that at least by the age of 7 months, infants reliably integrate and recognize emotional information across face and voice.

Part II: Identity information examines the integration of identity information from faces and voices, which play a central role in our social interactions. Indeed, both faces and voices are rich in information on a person's identity and gender. *Stefan R. Schweinberger* reviews evidence illustrating these audiovisual interactions during familiar speaker recognition. *Rebecca Watson* and *colleagues* focus on face–voice integrative processes related to gender, using dynamic, ecological "morphed video" stimuli. *Frederic Joassin* and *Salvatore Campanella* argue that the cross-modal integration of identity and gender information through faces and voices involve similar networks. While the above chapters report results obtained using functional magnetic resonance imaging, *Aina Puce* reviews evidence on face–voice integrative processes obtained using neurophysiological techniques.

Part III: Affective Information is dedicated to the integration of emotional information in voice (affective prosody) and face (emotional facial expressions). *Gilles Pourtois* and *Monica Dhar* review theoretical models and provide behavioural data arguing that perception of emotion can be conceptualized as an object-based multisensory phenomenon. *Tobias Brosch* and *Didier Grandjean* discuss cross-modal influences of emotion on spatial attention and their neural correlates as determined by electrophysiological and brain imaging methods. *Benjamin Kreifelts, Dirk Wildgruber* and *Thomas Ethofer* provide an overview on methodological approaches for studying audiovisual integration of emotion using functional magnetic resonance imaging and discuss the advantages and weaknesses of different analysis strategies. *Beatrice de Gelder, Bernard M.C. Stienen* and *Jan Van den Stock* review the explosion of research on emotional face–voice integration since their pioneering innovative work in this domain. They extend the review to abnormal affective processing in schizophrenia and autism—a perfect transition to the last part of the book.

Part IV: Impairment illustrates the importance of cross-modal face–voice interactions by showing their impact on people's life when these processes are altered. *Pierre Maurage, Scott Love* and *Fabien D'Hondt* provide evidence for a cross-modal deficit when chronic alcoholic patients are confronted with emotional stimuli. *Barbra Zupan* stresses the role of audition in processing of bimodal cues of speech and emotion in individuals with hearing loss. *Dyna Delle-Vigne, Charles Konreich, Paul Verbank* and *Salvatore Campanella* suggest that emotional cross-modal stimulations, through the use of cognitive event-related potentials, may help discriminate more clearly between different psychiatric populations.

These contributions reflect a dynamic, emerging research field situated at the conjunction between two active currents of neuroscience and psychology: Social Neuroscience and Multimodal Integration. We hope they illustrate important recent advances in these exciting domains and constitute an interesting reading.

Glasgow, UK — Pascal Belin
Brussels, Belgium — Salvatore Campanella
Tubingen, Germany — Thomas Ethofer

References

Belin, P., Fecteau, S., & Bedard, C. (2004) Thinking the voice: neural correlates of voice perception. *Trends in Cognitive Sciences*, 8, 129–135.

Bruce, V., & Young, A. (1986) Understanding face recognition. *British Journal of Psychology,* 77, 305–327.

Campanella, S., & Belin, P. (2007) Integrating face and voice in person perception. *Trends in Cognitive Sciences*, 11, 535–543.

Contents

Part I Evolution and Development

1 **Multisensory Recognition in Vertebrates (Especially Primates)** 3
Ipek G. Kulahci and Asif A. Ghazanfar

2 **Cross-Modal Representation in Humans and Nonhuman Animals: A Comparative Perspective** 29
Akihiro Izumi

3 **Representation and Integration of Faces and Vocalizations in the Primate Ventral Prefrontal Cortex** 45
Maria M. Diehl and Lizabeth M. Romanski

4 **Intersensory Perception of Faces and Voices in Infants** 71
Ross Flom

5 **The Early Development of Processing Emotions in Face and Voice** 95
Tobias Grossman

Part II Identity Information

6 **Audiovisual Integration in Speaker Identification** 119
Stefan R. Schweinberger

7 **Audiovisual Integration of Face–Voice Gender Studied Using "Morphed Videos"** 135
Rebecca Watson, Ian Charest, Julien Rouger, Christoph Casper, Marianne Latinus, and Pascal Belin

8 **Cross-Modal Integration of Identity and Gender Information Through Faces and Voices Involves a Similar Cortical Network** 149
Salvatore Campanella and Frédéric Joassin

9 **Neurophysiological Correlates of Face and Voice Integration** 163
Aina Puce

Part III Affective Information

10 **Integration of Face and Voice During Emotion Perception: Is There Anything Gained for the Perceptual System Beyond Stimulus Modality Redundancy?** 181
Gilles Pourtois and Monica Dhar

11 **Cross-Modal Modulation of Spatial Attention by Emotion** 207
Tobias Brosch and Didier Grandjean

12 **Audiovisual Integration of Emotional Information from Voice and Face** 225
Benjamin Kreifelts, Dirk Wildgruber, and Thomas Ethofer

13 **Emotions by Ear and by Eye** 253
Beatrice de Gelder, Bernard M.C. Stienen, and Jan Van den Stock

Part IV Impairment

14 **Crossmodal Integration of Emotional Stimuli in Alcohol Dependence** 271
Pierre Maurage, Scott Love, and Fabien D'Hondt

15 **The Role of Audition in Audiovisual Perception of Speech and Emotion in Children with Hearing Loss** 299
Barbra Zupan

16 **Searching for a Greater Sensitivity of Cognitive Event-Related Potentials Through a Crossmodal Procedure for a Better Clinical Use in Psychiatry** 325
D. Delle- Vigne, C. Kornreich, P. Verbanck, and S. Campanella

Index 369

Contributors

Pascal Belin Voice Neurocognition Laboratory, Institute of Neuroscience and Psychology, College of Medical, Veterinary and Life Sciences, University of Glasgow, Glasgow, UK

International Laboratories for Brain, Music and Sound (BRAMS), Université de Montréal & McGill University, Montreal, Quebec, Canada

Tobias Brosch New York University, Department of Psychology, New York, NY, USA

Salvatore Campanella Laboratory of Psychological Medicine, Free University of Brussels, Brussels, Belgium

CHU Brugmann, Psychiatry Department (EEG), The Belgian Fund for Scientific Research (FNRS), Brussels, Belgium

Christoph Casper Department of Business Administration and Human Resource Management, University of Cologne, Cologne, Germany

Ian Charest MRC Cognition and Brain Sciences Unit, Cambridge, UK

D. Delle-Vigne Laboratory of Psychological Medicine, Free University of Brussels, Brussels, Belgium

Fabien D'Hondt Lille Nord de France University, Lille, France

Functional Neurosciences and Pathology Lab, UDSL, Lille, France

Monica Dhar Department of Experimental-Clinical and Health Psychology, Ghent University, Ghent, Belgium

Maria M. Diehl Department of Neurobiology & Anatomy and Center for Navigation and Communication Sciences, University of Rochester, Rochester, NY, USA

Thomas Ethofer Department of General Psychiatry, University of Tübingen, Tübingen, Germany

Ross Flom Department of Psychology, Brigham Young University, Provo, UT, USA

Beatrice de Gelder Cognitive and Affective Neuroscience Laboratory, Tilburg University, Tilburg, The Netherlands

Asif A. Ghazanfar Department of Ecology & Evolutionary Biology, Princeton University, Princeton, NJ, USA

Department of Psychology, Princeton University, Princeton, NJ, USA

Neuroscience Institute, Princeton University, Princeton, NJ, USA

Didier Grandjean Department of Psychology, Centre Interfacultaire en Sciences Affectives (NCCR), University of Geneva, Geneva, Switzerland

Tobias Grossman Centre for Brain and Cognitive Development, Birkbeck, University of London, Bloomsbury, UK

Akihiro Izumi Primate Research Institute, Kyoto University, Inuyama, Japan

Frédéric Joassin Clinique de la mémoire, CHU Ambroise Paré, Mons, Belgium

C. Kornreich Laboratory of Psychological Medicine, Free University of Brussels, Brussels, Belgium

Benjamin Kreifelts Department of General Psychiatry, University of Tübingen, Tübingen, Germany

Ipek G. Kulahci Department of Ecology & Evolutionary Biology, Princeton University, Princeton, NJ, USA

Marianne Latinus Voice Neurocognition Laboratory, Institute of Neuroscience and Psychology, College of Medical, Veterinary and Life Sciences, University of Glasgow, Glasgow, UK

Scott Love Department of Psychological and Brain Sciences, Indiana University, Bloomington, Indiana, USA

Pierre Maurage Neuroscience, Systems and Cognition (NEUROCS), and Health and Psychological Development (CSDP) Research Units, Institute of Psychology, Catholic University of Louvain, Louvain-la-Neuve, Belgium

Gilles Pourtois Department of Experimental-Clinical and Health Psychology, Ghent University, Ghent, Belgium

Aina Puce Department of Psychological and Brain Sciences, Indiana University, Bloomington, IN, USA

Lizabeth M. Romanski Department of Neurobiology & Anatomy and Center for Navigation and Communication Sciences, University of Rochester, Rochester, NY, USA

Julien Rouger Brain Innovation BrainVoyager, Maastricht, The Netherlands

Stefan R. Schweinberger Department of General Psychology and Cognitive Neuroscience, DFG Research Unit Person Perception, Friedrich-Schiller-University of Jena, Jena, Germany

Bernard M.C. Stienen Cognitive and Affective Neuroscience Laboratory, Tilburg University, Tilburg, The Netherlands

Jan Van den Stock Cognitive and Affective Neuroscience Laboratory, Tilburg University, Tilburg, The Netherlands

P. Verbanck Laboratory of Psychological Medicine, Free University of Brussels, Brussels, Belgium

D. Delle Vigne Laboratory of Psychological Medicine, Free University of Brussels, Brussels, Belgium

Rebecca Watson Voice Neurocognition Laboratory, Institute of Neuroscience and Psychology, College of Medical, Veterinary and Life Sciences, University of Glasgow, Glasgow, UK

Dirk Wildgruber Department of General Psychiatry, University of Tübingen, Tübingen, Germany

Barbra Zupan Department of Applied Linguistics, Brock University, St. Catharines, ON, Canada

Part I
Evolution and Development

Chapter 1
Multisensory Recognition in Vertebrates (Especially Primates)

Ipek G. Kulahci and Asif A. Ghazanfar

A monkey wakes up next to her group mates as the sun rises. Throughout the day, she needs to make a number of decisions. Who should she forage with? Who should she cooperate with in order to chase away unfamiliar monkeys? Are there any particular individuals that she should avoid interacting with? When she is not foraging or defending her territory, she can usually be seen grooming another individual. However, choosing whom to groom presents yet another decision she needs to make. On this particular day, she may even end up deciding with whom she is going to mate. This is a complex but important decision, requiring the selection of a high quality male among many others based on a set of physical characteristics. All these myriad decisions require her to know the individuals in the group, recognize specific individuals among others, and remember past interactions with group members.

This monkey, like us and most other animals, constantly makes these decisions based on the evaluation of the signals from the environment. In a dynamic social environment, it is essential that animals are well equipped for detecting, learning, and discriminating communicatively relevant information including the identities and status of others. A question of great interest to biologists interested in communication and social decision-making is how individuals integrate information from more than one signal in order to identify individuals in a group. The signal receiver's ability to integrate two signals evolves through the interactions between signal senders and receivers (Rowe, 1999). To understand the evolution of cross-modal

I.G. Kulahci
Department of Ecology & Evolutionary Biology,
Princeton University, Princeton, NJ 08540, USA

A.A. Ghazanfar (✉)
Department of Ecology & Evolutionary Biology,
Princeton University, Princeton, NJ 08540, USA

Department of Psychology, Princeton University, Princeton, NJ 08540, USA

Neuroscience Institute, Princeton University, Princeton, NJ 08540, USA
e-mail: asifg@princeton.edu

P. Belin et al. (eds.), *Integrating Face and Voice in Person Perception*,
DOI 10.1007/978-1-4614-3585-3_1, © Springer Science+Business Media New York 2013

(or *multisensory*) recognition and integration, we need to consider the receivers' perspective, such as the physiological, morphological, behavioral, and neural characteristics that assist the ability to detect, discriminate, and remember signaler's output (Guilford & Dawkins, 1991).

Our main goals in this chapter are to show that the ability to combine signals across different sensory modalities to learn about others' identity and affective states is not unique to humans, and to demonstrate that studies on animal behavior (nonhuman primates, in particular) can provide valuable insights into our understanding of human multisensory processing. First, we focus on a few examples from different vertebrate species and behavioral contexts to address why the combination of acoustic and visual signals is essential for survival and reproductive fitness. With these examples, we aim to outline some of the mechanisms through which signal structure influences receivers' ability to detect and process multisensory signals. The visual components of these signals are produced not only by the face of the signaler, but in many cases also by animals' body postures, body size, and movements. These all convey important information about individual identity and characteristics. In the second part of this chapter, we focus on what we know about both the behavior and the neurobiology of multisensory processing of faces and voices in nonhuman primates. We hope that this will help bridge animal studies with our knowledge of human recognition of individuals across modalities.

Even though studying animal behavior as it applies to multisensory signal processing presents some challenges (Hoy, 2005), animal studies can provide valuable insight into how humans use multisensory signals. By studying multiple species, we will achieve a more complete framework for the evolution of multisensory integration. Furthermore, by comparing closely related species that differ in whether or how they integrate signals, we can learn about the role of communication in the emergence of different species (Bro-Jorgensen, 2010) as well as the environmental conditions that may have favored the emergence of this ability. In addition, working with animals offers some experimental advantages over working with humans. For example, in the laboratory, an individual's experiences since birth can be documented, manipulated, and controlled. Large sample sizes over multiple generations can be achieved when working with certain taxa such as arthropods, as we illustrate in the next section. Finally, the functional aspects of multisensory integration can be experimentally tested either by modifying signals (such as presenting 3D animal models), or by modifying receivers' access to the signals in a natural setting (such as studying species in different environments with varying levels of environmental noise).

1 Audiovisual Recognition for Multiple Behaviors in Multiple Species

A variety of nonhuman animals, ranging from insects to mammals, use signals in more than one modality for communication (Rowe, 1999). Acoustic and visual signals are frequently combined for distinguishing conspecifics (members of same species as the signal receiver) from heterospecifics (members of a different species), recognizing individuals, detecting particular individuals in noisy or unreliable environments, and learning about characteristics of others. These abilities are essential for survival and reproduction, and enhance many behaviors such as territorial defense, predator avoidance, prey detection, foraging, mate choice, and even parental care. Providing a comprehensive overview of why animals signal in more than one modality is beyond the scope of this chapter, we therefore refer interested readers to the following references for more in-depth approaches to this topic: Bro-Jorgensen (2010); Candolin (2003); Hebets and Papaj (2005); Rowe (1999).

1.1 Aggression and Territorial Defense

Individuals in most species hold territories in which they have exclusive access to resources essential for survival and reproduction. Some of the critical resources include food, mates and shelter; therefore, territories need to be protected against intruders (Wilson, 1975). Errors in individual recognition can be devastating since territorial defense is usually energetically costly, it reduces the time available for other activities, and it may attract the attention of predators (Brown, 1964). Because of these costs, territory holders need to be certain that they are displaying aggression towards an intruder rather than towards a group member, a potential mate, or even a predator.

The accuracy of territorial response would be higher when more than one cue can be used to detect the presence and the identity of an intruder. Two signals are considered to be redundant when they transmit the same information and increase receivers' response accuracy (Partan & Marler, 1999). Experimental studies of the role of visual–acoustic integration in intruder detection in pied currawong (*Strepera graculina*; Fig. 1.1a) reveals that males of this species display higher levels of territorial behavior—characterized by movement towards the speaker—when the playback vocalization is accompanied by an artificial currawong model. Furthermore, placing the model and the speaker close to each other leads to spatially less variable territorial defense response, suggesting that the males integrate the two cues not only to detect the presence of intruders, but also to determine location of intruders (Lombardo, Mackey, Tang, Smith, & Blumstein, 2008).

An animal's recognition abilities in variable environments are also enhanced through multisensory signals (Hebets, 2005; Hebets & Papaj, 2005). Most species utilize that fact that acoustic signals and visual signals propagate under different environmental conditions and distances. For example, males of the territorial Bornean ranid frog (*Staurois guttatus*) signal their presence to intruder males through vocalizations and multiple visual displays including foot-flagging, upright posture, and vocal-sac inflation. Calls are usually detected from a longer distance than visual displays, so the calls direct attention of the intruders towards the visual signals. By combining the two modalities, intruders are able to quickly and accurately detect the presence and the location of the territory holders (Grafe & Wanger, 2007). Similarly, vocalizations of barking geckos (*Ptenopus garrulus garrulus*), a nocturnal

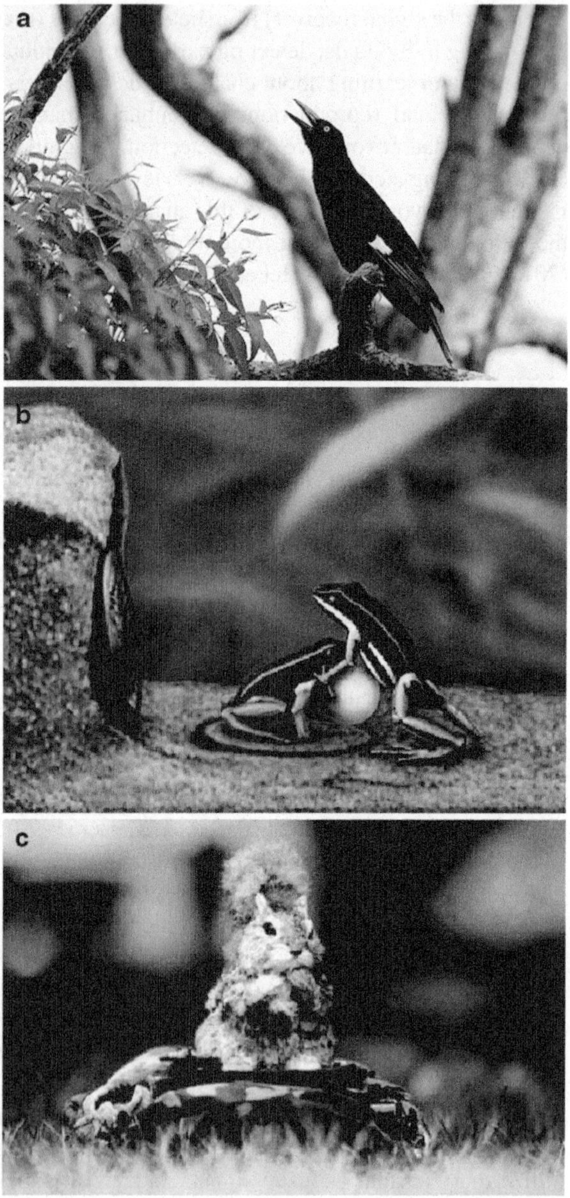

Fig. 1.1 (**a**) A singing pied currawong. (Photo by Steve Happ.) (**b**) Dart-poison frog electromechanical model receiving an attack from a real male frog (Narins et al., 2003). (**c**) Robot squirrel whose tail flagging combined with alarm calls elicits increased alarm behavior in real squirrels (Partan et al., 2009)

territorial species, convey several individual characteristics (Hibbitts, Whiting, & Stuart-Fox, 2007). Larger males of this species have a low frequency call; larger males are also stronger than smaller males, and they have a higher chance of winning in an aggressive context. By inferring the size of each other through the frequency of calls, males can evaluate whether or not to participate in a fight even when they cannot see their opponent. The ability to infer characteristics of others through vocalizations is important in many other behavioral contexts, and we continue to explore this ability in the next section.

Among amphibians, dart-poison frogs (*Epipedobates femoralis*) attack territorial intruders only if the intruders produce a dynamic, bimodal display (Narins, Grabul, Soma, Gaucher, & Hodl, 2005; Narins, Hodl, & Grabul, 2003). By using an electromechanical frog model, experiments in the wild revealed that neither unimodal cues presented in isolation nor static bimodal stimuli elicit attacks (Fig. 1.1b). These results suggest that integration of dynamic bimodal cues is necessary to elicit aggression in this species. Territorial males presented with visual and auditory cues separated by experimentally introduced temporal delays or spatial disparities will reduce their frequency of attack on the frog model (Narins et al., 2005). In the temporal integration experiments, bimodal stimuli with temporal overlap during calling bouts consistently evoke aggressive behavior, but stimuli lacking bimodal temporal overlap are relatively ineffective at the same task. In the spatial integration studies, despite presenting the components of the bimodal stimulus with an initial spatial disparity of up to 12 cm, attack behavior persists.

1.2 Mate Choice

One of the main reasons for defending a territory is to have access to mates (Brown, 1964). Successful mate choice, which directly influences reproductive success and therefore evolution, requires the ability to combine multiple signals from potential mates. In majority of the animal species, females choose the males to mate with, and their choice requires them to avoid heterospecific males and to choose the best male among conspecifics. For group-living species in which both mates participate in parental care, it is also essential that females are able to distinguish their mates from other males in the group (Sherman, Reeve, & Pfennig, 1997).

Males have to attract females' attention towards the signals that convey their quality; therefore, they frequently use redundant multisensory signaling to increase the likelihood that females will detect their signals across variable environments (Candolin, 2003). For example, Galapagos finches (*Geospiza* spp.), commonly known among biologists as "Darwin's finches," have been studied since 1972 for their behavior and speciation patterns (Grant & Grant, 2002). Some species, such as the medium ground finch (*Geospiza fortis*), have two morphological variants that co-occur in a single population. The morphs differ in their beak shape, and females prefer to mate with a male from their own morph (Huber, Leon, Hendry, Bermingham, & Podos, 2007). This assortative mating is acquired through the use of visual cues at closer distances and vocal cues at longer distances (Grant & Grant, 1989; Podos, 2010). The frequency bandwidth of males' songs is strongly correlated with their beak morphology (Huber & Podos, 2006). It turns out that the territorial males are able to predict the beak shape and therefore the morph of intruder males just by listening to their song, and direct their territorial defense only towards males of their own morph, since they are perceived as competition for mates and food (Huber et al., 2007; Podos, 2010). Most likely, females are also able to use song properties to infer the visual morph of the singing males, and use this knowledge to choose a mate (Podos, 2010).

While the acoustic signals allow receivers to predict the visual properties of senders in some species, in other species, the visual and the acoustic signals can convey information about different characteristics of the senders. Males of Barbary doves (*Streptopelia risoria*) display to females using a complex signal composed of a visual bow and a call. The visual signal, the bow itself, is stereotypic across individuals and is informative about the sex of the displaying individual, while the call, which shows interindividual variation in frequency, duration and intercall interval, is informative about the quality of the singing male. The repetition rates of the bows and the calls are strongly correlated with each other. As a result, females combine these two signals to choose between males (Fusani, Hutchison, & Hutchison, 1997).

Multisensory integration ability is important for many species' mating behaviors; however, it is critical for mate choice by females in lekking species. Leks are large aggregations formed by males in order to attract females who visit these groups to choose a mate (Bradbury, 1981). The males of some amphibians, for instance, will gather together and sing, forming a chorus. This coordinated behavior attracts the attention of females from a distance. However, it also presents significant challenges for both sexes. The males need to distinguish themselves from others and the females need to locate the highest quality males in the throng (Roberts, Taylor, & Uetz, 2007; Taylor, Buchanan, & Doherty, 2007; Wollerman, 1999). A male has a higher chance of being detected and distinguished by a female if she can see him as he calls. Using a robotic male frog model, it's been shown that female Tungara frogs (*Physalaemus pustulosus*), for example, prefer the calls of males that are synchronized with a visual signal, such as vocal sac inflation, versus only the calls. This preference for the acoustic-visual display becomes especially important when females cannot acoustically distinguish between different males in a chorus. Oddly, coupling a vocalization with only the vocal sac (without the body of the robotic frog) is enough to trigger this preference (Taylor, Klein, Stein, & Ryan, 2008). Similarly, when choosing between two songs that are equally good, females of squirrel tree frogs (*Hyla squirella*) prefer the male calls that are coupled with a frog model (Roberts et al., 2007).

1.3 Predator and Prey Interactions

One of the critical situations in which correct identification and interpretation of signals has a direct influence on survival is predator detection. Individuals can detect a predator either by encountering the predator, or by receiving alarm signals from other individuals. However, despite the fact that alarm signals are usually multisensory, only a small number of studies have addressed multisensory processing in predator detection and alarm signals (Partan, Larco, & Owens, 2009; Partan,

Yelda, Price, & Shimizu, 2005). Animals signal presence of predators through changes in body posture as well as through increased movement and vocalization. Most of us living in areas frequented by squirrels have witnessed squirrel alarm behaviors; when disturbed, they usually flag their tails and bark. During our encounters with the alarmed squirrels, we may have also witnessed a phenomenon common in rodents and birds; when an individual encounters a conspecific who is displaying alarm behavior, s/he is likely to repeat the alarm behavior (Partan et al., 2009). To understand the role of acoustic-visual integration in how conspecifics detect alarm behavior of others, eastern gray squirrels (*Sciurus carolinensis*) were presented with a robotic squirrel that displayed alarm behavior (Partan et al., 2009) (Fig. 1.1c). When tail-flagging of the robotic squirrel is accompanied with the playback of a bark, wild squirrels repeated the alarm behavior at higher rates in comparison to when only barking or tail-flagging was present.

2 Audiovisual Processing in Nonhuman Primates

Given the ubiquity of multisensory processes in the animal kingdom, it is somewhat odd that anyone would think that humans were anything special in this regard. Yet, using data from neuroanatomical studies, it was once hypothesized that humans were unique in their ability to form multisensory associations. This followed from the basic tenet of neocortical organization: different regions of the cortex have different functions. Some regions receive visual, auditory, tactile, olfactory and gustatory sensations. Each of these sensory regions is thought to send projections which converge on an "association area" which then enables the association of between the different senses and between the senses and movement. According to a highly influential two-part review by Norman Geschwind, entitled, "Disconnexion syndromes in Animals and Man" (Geschwind, 1965a, 1965b), the connections between sensory modalities via their convergence in association areas are not robust in nonhuman animals, limiting their ability to make multisensory associations. In contrast, humans can readily make such associations, for example, between the sight of a lion and the sounds of its roar, but, apparently, a lion cannot.

This picture of human versus nonhuman multisensory abilities based on anatomy led to the idea that human speech and language evolved in parallel with robust multisensory connections within the neocortex. Geschwind claimed that the "ability to acquire speech has as a prerequisite the ability to form cross-modal associations" (Geschwind, 1965a, 1965b). This view of cross-modal associations as a potentially uniquely human capacity remains present even in more current ideas about the evolution of speech and language. For example, it's been suggested that human language depends upon our unique ability to imitate in multiple modalities which in turn relies on a "substantial change in neural organization, one that affects not only imitation but also communication" (Hauser, Chomsky, & Fitch, 2002) (page 1,575). For the remainder of this chapter, we focus on audiovisual communication in nonhuman primates. Our purpose is twofold: (1) to debunk the view that human communication is uniquely multisensory; (2) to show that the neural mechanisms of multisensory processes extend beyond neocortical association areas and do not require a uniquely human brain architecture.

2.1 Monkeys Match Facial Expressions to Vocal Expressions

It is widely accepted that human speech is fundamentally a multisensory behavior, with face-to-face communication perceived through both the visual and auditory channels. What is true for human speech is also true for vocal communication in nonhuman primates: vision and audition are inextricably linked. Human and primate vocalizations are produced by coordinated movements of the lungs, larynx (vocal folds), and the supralaryngeal vocal tract (Ghazanfar & Rendall, 2008). The vocal tract consists of the column of air derived from the pharynx, mouth, and nasal cavity. In humans, speech-related vocal tract motion results in the predictable deformation of the face around the oral aperture and other parts of the face (Jiang, Alwan, Keating, Auer, & Bernstein, 2002; Yehia, Kuratate, & Vatikiotis-Bateson, 2002; Yehia, Rubin, & Vatikiotis-Bateson, 1998). For example, human adults automatically link high-pitched sounds to facial postures producing an/i/sound and low-pitched sounds to faces producing an/a/sound (Kuhl, Williams, & Meltzoff, 1991). In primate vocal production, there is a similar link between acoustic output and facial dynamics. Different macaque monkey vocalizations are produced with unique lip configurations and mandibular positions and the motion of such articulators influences the acoustics of the signal (Hauser, Evans, & Marler, 1993; Hauser & Ybarra, 1994). Coo calls, like/u/in speech, are produced with the lips protruded, while screams, like the/i/in speech, are produced with the lips retracted (Fig. 1.2). Thus, it is likely that many of the facial motion cues that humans use for speech-reading are present in other primates as well.

Given that both humans and other extant primates use both facial and vocal expressions as communication signals, it is perhaps not surprising that many primates other than humans recognize the correspondence between the visual and auditory components of vocal signals. Macaque monkeys (*Macaca mulatta*), capuchins (*Cebus apella*), and chimpanzees (*Pan troglodytes*) all recognize auditory–visual correspondences between their various vocalizations (Evans, Howell, & Westergaard, 2005; Ghazanfar & Logothetis, 2003; Izumi & Kojima, 2004; Parr, 2004). For example, rhesus monkeys tested in a preferential looking paradigm readily match the facial expressions of "coo" and "threat" calls with their associated vocal components (Ghazanfar & Logothetis, 2003). Perhaps more pertinent, rhesus monkeys can also segregate competing voices in a chorus of coos, much as humans might with speech in a cocktail party scenario, and match them to the correct number of individuals seen cooing on a video screen (Jordan, Brannon, Logothetis, & Ghazanfar, 2005) (Fig. 1.3a). Finally, macaque monkeys use formants (i.e., vocal tract

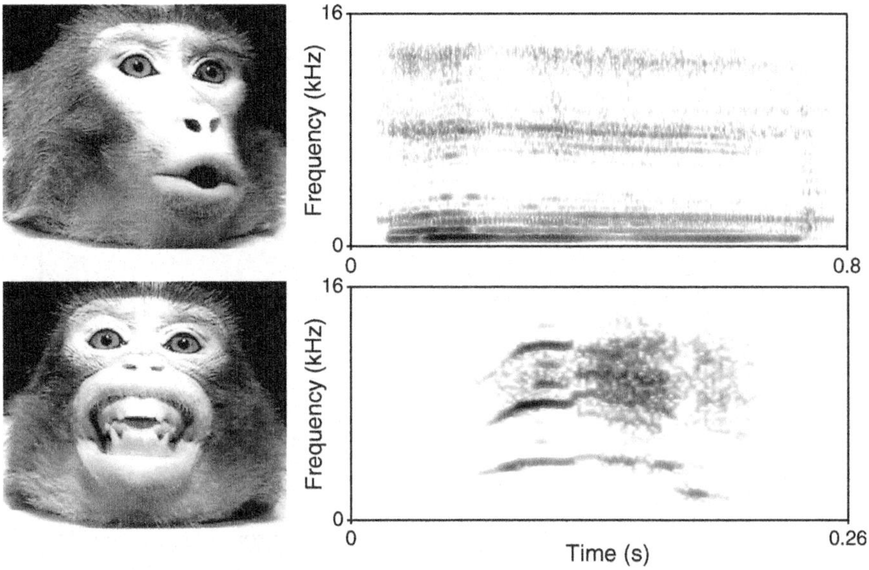

Fig. 1.2 Exemplars of the facial expressions produced concomitantly with vocalizations. Rhesus monkey coo and scream calls taken at the midpoint of the expressions with their corresponding spectrograms

resonances) as acoustic cues to assess age-related body size differences among conspecifics (Fig. 1.3b) (Ghazanfar et al., 2007). They do so by linking across modalities the body size information embedded in the formant spacing of vocalizations (Fitch, 1997) with the visual size of animals who are likely to produce such vocalizations (Ghazanfar et al., 2007)—a capacity that is likely useful in assessing the identity of competitors.

In a recent experiment, macaque monkeys demonstrated that they could recognize familiar individuals of different species (monkey and human) and across modalities (Sliwa, Duhamel, Pascalis, & Wirth, 2011). The monkeys had daily exposure to both conspecifics and human individuals from infancy and were familiarized with both the humans and other rhesus monkeys serving as stimuli in the experiment via recent real life daily exposure (housing "roommates," caregivers, and researchers). In a free preferential looking time paradigm, monkeys spontaneously matched the faces of known individuals to their voices, regardless of species. Their known preferences for interacting with particular individuals were also apparently in the strength of their multisensory recognition. Overall, this experiment demonstrates the existence of individual recognition in rhesus monkeys comprising at least two elements of identity (vocal and visual). It also shows that individual recognition extends adaptably from conspecifics to personally known humans.

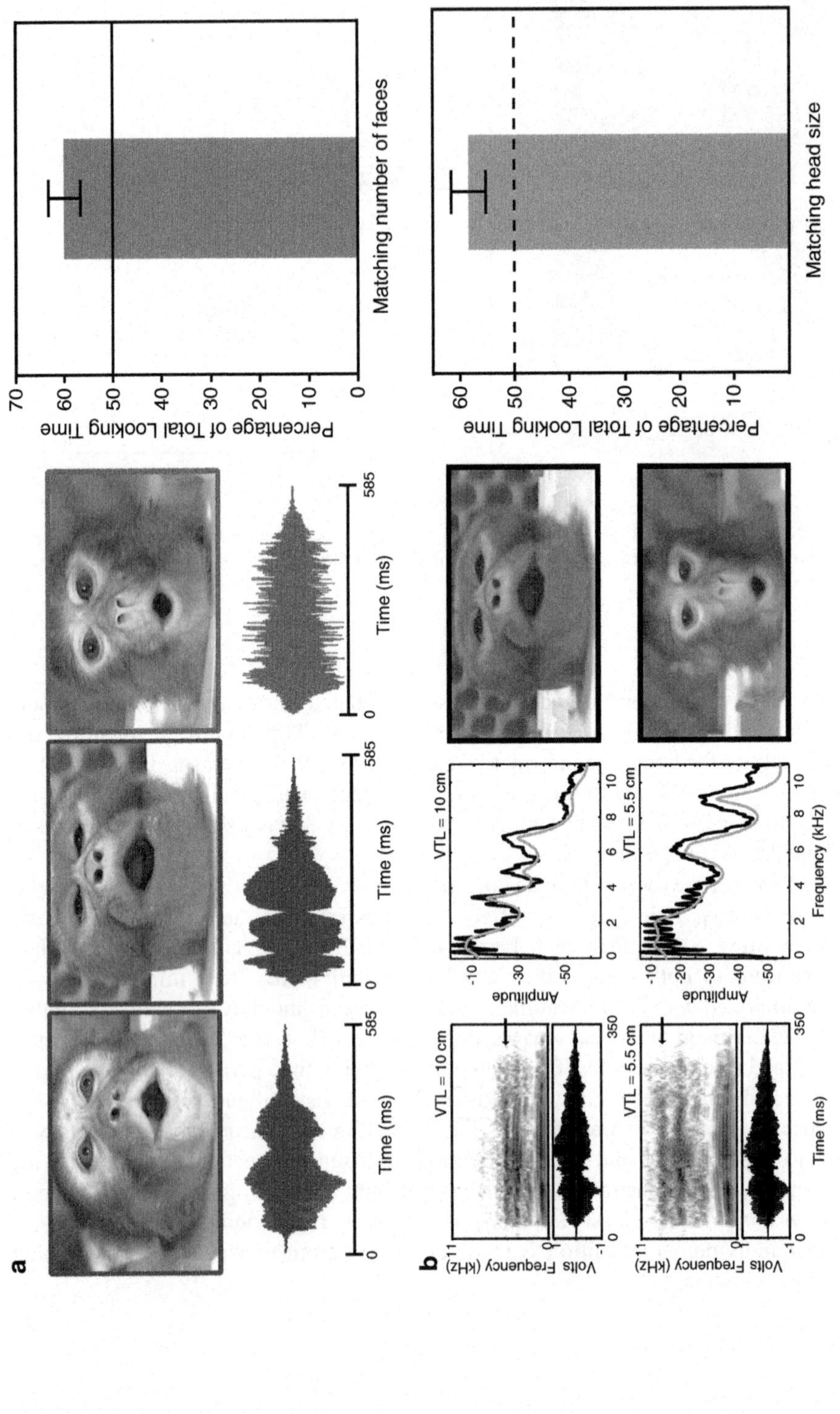
a
Time (ms)
0
585
Percentage of Total Looking Time
Matching number of faces
b
VTL = 10 cm
VTL = 5.5 cm
Volts Frequency (kHz)
Amplitude
Frequency (kHz)
Time (ms)
Matching head size

Fig. 1.3 Monkeys can match across modalities. To test this, we adopted the preferential-looking paradigm which does not require training or reward. In the paradigm, subjects are seated in front of two LCD monitors and shown two side-by-side digital videos, only one of which corresponds to the sounds track heard through a centrally located speaker. A trial consists of the two videos played in a continuous loop with one of the two sound tracks also played in a loop through the speaker. The dependent measure is percentage of total looking time to the match video. (**a**) Monkeys segregate coo vocalizations from different individuals and look to correct number of conspecific individuals displayed on the screen. Still frames extracted from a stimulus set along with their acoustic counterparts below. The *bar graph* shows the mean percentage (±SEM) of total looking time to the matching video display; chance is 50%. (**b**) A single coo vocalizations were synthesized to mimic large and small sounding individuals. Diagrams in the *left panels* show the spectrograms and waveforms of a coo vocalization resynthesized with two different vocal tract lengths. The *arrow* in the spectrogram indicates the position of an individual formant which increases in frequency as the apparent vocal tract length decreases. In the *middle panels*, power spectra (*black line*) and linear predictive coding spectra (*gray lines*) for the long vocal tract length (10 cm, *top panel*) and short vocal tract length (5.5 cm, *bottom panel*). Still frames show the visual components of a large and small monkey. The *bar graph* shows the mean percentage of total looking time spent looking at the matching video display; the *dotted line* indicates chance expectation. *Error bars* are SEM

2.2 Monkeys Integrate Faces and Voices

All of the above experiments with monkeys and apes test whether or not they can *match* faces to voices. None provide direct evidence as to whether multisensory communication signals provide a behavioral advantage. To bridge this gap, monkeys were trained to detect auditory, visual, or audiovisual vocalizations embedded in noise as fast and as accurately as possible (Chandrasekaran, Lemus, Trubanova, Gondan, & Ghazanfar, 2011). A free-response task was designed to approximate a natural face-to-face vocal communication event. In such settings, the vocal components of the communication signals are degraded by environmental noise. The face and its motion, on the other hand, are usually perceived clearly. In the task, monkeys responded to "coo" calls that are affiliative vocalizations commonly produced by macaque monkeys in a variety of contexts (Fig. 1.4). All vocalizations had five different levels of sound intensity and were embedded in a constant background noise. For dynamic faces, we used computer-generated monkey avatars (Fig. 1.4a). The use of avatars allowed us to restrict facial motion to the mouth region, ensure constant lighting and background, and to parameterize the size of the mouth opening while keeping eye and head positions constant. The degree of mouth-opening was in accordance with the intensity of the associated vocalization: greater sound intensity was coupled to larger mouth openings by the dynamic face. Two coos were paired with two monkey avatars, respectively, and this pairing was kept constant (Fig. 1.4a).

During the task, one avatar face would be continuously on the screen for a block of trials; the background noise was also continuous (Fig. 1.4b). In the "visual only (V)" condition, this avatar would move its mouth without any corresponding auditory component; that is, it silently produced a coo. In the "auditory-only (A)" condition, the vocalization normally paired with the *other* avatar (which is not on the screen) is presented with the *static* face of the avatar. Finally, in the "audiovisual (AV)" condition, the avatar moves its mouth accompanied by the corresponding vocalization and in accordance (degree of mouth opening) with its intensity. Each condition (V, A, or AV) was presented after a variable (drawn from a uniform distribution) interval between 1 and 3 s, and subjects indicated the detection of an event (visible mouth motion, auditory signal or both) by pressing a lever. Under these task conditions, monkeys (like humans in similar experiments) integrated faces and voices, allowing them to significantly decrease their reaction times relative to unimodal conditions (Chandrasekaran, Lemus, Trubanova et al., 2011) (Fig. 1.4c). This is the first evidence for a behavioral advantage for combining faces and voices in a nonhuman primate species.

Taken together, these behavioral data suggest that humans are not at all unique among primates in their ability to perceive and integrate communication signals across modalities. As we describe below, the neural mechanisms by which monkeys process these signals are also similar to those used by humans.

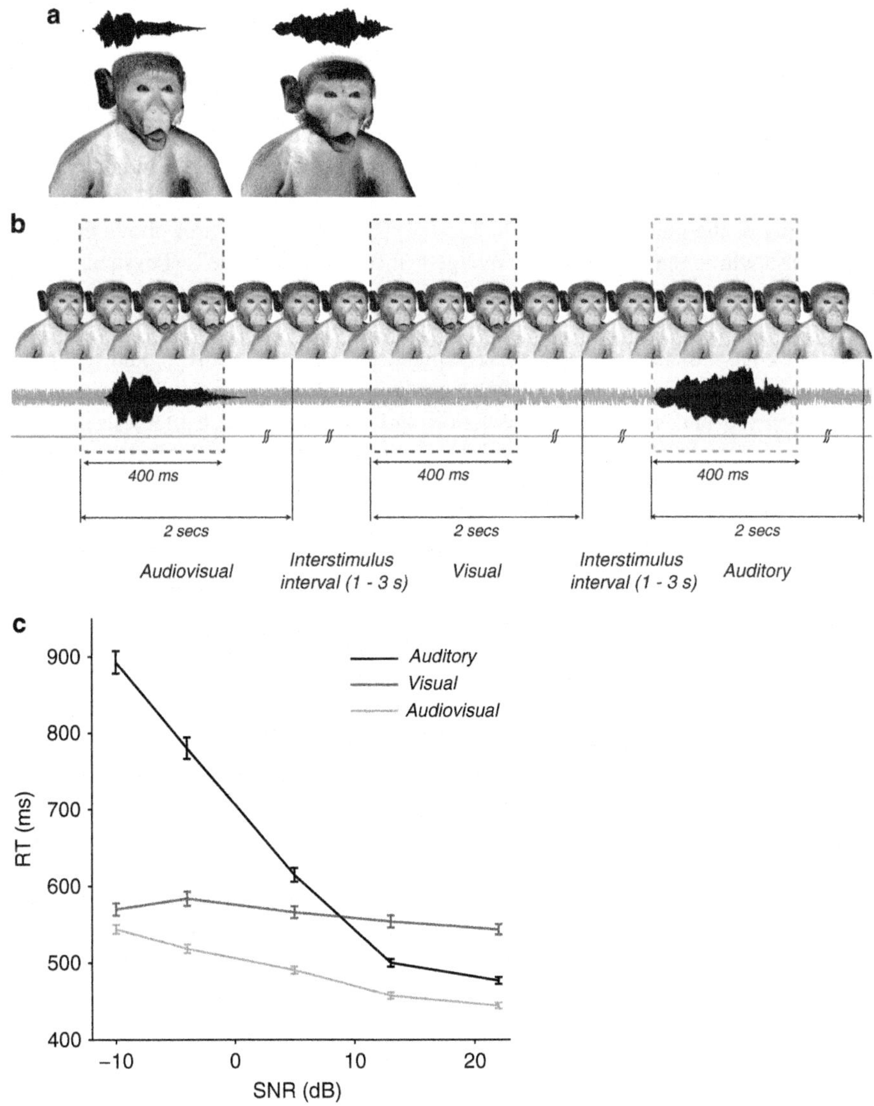

Fig. 1.4 (**a**) Waveform of coo vocalizations detected by the monkeys and their respective avatars below. (**b**). Free-response paradigm task structure. An avatar face was always on the screen. Visual, auditory, and audiovisual stimuli were randomly presented with an interstimulus interval of 1–3 s drawn from a uniform distribution. Responses within a 2 s window after stimulus onset were considered to be hits. Responses in the interstimulus interval are considered to be false alarms and led to timeouts. (**c**) Mean RTs obtained by pooling across all sessions as a function of SNR for the unisensory and multisensory conditions for one monkey. *Error bars* denote standard error of the mean estimated using bootstrapping. *X*-axes denote SNR in dB. *Y*-axes depict RT in milliseconds

3 Neocortical Processing of Face–Voice Signals in Monkeys

Although it is generally recognized that we and other animals use our different senses in an integrated fashion, we assume that, at the neural level, these senses are, for the most part, processed independently but then converge at critical nodes. This idea extends as far back as Leonardo da Vinci's (1452–1519) research into the neuroanatomy of the human brain. He suggested that there was an area above the pituitary fossa where the five senses converged (the "sensu comune") (Pevsner, 2002). Until recently, this basic tenet of neocortical organization has not changed to a large degree since da Vinci's time, as it has long been argued that different regions of the cortex have different functions segregated according to sense modality. Some regions receive visual sensations, others auditory sensations and still others tactile sensations (so and so forth, for olfaction and gustation). Each of these sensory regions is thought to send projections which converge on an "association area" which then enables the association of between the different senses and between the senses and movement.

Thus, according to this traditional line of thinking, the linking of vision with audition in the multisensory vocal perception described above would be attributed to the functions of association areas such as the superior temporal sulcus in the temporal lobe or the principal and intraparietal sulci located in the frontal and parietal lobes, respectively. Although these regions may certainly play important roles (see below), they are certainly not necessary for all types of multisensory behaviors (Ettlinger & Wilson, 1990), nor are they the sole regions for multisensory convergence (Driver & Noesselt, 2008; Ghazanfar & Schroeder, 2006). The auditory cortex, in particular, has many potential sources of visual inputs (Ghazanfar & Schroeder, 2006) and this is borne out in the increasing number of studies demonstrating visual modulation of auditory cortical activity (Bizley, Nodal, Bajo, Nelken, & King, 2007; Ghazanfar, Chandrasekaran, & Logothetis, 2008; Ghazanfar, Maier, Hoffman, & Logothetis, 2005; Kayser, Petkov, Augath, & Logothetis, 2007; Kayser, Petkov, C.I. & Logothetis, N.K, 2008; Schroeder & Foxe, 2002). Here we focus on those auditory cortical studies investigating face–voice integration specifically.

Recordings from both primary and lateral belt auditory cortex reveal that responses to the voice are influenced by the presence of a dynamic face (Ghazanfar et al., 2005, 2008). Monkey subjects viewing unimodal and bimodal versions of two different species-typical vocalizations ("coos" and "grunts") show both enhanced and suppressed local field potential (LFP) responses in the bimodal condition relative to the unimodal auditory condition (Ghazanfar et al., 2005). Consistent with evoked potential studies in humans (Besle, Fort, Delpuech, & Giard, 2004; van Wassenhove, Grant, & Poeppel, 2005), the combination of faces and voices led to integrative responses (significantly different from unimodal responses) in the vast majority of auditory cortical sites—both in primary auditory cortex and the lateral belt auditory cortex. The data demonstrated that LFP signals in the auditory cortex are capable of multisensory integration of facial and vocal signals in monkeys (Ghazanfar et al., 2005) and have subsequently been confirmed at the single unit level in the lateral belt cortex as well (Ghazanfar et al., 2008) (Fig. 1.5).

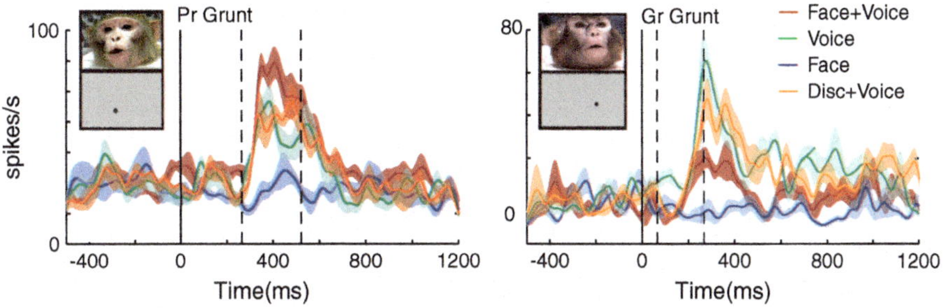

Fig. 1.5 Single neuron examples of multisensory integration of Face + Voice stimuli compared with Disk + Voice stimuli in the lateral belt area. The *left panel* shows an enhanced response when voices are coupled with faces, but no similar modulation when coupled with disks. The *right panel* shows similar effects for a suppressed response. *X*-axes show time aligned to onset of the face (*solid line*). *Dashed lines* indicate the onset and offset of the voice signal. *Y*-axes depict the firing rate of the neuron in spikes per second. *Shaded regions* denote the SEM

The specificity of face–voice integrative responses was tested by replacing the dynamic faces with dynamic disks that mimicked the aperture and displacement of the mouth. In human psychophysical experiments, such artificial dynamic stimuli can still lead to enhanced speech detection, but not to the same degree as a real face (Bernstein, Auer, & Takayanagi, 2004; Schwartz, Berthommier, & Savariaux, 2004). When cortical sites or single units were tested with dynamic disks, far less integration was seen when compared to the real monkey faces (Ghazanfar et al., 2005, 2008) (Fig. 1.5). This was true primarily for the lateral belt auditory cortex (LFPs and single units) and was observed to a lesser extent in the primary auditory cortex (LFPs only). This suggests that there may be increasingly specific influences of "extra" sensory modalities as one moves away from the primary sensory regions.

Unexpectedly, grunt vocalizations were over-represented relative to coos in terms of enhanced multisensory LFP responses (Ghazanfar et al., 2005). As coos and grunts are both produced frequently in a variety of affiliative contexts and are broadband spectrally, the differential representation cannot be attributed to experience, valence or the frequency tuning of neurons. One remaining possibility is that this differential representation may reflect a behaviorally relevant distinction, as coos and grunts differ in their direction of expression and range. Coos are generally contact calls rarely directed toward any particular individual. In contrast, grunts are often directed towards individuals in one-on-one situations, often during social approaches as in baboons and vervet monkeys (Cheney & Seyfarth, 1982; Palombit, Cheney, & Seyfarth, 1999). Given their production at close range and context, grunts may produce a stronger face–voice association than coo calls. This distinction appeared to be reflected in the pattern of significant multisensory responses in auditory cortex; that is, this multisensory bias towards grunt calls may be related to the fact the grunts (relative to coos) are often produced during intimate, one-to-one social interactions.

3.1 The Superior Temporal Sulcus is a Source of Face-Sensitive Input to the Auditory Cortex

The face-specific visual influence on the lateral belt auditory cortex begs the question as to its anatomical source. Although there are multiple possible sources of visual input to auditory cortex (Ghazanfar & Schroeder, 2006), the STS is likely to be a prominent one, particularly for integrating faces and voices, for the following reasons. First, there are reciprocal connections between the STS and the lateral belt and other parts of auditory cortex (Barnes & Pandya, 1992; Seltzer & Pandya, 1994). Second, neurons in the STS are sensitive to both faces and biological motion (Harries & Perrett, 1991; Oram & Perrett, 1994). Finally, the STS is known to be multisensory (Barraclough, Xiao, Baker, Oram, & Perrett, 2005; Benevento, Fallon, Davis, & Rezak, 1977; Bruce, Desimone, & Gross, 1981; Chandrasekaran & Ghazanfar, 2009; Schroeder & Foxe, 2002). One mechanism for establishing whether auditory cortex and the STS interact at the functional level is to measure their temporal correlations as a function of stimulus condition. Concurrent recordings LFPs and spiking activity in the lateral belt of auditory cortex and the upper bank of the STS revealed that functional interactions, in the form of gamma band correlations, between these two regions increased in strength during presentations of faces and voices together relative to the unimodal conditions (Ghazanfar et al., 2008) (Fig. 1.6a). Furthermore, these interactions were not solely modulations of response strength, as phase relationships were significantly less variable (tighter) in the multisensory conditions (Fig. 1.6b).

The influence of the STS on auditory cortex was not merely on its gamma oscillations. Spiking activity seems to be *modulated*, but not "driven," by ongoing activity arising from the STS. Three lines of evidence suggest this scenario. First, visual influences on single neurons were most robust when in the form of dynamic faces and were only apparent when neurons had a significant response to a vocalization (i.e., there were no overt responses to faces alone). Second, these integrative responses were often "face-specific" and had a wide distribution of latencies, which suggested that the face signal was an ongoing signal that influenced auditory responses (Ghazanfar et al., 2008). Finally, this hypothesis for an ongoing signal is supported by the sustained gamma band activity between auditory cortex and STS and by a spike-field coherence analysis. This analysis reveals that just prior to spiking activity in the auditory cortex, there is an increase in gamma band power in the STS (Ghazanfar et al., 2008) (Fig. 1.6c).

Both the auditory cortex and the STS have multiple bands of oscillatory activity generated in responses to stimuli that may mediate different functions (Chandrasekaran & Ghazanfar, 2009; Lakatos et al., 2005). Thus, interactions between the auditory cortex and the STS are not limited to spiking activity and high frequency gamma oscillations. Below 20 Hz, and in response to naturalistic audiovisual stimuli, there are directed interactions from auditory cortex to STS, while above 20 Hz (but below the gamma range), there are directed interactions from STS to auditory cortex (Kayser & Logothetis, 2009). Given that different frequency bands in the STS integrate faces and voices in distinct ways (Chandrasekaran & Ghazanfar, 2009),

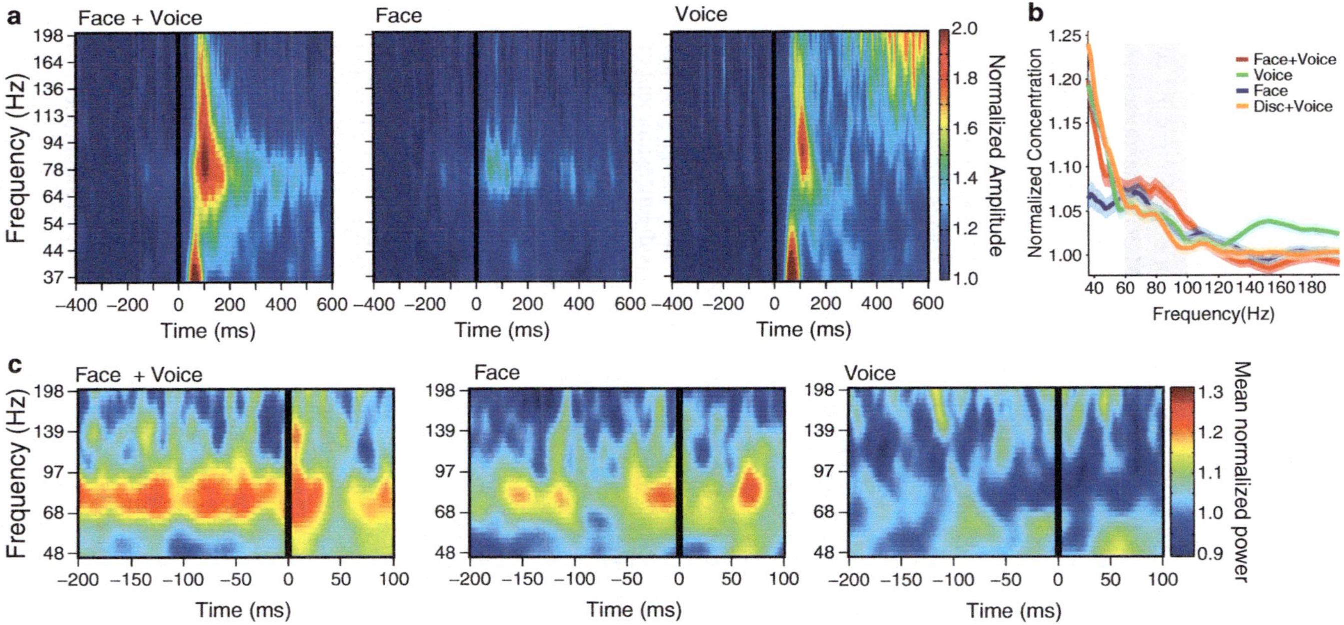

Fig. 1.6 (**a**) Time-frequency plots (cross-spectrograms) illustrate the modulation of functional interactions (as a function of stimulus condition) between the lateral belt auditory cortex and the STS for a population of cortical sites. *X*-axes depict the time in milliseconds as a function of onset of the auditory signal (*solid black line*). *Y*-axes depict the frequency of the oscillations in Hz. *Color-bar* indicates the amplitude of these signals normalized by the baseline mean. (**b**) Population phase concentration from 0 to 300 ms after voice onset. *X*-axes depict frequency in Hz. *Y*-axes depict the average normalized phase concentration. *Shaded regions* denote the SEM across all electrode pairs and calls. All values are normalized by the baseline mean for different frequency bands. (**c**) Spike-field cross-spectrogram illustrates the relationship between the spiking activity of auditory cortical neurons and the STS local field potential across the population of cortical sites. *X*-axes depict time in milliseconds as a function of the onset of the multisensory response in the auditory neuron (*solid black line*). *Y*-axes depict the frequency in Hz. *Color-bar* denotes the cross-spectral power normalized by the baseline mean for different frequencies

it's possible that these lower frequency interactions between the STS and auditory cortex also represent distinct multisensory processing channels.

Two things should be noted here. The first is that functional interactions between STS and auditory cortex are not likely to occur solely during the presentation of faces with voices. Other congruent, behaviorally salient audiovisual events such as looming signals (Cappe, Thut, Romei, & Murray, 2009; Gordon & Rosenblum, 2005; Maier, Neuhoff, Logothetis, & Ghazanfar, 2004) or other temporally coincident signals may elicit similar functional interactions (Maier, Chandrasekaran, & Ghazanfar, 2008; Noesselt et al., 2007). The second is that there are other areas that, consistent with their connectivity and response properties (e.g., sensitivity to faces and voices), could also (and very likely) have a visual influence on auditory cortex. These include the ventrolateral prefrontal cortex (Romanski, Averbeck, & Diltz, 2005; Sugihara, Diltz, Averbeck, & Romanski, 2006) and the amygdala (Gothard, Battaglia, Erickson, Spitler, & Amaral, 2007; Kuraoka & Nakamura, 2007).

3.2 Viewing Vocalizing Conspecifics: Eye Movements and the Auditory Cortex

Humans and other primates readily link facial expressions with appropriate, congruent vocal expressions. What cues they use to make such matches are not known. One method for investigating such behavioral strategies is the measurement of eye movement patterns. When human subjects are given *no* task or instruction regarding what acoustic cues to attend, they will consistently look at the eye region more than the mouth when viewing videos of human speakers (Klin, Jones, Schultz, Volkmar, & Cohen, 2002). Macaque monkeys exhibit the exact same strategy. The eye movement patterns of monkeys viewing conspecifics producing vocalizations reveal that monkeys spend most of their time inspecting the eye region relative to the mouth (Ghazanfar, Nielsen, & Logothetis, 2006) (Fig. 1.7a). When they did fixate on the mouth, it was highly correlated with the onset of mouth movements (Fig. 1.7b). This, too, was highly reminiscent of human strategies: subjects asked to identify words increased their fixations onto the mouth region with the onset of facial motion (Lansing & McConkie, 2003).

Somewhat surprisingly, activity in both primary auditory cortex and belt areas is influenced by eye position. When the spatial tuning of primary auditory cortical neurons is measured with the eyes gazing in different directions, ~30% of the neurons are affected by the position of the eyes (Werner-Reiss, Kelly, Trause, Underhill, & Groh, 2003). Similarly, when LFP-derived current-source density activity was measured from auditory cortex (both primary auditory cortex and caudal belt regions), eye position significantly modulated auditory-evoked amplitude in about 80 % of sites (Fu et al., 2004). These eye-position effects occurred mainly in the upper cortical layers, suggesting that the signal is fedback from another cortical area. A possible source includes the frontal eye field (FEF) located in the frontal

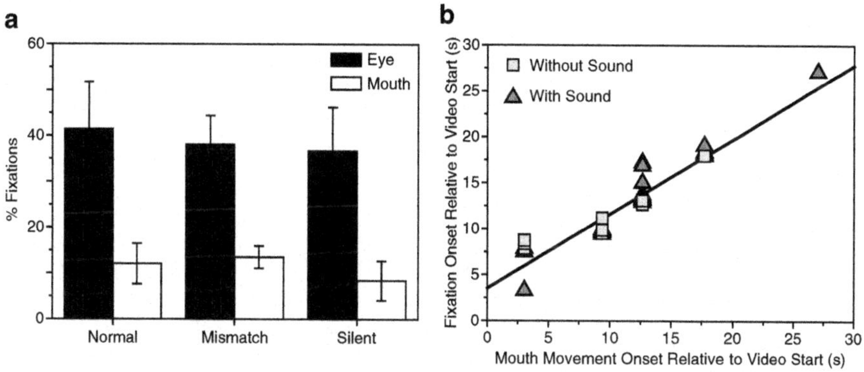

Fig. 1.7 (**a**) The average fixation on the eye region versus the mouth region across three subjects while viewing a 30-s video of vocalizing conspecific. The audio track had no influence on the proportion of fixations falling onto the mouth or the eye region. *Error bars* represent SEM. (**b**) We also find that when monkeys do saccade to the mouth region, it is tightly correlated with the onset of mouth movements ($r = 0.997$, $p < 0.00001$)

lobes, the medial portion of which generates relatively long saccades (Robinson & FUchs, 1969), is interconnected with both the STS (Schall, Morel, King, & Bullier, 1995; Seltzer & Pandya, 1989) and multiple regions of the auditory cortex (Hackett, Stepniewska, & Kaas, 1999; Romanski, Bates, & Goldman-Rakic, 1999; Schall et al., 1995).

It does not take a huge stretch of the imagination to link these auditory cortical processes to the oculomotor strategy for looking at vocalizing faces. A dynamic, vocalizing face is a complex sequence of sensory events, but one that elicits fairly stereotypical eye movements: we and other primates fixate on the eyes but then saccade to mouth when it moves before saccading back to the eyes. Is there a simple scenario that could link the proprioceptive eye position effects in the auditory cortex with its face–voice integrative properties (Ghazanfar & Chandrasekaran, 2007)? Reframing (ever so slightly) the hypothesis of Schroeder and colleagues (Lakatos, Chen, O'Connell, Mills, & Schroeder, 2007; Schroeder, Lakatos, Kajikawa, Partan, & Puce, 2008), one possibility is that the fixations at the onset of mouth movements send a signal to the auditory cortex which resets the phase of an ongoing oscillation. This proprioceptive signal thus primes the auditory cortex to amplify or suppress (depending on the timing) of a subsequent auditory signal originating from the mouth. Given that mouth movements precede the voiced components of both human (Chandrasekaran, Trubanova, Stillittano, Caplier, & Ghazanfar, 2009) and monkey vocalizations (Chandrasekaran & Ghazanfar, 2009; Ghazanfar et al., 2005), the temporal order of visual to proprioceptive to auditory signals is consistent with this idea. This hypothesis is also supported (though indirectly) by the finding that sign of face–voice integration in the auditory cortex and the STS is influenced by the timing of mouth movements relative to the onset of the voice (Chandrasekaran & Ghazanfar, 2009; Ghazanfar et al., 2005).

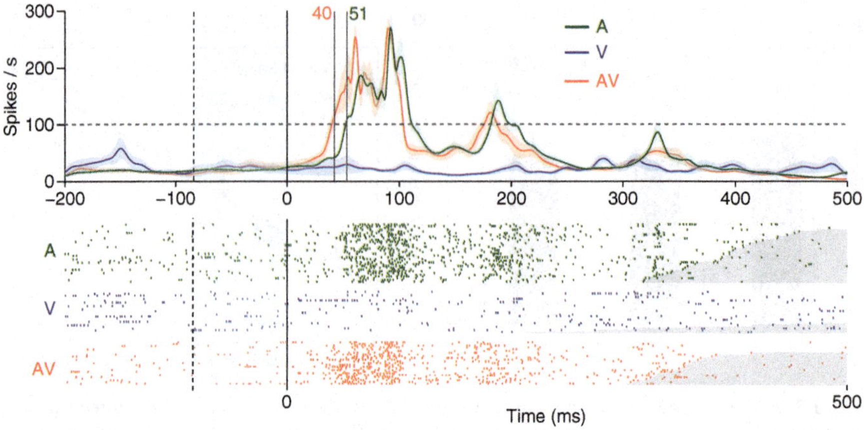

Fig. 1.8 Example of an audiovisual integrative response in an auditory cortical neuron during an audiovisual detection task (as shown in Fig. 1.4). *Top*, Smoothed peristimulus time histogram (gaussian, $\sigma = 8$ ms) of the cortical neuron for the loudest SNR for the audiovisual, visual and auditory components of the vocalization. *X*-axes depicts time in millisecond. *Y*-axes depicts firing rate in spikes/s. Note, the speedup in latency for audiovisual compared to auditory vocalizations. *Bottom*, the spiking rasters for this multiunit cortical site showing a general shift in the latency of the response for audiovisual compared to auditory responses. *X*-axes depict time in milliseconds. *Solid lines* denote auditory onset, *dotted line* and *blue shaded* region denote the region of visual motion

3.3 Linking Multisensory Behavior to Neurophysiology

None of the neurophysiological studies described above required the monkeys to concurrently perform a multisensory behavioral task. Thus, we do not know the relationship between multisensory neural activity and multisensory behaviors. To remedy this situation, the dynamics of audiovisual integration was investigated in the auditory cortex of monkeys detecting "coo" vocalizations in the free response paradigm described above ("Monkeys integrate faces and voices") (Chandrasekaran, Lemus, & Ghazanfar, 2011). Macaque monkeys detected visual, auditory, or audiovisual coo vocalizations by monkey avatars in a background of noise as fast and as accurately as possible. Audiovisual RTs were faster than RTs to both auditory- and visual-only conditions. During these behaviors, spiking activity and local field potential (LFPs) were recorded from the core and lateral belt regions of auditory cortex. Activity in auditory cortex covaried with behavior. First, a decrease in signal-to-noise ratio (SNR) of the auditory-only vocalization increased latency and decreased magnitude of spiking responses. Second, spiking responses were faster for audiovisual compared to auditory vocalizations—a parallel with decreasing reaction times (Fig. 1.8). Spiking responses were however absent to visual-only vocalizations—suggesting subthreshold effects of visual input. In keeping with the profiles of spiking activity, LFP responses to audiovisual vocalizations were faster than auditory vocalizations. In addition, evoked and induced power, as well as intertrial phase coherence, in the 10–30 Hz band of the LFP was suppressed for audiovisual compared to auditory vocalizations. Suppression was maximal for the largest SNR and decreased with the decrease in SNR. This suggests that visual input into auditory cortex changes the state of network dynamics in this frequency band. We also found that the onset of visual mouth motion led to an increase in the intertrial phase coherence in the 10–30 Hz band immediately after visual onset. This is consistent with the phase-resetting hypothesis described above. Taken together, these results suggest that during the detection of audiovisual vocalizations, visual cues speed up and alter the dynamics of the circuits in auditory cortex processing vocalizations.

4 Conclusions

Communication is, by default, a multisensory phenomenon. This is evident in the automatic integration of the senses during vocal perception in humans, monkeys, and numerous other species. Multisensory processes enhance multiple types of behavior, including territorial defense, mate choice, and predator recognition. The overwhelming evidence from the primate studies reviewed here, and numerous other studies from different domains of neuroscience, all converge on the idea that the neocortex is fundamentally multisensory (Ghazanfar & Schroeder, 2006). It is not confined to a few "sensu comune" in the association cortices. It is all over. This does not mean, however, that every cortical area is uniformly multisensory, but rather that cortical areas maybe weighted differently by "extra"-modal inputs depending on the task at hand and its context.

Acknowledgments The authors gratefully acknowledge the scientific contributions and numerous discussions with the following people: Chand Chandrasekaran, Luis Lemus, Joost Maier, Darshana Narayanan, Stephen Shepherd, and Daniel Takahashi. This work was supported by NIH R01NS054898, NSF BCS-0547760 CAREER Award, and the James S. McDonnell Scholar Award.

References

Barnes CL, Pandya DN (1992) Efferent cortical connections of multimodal cortex of the superior temporal sulcus in the rhesus-monkey. The Journal of Comparative Neurology 318:222–244

Barraclough NE, Xiao D, Baker CI, Oram MW, Perrett DI (2005) Integration of visual and auditory information by superior temporal sulcus neurons responsive to the sight of actions. Journal of Cognitive Neuroscience 17:377–391

Benevento LA, Fallon J, Davis BJ, Rezak M (1977) Auditory-visual interactions in single cells in the cortex of the superior temporal sulcus and the orbital frontal cortex of the macaque monkey. Experimental Neurology 57:849–872

Bernstein LE, Auer ET, Takayanagi S (2004) Auditory speech detection in noise enhanced by lipreading. Speech Communication 44:5–18

Besle J, Fort A, Delpuech C, Giard MH (2004) Bimodal speech: early suppressive visual effects in human auditory cortex. The European Journal of Neuroscience 20:2225–2234

Bizley JK, Nodal FR, Bajo VM, Nelken I, King AJ (2007) Physiological and anatomical evidence for multisensory interactions in auditory cortex. Cerebral Cortex 17:2172–2189

Bradbury JW (1981) The evolution of leks. In: Alexander RD, Tinkle DW (eds) Natural selection and social behavior. Chiron Press, New York, NY

Bro-Jorgensen J (2010) Dynamics of multiple signalling systems: animal communication in a world in flux. Trends in Ecology & Evolution 25:292–300

Brown JL (1964) The evolution of diversity in avian territorial systems. The Wilson Bulletin 76: 160–169

Bruce C, Desimone R, Gross CG (1981) Visual properties of neurons in a polysensory area in superior temporal sulcus of the macaque. Journal of Neurophysiology 46:369–384

Candolin U (2003) The use of multiple cues in mate choice. Biological Reviews 78:575–595

Cappe C, Thut G, Romei V, Murray MM (2009) Selective integration of auditory-visual looming cues by humans. Neuropsycholgia 47:1045–1052

Chandrasekaran, C., Lemus, L., & Ghazanfar, A. A. (2011) Dynamic faces speed up vocal processing in the auditory cortex of behaving monkeys. Washington, D.C: Society for Neuroscience.

Chandrasekaran C, Ghazanfar AA (2009) Different neural frequency bands integrate faces and voices differently in the superior temporal sulcus. Journal of Neurophysiology 101:773–788

Chandrasekaran C, Lemus L, Trubanova A, Gondan M, Ghazanfar AA (2011b) Monkeys and humans share a common computation for face/voice integration. PLoS Computational Biology 7(9):e1002165

Chandrasekaran C, Trubanova A, Stillittano S, Caplier A, Ghazanfar AA (2009) The natural statistics of audiovisual speech. PLoS Computational Biology 5:e1000436

Cheney DL, Seyfarth RM (1982) How vervet monkeys perceive their grunts - field playback experiments. Animal Behaviour 30:739–751

Driver J, Noesselt T (2008) Multisensory interplay reveals crossmodal influences on 'sensory-specific' brain regions, neural responses, and judgments. Neuron 57:11–23

Ettlinger G, Wilson WA (1990) Cross-modal performance: behavioural processes, phylogenetic considerations and neural mechanisms. Behavioural Brain Research 40:169–192

Evans TA, Howell S, Westergaard GC (2005) Auditory-visual cross-modal perception of communicative stimuli in tufted capuchin monkeys (Cebus apella). Journal of Experimental Psychology Animal Behavior Processes 31:399–406

Fitch WT (1997) Vocal tract length and formant frequency dispersion correlate with body size in rhesus macaques. The Journal of the Acoustical Society of America 102:1213–1222

Fu KMG, Shah AS, O'Connell MN, McGinnis T, Eckholdt H, Lakatos P et al (2004) Timing and laminar profile of eye-position effects on auditory responses in primate auditory cortex. Journal of Neurophysiology 92:3522–3531

Fusani L, Hutchison RE, Hutchison JB (1997) Vocal-postural Co-ordination of a sexually dimorphic display in a monomorphic species: The Barbary dove. Behaviour 134:321–335

Geschwind N (1965a) Disconnexion syndromes in animals and man, part I. Brain 88:237–294

Geschwind N (1965b) Disconnexion syndromes in animals and man, part II. Brain 88:585–644

Ghazanfar AA, Chandrasekaran CF (2007) Paving the way forward: integrating the senses through phase-resetting of cortical oscillations. Neuron 53:162–164

Ghazanfar AA, Chandrasekaran C, Logothetis NK (2008) Interactions between the superior temporal sulcus and auditory cortex mediate dynamic face/voice integration in rhesus monkeys. Journal of Neuroscience 28:4457–4469

Ghazanfar AA, Logothetis NK (2003) Facial expressions linked to monkey calls. Nature 423: 937–938

Ghazanfar AA, Maier JX, Hoffman KL, Logothetis NK (2005) Multisensory integration of dynamic faces and voices in rhesus monkey auditory cortex. Journal of Neuroscience 25:5004–5012

Ghazanfar AA, Nielsen K, Logothetis NK (2006) Eye movements of monkeys viewing vocalizing conspecifics. Cognition 101:515–529

Ghazanfar AA, Rendall D (2008) Evolution of human vocal production. Current Biology 18: R457–R460

Ghazanfar AA, Schroeder CE (2006) Is neocortex essentially multisensory? Trends in Cognitive Sciences 10:278–285

Ghazanfar AA, Turesson HK, Maier JX, van Dinther R, Patterson RD, Logothetis NK (2007) Vocal tract resonances as indexical cues in rhesus monkeys. Current Biology 17:425–430

Gordon MS, Rosenblum LD (2005) Effects of intrastimulus modality change on audiovisual time-to-arrival judgments. Perception & Psychophysics 67:580–594

Gothard KM, Battaglia FP, Erickson CA, Spitler KM, Amaral DG (2007) Neural responses to facial expression and face identity in the monkey amygdala. Journal of Neurophysiology 97: 1671–1683

Grafe TU, Wanger TC (2007) Multimodal signaling in male and female foot-flagging frogs *Staurois guttatus* (ranidae): An alerting function of calling. Ethology 113:772–781

Grant BR, Grant PR (1989) Evolutionary dynamics of a natural population: the large cactus finch of the Galapagos. Chicago University Press, Chicago, IL

Grant PR, Grant BR (2002) Unpredictable evolution in a 30-year study of Darwin's finches. Science 296:707–711

Guilford T, Dawkins MS (1991) Receiver psychology and the evolution of animal signals. Animal Behaviour 42:1–14

Hackett TA, Stepniewska I, Kaas JH (1999) Prefrontal connections of the parabelt auditory cortex in macaque monkeys. Brain Research 817:45–58

Harries MH, Perrett DI (1991) Visual processing of faces in temporal cortex - physiological evidence for a modular organization and possible anatomical correlates. Journal of Cognitive Neuroscience 3:9–24

Hauser MD, Chomsky N, Fitch W (2002) The faculty of language: What is it, who has it, and how did it evolve? Science 298:1569–1579

Hauser MD, Evans CS, Marler P (1993) The role of articulation in the production of rhesus-monkey, macaca-mulatta, vocalizations. Animal Behaviour 45:423–433

Hauser MD, Ybarra MS (1994) The role of lip configuration in monkey vocalizations - experiments using xylocaine as a nerve block. Brain and Language 46:232–244

Hebets EA (2005) Attention-altering signal interactions in the multimodal courtship display of the wolf spider *Schizocosa uetzi*. Behavioral Ecology 16:75–82

Hebets EA, Papaj DR (2005) Complex signal function: developing a framework of testable hypotheses. Behavioral Ecology and Sociobiology 57:197–214

Hibbitts T, Whiting M, Stuart-Fox D (2007) Shouting the odds: Vocalization signals status in a lizard. Behavioral Ecology and Sociobiology 61:1169–1176

Hoy R (2005) Animal awareness: The (un)binding of multisensory cues in decision making by animals. Proceedings of the National Academy of Sciences of the United States of America 102:2267–2268

Huber SK, Leon LFD, Hendry AP, Bermingham E, Podos J (2007) Reproductive isolation of sympatric morphs in a population of Darwin's finches. Proceedings of the Royal Society B: Biological Sciences 274:1709–1714

Huber SK, Podos J (2006) Beak morphology and song features covary in a population of Darwin's finches (*Geospiza Fortis*). Biological Journal of the Linnean Society 88:489–498

Izumi A, Kojima S (2004) Matching vocalizations to vocalizing faces in a chimpanzee (Pan troglodytes). Animal Cognition 7:179–184

Jiang JT, Alwan A, Keating PA, Auer ET, Bernstein LE (2002) On the relationship between face movements, tongue movements, and speech acoustics. Eurasip Journal on Applied Signal Processing 2002:1174–1188

Jordan KE, Brannon EM, Logothetis NK, Ghazanfar AA (2005) Monkeys match the number of voices they hear with the number of faces they see. Current Biology 15:1034–1038

Kayser C, Logothetis NK (2009) Directed interactions between auditory and superior temporal cortices and their role in sensory integration. Frontiers in Integrative Neuroscience 3:7

Kayser C, Petkov CI, Augath M, Logothetis NK (2007) Functional imaging reveals visual modulation of specific fields in auditory cortex. Journal of Neuroscience 27:1824–1835

Kayser C, Petkov CI, Logothetis NK (2008) Visual modulation of neurons in auditory cortex. Cerebral Cortex 18:1560–1574

Klin A, Jones W, Schultz R, Volkmar F, Cohen D (2002) Visual fixation patterns during viewing of naturalistic social situations as predictors of social competence in individuals with autism. Archives of General Psychiatry 59:809–816

Kuhl PK, Williams KA, Meltzoff AN (1991) Cross-modal speech perception in adults and infants using nonspeech auditory stimuli. Journal of Experimental Psychology Human Perception and Performance 17:829–840

Kuraoka K, Nakamura K (2007) Responses of single neurons in monkey amygdala to facial and vocal emotions. Journal of Neurophysiology 97:1379–1387

Lakatos P, Chen C-M, O'Connell MN, Mills A, Schroeder CE (2007) Neuronal oscillations and multisensory interaction in primary auditory cortex. Neuron 53:279–292

Lakatos P, Shah AS, Knuth KH, Ulbert I, Karmos G, Schroeder CE (2005) An oscillatory hierarchy controlling neuronal excitability and stimulus processing in the auditory cortex. Journal of Neurophysiology 94:1904–1911

Lansing IR, McConkie GW (2003) Word identification and eye fixation locations in visual and visual-plus-auditory presentations of spoken sentences. Perception & Psychophysics 65: 536–552

Lombardo S, Mackey E, Tang L, Smith B, Blumstein D (2008) Multimodal communication and spatial binding in pied currawongs (*Strepera graculina*). Animal Cognition 11:675–682

Maier JX, Chandrasekaran C, Ghazanfar AA (2008) Integration of bimodal looming signals through neuronal coherence in the temporal lobe. Current Biology 18:963–968

Maier JX, Neuhoff JG, Logothetis NK, Ghazanfar AA (2004) Multisensory integration of looming signals by rhesus monkeys. Neuron 43:177–181

Narins PM, Grabul DS, Soma KK, Gaucher P, Hodl W (2005) Cross-modal integration in a dart-poison frog. Proceedings of the National Academy of Sciences of the United States of America 102:2425–2429

Narins PM, Hodl W, Grabul DS (2003) Bimodal signal requisite for agonistic behavior in a dart-poison frog, epipedobatesfemoralis. Proceedings of the National Academy of Sciences of the United States of America 100:577–580

Noesselt T, Rieger JW, Schoenfeld MA, Kanowski M, Hinrichs H, Heinze H-J et al (2007) Audiovisual temporal correspondence modulates human multisensory superior temporal sulcus plus primary sensory cortices. Journal of Neuroscience 27:11431–11441

Oram MW, Perrett DI (1994) Responses of anterior superior temporal polysensory (Stpa) neurons to biological motion stimuli. Journal of Cognitive Neuroscience 6:99–116

Palombit RA, Cheney DL, Seyfarth RM (1999) Male grunts as mediators of social interaction with females in wild chacma baboons (papio cynocephalus ursinus). Behaviour 136:221–242

Parr LA (2004) Perceptual biases for multimodal cues in chimpanzee (pan troglodytes) affect recognition. Animal Cognition 7:171–178

Partan SR, Larco CP, Owens MJ (2009) Wild tree squirrels respond with multisensory enhancement to conspecific robot alarm behaviour. Animal Behaviour 77:1127–1135

Partan S, Marler P (1999) Communication goes multimodal. Science 283:1272–1273

Partan S, Yelda S, Price V, Shimizu T (2005) Female pigeons, *Columba livia,* respond to multisensory audio/video playbacks of male courtship behaviour. Animal Behaviour 70:957–966

Pevsner J (2002) Leonardo da Vinci's contributions to neuroscience. Trends in Neurosciences 25:217–220

Podos J (2010) Acoustic discrimination of sympatric morphs in Darwin's finches: a behavioural mechanism for assortative mating? Philosophical Transactions of the Royal Society B: Biological Sciences 365:1031–1039

Roberts JA, Taylor PW, Uetz GW (2007) Consequences of complex signaling: Predator detection of multimodal cues. Behavioral Ecology 18:236–240

Robinson DA, FUchs AF (1969) Eye movements evoked by stimulation of frontal eye fields. Journal of Neurophysiology 32:637–648

Romanski LM, Averbeck BB, Diltz M (2005) Neural representation of vocalizations in the primate ventrolateral prefrontal cortex. Journal of Neurophysiology 93:734–747

Romanski LM, Bates JF, Goldman-Rakic PS (1999) Auditory belt and parabelt projections to the prefrontal cortex in the rhesus monkey. The Journal of Comparative Neurology 403:141–157

Rowe C (1999) Receiver psychology and the evolution of multicomponent signals. Animal Behaviour 58:921–931

Schall JD, Morel A, King DJ, Bullier J (1995) Topography of visual cortex connections with frontal eye field in macaque: Convergence and segregation of processing streams. Journal of Neuroscience 15:4464–4487

Schroeder CE, Foxe JJ (2002) The timing and laminar profile of converging inputs to multisensory areas of the macaque neocortex. Cognitive Brain Research 14:187–198

Schroeder CE, Lakatos P, Kajikawa Y, Partan S, Puce A (2008) Neuronal oscillations and visual amplification of speech. Trends in Cognitive Sciences 12:106–113

Schwartz J-L, Berthommier F, Savariaux C (2004) Seeing to hear better: Evidence for early audio-visual interactions in speech identification. Cognition 93:B69–B78

Seltzer B, Pandya DN (1989) Frontal-lobe connections of the superior temporal sulcus in the rhesus-monkey. The Journal of Comparative Neurology 281:97–113

Seltzer B, Pandya DN (1994) Parietal, temporal, and occipital projections to cortex of the superior temporal sulcus in the rhesus monkey: A retrograde tracer study. The Journal of Comparative Neurology 343:445–463

Sherman PW, Reeve HK, Pfennig DW (1997) Recognition systems. In: Krebs JR, Davies NB (eds) Behavioural ecology: an evolutionary approach. Cambridge University Press, Cambridge, pp 69–96

Sliwa J, Duhamel JR, Pascalis O, Wirth S (2011) Spontaneous voice-face identity matching by rhesus monkeys for familiar conspecifics and humans. Proceedings of the National Academy of Sciences of the United States of America 108:1735–1740

Sugihara T, Diltz MD, Averbeck BB, Romanski LM (2006) Integration of auditory and visual communication information in the primate ventrolateral prefrontal cortex. Journal of Neuroscience 26:11138–11147

Taylor RC, Buchanan BW, Doherty JL (2007) Sexual selection in the squirrel treefrog *Hyla squirella*: the role of multimodal cue assessment in female choice. Animal Behaviour 74:1753–1763

Taylor RC, Klein BA, Stein J, Ryan MJ (2008) Faux frogs: multimodal signalling and the value of robotics in animal behavior. Animal Behaviour 76:1089–1097

van Wassenhove V, Grant KW, Poeppel D (2005) Visual speech speeds up the neural processing of auditory speech. Proceedings of the National Academy of Sciences of the United States of America 102:1181–1186

Werner-Reiss U, Kelly KA, Trause AS, Underhill AM, Groh JM (2003) Eye position affects activity in primary auditory cortex of primates. Current Biology 13:554–562

Wilson EO (1975) Sociobiology the new synthesis. Harvard University Press, Cambridge

Wollerman L (1999) Acoustic interference limits call detection in a neotropical frog, *Hyla ebraccata*. Animal Behaviour 57:529–536

Yehia HC, Kuratate T, Vatikiotis-Bateson E (2002) Linking facial animation, head motion and speech acoustics. Journal of Phonetics 30:555–568

Yehia H, Rubin P, Vatikiotis-Bateson E (1998) Quantitative association of vocal-tract and facial behavior. Speech Communication 26:23–43

Chapter 2
Cross-Modal Representation in Humans and Nonhuman Animals: A Comparative Perspective

Akihiro Izumi

Abstract Auditory–visual representation provide redundant information about vocal individuals (i.e., who is vocalizing), and studies have reported such an ability in various vertebrate species. I introduce behavioral evidences of such abilities in animals and characterize the experimental paradigms that have been used in this field of study. I then compare vocal-type representation in nonhuman primates with that in humans, and discuss the evolution of human-specific phoneme representation (representation of articulatory gestures) that might relate to the faculty of language.

1 Introduction

The ability to integrate information cross-modally has been regarded as essential for the emergence of language (Geschwind, 1965). To determine whether this ability is not limited to humans, previous studies examined whether nonhuman primates could perform a tactile–visual matching-to-sample task. Although humans showed tactile–visual representation of object shapes from an early stage of development (Meltzoff & Borton, 1979), rhesus monkeys failed to perform the task, suggesting that cross-modal performance is unique to humans (Ettlinger, 1967; Ettlinger & Blakemore, 1967).

A pioneering study by Davenport and Rogers (1970) demonstrated that a cross-modal ability is shared by nonhuman primates. They trained apes (chimpanzees, orangutans, and gorillas) to perform a visual–tactile matching-to-sample task. In a trial, the following three objects were presented as stimuli: a sample stimulus which the subject apes could see but could not touch, and two comparison stimulus which the apes could touch but could not see. Only one of the two tactually presented

A. Izumi (✉)
Primate Research Institute, Kyoto University,
Kanrin, Inuyama, Aichi 484-8506, Japan
e-mail: izumi@pri.kyoto-u.ac.jp

P. Belin et al. (eds.), *Integrating Face and Voice in Person Perception*,
DOI 10.1007/978-1-4614-3585-3_2,

comparison stimuli was identical to the sample stimulus, and the task was to identify this identical stimulus. Some of the apes (including chimpanzees and orangutans) acquired the task and could transfer their performances to novel stimuli that had not been used for training. A follow-up study showed that the apes immediately transferred their performances when the sample stimulus was replaced with a picture (color or black and white) of the object (Davenport & Rogers, 1971).

In addition to this visual–tactile matching-to-sample study, Davenport, Rogers, and Russell (1973) demonstrated tactile to visual matching in six experimentally naive chimpanzees that were different from those included in the previous studies. Later, Cowey and Weiskrantz (1975) showed similar visual–tactile matching in rhesus monkeys. Cross-modal performance has now been demonstrated in a number of species, especially primates, and recently many attempts have been made to reveal neural mechanisms underlying such an ability (for review, see Ghazanfar & Schroeder, 2006; Stein & Stanford, 2008).

The ability to match novel auditory stimuli to visual stimuli seems to be essential for language acquisition in human infants (e.g., Gogate & Bahrick, 1998). To reveal evolutional specializations in humans, it is important to compare humans with other primate species that are of evolutional proximity. Despite the importance of such studies from the comparative perspective, investigations on auditory–visual representations have not been done extensively possibly because of the difficulty faced by nonhuman primates to acquire auditory association tasks.

Here I focus on auditory–visual conceptual representations of various types, including stimulus identity, because of their possible relevance to language ability in humans. Besides conceptual representations, auditory–visual effects related to sound localization have been reported. In humans the perceived spatial location of a sound source is influenced by the location of a visual stimulus that is presented in close temporal proximity (Howard & Templeton, 1966). These auditory and visual stimuli are perceived as originating from a common source, and the superiority of vision in terms of spatial resolution has a dominant effect on source localization. This phenomenon is known as the ventriloquist effect and has also been reported in rhesus monkeys (Woods & Recanzone, 2004). The auditory–visual processes concerned with source localization (i.e., where the sound comes from) seem to differ from those concerned with source identification (what the source is) in their behavioral properties. These two types of processes might also differ in the neural mechanisms; they correspond to information streams on stimulus location ("where") and identity ("what") in auditory and visual modalities (for review, see Calvert, Brammer, & Iversen, 1998).

I first introduce behavioral evidences of auditory–visual representations in nonhuman animals and characterize the experimental paradigms that are used in this field of study. The methodology of acquiring behavioral evidence is a critical issue in the examination of mental representation in both nonhumans and nonlinguistic infants. Auditory–visual representations of concepts are classified into two types: representations of sound source identity (e.g., identification of vocal individuals) and representations of vocal types (e.g., phoneme representations in humans). I compare these representations in nonhumans with those in humans, and discuss the evolution of human specific representations of phonemes and articulatory gestures that might relate to the faculty of language.

2 Methodology for Examining Auditory–Visual Representations in Animals

Table 2.1 summarizes studies that demonstrated auditory–visual representations in nonhuman animals. These studies used four types of experimental paradigms. Two paradigms (preferential looking and expectancy violation) were developed in studies with human infants; both of them use spontaneous looking behavior and do not require task training. The remaining two paradigms (cross-modal interference and matching-to-sample) require subjects to perform behavioral tasks.

2.1 Preferential Looking

The preferential looking paradigm is advantageous when examining the subject's spontaneous responses, because it does not require task training and involves minimum habituation to the experimental setting. This paradigm has been widely used in studies with preverbal infants, including studies examining auditory–visual representations of vowels (e.g., Kuhl & Meltzoff, 1982; Patterson & Werker, 2003). Ghazanfar and Logothetis (2003) used the paradigm to examine auditory–visual representations of species-specific vocalizations in rhesus monkeys. A subject monkey was restrained on a monkey chair and was presented with two movies showing a monkey articulating two types of vocalization (coo and threat). One of the movies was accompanied by sound playback and the other was silent. The monkeys spent more time looking at the movie with sound playback than at the silent movie, suggesting auditory–visual representation of vocalizations. Using a similar method, Evans, Howell, and Westergaard (2005) demonstrated auditory–visual representation of conspecific vocalizations in capuchin monkeys.

This paradigm has also been used to examine other aspects of cross-modal representation, possibly because of its simplicity and widespread use. Maier, Neuhoff, Logothesis, and Ghazanfar (2004) showed that rhesus monkeys spent more time looking at coincident visual and auditory looming stimuli, suggesting auditory–visual representation of these stimuli. Jordan, Brannon, Logothetis, and Ghazanfar (2005) found that rhesus monkeys preferred to look at a movie in which the number of conspecifics shown matched the number of vocalizations heard, suggesting the monkeys had representation of the number of voices.

Basically, this paradigm uses spontaneous preference for looking at the visual stimulus that corresponds to the sound played at the same time. The difficulty with this paradigm is that such a preference might not always exist. To examine cross-species representations of vocalizations, Zangenehpour, Ghazanfar, Lewkowicz, and Zatorre (2009) conducted a similar preferential looking experiment in vervet monkey infants, using rhesus vocalizations as stimuli. The vervet monkeys showed a cross-modal effect, but spent more time looking at incongruent stimuli than at congruent stimuli (a result opposite to that previously observed in rhesus monkeys by

Table 2.1 Behavioral studies on auditory–visual cross-modal representation in nonhuman animals

Cue (auditory–visual correspondence)	Subject animal	Stimulus	Paradigm	Article
Arbitrary	Rhesus monkey	Geometric	Cross-modal matching	Gaffan and Harrison (1991), Murray and Gaffan (1994)
	Cynomolgus monkey	Geometric	Cross-modal matching	Colombo and Graziano (1994)
	Bonobo	Words, lexigrams and pictures	Cross-modal matching	Savage-Rumbaugh et al. (1988)
	Domestic dog	Words and objects	Cross-modal matching	Kaminski et al. (2004)
Species	Guinea baboon	Humans and baboons	Cross-modal interference (priming)	Martin-Malivel and Fagot (2001)
	Japanese monkey	Humans and monkeys	Expectancy violation	Adachi, Kuwahata, Fujita, Tomonaga, and Matsuzawa (2006)
	Chimpanzee	Humans and objects	Cross-modal matching	Hashiya and Kojima (1997, 2001a, 2001b)
Individuality	Chimpanzee	Humans	Cross-modal matching	Hashiya and Kojima (2001b)
	Chimpanzee	Conspecifics	Cross-modal matching	Kojima et al. (2003)
	Gray-cheeked mangabey	Conspecifics	Preferential looking	Bovet and Deputte (2009)
	Squirrel monkey	Humans	Cross-modal interference	Adachi and Fujita (2007)
	Domestic dog	Humans	Expectancy violation	Adachi et al. (2007)
	Domestic horse	Conspecifics	Expectancy violation	Proops et al. (2009)
Vocal type	Chimpanzee	Conspecifics	Cross-modal matching	Izumi and Kojima (2004)
	Rhesus monkey	Conspecifics	Preferential looking	Ghazanfar and Logothetis (2003)
	Capuchin monkey	Conspecifics	Preferential looking	Evans et al. (2005)
	Vervet monkey	Rhesus monkeys	Preferential looking	Zangenehpour et al. (2009)
Vocal number	Rhesus monkey	Conspecifics	Preferential looking	Jordan et al. (2005)
Body size (formant)	Rhesus monkey	Conspecifics	Preferential looking	Ghazanfar et al. (2007)
Looming/receding	Rhesus monkey	Looming/receding stimuli	Preferential looking	Maier et al. (2004)

Ghazanfar and Logothesis (2003)). The authors explained their results by suggesting that infant vervet monkeys showed more fear and anxiety when exposed to congruent stimuli than to incongruent stimuli. Bovet and Deputte (2009) conducted a similar experiment with six mangabeys (a species of Old World monkey) to examine cross-modal representation of vocal individuality. Two of the mangabeys preferred to look at congruent stimuli, but the other four preferred to look at incongruent stimuli. Still pictures were used in that study but other studies (e.g., Ghazanfar & Logothetis, 2003) used movies as visual stimuli. A movie and its accompanying soundtrack share characteristics other than the conceptual content. Rather than conceptual representations of vocal types, monkeys might use cues such as temporal synchronization to detect the congruency of auditory–visual stimuli.

2.2 *Expectancy Violation*

Another experimental paradigm requiring minimal training or habituation is expectancy violation. This paradigm is based on the assumption that subjects will spend more time looking at events that violate their expectations, and has been used in studies with preverbal infants (e.g., Spelke, 1985; Wynn, 1992) and animals (e.g., Santos & Hauser, 1999).

Using this paradigm, Adachi, Kuwahata, and Fujita (2007) demonstrated that domestic dogs possess auditory–visual representations of their owners. In a trial, a dog was presented with the voice of its owner or that of a stranger, followed by a presentation of either the owner's or the stranger's face. Dogs spent more time looking at a face when the depicted individual was different from the vocal individual (i.e., the incongruent condition). Similarly, Proops, McComb, and Reby (2009) showed that domestic horses possess cross-modal representations of familiar horses. After visual presentation of a familiar horse, a vocalization of either the visually presented horse (congruent vocalization) or another familiar horse (incongruent vocalization) was played. After the incongruent vocalization was played, the horses responded more quickly and looked in the direction of the stimulus horse more often and for a longer time. In both the dogs and horses, the presentation of the preceding stimuli induced the subjects' cross-modal representation of individuals, and subsequent presentation of cues from different individuals seemed to violate their expectations.

2.3 *Cross-Modal Interference*

In the cross-modal interference paradigm, the subject is first trained to perform a behavioral task with one sensory modality (e.g., a discrimination task in the visual modality). After acquisition of the task, a stimulus of another modality is introduced in a part or trials. Although the newly introduced modality does not serve as a task cue, it is assumed to have an effect on task performance if the subject performs the trained task using a cross-modal concept of some types.

Martin-Malivel and Fagot (2001) trained two Guinea baboons to perform an auditory go/no-go task. One baboon was required to respond (a go response) to human vocalizations and not respond (a no-go response) to baboon vocalizations, and the other baboon was trained vice versa. Before the presentation of the vocal stimulus, a picture of either a human or baboon was presented as a prime stimulus. One of the two baboons showed quicker responses when the stimulus category (i.e., human or baboon) of the prime stimulus matched the vocal stimulus, suggesting cross-modal priming. The other baboon did not show such a cross-modal priming effect.

Adachi and Fujita (2007) trained two squirrel monkeys to perform a visual matching-to-sample task. In a trial, a picture of one of two familiar humans (primary and secondary caretakers) was presented as a sample stimulus and then two figures (heart and moon) were presented as choice stimuli. Each of the two figures was corresponded to the primary or the secondary caretakers, respectively. The monkeys had to select the corresponding figure in response to the sample stimulus. In the test trials, the voice of one of the caretakers was presented just before presenting the choice stimuli. Matching performances in trials with the secondary caretaker's face were reduced by presentation of the primary caretaker's voice, showing a cross-modal interference effect of individuality. On the other hand, presentations of the secondary caretaker's voice had no cross-modal effect on matching performances. Although the monkeys might possess cross-modal representations of the primary caretaker alone, why matching performances with the primary caretaker's face were not affected by the (incongruent) voice of the secondary caretaker remains unknown. Although Martin-Malivel and Fagot (2001) and Adachi and Fujita (2007) trained monkeys to perform behavioral tasks with a sensory modality, the experimental paradigm used by these authors did not involve training the subject animals to perform a cross-modal task per se. Similar to the preferential looking and expectancy violation paradigms, the suggested cross-modal representations in these studies did not seem to be a result of training. The difficulties with this paradigm seem to be that the cross-modal effect might not always be robust and it depends on subjects and conditions.

2.4 Cross-Modal Matching-to-Sample

The classical and most direct method to examine cross-modal representation is by training animals to perform cross-modal matching. The pioneering study by Davenport and Rogers (1970) used a tactile–visual matching-to-sample task. Studies have confirmed that macaque monkeys can associate auditory and visual stimuli (Colombo & Graziano, 1994; Gaffan & Harrison, 1991; Murray & Gaffan, 1994). For example, Gaffan and Harrison (1991) successfully trained six rhesus monkeys to perform a matching-to-sample task with six arbitrary pairs of auditory–visual stimuli. However, this type of auditory association task is generally difficult to acquire by animals including nonhuman primates, and many trials are required to acquire this task.

Although studies with a bonobo (a species of ape; Savage-Rumbaugh, Sevcik, & Hopkins, 1988) and a domestic dog (Kaminski, Call, & Fischer, 2004) used a sort of auditory–visual matching-to-sample task to demonstrate their symbolic representations, the process of task acquisition and the degree of transfer to novel stimuli seemed to be substantially different from those observed in monkeys in previous studies (e.g., Gaffan & Harrison, 1991). The bonobo studied by Savage-Rumbaugh et al. (1998) was trained to use lexigrams (visual symbols that represents objects and other concepts) and exposed to human spoken English prior to the experiment. Without additional training to perform cross-modal matching, the bonobo correctly identified corresponding lexigrams and pictures in response to spoken English. The dog studied by Kaminski et al. (2004) learned to retrieve more than 200 items in response to humans' vocal requests during everyday interactions with humans. The dog showed an ability of "fast mapping" (Carey & Bartlett, 1978) based on the principle of exclusion; it could inferentially link novel words to novel items. The remarkable performances of the bonobo and dog in these studies are interesting in terms of the evolution of human symbolic representation, and further research is needed to examine whether such abilities are shared by conspecific and heterospecific animals.

An arbitrary relationship between auditory and visual stimuli (such as white noise matched with a red square) needs to be acquired during training. Although the successful acquisition of an arbitrary cross-modal matching task suggests that an animal has such a learning potential, whether this potential is realized in the animal's natural environment is unclear. On the other hand, natural associations between stimuli might be acquired during everyday life. For example, humans acquire association between the voices and faces of familiar persons during everyday life. In case of matching-to-sample paradigms, determining whether the subject possessed cross-modal representation before the experimental training or whether they acquired this association through intensive training is important.

If matching-to-sample performance is immediately transferred to novel untrained stimuli, the performance is not mere an association between trained auditory and visual stimuli. For example, Hashiya and Kojima (2001a) trained a chimpanzee to perform an auditory–visual matching-to-sample task. Sounds of various familiar objects (e.g., bell, whistle) were presented as sample stimuli, and the task was to select the corresponding picture (i.e., a picture of the sound source). Intensive training was necessary for the chimpanzee to master the task. After acquisition of the task with the initial six objects, the chimpanzee immediately transferred its cross-modal performance to novel stimuli. The results suggested that the chimpanzee possessed some type of cross-modal concept of sound-producing objects during everyday life.

Izumi and Kojima (2004) investigated chimpanzees for cross-modal representations of their vocal types (e.g., pant hoots, screams) using a cross-modal individual recognition task similar to that described by Kojima, Izumi, and Ceugniet (2003). The sample stimulus was a chimpanzee (sample individual) vocalization, and the test stimuli were two choice movies. In a training trial, one of the movies showed the sample individual and the other showed a different chimpanzee, and the correct response was to select the former movie. In a test trial, one of the movies depicted the sample individual vocalizing the same type of vocalization as the sample vocalization (congruent movie), while the same sample individual vocalized another type of vocalization in the other movie (incongruent movie). Because both movies depicted the sample individual, a response to either movie was rewarded (i.e., probe test trials with nondifferential reinforcement). The chimpanzee preferentially selected the congruent movie, suggesting cross-modal representation of vocal types. Although the task itself was cross-modal matching of vocal individuals, the results suggested cross-modal interference of vocal types that reflected the chimpanzee's spontaneous responses.

Because monkeys show auditory–visual abilities in other experimental paradigms without training (e.g., Ghazanfar & Logothetis, 2003), it seems strange that nonhuman primates need intensive training to acquire auditory–visual matching-to-sample tasks (e.g., Hashiya & Kojima, 2001a). Ironically, the cause of the difficulty in acquiring the task might be related to the intensive training itself, in which animal subjects usually perform several tens or more trials in a day. For example, in a study using the expectancy violation paradigm, a horse performed only one trial a day, and there was an interval of 4 or more days between trials to prevent habituation (Proops et al., 2009). These animals might be expected to possess auditory–visual representations before training, but habituation to the experimental settings especially to the auditory stimuli might influence their performance in matching-to-sample tasks. In practice, the introduction of novel auditory stimuli, or using trial-unique stimuli, seems to facilitate the acquisition of auditory tasks in nonhuman primates (Hashiya & Kojima, 1997; Wright, Shyan, & Jitsumori, 1990). Humans can perform auditory–visual matching without explicit training. Similarly, the bonobo studied by Savage-Rumbaugh et al. (1988) and the dog studied by Kaminski et al. (2004) acquired remarkable auditory–visual performance without any specific cross-modal training. These impressive animals might somehow have learned to overcome the effects of habituation to auditory cues during everyday interactions with humans.

3 What Is Special About Human Auditory–Visual Representation?

As discussed above, studies during the last decade have demonstrated auditory–visual conceptual representations in nonhuman animals. Auditory–visual representations can be classified into two types: representations of sound source identity (e.g., vocal individuality) and representations of vocal types. Below I discuss whether these representations are specialized in human evolution.

3.1 Sound-Source Identification

Among the previous studies (Table 2.1), auditory–visual representations of vocal individuals have been demonstrated in various species including dogs (Adachi et al.,

2007) and horses (Proops et al., 2009). The identification of species and that of individuals constitute sound-source identification (i.e., what the source is). Various species might possess such a cross-modal identity representation because it confers the advantage providing redundant information on individual or object identity from multiple modalities.

Nonhuman primates show less robust auditory short-term memory compared with auditory and visual memory in humans and visual memory in monkeys (Japanese monkey: Kojima, 1985; cebus monkey: Colombo & D'Amato, 1986; chimpanzee: Hashiya & Kojima, 2001a). Cross-modal identification of sound sources is hypothesized to help maintain auditory memory by using visual memory. Colombo and Graziano (1994) trained two cynomolgus monkeys to perform an auditory–visual delayed matching-to-sample task with two arbitrary pairs of auditory–visual stimuli. During the delay period of 3 s or 9 s, interference was provided auditorily (by playing music) or visually (by switching on the house light). The performance of both monkeys was strongly affected by the visual interference, suggesting that the monkeys remembered visual information during the delay period. Monkeys might use visual information because of its advantages in memory retention.

A prerequisite for performing cross-modal identification is categorization of the stimuli within each modality. In case of auditory–visual individual recognition, identification of individuals using each of these modalities is necessary. Vocal recognition of individuals has been well investigated in avian species (e.g., Jouventin, Aubin, & Lengagne, 1999). Studies have also demonstrated vocal recognition of individuals in a wide range of mammals including primates (e.g., Ceugniet & Izumi, 2004; Weiss, Garibaldi, & Hauser, 2001), dolphins (Sayigh et al., 1998), and rodents (e.g., Blumstein & Daniel, 2004). Although there have been relatively few studies of the visual recognition of individuals, identification of the faces of individual conspecifics has been reported in cattles (Coulon, Deputte, van Heyman, & Baudoin, 2009), horses (Proops et al., 2009), and primates (e.g., Parr, Winslow, Hopkins, & de Waal, 2000; Pokorny & de Waal, 2009).

Perhaps animals that can identify individuals by both of the two modalities are able to perform cross-modal identification. What might differ among species is the degree to which animals use abstract information from the sound source. Few studies have been conducted on this aspect: however, Hashiya and Kojima (2001b) reported that a chimpanzee acquired the task of matching human voices and faces on the basis of individual identity, and transferred the performance to novel stimuli from unfamiliar humans. When the voice and both of the two faces belonged to unfamiliar humans, the chimpanzee tended to select a sex-matched picture in response to the sample vocalization. The chimpanzee seemed to possess some knowledge on sex differences in terms of human voices and faces and could use this knowledge to identify individuals by their voices. Using a preferential looking paradigm in rhesus monkeys, Ghazanfar et al. (2007) showed that the monkeys perceived the relationship between age-related body growth and acoustic characters of voices (formants).

3.2 Representation of Vocal Types

Unlike cross-modal representation of sound sources, representation of vocal types has been reported only in primates. Two studies using the preferential looking paradigm demonstrated auditory–visual representation of conspecific vocal types in monkeys (rhesus monkeys: Ghazanfar & Logothetis, 2003; capuchin monkeys: Evans et al., 2005). Using a similar method, Zangenehpour et al. (2009) demonstrated that vervet monkey infants possessed auditory–visual representation of rhesus monkey vocalizations. Results showing cross-species vocal representations can be considered in terms of perceptual narrowing; infant monkeys have cross-modal sensitivity to a broad range of monkey vocalizations, and such sensitivity shows specialization through experience with conspecifics (Lewkowicz & Ghazanfar, 2009). At the same time, these results also raise the possibility that such cross-modal effects rely on relatively general processing of auditory–visual events and do not necessarily require conceptual representations of vocal types. Specifically, temporal synchronization of sounds and fine facial movements could provide monkeys with cues that they can use to detect the congruency of auditory–visual stimuli. Although whether the looking preferences of the monkeys were actually affected by auditory-visual synchrony is unknown, human infants perceive such synchronization (e.g., Dodd, 1979; Spelke, 1979). Spelke (1979) showed that 4-month-old infants preferred to look at the bouncing object that accompanied temporally synchronized sounds. Human infants were reported to perceive both temporal synchrony and phonetic correspondence of auditory–visual speech (Kuhl & Meltzoff, 1984; Kuhl, Williams, & Meltzoff, 1991). Further studies with nonhuman species are required to separate the effects of temporal synchrony from those of conceptual correspondence (i.e., matching of vocal types) between vocal sounds and faces.

Izumi and Kojima (2004) used a vocal-movie matching-to-sample task and showed that a chimpanzee preferred to select movies that were congruent in vocal type with sample vocalizations. To avoid the effects of synchronization, the sounds and movies always came from different utterances; they were recorded in different occasion so they did not synchronize even if they were congruent in vocal type. The spontaneous preference for the congruent movie suggested that the chimpanzee used a representation that was related to auditory and visual stimuli of the same vocal type. Humans perceive phonetic information by observing mouse and lip movements (speechreading or lipreading: e.g., Bernstein, Demorest, & Tucker, 2000), and visual observation of the speaker's facial expression and lip movements improves the intelligibility of speech especially in noisy environments (e.g., Munhall & Vatikiotis-Bateson, 1998; Sumby & Pollack, 1954). The results of Izumi and Kojima (2004) apparently demonstrate a phenomenon similar to human speechreading, but the authors did not suggest that the chimpanzee possessed phoneme representations similar to those of humans. Together with phoneme information, humans perceive the cross-modal correspondence of affective states (e.g., Walker, 1982). Movie clips showing chimpanzee might contain various contextual cues such as gaze directions, head movements, and facial expressions other than mouse or lip movements. Vocalizations of nonhuman animals are closely related to the animals' emotional state in nature, and the vocal-type matching in the chimpanzee might be mediated by affective state properties that are shared by auditory and movie stimuli (i.e., what the affective context is). Parr (2001) demonstrated that chimpanzees matched emotional movies and pictures of chimpanzees' facial expressions on the basis of their emotional meaning. For example, the chimpanzees matched a hypodermic needle and the bared-teeth face, both of which are related to negative emotions.

Both faces and voices contain rich information about individual identity, affective state, and phonemes. Humans seem to integrate these types of information cross-modally (for review, see Campanella & Belin, 2007). To further examine whether nonhuman primates integrate information beyond individual identity, more controlled stimuli must be used in particular to separate the effects of affective state and phonemic properties. One solution is to use human speech. Using human phonemes as stimuli, phonological phenomena such as the phoneme boundary effect of consonants have been examined in nonhuman primates (e.g., Kojima, 2003). A problem in using human speech is that it is unknown whether subject animals process such stimuli in the same way that they process their conspecific vocalizations or whether they process them only as sound sequences. Another approach is to use synthesized faces and voices that imitate those of conspecifics but do not include contextual cues. Such a technique will enable us to manipulate lip shapes of chimpanzees in order to represent different vocalizations while preserving other facial expressions. Although no study has examined whether nonhuman animals show the McGurk effect (McGurk & MacDonald, 1976), it seems possible to examine such an effect with synthesized auditory–visual stimuli.

I hypothesize that auditory–visual representations of vocal types in nonhuman primates are different from phoneme representations in humans, both in their function and mechanisms. In chimpanzees, such cross-modal representation seems to have the function of providing redundant information with which the animal can infer the other individual's affective state and its background social context. According to this notion, chimpanzees probably do not show McGurk effect because they do not integrate auditory–visual phoneme information. On the other hand, perceptual representations of phonemes in humans seem to be closely related to representations of articulatory gestures (Liberman & Whalen, 2000). Kuhl and Meltzoff (1982) reported that human infants aged 18–20 weeks tried to mimic stimulus vowels in an experiment for examining auditory–visual representations. Cross-modal representations of phonemes in humans seem to be relevant to the acquisition of articulatory gestures. In nonhuman primates, cross-modal representations of vocalizations seem to be irrelevant to vocal learning; these animals do not mimic vocalizations and they rarely show an evidence of vocal production learning (i.e., learning how to produce vocal sounds; Janik & Slater, 1997; Yamaguchi & Izumi, 2008). Izumi, Kuraoka, Kojima, and Nakamura (2001) reported that rhesus monkeys could be conditioned to express three facial actions (tongue protrusion, mouth opening, and mouth distortion) in response to arbitrary visual cues. Interestingly, vocalizations seem to be difficult for monkeys to control, but not mouth movements. Unlike humans, nonhuman primates seem to lack phoneme representation that can transfer information between perception and action.

4 Conclusion

Cross-modal representation was once believed to be unique to humans, but now various animals, especially primates, are known to possess this ability including auditory and visual modalities. Cross-modal identity representations (e.g., auditory–visual individual representation) provide redundant information about individual and object identity (i.e., what the source is) from multiple modalities, and seem to be shared by various vertebrate species. Auditory–visual representations of vocal types have been reported only in primates. Although this type of representation is apparently similar to phoneme representation in humans, these are hypothesized to differ in both their mechanisms and functions. Human phoneme representations are closely related to representations of articulatory gestures (i.e., "how" the phonemes are pronounced), and seem to be relevant to the acquisition of speech. Instead of articulatory gestures, vocal type representations in nonhumans might be mediated by cues such as affective properties that are shared by auditory and visual stimuli (e.g., "what" the affective context is). Similar to identity representations such vocal type representations seem to provide redundant information enabling animals to infer another individual's affective state and its background social context.

In future, whether nonhuman animals truly lack phoneme representations similar to those in humans must be examined. One promising way of examining the effects of mouth movements on acoustic perception, such as the McGurk effect, is to use synthesized faces and voices that imitate those of conspecifics while excluding contextual cues. Another important question for investigation is why nonhumans usually show difficulty in acquiring cross-modal tasks, whereas humans do not need explicit training. A bonobo (Savage-Rumbaugh et al., 1988) and domestic dog (Kaminski et al., 2004) were found to show exceptional performance in task acquisition and extensive transfer of their performances. Further study is necessary to determine whether other bonobos, dogs, and animals belonging to other species also have such abilities, and how such remarkable performance is acquired during everyday interactions with humans.

References

Adachi, I., & Fujita, K. (2007). Cross-modal representation of human caretakers in squirrel monkeys. *Behavioural Processes, 74*, 27–32.

Adachi, I., Kuwahata, H., & Fujita, K. (2007). Dogs recall their owner's face upon hearing the owner's voice. *Animal Cognition, 10*, 17–21.

Adachi, I., Kuwahata, H., Fujita, K., Tomonaga, M., & Matsuzawa, T. (2006). Japanese macaques form a cross-modal representation of their own species in their first year of life. *Primates, 47*, 350–354.

Bernstein, L. E., Demorest, M. E., & Tucker, P. E. (2000). Speech perception without hearing. *Perception & Psychophysics, 62*, 233–252.

Blumstein, D. T., & Daniel, J. C. (2004). Yellow-belled marmots discriminate between the alarm calls of individuals and are more responsive to calls from juveniles. *Animal Behaviour, 68*, 1257–1265.

Bovet, D., & Deputte, B. L. (2009). Matching vocalizations to faces of familiar conspecifics in grey-cheeked mangabeys (*Lophocebus albigena*). *Folia Primatologica, 80*, 220–232.

Calvert, G. A., Brammer, M. J., & Iversen, S. D. (1998). Crossmodal identification. *Trends in Cognitive Sciences, 2*, 247–253.

Campanella, S., & Belin, P. (2007). Integrating face and voice in person perception. *Trends in Cognitive Sciences, 11*, 535–543.

Carey, S., & Bartlett, E. (1978). Acquiring a single new word. *Papers and Reports on Child Language Development, 15*, 17–29.

Ceugniet, M., & Izumi, A. (2004). Vocal individual discrimination in Japanese monkeys. *Primates, 45*, 119–128.

Colombo, M., & D'Amato, M. R. (1986). A comparison of visual and auditory short-term memory in monkeys (*Cebus apella*). *Quarterly Journal of Experimental Psychology, 38B*, 425–448.

Colombo, M., & Graziano, M. (1994). Effects of auditory and visual interference on auditory-visual delayed matching to sample in monkey (*Macaca fascicularis*). *Behavioral Neuroscience, 108*, 636–639.

Coulon, M., Deputte, B. L., van Heyman, Y., & Baudoin, C. (2009). Individual recognition in domestic cattle (*Bos taurus*): Evidence from 2D-images of heads from different breeds. *PLoS One, 4*, e4441.

Cowey, A., & Weiskrantz, L. (1975). Demonstration of cross-modal matching in rhesus monkeys, *Macaca mulatta*. *Neuropsychologia, 13*, 117–120.

Davenport, R. K., & Rogers, C. M. (1970). Intermodal equivalence of stimuli in apes. *Science, 168*, 279–280.

Davenport, R. K., & Rogers, C. M. (1971). Perception of photographs by apes. *Behaviour, 39*, 318–320.

Davenport, R. K., Rogers, C. M., & Russell, I. S. (1973). Cross modal perception in apes. *Neuropsychologia, 11*, 21–28.

Dodd, B. (1979). Lip reading in infants: Attention to speech presented in- and out-of-synchrony. *Cognitive Psychology, 11*, 478–784.

Ettlinger, G. (1967). Analysis of cross-modal effects and their relationship to language. In F. L. Darley & C. H. Millikan (Eds.), *Brain mechanisms underlying speech and language* (pp. 53–60). New York, Grune & Stratton.

Ettlinger, G., & Blakemore, C. B. (1967). Cross-modal matching in the monkey. *Neuropsychologia, 5*, 147–154.

Evans, T. A., Howell, S., & Westergaard, G. C. (2005). Auditory-visual cross-modal perception of communicative stimuli in tufted capuchin monkeys (*Cebus apella*). *Journal of Experimental Psychology: Animal Behavior Processes, 31*, 399–406.

Gaffan, D., & Harrison, S. (1991). Auditory-visual associations, hemispheric specialization and temporal-frontal interaction in the rhesus monkey. *Brain, 114*, 2133–2144.

Geschwind, N. (1965). Disconnection syndrome in animals and man. *Brain, 88*, 237–294.

Ghazanfar, A. A., & Logothetis, N. K. (2003). Facial expressions linked to monkey calls. *Nature, 423*, 937–938.

Ghazanfar, A. A., & Schroeder, C. E. (2006). Is neocortex essentially multisensory? *Trends in Cognitive Sciences, 10*, 278–285.

Ghazanfar, A. A., Turesson, H. K., Maier, J. X., van Dinther, R., Patterson, R. D., & Logothetis, N. K. (2007). Vocal-tract resonances as indexical cues in rhesus monkeys. *Current Biology, 17*, 425–430.

Gogate, L. J., & Bahrick, L. E. (1998). Intersensory redundancy facilitates learning of arbitrary relations between vowel sounds and objects in seven-month-old infants. *Journal of Experimental Child Psychology, 69*, 133–149.

Hashiya, K., & Kojima, S. (1997). Auditory-visual intermodal matching by a chimpanzee (*Pan troglodytes*). *Japanese Psychological Research, 39*, 182–190.

Hashiya, K., & Kojima, S. (2001a). Acquisition of auditory-visual intermodal matching-to-sample by a chimpanzee (*Pan troglodytes*): Comparison with visual-visual intermodal matching. *Animal Cognition, 4*, 231–239.

Hashiya, K., & Kojima, S. (2001b). Hearing and auditory-visual intermodal recognition in the chimpanzee. In T. Matsuzawa (Ed.), *Primate origins of human cognition and behavior* (pp. 155–189). Tokyo: Springer.

Howard, I. P., & Templeton, W. B. (1966). *Human spatial orientation*. New York, NY: Willey.

Izumi, A., & Kojima, S. (2004). Matching vocalizations to vocalizing faces in a chimpanzee (*Pan troglodytes*). *Animal Cognition, 7*, 179–184.

Izumi, A., Kuraoka, K., Kojima, S., & Nakamura, K. (2001). Visually guided facial actions in rhesus monkeys. *Cognitive, Affective, & Behavioral Neuroscience, 1*, 266–269.

Janik, V. M., & Slater, P. J. B. (1997). Vocal learning in mammals. *Advances in the Study of Behavior, 26*, 59–99.

Jordan, K. E., Brannon, E. M., Logothetis, N. K., & Ghazanfar, A. A. (2005). Monkeys match the number of voices they hear to the number of faces they see. *Current Biology, 15*, 1034–1038.

Jouventin, P., Aubin, T., & Lengagne, T. (1999). Finding a parent in a king penguin colony: The acoustic system of individual recognition. *Animal Behaviour, 57*, 1175–1183.

Kaminski, J., Call, J., & Fischer, J. (2004). Word learning in a domestic dog: Evidence for "fast mapping". *Science, 304*, 1682–1683.

Kojima, S. (1985). Auditory short-term memory in the Japanese monkey. *International Journal of Neuroscience, 25*, 255–262.

Kojima, S. (2003). *A search for the origins of human speech: Auditory and vocal functions of the chimpanzee*. Kyoto: Kyoto University Press.

Kojima, S., Izumi, A., & Ceugniet, M. (2003). Identification of vocalizers by pant hoots, pant grunts and screams in a chimpanzee. *Primates, 44*, 225–230.

Kuhl, P. K., & Meltzoff, A. N. (1982). The bimodal development of speech in infancy. *Science, 218*, 1138–1141.

Kuhl, P. K., & Meltzoff, A. N. (1984). The intermodal representation of speech in infants. *Infant Behavior & Development, 7*, 361–381.

Kuhl, P. K., Williams, K. A., & Meltzoff, A. N. (1991). Cross-modal speech perception in adults and infants using nonspeech auditory stimuli. *Journal of Experimental Psychology: Human Perception & Performance, 17*, 829–840.

Lewkowicz, D. J., & Ghazanfar, A. A. (2009). The emergence of multisensory systems through perceptual narrowing. *Trends in Cognitive Sciences, 13*, 470–478.

Liberman, A. M., & Whalen, D. H. (2000). On the relation of speech to language. *Trends in Cognitive Sciences, 4*, 187–196.

Maier, J. X., Neuhoff, J. G., Logothetis, N. K., & Ghazanfar, A. A. (2004). Multisensory integration of looming signals by rhesus monkeys. *Neuron, 43*, 177–181.

Martin-Malivel, J., & Fagot, J. (2001). Cross-modal integration and conceptual categorization in baboons. *Behavioural Brain Research, 122*, 209–213.

McGurk, H., & MacDonald, J. (1976). Hearing lips and seeing voices. *Nature, 264*, 746–748.

Meltzoff, A. N., & Borton, R. W. (1979). Intermodal matching by human neonates. *Nature, 282*, 403–404.

Munhall, K. G., & Vatikiotis-Bateson, E. (1998). The moving face during speech communication. In R. Campbell, B. Dodd, D. Burnham (Eds.), *Hearing by Eye 2: Advances in the psychology of speechreading and auditory-visual speech* (pp. 123–139). Hove, UK: Psychology Press.

Murray, E. A., & Gaffan, D. (1994). Removal of the amygdala plus subjacent disrupts the retention of both intramodal and crossmodal associative memories in monkeys. *Behavioral Neuroscience, 108*, 494–500.

Parr, L. A. (2001). Cognitive and physiological markers of emotional awareness in chimpanzees (*Pan troglodytes*). *Animal Cognition, 4*, 223–229.

Parr, L. A., Winslow, J. T., Hopkins, W. D., & de Waal, F. B. M. (2000). Recognizing facial cues: Individual discrimination by chimpanzees (*Pan troglodytes*) and rhesus monkeys (*Macaca mulatta*). *Journal of Comparative Psychology, 114*, 47–60.

Patterson, M. L., & Werker, J. F. (2003). Two-month-old infants match phonetic information in lips and voice. *Developmental Science, 6*, 191–196.

Pokorny, J. J., & de Waal, F. B. M. (2009). Monkeys recognize the faces of group mates in photographs. *Proceedings of the National Academy of Sciences of the United States of America, 106*, 21539–21543.
Proops, L., McComb, K., & Reby, D. (2009). Cross-modal individual recognition in domestic horses (*Equus caballus*). *Proceedings of the National Academy of Sciences of the United States of America, 106*, 947–951.
Santos, L. S., & Hauser, M. D. (1999). How monkeys see the eyes: Cotton-top tamarins' reaction to changes in visual attention and action. *Animal Cognition, 2*, 131–139.
Savage-Rumbaugh, S., Sevcik, R. A., & Hopkins, W. D. (1988). Symbolic cross-modal transfer in two species of chimpanzees. *Child Development, 59*, 617–625.
Sayigh, L. S., Tyack, P. L., Wells, R. S., Solow, A. R., Scott, M. D., & Irvine, A. B. (1998). Individual recognition in wild bottlenose dolphins: A field test using playback experiments. *Animal Behaviour, 57*, 41–50.
Spelke, E. S. (1979). Perceiving bimodally specified events in infancy. *Developmental Psychology, 15*, 626–636.
Spelke, E. S. (1985). Preferential-looking methods as tools for the study of cognition in infancy. In G. Gottlieb & N. Krasnegor (Eds.), *Measurement of audition and vision in the first year of postnatal life* (pp. 323–363). Norwood, NJ: Ablex.
Stein, B. E., & Stanford, T. R. (2008). Multisensory integration: Current issues from the perspective of the single neuron. *Nature Reviews Neuroscience, 9*, 255–266.
Sumby, W. H., & Pollack, I. (1954). Visual contribution to speech intelligibility in noise. *Journal of the Acoustical Society of America, 26*, 212–215.
Walker, A. S. (1982). Intermodal perception of expressive behaviors by human infants. *Journal of Experimental Child Psychology, 33*, 514–535.
Weiss, D. J., Garibaldi, B. T., & Hauser, M. D. (2001). The production and perception of long calls by cotton-top tamarins (*Saguinus oedipus*): Acoustic analyses and playback experiments. *Journal of Comparative Psychology, 115*, 258–271.
Woods, T. M., & Recanzone, G. H. (2004). Visually induced plasticity of auditory spatial perception in macaques. *Current Biology, 14*, 1559–1564.
Wright, A. A., Shyan, M. R., & Jitsumori, M. (1990). Auditory same/different concept learning by monkeys. *Animal Learning & Behavior, 18*, 287–294.
Wynn, K. (1992). Addition and subtraction by human infants. *Nature, 358*, 749–750.
Yamaguchi, C., & Izumi, A. (2008). Vocal learning in nonhuman primates: Importance of vocal contexts. In N. Masataka (Ed.), *The origins of language: Unrevealing evolutionary forces* (pp. 75–84). Tokyo: Springer.
Zangenehpour, S., Ghazanfar, A. A., Lewkowicz, D. J., & Zatorre, R. J. (2009). Heterochrony and cross-species intersensory matching by infant vervet monkeys. *PLoS One, 4*, e4302.

Chapter 3
Representation and Integration of Faces and Vocalizations in the Primate Ventral Prefrontal Cortex

Maria M. Diehl and Lizabeth M. Romanski

1 Introduction

The integration of facial gestures and vocal signals is an essential process in social communication. Facial and vocal signals provide an abundant source of information that we use in our everyday interactions to communicate our intentions and obtain emotional and cognitive information from others. Face–voice integration relies on several brain regions, including language regions in the ventral frontal lobe. Neuroimaging has made great strides in describing activity in temporal and frontal regions during speech processing, but we have relatively little understanding of the cellular mechanisms that underlie face–voice integration in the frontal lobe. Much of the neurophysiology research into the cellular details of face and voice processing has been focused on nonhuman primates in an attempt to characterize the neural circuit involved in social communication. While much of this research has elaborated on these sensory processes in primary and secondary cortical areas, more recent research has embarked upon how higher order cortical areas like the prefrontal cortex (PFC) process face and voice information. This chapter will focus on the role of the ventrolateral prefrontal cortex in the processing and integration of face and vocal information in nonhuman primates. We will first describe studies on face-responsive cells in the nonhuman primate cortex, including inferotemporal cortex and the Superior Temporal Sulcus and finally face processing in PFC. This will be followed by auditory responses in PFC. Finally, we will examine the integration of faces and voices by single cells in the primate prefrontal cortex and their potential role in recognition and social communication.

M.M. Diehl • L.M. Romanski, Ph.D. (✉)
Department of Neurobiology & Anatomy and Center for Navigation and Communication Sciences, University of Rochester, Rochester, NY 14626, USA
e-mail: maria_diehl@URMC.Rochester.edu; liz_romanski@urmc.rochester.edu

P. Belin et al. (eds.), *Integrating Face and Voice in Person Perception*,
DOI 10.1007/978-1-4614-3585-3_3,

2 Face Processing in Nonhuman Primates

2.1 Behavioral Responses to Faces

Faces are among the most important social cues used by nonhuman primates (NHPs) for a variety of interactions including kin recognition, conspecific communication of danger, food finding, mating behaviors, and mother–infant interactions. There have been extensive systematic qualitative descriptions of facial and body gestures in multiple species by Darwin (1872), in NHPs (Andrew, 1963; Redican, 1975; van Hooff, 1962), including squirrel monkeys (Marriott & Salzen, 1978), capuchin monkeys (Weigel, 1979), and rhesus monkeys (Altmann, 1962; Hauser & Marler, 1993; Hinde & Rowell, 1962; Maestripieri & Wallen, 1997; Partan, 2002). Such studies have provided us with a rich catalog of vocal repertoires, facial and bodily gestures that NHPs use during specific contexts of social communication. In particular, Partan (2002) has investigated the interaction of facial and vocal signals used during social communication in the rhesus macaque, which is used in neurobiological research. This work has described the facial expressions and body postures that are associated with particular vocal signals. By examining the relationship between occurrences in specific body movements (such as ear, eye, head, and overall body posture) with occurrences in vocal behaviors, Partan (2002) found that while most behaviors were exclusively visual—such as an open-mouth stare with ears pointing forward—a significant proportion (30 %) of behaviors were multimodal in nature. Partan's work highlights the importance of faces and vocalizations in NHP social communication.

Studies have also addressed how NHPs process faces in controlled tasks in order to compare psychophysical responses in NHPs to those in human subjects. The eyes seem to be the most salient feature in face processing in both humans (Haith, Bergman, & Moore, 1977; Klin, Jones, Schultz, Volkmar, & Cohen, 2002; Vinette, Gosselin, & Schyns, 2004) and NHPs (Gothard et al., 2004; Guo, Robertson, Mahmoodi, Tadmor, & Young, 2003; Nahm, Perret, Amaral, & Albright, 1997). To determine how face perception is different from perception of other complex visual objects, studies have tested NHPs' recognition and discrimination of inverted and upright faces. Some studies find that NHPs process faces similarly to humans (Parr, Dove, & Hopkins, 1998; Parr & Heintz, 2006, 2009; Tomonaga, 1999, 2007), whereas others have found important differences in face processing mechanisms (Dittrich, 1990; Gothard, Erickson, & Amaral, 2004; Rosenfeld & Van Hoesen, 1979). Differences in testing could explain the disparate findings. Parr and colleagues have investigated NHPs' discrimination of faces and objects using a match-to-sample paradigm and found that chimpanzees, like humans, use 2nd order configural processing—using relational information about facial features such as the distance between the eyes and the mouth, whereas monkeys rely more on 1st order spatial relationships of facial features—using primary configural information such as the location of the eyes above the mouth (Parr, Winslow, & Hopkins, 1999; Parron & Fagot, 2007; Tomonaga, 2007). Studies using the Thatcher effect to specifically assess configural face processing, where only the eyes and mouth remain upright while the remaining facial features are inverted, have found that manipulating the eye and mouth orientation was more salient in upright compared to inverted faces similar to that in human face processing (Adachi, Chou, & Hampton, 2009; Dahl, Logothetis, Bulthoff, & Wallraven, 2010). While the previous two studies used a habituation–dishabituation paradigm, Parr and colleagues used the match to sample paradigm to investigate the differences in face perception using changes in expression across different identities in the behavioral paradigm. During the match to sample task, subjects undergo extensive training, preventing subjects from generalizing across categories of expression when combined with changes in identity. It would be interesting to see if similar results ensued if a nonmatch to sample task was employed to answer the same question. This task along with the habituation–dishabituation paradigm is easier to implement in NHPs because there is a natural tendency of attending to novel items in both humans and NHPs. Ongoing studies in our laboratory have demonstrated rhesus macaques' ability to discriminate audiovisual face–vocalization stimuli on the basis of emotional expression or caller identity in a nonmatch to sample task (unpublished results). In these studies, a change in facial identity is easier to discriminate compared to a change in emotional expression when the identity remains the same.

2.2 *Neuronal Responses to Faces in the Temporal Lobe*

Most of the literature on the cellular neurophysiology of face processing has focused on the temporal cortex of the macaque brain. Early studies established that the inferotemporal (IT) cortex receives afferents from prestriate and striate cortices that are involved in complex visual processing, and that these neurons respond differentially to complex visual stimuli (Gross, Bender, & Rocha-Miranda, 1969) as well as to particular features within images of objects (Tanaka, Saito, Fukada, & Moriya, 1991). IT neurons are not tuned to a specific stimulus; rather they respond to shape, color, texture, or combinations of these features (Albright, Desimone, & Gross, 1984) and may also be influenced by stimulus size and position changes (Ito, Tamura, Fujita, & Tanaka, 1995). More recent studies using awake behaving monkeys have examined responses in IT cortex to complex visual stimuli including faces during behavioral tasks. A notable study by Yamane, Kaji, and Kawano (1988) found that IT neurons may process faces by attending to specific features including the distance between the eyes, mouth, and hairline during a discrimination task, which can be summed in a population to identify particular features. Another group examined the effect of facial identity on IT neurons during a face discrimination task and found that anterior IT cortical neurons encode information about facial identity, whereas neurons located in the anterior superior temporal sulcus process perceptual information such as differences in facial views (Eifuku, De Souza, Tamura, Nishijo, & Ono, 2004). It was also found that neurons in anterior IT cortex can encode a face and visual pattern-paired associate, indicating the capability of these neurons to link and integrate facial features involved in identity processing and semantic associations (Eifuku, Nakata, Sugimori, Ono, & Tamura, 2010).

In contrast to the selectivity of IT neurons for features associated with identity, neurons within superior temporal sulcus (STS) are modulated by facial expression and features including face view and gaze direction (Perrett, Rolls, & Caan, 1982; Perrett, et al., 1985; Perrett, Mistlin, & Chitty, 1987). Such face-view responses are distributed rostrocaudally along the STS (DeSouza, Eifuku, Tamura, Nishijo, & Ono, 2005). Hasselmo, Rolls, and Baylis (1989) used a go/no-go task to examine the neural response to facial expression and identity in the STS and IT cortex. Using stimuli from three monkeys making three different expressions (neutral, slight threat, and full threat) and 35 unfamiliar faces, they found that most cells responsive to facial identity were found in IT cortex, whereas cells responsive to the facial expression tended to cluster in the STS. When the time course of IT neuronal response trains are examined, an interesting picture emerges. Sugase, Yamane, Ueno, and Kawano (1999) found that when viewing faces varying in identity and expression, global information such as stimulus category or identity (monkey, human, object) is represented in the first part of the spike train, while fine information such as expression within a category is encoded in the latter part of the spike train.

Collectively, these studies provide evidence for the complex processing that takes places in different regions of the temporal cortex during face perception. While there is detailed information on visual processing (Gross, 1994; Miyashita, 1993; Tanaka, 1996) and face processing (Perrett, Hietanen, Oram, & Benson, 1992) in the temporal lobe, fewer studies are available on face processing in prefrontal cortex, another essential but less-studied node in the face processing network.

2.3 *Prefrontal Cortex Processing of Faces*

Strong connections exist between the temporal lobe areas involved in complex visual processing and prefrontal cortex (PFC). The PFC is a heterogeneous region receiving information from sensory cortices and subcortical structures to execute complex tasks involving goal-directed behavior (Fuster, 2001; Goldman-Rakic, 1996a, 1996b; Miller & Cohen, 2001). Decades of research demonstrate its involvement in higher order cognitive functions including working memory, decision-making, and social communication processes such as language and face–voice processing. Early studies of prefrontal function focused on PFC's visual properties by testing neurons with a variety of visual stimuli to characterize the response properties (Pigarev, Rizzolatti, & Schandolara, 1979; Rizzolatti, Scandolara, Matelli, & Gentilucci, 1981; Thorpe, Rolls, & Maddison, 1983). Visual responses in the frontal eye fields located in dorsolateral PFC demonstrate that region encodes information about spatial location but not features of visual stimuli such as color or shape (see initial studies by Mohler, Goldberg, & Wurtz, 1973 and Bruce & Goldberg, 1985). Delay activity has also been extensively documented in PFC, demonstrating that this area of the brain holds relevant information online in order to make a decision during the maintenance of spatial locations (Bruce & Goldberg, 1985; Funahashi, Bruce, & Goldman-Rakic, 1989; Funahashi, Chafee, & Goldman-Rakic, 1993; Kojima & Goldman-Rakic, 1984; Niki & Watanabe, 1976; Quintana & Fuster, 1992; Rao, Rainer, & Miller, 1997) and various visual stimuli (Fuster, Bauer, & Jervey, 1982; Miller, Erickson, & Desimone, 1996; Quintana & Fuster, 1992; Quintana, Yajeya, & Fuster, 1988; Rao et al., 1997; Watanabe, 1986) during working memory tasks. While there are an abundance of studies in the human literature documenting the BOLD activation of PFC during working memory and decision-making tasks using fMRI, a subset of such studies have demonstrated activity of PFC during face processing (Dolan et al., 1996; Ishai, Pessoa, Bikle, & Ungerleider, 2004; Ishai, Schmidt, & Boesiger, 2005; Kesler-West et al., 2001; LoPresti et al., 2008; Nomura et al., 2004; Sergerie, Lepage, & Armony, 2005; Vuilleumier, Armony, Driver, & Dolan, 2001). It is typically orbital and ventral prefrontal cortex that are activated during working memory tasks using face stimuli (Dolan et al., 1996) as well as during the perception of emotional faces (Iidaka et al., 2001; Ishai et al., 2005; Kesler-West et al., 2001; Pourtois, Schwartz, Seghier, Lazeyras, & Vuilleumier, 2006). Neural recordings in the human brain have also shown activity of ventrolateral PFC during face processing (Marinkovic, Trebon, Chauvel, & Halgren, 2000). In order to understand the neuronal mechanisms that underlie face processing, it is necessary to examine such questions about face processing using a single unit neurophysiological approach in animals performing similar behavioral tasks that are both passive and active in nature.

Face responsive neurons in PFC were first documented in 1983 when Thorpe and colleagues tested neurons in orbitofrontal cortex (OFC) for responses to complex visual and gustatory stimuli and found a small population of cells responsive to face stimuli, especially when paired with a stimulus reinforcer. A landmark study by O'Scalaidhe and colleagues documented face-responsive neurons in both OFC and ventrolateral prefrontal cortex (VLPFC) of the rhesus monkey during both passive fixation and working memory tasks (1997, 1999). In this study it was determined that some neurons in ventral prefrontal cortex responded selectively to faces compared to object and other nonface stimuli (Fig. 3.1). Cells that showed a twofold increase in firing rate to face stimuli versus nonface stimuli were considered face-selective. Control stimuli, which included scrambled and inverted faces–stimuli that contain many of the same features as natural faces, did not evoke responses in these prefrontal "face cells". Therefore, it was not simply the presence of facial features alone that drove the cells' responses; rather, face configuration was an important feature. Furthermore, some prefrontal "face cells" showed sustained firing which lasted over 200 ms after stimulus presentation which suggested mnemonic processing beyond the stimulus period (O'Scalaidhe, Wilson, & Goldman-Rakic, 1999) even when working memory was not explicitly required.

These sustained responses to face stimuli during passive presentation demonstrate that prefrontal activation by faces is not dependent upon task contexts but upon the salience of the stimulus itself. Nonetheless, O'Scalaidhe and colleagues further examined VLPFC neurons during a conditional delayed response task using face and nonface stimuli as cues to make leftward or rightward saccades. While some

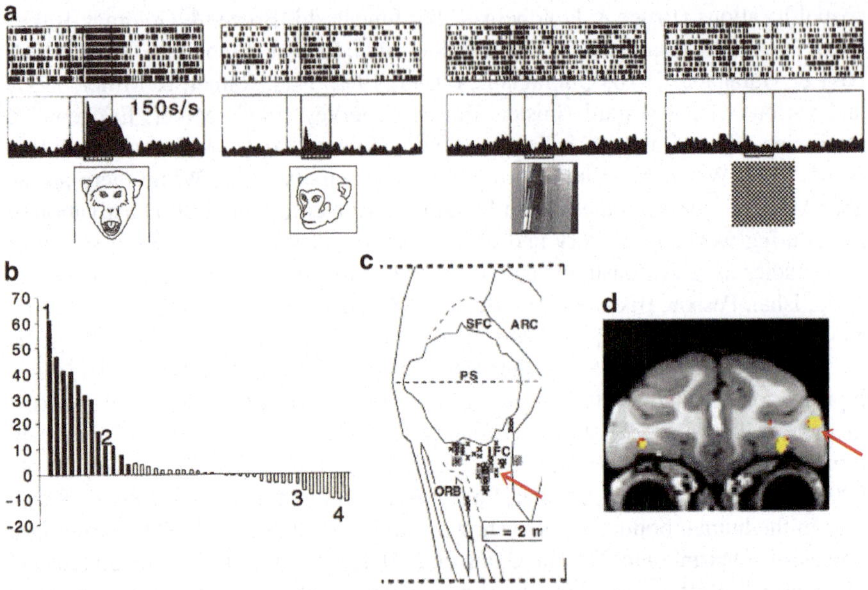

Fig. 3.1 Responses to faces in the ventral prefrontal cortex. (**a–c**) Responses of VLPFC neurons to face stimuli from O'Scalaidhe et al. (1997). (**a**) A single neuron had a selective response to a forward-view threat face (1), a neutral profile monkey face (2), and nonface visual stimuli (3, 4). (**b**) A selectivity graph of the neuronal data from the cell in (**a**) tested with 40 face and nonface stimuli including 1–4 from (**a**). The *bars* in *black* were face stimuli and elicited a response that was greater than the nonface stimuli. (**c**) A flat map of the prefrontal cortex depicting the locations of face cells in the ventrolateral prefrontal and orbital areas. A *red arrow* indicates the ventral prefrontal face cells. (**d**) Data from Tsao et al. (2008) showing activations of face patches in the ventrolateral prefrontal cortex and the orbitofrontal cortex. The *red arrow* indicates the same location as in O'Scalaidhe figure (**c**) (adapted from O'Scalaidhe et al., 1997 and Tsao et al., 2008)

neurons demonstrated delay activity, there were also responses in the sample period to the face stimuli themselves. Moreover, the delay activity was unique to faces since neurons failed to show delay activity with nonface stimuli regardless of the motor response. Upon examination of the recording locations, O'Scalaidhe and colleagues pinpointed several areas in VLFPC where face responsive neurons clustered: (1) on the inferior frontal convexity several mm below the principle sulcus, just behind the midpoint of the principal sulcus; (2) within the inferior frontal sulcus, located between the principle sulcus and lower limb of the arcuate sulcus (also known as the prefrontal dimple); and (3) in the lateral orbital cortex, at the anterior–posterior level of the inferior prefrontal sulcus (Fig. 3.7). Fewer recordings were completed in the lateral orbital cortex, with the idea that further examination of this area would demonstrate additional face-selective neurons, since previous OFC recordings have localized face-responsive neurons (Rolls, 1996). A more recent study by Rolls, Critchley, Browning, and Inoue (2006) describes a similar population of face-responsive cells in OFC that are differentially responsive to monkey and human face categories, facial expression and identity as well as changes in the angle of gaze during a go/no-go task. It is important to note that extensive searching was conducted to find such cells in OFC. In this study, 812 out of 3,168 OFC neurons were found to be visually responsive, and only 32 responded to face stimuli. Additional investigation will clarify the function and purpose of these responses and how they fit into the bigger picture of face processing and social communication.

Importantly, data from the single unit recordings have been confirmed with fMRI in the macaques which have demonstrated activation of face-responsive "patches" in similar VLPFC locations shown by O'Scalaidhe, Wilson, and Goldman-Rakic (1997) and O'Scalaidhe et al. (1999). In their studies, Tsao, Schweers, Moeller, and Freiwald (2008) identified three distinct patches in the frontal lobe which were more active during face than nonface stimuli. The locations of these patches in the ventrolateral part of the inferior convexity (PL), just anterior to the inferior limb of the arcuate sulcus (PA) and the lateral orbital cortex (PO) correspond to areas in VLPFC and OFC where face-responsive neurons were found during neurophysiological recordings by O'Scalaidhe et al. (1997, 1999). This group also found face "patches" in the temporal lobe and linked their activity to a circuit that processes face information (Moeller, Freiwald, & Tsao, 2008). There are a number of different brain areas that differentially respond to various aspects of faces and may constitute a large network for the processing of social information.

3 Vocalization Processing in the Primate Prefrontal Cortex

The human frontal lobe has been associated with language processing and communication for more than a century (Broca, 1861). Lesion, psychophysical and neuroimaging studies have described the role of the inferior frontal gyrus with speech, language and higher auditory functions. Neuroimaging studies of the human brain have shown activation of ventrolateral frontal lobe areas such as Brodmann's areas 44, 45, and 47 in a variety of communication related processes including auditory working memory, phonological processing, comprehension, semantic judgment, and speech–gesture integration (Buckner, Raichle, & Petersen, 1995; Demb et al., 1995; Fiez et al., 1996; Friederici, Ruschemeyer, Hahne, & Fiebach, 2003; Gabrieli, Poldrack, & Desmond, 1998; Stevens, Goldman-Rakic, Gore, Fulbright, & Wexler, 1998; Stromswold, 1996; Xu et al., 2010; Zatorre, Meyer, Gjedde, & Evans, 1996). Nonetheless, our understanding of the neuronal processes which occur during these higher auditory processes is limited due to the lack of a suitable animal model. Of the animals that engage in productive and receptive communication, only NHPs have an enlarged frontal lobe, with regions that are anatomically similar to human frontal lobe regions (Petrides & Pandya, 2002). Studies which have assessed the role of the NHP frontal lobe in higher auditory processing are far fewer than those that have examined the frontal lobes' role in visual processing, despite the obvious involvement of the human frontal lobe in language. A suitable animal model in which to examine communication processes would further our understanding of the cellular mechanisms which underlie communication.

3.1 Early Studies of Prefrontal Cortex and Auditory Processing in Nonhuman Primates

Early behavioral studies suggested that the PFC might play a role in auditory cognition in NHPs. Large lesions of lateral PFC in primates were shown by some to disrupt performance of auditory discrimination tasks (Goldman & Rosvold, 1970; Gross, 1963; Gross & Weiskrantz, 1962; Weiskrantz & Mishkin, 1958). A precise location for the effect of these lesions was not clear, however, and differences in tasks used to assess the effect of lesions on auditory memory differed making comparisons across studies difficult.

Direct assessment of the response of prefrontal neurons to complex auditory stimuli was somewhat more effective. Several studies demonstrated that neurons in the PFC respond to auditory stimuli or are active during auditory tasks in Old and New World primates (Azuma & Suzuki, 1984; Bodner, Kroger, & Fuster, 1996; Ito, 1982; Newman & Lindsley, 1976; Tanila, Carlson, Linnankoski, & Kahila, 1993; Tanila, Carlson, Linnankoski, Lindroos, & Kahila, 1992; Vaadia, Benson, Hienz, & Goldstein, 1986; Watanabe, 1986, 1992). There was evidence of weakly responsive auditory neurons that were distributed across a wide region of the PFC or single cells that were active in tasks that used an auditory stimulus to signal a specific event (Bodner et al., 1996; Newman & Lindsley, 1976; Tanila et al., 1992, 1993; Watanabe, 1986, 1992). Several studies suggested that dorsolateral prefrontal cortex (DLPFC) neurons were active during auditory localization. For example, Azuma and Suzuki (1984) demonstrated that single neurons in the DLFPC were selectively activated by auditory stimuli presented in the contralateral field while a Vaadia et al. (1986) showed that caudal principal sulcus and arcuate neurons are most active during auditory localization rather than during passive fixation. A single study noted phasic responses to click stimuli in the lateral orbital cortex, area 12o (Benevento, Fallon, Davis, & Rezak, 1977). Few studies, however, noted a specific location for auditory processing or demonstrated robust activation with communication stimuli. This may be due, in part, to the fact that studies have often confined electrode penetrations to caudal and dorsolateral prefrontal cortex (Azuma & Suzuki, 1984; Bodner et al., 1996; Ito, 1982; Tanila et al., 1992, 1993), where presumptive auditory inputs to the frontal lobe are more dispersed and might be related to spatial localization rather than to vocalizations (Romanski, 2004). Few studies have focused neurophysiological recordings to prefrontal areas that selectively receive a wealth of auditory afferents.

3.2 Auditory Projections to the Prefrontal Cortex

Early anatomical studies indicated that a rostrocaudal topography exists such that the caudal superior temporal gyrus (STG) and caudal PFC (areas 8a and caudal area 46) are reciprocally connected (Barbas, 1992; Chavis & Pandya, 1976; Pandya & Kuypers, 1969; Petrides & Pandya, 1988; Petrides & Pandya, 2002; Romanski,

Bates, & Goldman-Rakic, 1999) while the rostral STG is reciprocally connected with rostral principalis (rostral 46 and 10) and orbitofrontal areas (areas 11 and 12) (Chavis & Pandya, 1976; Pandya, Hallett, & Kmukherjee, 1969; Pandya & Kuypers, 1969). In the last decade, studies have characterized these temporoprefrontal connections in greater detail utilizing the core–belt organization (Kaas & Hackett, 1998; Morel, Garraghty, & Kaas, 1993). These studies include the analysis of injections into the auditory belt (Romanski, Tian et al., 1999), the parabelt (Hackett, Stepniewska, & Kaas, 1998), and the PFC (Hackett et al., 1998; Romanski, Bates et al., 1999). Together these studies have refined the direct rostral–caudal topography that was previously noted showing that the rostral frontal lobe areas are densely connected with anterior belt and parabelt regions (Hackett et al., 1998; Romanski, Bates et al., 1999). Moreover, the caudal parabelt and belt are reciprocally connected with caudal and dorsolateral frontal lobe. The densest projections to the frontal lobe originate in higher order auditory processing regions including the parabelt as well as the rostral STG and the dorsal bank of the STS including multisensory area TPO and area TAa (Petrides & Pandya, 1988; Romanski, 2007; Romanski, Bates et al., 1999).

When anatomical injections are combined with auditory physiology more specificity regarding the frontal lobe auditory projection field emerges. While projections from the temporal lobe to the prefrontal cortex are present, it is not know which of these projections are acoustic in nature. In Romanski, Tian et al. (1999) the lateral belt auditory areas AL, ML, and CL, were physiologically identified, as in previous studies by Rauschecker, Tian, and Hauser (1995) and injections of anterograde and retrograde anatomical tracers were placed into each of the belt areas. Analysis of the anatomical connections revealed that five specific regions of the frontal lobes received input from as early as the lateral belt including the frontal pole (area 10), the principal sulcus (area 46), VLPFC (areas 12vl and 45), the lateral orbital cortex (areas 11 and 12o), and the dorsal periarcuate region (area 8a) (Fig. 3.2). Moreover, these connections were topographically organized such that projections from AL typically involved the frontal pole, the rostral principal sulcus, anterior VLPFC and the lateral orbital cortex while projections from area CL targeted the dorsal periarcuate cortex and the caudal principal sulcus. The topographic specificity of this rostrocaudal, frontal–temporal connectivity is indicative of separate streams of auditory information that target distinct functional domains of the frontal lobe, similar to the spatial and nonspatial visual streams which target dorsal–spatial and ventral–object prefrontal regions (Barbas, 1988; Ungerleider & Mishkin, 1982; Webster, Bachevalier, & Ungerleider, 1994; Wilson, O'Scalaidhe, & Goldman-Rakic, 1993). One auditory pathway, originating in CL, targets caudal DLPFC, which has been shown to be involved in visuospatial processing; the other pathway, originating in AL, targets rostral and ventral prefrontal areas, which, in visual neurophysiology studies, appears to process objects and faces (O'Scalaidhe et al., 1997, 1999; Wilson et al., 1993). The identification of a ventral auditory stream which targets the VLPFC just as visual extrastriate areas target face-processing areas in VLPFC is especially interesting for face and voice integration.

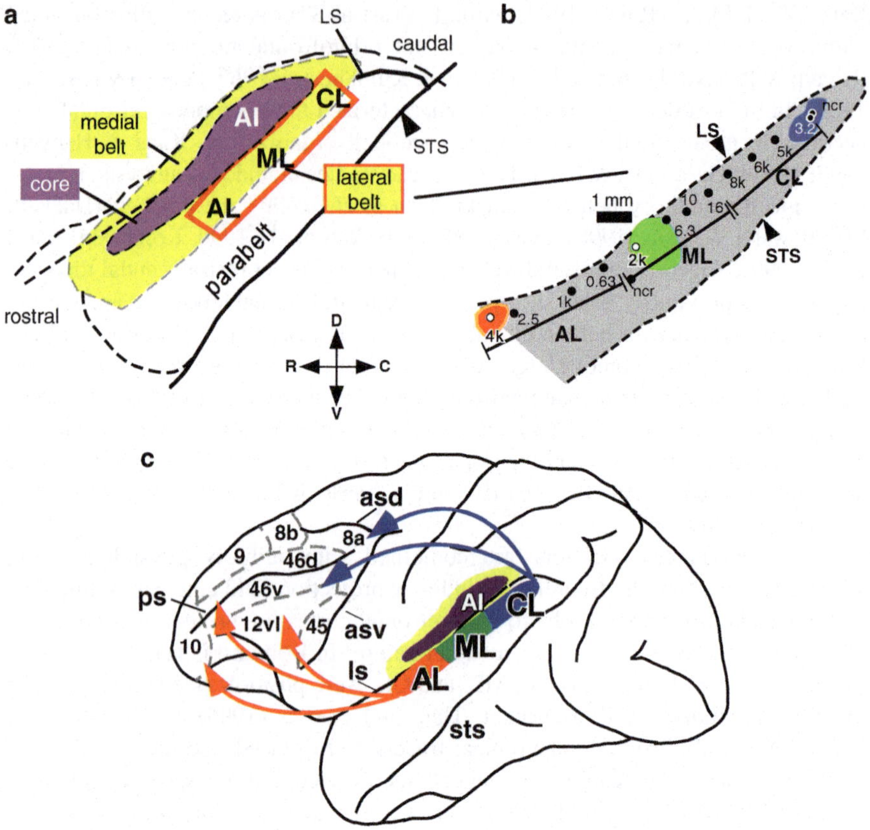

Fig. 3.2 Auditory projections to the prefrontal cortex. Injections placed into the auditory belt resulted in topographically organized projections to the ventral and dorsal prefrontal cortex. (**a**) Depicts the location of the auditory belt cortex outlined in *red*. The belt auditory cortex is shown enlarged in (**b**), with tracer injections placed into AL (*red*), ML (*green*), and CL (*blue*). The numbers placed at each electrode track location represent the characteristic frequency average for the track. Reversals of the frequency map occur at the borders of these areas. (**c**) A schematic of the projections to the prefrontal cortex from the belt indicate that anterior belt projects to ventral and anterior prefrontal areas while the caudal belt projects more densely to caudal, dorsal prefrontal areas, indicated a dorsal–ventral stream for auditory projections similar to that of the visual system (adapted from Romanski, Tian et al., 1999)

3.3 Auditory Responsive Domain in VLPFC

Using the information provided by anatomical studies, an auditory responsive domain has been defined within the VLPFC of the macaque monkey (Romanski & Goldman-Rakic, 2002) in an area that has been shown to receive acoustic afferents from ventral stream auditory neurons in the anterior belt, parabelt, and the dorsal bank of the STS (Diehl, Bartlow-Kang, Sugihara, & Romanski, 2008; Hackett, Stepniewska, & Kaas, 1999; Romanski, Bates et al., 1999; Romanski, Tian et al., 1999).

Neurons in VLPFC are responsive to complex acoustic stimuli including, but not limited to, species-specific vocalizations (Romanski & Goldman-Rakic, 2002). The discovery of complex auditory responses in the macaque VLPFC is in line with human fMRI studies indicating that a homologous region of the human brain, area 47 (pars orbitalis) is specifically activated by human vocal sounds compared with animal and nonvocal sounds (Fecteau, Armony, Joanette, & Belin, 2005). The initial study reporting auditory responses in the NHP VLPFC characterized the auditory responsive cells as being responsive to several types of complex sounds including species-specific vocalizations, human speech sounds, environmental sounds, and other complex acoustic stimuli (Romanski & Goldman-Rakic, 2002, Fig. 3.3). More than 74 % of the auditory neurons in this study responded to vocalizations, while fewer than 10 % of cells responded to pure tones or noise stimuli, prompting further investigation of this area in vocalization processing.

3.4 Representation of Vocalizations in VLPFC

Studies which have followed up on the localization of a discrete sound processing region in the PFC of NHPs have focused on determining what the neurons in this prefrontal area encode. Perhaps neurons at higher levels of the auditory hierarchy process complex stimuli in a more abstract manner than lower order sensory neurons or show evidence of greater selectivity. As mentioned above, studies have shown that VLPFC auditory neurons do not readily respond to simple acoustic stimuli such as pure tones (Romanski & Goldman-Rakic, 2002). Several studies have suggested that VLPFC neurons are robustly responsive to vocalizations and other complex sounds (Averbeck & Romanski, 2004; Gifford, Maclean, Hauser, & Cohen, 2005; Romanski, Averbeck, & Diltz, 2005; Russ, Ackelson, Baker, & Cohen, 2008). This is true not only of Old World Primates but also of New World Primates, since it has recently been shown that the ventral frontal lobe of marmosets shows C-fos activity during the perception of vocalizations in an antiphonal calling paradigm (Miller, Diauro, Pistorio, Hendry, & Wang, 2010; Fig. 3.4). Would these higher order auditory neurons be more likely to process the referential meaning within communication sounds or complex acoustic features that are a part of these and other sounds? PET and fMRI studies have suggested that the human inferior frontal gyrus, or ventral frontal lobe, plays a role in semantic processing and but also in phonological encoding (Buckner et al., 1995; Demb et al., 1995; Fiez et al., 1996; Friederici et al., 2003; Gabrieli et al., 1998; Poldrack et al., 1999; Stevens et al., 1998; Stromswold, Caplan, Alpert, & Rauch, 1996; Zatorre et al., 1996). If macaque VLPFC is a functional homologue of human inferior frontal gyrus then one might expect NHP ventral prefrontal neurons to encode the semantic meaning of particular vocalizations or of their phonological representation.

In a series of studies (Averbeck & Romanski, 2006; Romanski et al., 2005), VLPFC neurons were tested with a behaviorally and acoustically categorized library of *Macaca mulatta* calls (Hauser, 1996; Hauser & Marler, 1993), which contained

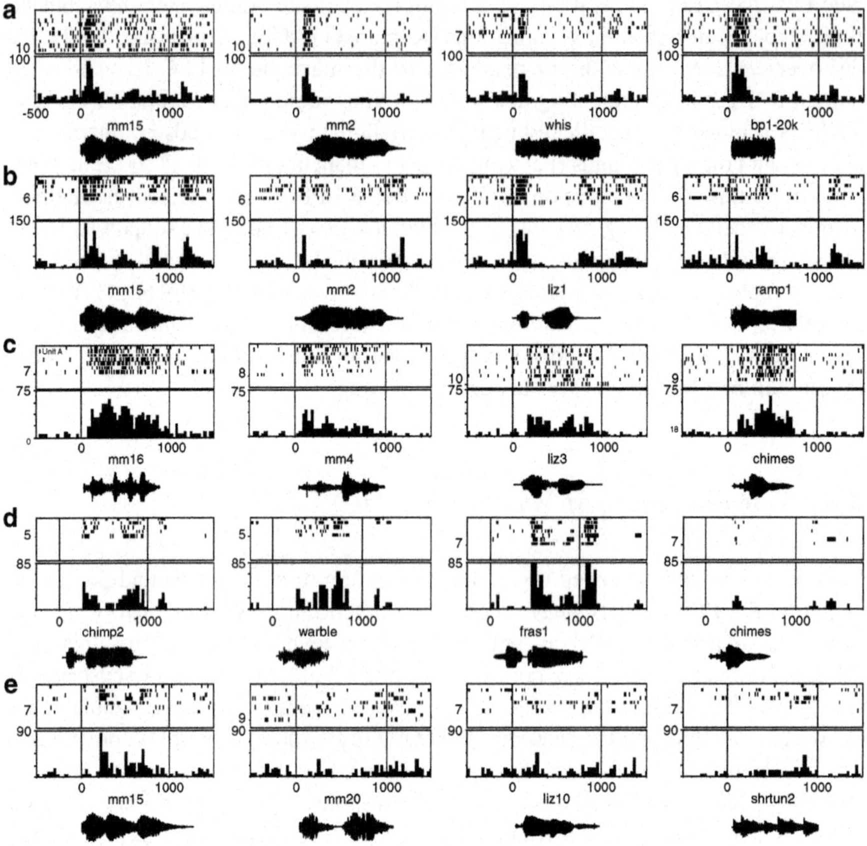

Fig. 3.3 Prefrontal auditory neuron response profiles from Romanski and Goldman-Rakic (2002). Responses of five cells (**a–e**) to auditory stimuli are shown as raster (*top panels*) and post-stimulus time histograms (*bottom panels*). The sounds used are shown as waveforms below each response panel. Cell (**a**) gave a nonspecific phasic onset response to all auditory stimuli tested, whereas monkey and human vocalizations (first three panels) elicited a greater response than nonvocal stimuli (*last panel*) in (**b**). Some cells had sustained responses to auditory stimuli (**c**) that lasted the length of the auditory stimuli. Cells in (**d**) and (**e**) were selective for vocalization stimuli. In (**d**) both animal and human vocalizations drove the cell while in (**e**) a single species-specific vocalization drove the cell. Adapted from Romanski and Goldman-Rakic (2002)

exemplars from each of ten identified call categories. Prefrontal neurons exhibited similar call selectivity as lateral belt auditory cortex neurons responding to 1–4 call types (Romanski et al., 2005; Tian, Reser, Durham, Kustov, & Rauschecker, 2001). Decoding analyses and associated information theoretic techniques showed that single cells, on average, could correctly classify their best call in about 55 % of individual trials and the second and third best calls in about 32 and 22 % of trials (Romanski et al., 2005). This is similar to the encoding of faces by temporal lobe "face" cells (Rolls & Tovee, 1995) where cells respond optimally to a few stimuli but in a decreasing gradient to others.

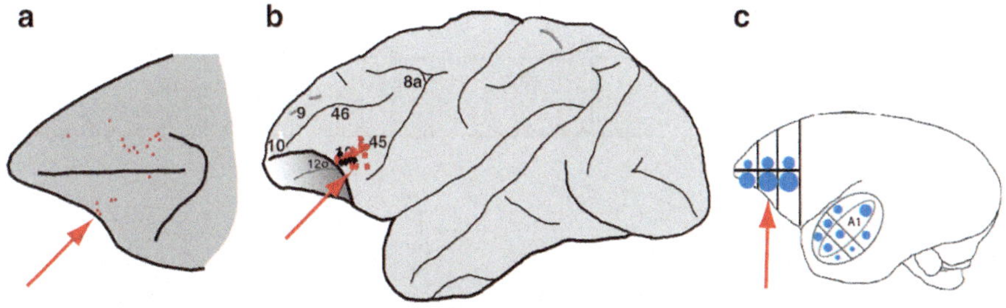

Fig. 3.4 Locations of auditory responsive regions in the prefrontal cortex of nonhuman primates. Several studies have identified auditory responsive regions in NHPs. In (**a**), auditory responsive neurons were identified in Tanila et al. (1993, 1992), from recordings in the *Macaca fascicularis* using a broad range of complex sounds. (**b**) Recordings from Romanski et al. (2005) using only species-specific vocalizations found vocalization responsive neurons in the ventral prefrontal cortex of *Macaca mulatta*. (**c**) Localization of vocalization receptive areas in the marmoset frontal lobe with C-fos during an antiphonal calling paradigm (Miller et al., 2010)

How VLPFC neurons classify different vocalizations could elucidate the defining feature that prefrontal neurons encode. In a system that encodes referential meaning neurons might respond in a similar manner to calls with similar meaning but different acoustic morphology. Further assessment of VLPFC neuronal responses with a hierarchical cluster analysis showed that prefrontal neurons respond similarly to call types that were identical in acoustic morphology, suggesting phonological rather than semantic encoding (Romanski & Averbeck, 2009; Romanski et al., 2005; Fig. 3.5a–c). In the rhesus macaque repertoire there are ~10–12 call types that are given during behaviorally distinct contexts including high value and low value food calls, agonistic calls, affiliative calls, and mating calls. One must compare calls with similar semantic meaning that are acoustically different with calls that are acoustically similar but semantically different. The harmonic arch and warble are semantically similar in referring to the presence of high value food but acoustically dissimilar in several aspects, while the warble and coo are acoustically similar but semantically different in that the coo call can be affiliative and is often given in the context of low value food. Our analysis of prefrontal neuronal responses to calls from the rhesus repertoire determined that few cells responded in a similar manner to the semantically similar but acoustically different harmonic arch/warble. In contrast ~20 % of the population responded to warbles and coos, which are acoustically similar but semantically different (Romanski et al., 2005; Fig. 3.5d, e). Similar results have been found in lateral belt auditory cortex neurons which respond in a similar manner to calls which had similar acoustic morphology (Romanski & Averbeck, 2009; Tian et al., 2001).

The data on which the cluster analysis of prefrontal neuronal responses was based was gathered while awake animals were passively presented with vocalizations but did not explicitly perform a mnemonic or discrimination task. Thus it is possible that if active discrimination of a semantic referential was required, e.g., detecting only food calls with a button press, then prefrontal neuronal activity might reflect this semantic feature, since many studies have revealed the importance of task demands in dictating prefrontal activity (Miller & Cohen, 2001; Rainer, Asaad,

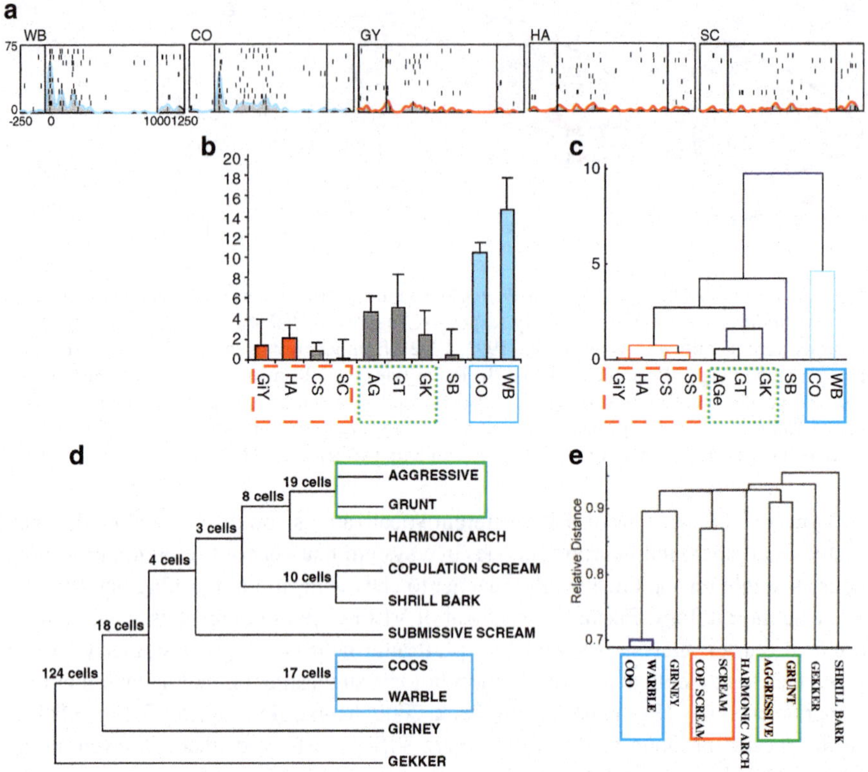

Fig. 3.5 Cluster analysis of vocalizations and auditory responsive prefrontal neurons. (**a**) A single prefrontal neuron's response to vocalizations is shown as raster and spike density plot, with the highest response to the warble (WB) and coo (CO) stimuli. (**b**) The bar chart of the average firing rate response to ten vocalization types for this same neuron and the (**c**), dendrogram of the spike rate to these vocalizations is shown. The dendrogram and bar chart show how the best responses to the warble and coo cluster together (*dark solid rectangle*) while the lowest response to the girney (GY), harmonic arch (HA), copulation scream (CS), and submissive scream (SC) also cluster together (*dashed line rectangle*). (**d**) When all the dendrograms for the population of VLPFC cells are summed together particular clusters occurred more frequently including Aggressive calls/Grunts and Warble/Coos. (**e**) The prefrontal neuronal clusters are similar to the clusters that emerge when the vocalizations are analyzed to determine acoustic similarities, suggesting that prefrontal neurons may encode acoustic features of the calls

& Miller, 1998). Without a behavioral requirement however, similarities in acoustic structure may drive the neuronal response (Romanski et al., 2005). In fact, one study suggested that prefrontal neuronal responses discriminate changes in semantic categories versus acoustic categories when calls are contrasted (Gifford et al., 2005). In their study, when a food call was presented several times and followed by either a nonfood call or another food-call, population neural responses differentiated transitions between semantically different categories, but not between semantically similar categories. However, the study did not include the case when acoustically similar calls from different semantic categories were contrasted (i.e., warble and coo) making it difficult to conclusively determine whether prefrontal neurons encode semantic or acoustic/phonological information. Furthermore, the paradigms used in traditional neurophysiology experiments are unlike the natural behavior used when listening or reacting to calls from other conspecifics. Experiments are progressing towards automated recordings during natural calling behaviors that would answer these questions and many more (Eliades & Wang, 2005; Miller et al., 2010).

4 Integration of Faces and Voices in the Prefrontal Cortex

We have discussed electrophysiological and anatomical data showing that neurons in ventral prefrontal cortex receive information about and respond to faces and vocalizations. The areas of PFC that have been shown to be face-responsive and vocalization-responsive are adjacent and overlapping (Fig. 3.7). In particular, anterolateral area 12vl has face-responsive neurons (O'Scalaidhe et al., 1997, 1999) and in a separate study neurons in this area responded to vocalizations (Romanski et al., 2005). It is clear that this location should have neurons that might be multisensory and integrate face and vocalization information. Moreover, projections from the auditory association cortex and visual extrastriate terminate in VLPFC and could easily synapse on prefrontal neurons providing the necessary auditory and visual information to convey a multisensory response (Stein & Meredith, 1993; Sugihara, Diltz, Averbeck, & Romanski, 2006). In addition, projections from polymodal regions of the STS could provide already integrated information to prefrontal neurons to be used in mnemonic or recognition processes. Thus, the substrates for audiovisual integration are clearly present in VLPFC, so it is not surprising that recent studies have documented neurons that are responsive to combinations of faces and vocalizations in VLPFC neurons (Romanski, 2007).

In Sugihara et al. (2006) VLPFC neurons were tested with faces, vocalizations, and their combination using both dynamic movie clips of monkeys vocalizing and static pictures paired with sound. While some neurons exhibited strong unimodal responses, more than half the recorded population exhibited a significant change in activity when face and vocalizations were presented simultaneously. Many neurons demonstrated multisensory enhancement, that is, an increase during bimodal stimulation that is significantly greater than the best unimodal response, while a slightly larger number of cells exhibited multisensory suppression, where the bimodal response was less than the unimodal response (Fig. 3.6). While there were some linear multisensory neurons which responded to the unimodal auditory and to the unimodal visual stimulus, most prefrontal multisensory neurons demonstrated a nonlinear multisensory response in that the neuronal response to the combined audiovisual stimulus was significantly different than the simple linear combination of auditory and visual responses. It was also found that face–vocalization stimuli evoked multisensory responses more frequently than nonface–nonvocalization combinations when both were tested. This adds support to the notion that VLPFC may be specialized for integrating face and vocalization information during communication

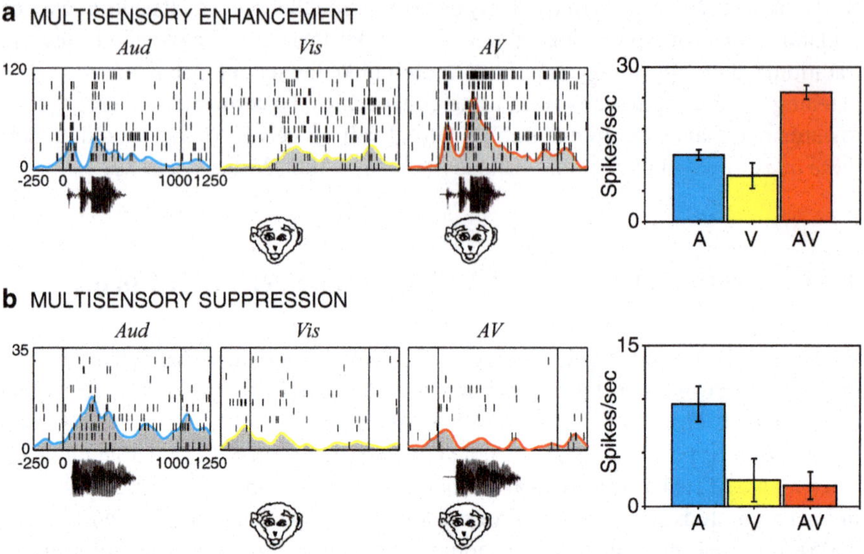

Fig. 3.6 Multisensory interactions in the prefrontal cortex. The types of multisensory responses that were seen when auditory and visual stimuli were presented separately and combined in the prefrontal cortex included (**a**), multisensory enhancement, where the response to combined stimuli is greater than the best unimodal response or (**b**), multisensory suppression where the response to the combined stimuli was less than the best unimodal response. Adapted from Sugihara et al. (2006)

rather than general auditory and visual stimuli, and setting it apart from other areas which integrate sensory stimuli in a more general sense.

In localizing these multisensory responses to the prefrontal cortex there appeared to be two somewhat separate VLPFC zones for multisensory processing. First, there is a large pool of multisensory and visual neurons covering most of the middle and posterior VLPFC (areas 12/47 and 45). These neurons are robustly responsive to visual stimuli with weak modulation by auditory stimuli. Unimodal neurons in this area are mostly visual and respond to faces but also to nonsocial stimuli such as objects, shapes, and patterns. Neurons in this region, which lie closer the arcuate sulcus (Fig. 3.7), receive their greatest input from the polymodal STS and IT cortex. Previous studies in nonhuman primates of visual working memory, decision-making, sequence planning, and visual search (Freedman & Miller, 2008; Kim & Shadlen, 1999; Rao et al., 1997; Wilson et al., 1993) may have recorded neurons within pool of VLPFC neurons since they lie below the caudal principal sulcus and are easier to reach than their anterolateral counterparts.

A smaller, potentially more specialized pool of vocalization-responsive neurons is located in VLPFC (area 12vl), anterior and lateral to the first pool. This is the region where vocalization responsive neurons were predominantly localized in previous studies (Romanski et al., 2005). These anterolateral VLPFC neurons respond vocalizations and to faces, but weakly to other visual stimuli. This area receives afferents from mainly polymodal and auditory association cortex. Multisensory

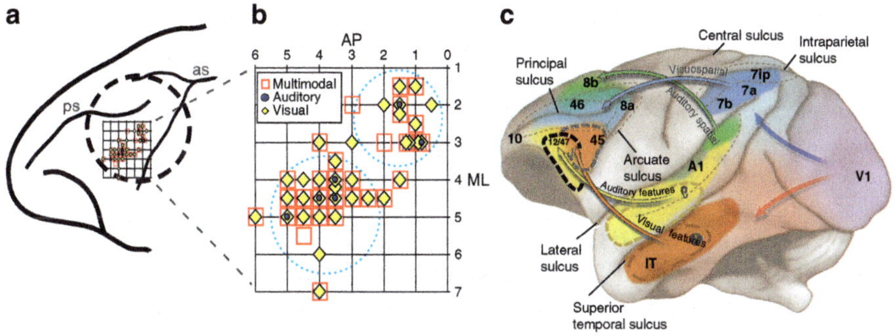

Fig. 3.7 Location of face–vocalization cells in the primate prefrontal cortex. (**a**) Schematic of the macaque brain which indicates the approximate locations of auditory, visual, and audiovisual responsive cells. The enlarged inset (**b**) shows these locations on the recording grid and two pools of multisensory neurons are noted with *blue dashed line circles* (adapted from Sugihara et al., 2006). (**c**) A pictorial representation of auditory and visual domains in the ventral prefrontal cortex that is color coded to indicate projections from the temporal and parietal lobes and their innervation of the prefrontal cortex. VLPFC receives input from both auditory cortical regions (*yellow*) and extrastriate temporal lobe visual regions (*orange*)

responses here favor faces and vocalizations, suggesting a more specialized role in the integration of social communication information. In contrast, the larger posterior pool might integrate social audiovisual as well as nonsocial audiovisual stimuli while the anterior vocalization responsive cells may be specialized for integration of social communication sounds with facial gestures.

Exactly what function does VLPFC serve in integrating face and vocal information? The similarities in anatomical location of this area with the human inferior frontal gyrus imply a role in social communication. There are of course a number of functions that integration might affect social communication. Clarification of semantic meaning and enhanced recognition of identity are both processes in which the integration of facial and vocal information would benefit social communication since integration of cross-modal stimuli can enhance accuracy or decrease reaction time (Stein & Meredith, 1993). Adding a facial expression to spoken words can clarify or even alter the meaning of an utterance (McGurk & MacDonald, 1976). Enhanced accuracy and comprehension by the addition of a corresponding visual gesture has been demonstrated in many studies, though in a direct assessment of speech with corresponding symbolic gestures, only the anterior VLPFC (area 47) was shown to be activated by both in an fMRI study (Xu, Gannon, Emmorey, Smith, & Braun, 2009). This area in the human brain is homologous with the VLPFC region which has been shown to integrate face and vocalizations (Sugihara et al., 2006). Thus, prefrontal neurons in the NHP may be assessing semantic meaning by integrating a vocalization with the corresponding facial gesture. Furthermore, VLPFC receives a robust innervation from the polymodal processing regions in the STS, which have been associated with the processing of facial expression and from IT cortex which could convey identity related feature information.

An area involved in semantic processing or in identity recognition would show a change between the original stimulus and an incongruent one. Human neuroimaging studies of the VLPFC have noted disparate findings on the processing of congruent versus incongruent speech–gesture or speech–face stimuli with some studies noting a decrease in ventral prefrontal activity for incongruent faces and voices (Calvert, Hansen, Iversen, & Brammer, 2001; Homae, Hashimoto, Nakajima, Miyashita, & Sakai, 2002; Jones & Callan, 2003) and others reporting increased activations during incongruent stimuli (Hein et al., 2007; Miller & D'Esposito, 2005; Naumer et al., 2009; Ojanen et al., 2005; Werner & Noppeney, 2010). The recordings in macaque VLPFC using face with corresponding vocalizations indicate that cells do not encode incongruence with a universal increase or decrease to all incongruent events. Rather, a change from the congruent can be encoded with either an increase or a decrease in neuronal activity depending on the original response to bimodal stimuli. Overall, suppression is more common in bimodal responsive VLPFC neurons than enhancement (Sugihara et al., 2006). This suggests that different pools of neurons may display different but potentially cooperative activity in signifying the integration of audiovisual stimuli.

Investigation of the specific facial features that are encoded by VLPFC neurons could illuminate what aspect of recognition with which the PFC is most involved. One study examined the response of prefrontal neurons to different rotated views of a human and monkey face and found that ventral prefrontal neurons responded best to forward (0° and 30°) face-views (Romanski & Diehl, 2011). This is not surprising given that most of the information that conveys identity is present in these views. Nonetheless, further tests revealed that these forward face-view neurons were also auditory responsive while nonselective cells were not, suggesting a specialization of face-responsive cells for face-to-face communication (Romanski & Diehl, 2011). Neurons which are most responsive to facial features and to the pitch of vocal stimuli would be most appropriate in a system that is involved in the identification of individuals. These responses have not yet been assessed in prefrontal neurons, though some previous work has suggested their involvement in identity processing (O'Scalaidhe et al., 1997). Furthermore, it is possible that VLPFC cells may encode the mouth movements which underlie vocalization production in a similar manner to the speech production activations in Broca's area.

The accumulation of evidence to date shows that cells in the ventral prefrontal cortex of the NHP respond to and integrate audiovisual information. VLPFC cells respond optimally to face and vocalization stimuli and exhibit multisensory enhancement or suppression when face–vocalization stimuli are combined. There is evidence to suggest that the primate VLFPC is involved in identity processing but may also play a role in the encoding of semantic information during communication in a manner that may be homologous to the human inferior frontal gyrus and language processing. Further work aimed at understanding the mechanism of sensory integration in the frontal lobes of nonhuman primates may provide us with an understanding of the cellular mechanisms which underlie recognition and speech perception in the human brain, which critically depends on the integration of multiple types of sensory information.

References

Adachi, I., Chou, D. P., & Hampton, R. R. (2009). Thatcher effect in monkeys demonstrates conservation of face perception across primates. *Current Biology, 19*(15), 1270–1273.

Albright, T. D., Desimone, R., & Gross, C. G. (1984). Columnar organization of directionally selective cells in visual area MT of the macaque. *Journal of Neurophysiology, 51*, 16–31.

Altmann, S. A. (1962). A field study of the sociobiology of rhesus monkeys, *Macaca mulatta*. *Annals of the New York Academy of Sciences, 102*, 338–435.

Andrew, R. J. (1963). Evolution of facial expression. *Science (New York, NY), 142*, 1034–1041.

Averbeck, B. B., & Romanski, L. M. (2004). Principal and independent components of macaque vocalizations: Constructing stimuli to probe high-level sensory processing. *Journal of Neurophysiology, 91*, 2897–2909.

Averbeck, B. B., & Romanski, L. M. (2006). Probabilistic encoding of vocalizations in macaque ventral lateral prefrontal cortex. *The Journal of Neuroscience: The Official Journal of the Society for Neuroscience, 26*, 11023–11033.

Azuma, M., & Suzuki, H. (1984). Properties and distribution of auditory neurons in the dorsolateral prefrontal cortex of the alert monkey. *Brain Research, 298*, 343–346.

Barbas, H. (1988). Anatomic organization of basoventral and mediodorsal visual recipient prefrontal regions in the rhesus monkey. *Journal of Comparative Neurology, 276*, 313–342.

Barbas, H. (1992). Architecture and cortical connections of the prefrontal cortex in the rhesus monkey. *Advances in Neurology, 57*, 91–115 [review].

Benevento, L. A., Fallon, J., Davis, B. J., & Rezak, M. (1977). Auditory-visual interaction in single cells in the cortex of the superior temporal sulcus and the orbital frontal cortex of the macaque monkey. *Experimental Neurology, 57*, 849–872.

Bodner, M., Kroger, J., & Fuster, J. M. (1996). Auditory memory cells in dorsolateral prefrontal cortex. *Neuroreport, 7*, 1905–1908.

Broca, P. (1861). Remarques su le siege defaulte de langage articule suivies d'une observation d'aphemie (perte de la parole). *Bulletin De La Societe' d'Anthropologie, 2*, 330–337.

Bruce, C. J., & Goldberg, M. E. (1985). Primate frontal eye fields. I. Single neurons discharging before saccades. *Journal of Neurophysiology, 53*(3), 603–635.

Buckner, R. L., Raichle, M. E., & Petersen, S. E. (1995). Dissociation of human prefrontal cortical areas across different speech production tasks and gender groups. *Journal of Neurophysiology, 74*(5), 2163–2173.

Calvert, G. A., Hansen, P. C., Iversen, S. D., & Brammer, M. J. (2001). Detection of audio-visual integration sites in humans by application of electrophysiological criteria to the BOLD effect. *NeuroImage, 14*(2), 427–438.

Chavis, D. A., & Pandya, D. N. (1976). Further observations on corticofrontal connections in the rhesus monkey. *Brain Research, 117*, 369–386.

Dahl, C. D., Logothetis, N. K., Bulthoff, H. H., & Wallraven, C. (2010). The thatcher illusion in humans and monkeys. *Proceedings of the Royal Society: Biological Sciences, 277*(1696), 2973–2981.

Darwin, C. (1872). *The expression of the emotions in man and animals*. London: John Murray.

Demb, J. B., Desmond, J. E., Wagner, A. D., Vaidya, C. J., Glover, G. H., & Gabrieli, J. D. (1995). Semantic encoding and retrieval in the left inferior prefrontal cortex: A functional MRI study of task difficulty and process specificity. *Journal of Neuroscience, 15*(9), 5870–5878.

DeSouza, W. C., Eifuku, S., Tamura, R., Nishijo, H., & Ono, T. (2005). Differential characteristics of face neuron responses within the anterior superior temporal sulcus of macaques. *Journal of Neurophysiology, 94*, 1252–1266.

Diehl, M. M., Bartlow-Kang, J., Sugihara, T., & Romanski, L. M. (2008). Distinct temporal lobe projections to auditory and visual regions in the ventral prefrontal cortex support face and vocalization processing. *Society for Neuroscience Abstracts, 34*.

Dittrich, W. (1990). Representation of faces in longtailed macaques (*Macaca fascicularis*). *Ethology, 85*, 265–278.

Dolan, R. J., Fletcher, P., Morris, J., Kapur, N., Deakin, J. F., & Frith, C. D. (1996). Neural activation during covert processing of positive emotional facial expressions. *NeuroImage, 4*(3 Pt 1), 194–200.

Eifuku, S., De Souza, W. C., Tamura, R., Nishijo, H., & Ono, T. (2004). Neuronal correlates of face identification in the monkey anterior temporal cortical areas. *Journal of Neurophysiology, 91*, 358–371.

Eifuku, S., Nakata, R., Sugimori, M., Ono, T., & Tamura, R. (2010). Neural correlates of associative face memory in the anterior inferior temporal cortex of monkeys. *The Journal of Neuroscience: The Official Journal of the Society for Neuroscience, 30*(45), 15085–15096.

Eliades, S. J., & Wang, X. (2005). Dynamics of auditory-vocal interaction in monkey auditory cortex. *Cerebral Cortex (New York, NY: 1991), 15*(10), 1510–1523.

Fecteau, S., Armony, J. L., Joanette, Y., & Belin, P. (2005). Sensitivity to voice in human prefrontal cortex. *Journal of Neurophysiology, 94*, 2251–2254.

Fiez, J. A., Raife, E. A., Balota, D. A., Schwarz, J. P., Raichle, M. E., & Petersen, S. E. (1996). A positron emission tomography study of the short-term maintenance of verbal information. *Journal of Neuroscience, 16*, 808–822.

Freedman, D. J., & Miller, E. K. (2008). Neural mechanisms of visual categorization: Insights from neurophysiology. *Neuroscience and Biobehavioral Reviews, 32*(2), 311–329.

Friederici, A. D., Ruschemeyer, S. A., Hahne, A., & Fiebach, C. J. (2003). The role of left inferior frontal and superior temporal cortex in sentence comprehension: Localizing syntactic and semantic processes. *Cerebral Cortex, 13*, 170–177.

Funahashi, S., Bruce, C. J., & Goldman-Rakic, P. S. (1989). Mnemonic coding of visual space in the monkey's dorsolateral prefrontal cortex. *Journal of Neurophysiology, 61*, 1–19.

Funahashi, S., Chafee, M. V., & Goldman-Rakic, P. S. (1993). Prefrontal neuronal activity in rhesus monkeys performing a delayed anti-saccade task. *Nature, 365*(6448), 753–756.

Fuster, J. M. (2001). The prefrontal cortex—An update: Time is of the essence. *Neuron, 30*(2), 319–333.

Fuster, J. M., Bauer, R. H., & Jervey, J. P. (1982). Cellular discharge in the dorsolateral prefrontal cortex of the monkey in cognitive tasks. *Experimental Neurology, 77*, 679–694.

Gabrieli, J. D. E., Poldrack, R. A., & Desmond, J. E. (1998). The role of left prefrontal cortex in language and memory. *Proceedings of the National Academy of Sciences of the United States of America, 95*, 906–913.

Gifford, G. W., III, Maclean, K. A., Hauser, M. D., & Cohen, Y. E. (2005). The neurophysiology of functionally meaningful categories: Macaque ventrolateral prefrontal cortex plays a critical role in spontaneous categorization of species-specific vocalizations. *Journal of Cognitive Neuroscience, 17*, 1471–1482.

Goldman, P. S., & Rosvold, H. E. (1970). Localization of function within the dorsolateral prefrontal cortex of the rhesus monkey. *Experimental Neurology, 27*, 291–304.

Goldman-Rakic, P. S. (1996a). The prefrontal landscape: Implications of functional architecture for understanding human mentation and the central executive. *Philosophical Transactions of the Royal Society of London. Series B: Biological Sciences, 351*, 1445–1453 [review] [72 refs].

Goldman-Rakic, P. S. (1996b). Regional and cellular fractionation of working memory. *Proceedings of the National Academy of Sciences of the United States of America, 93*(24), 13473–13480.

Gothard, K. M., Erickson, C. A., & Amaral, D. G. (2004). How do rhesus monkeys (*Macaca mulatta*) scan faces in a visual paired comparison task? *Animal Cognition, 7*(1), 25–36.

Gross, C. G. (1963). A comparison of the effects of partial and total lateral frontal lesions on test performance by monkeys. *Journal of Comparative Physiological Psychology, 56*, 41–47.

Gross, C. G. (1994). How inferior temporal cortex became a visual area. *Cerebral Cortex, 5*, 455–469.

Gross, C. G., Bender, D. B., & Rocha-Miranda, C. E. (1969). Visual receptive fields of neurons in inferotemporal cortex of the monkey. *Science, 166*, 1303–1306.

Gross, C. G., & Weiskrantz, L. (1962). Evidence for dissociation of impairment on auditory discrimination and delayed response following lateral frontal lesions in monkeys. *Experimental Neurology, 5*, 453–476.

Guo, K., Robertson, R. G., Mahmoodi, S., Tadmor, Y., & Young, M. P. (2003). How do monkeys view faces? A study of eye movements. *Experimental Brain Research. Experimentelle Hirnforschung. Experimentation Cerebrale, 150*(3), 363–374.

Hackett, T. A., Stepniewska, I., & Kaas, J. H. (1998). Subdivisions of auditory cortex and ipsilateral cortical connections of the parabelt auditory cortex in macaque monkeys. *Journal of Comparative Neurology, 394*, 475–495.

Hackett, T. A., Stepniewska, I., & Kaas, J. H. (1999). Prefrontal connections of the parabelt auditory cortex in macaque monkeys. *Brain Research, 817*, 45–58.

Haith, M. M., Bergman, T., & Moore, M. J. (1977). Eye contact and face scanning in early infancy. *Science (New York, NY), 198*(4319), 853–855.

Hasselmo, M. E., Rolls, E. T., & Baylis, G. C. (1989). The role of expression and identity in the face-selective responses of neurons in the temporal visual cortex of the monkey. *Behavioural Brain Research, 32*, 203–218.

Hauser, M. D. (1996). *The evolution of communication*. Cambridge, MA: MIT Press.

Hauser, M. D., & Marler, P. (1993). Food associated calls in rhesus macaques (*Macaca mulatta*). I. Socioecological factors. *Behavioral Ecology, 4*, 194–205.

Hein, G., Doehrmann, O., Muller, N. G., Kaiser, J., Muckli, L., & Naumer, M. J. (2007). Object familiarity and semantic congruency modulate responses in cortical audiovisual integration areas. *The Journal of Neuroscience: The Official Journal of the Society for Neuroscience, 27*(30), 7881–7887.

Hinde, R. A., & Rowell, T. E. (1962). Communication by postures and facial expressions in the rhesus monkey (*Macaca mulatta*). *Proceedings of the Zoological Society of London, 138*, 1–21.

Homae, F., Hashimoto, R., Nakajima, K., Miyashita, Y., & Sakai, K. L. (2002). From perception to sentence comprehension: The convergence of auditory and visual information of language in the left inferior frontal cortex. *NeuroImage, 16*(4), 883–900.

Iidaka, T., Omori, M., Murata, T., Kosaka, H., Yonekura, Y., Okada, T., et al. (2001). Neural interaction of the amygdala with the prefrontal and temporal cortices in the processing of facial expressions as revealed by fMRI. *Journal of Cognitive Neuroscience, 13*(8), 1035–1047.

Ishai, A., Pessoa, L., Bikle, P. C., & Ungerleider, L. G. (2004). Repetition suppression of faces is modulated by emotion. *Proceedings of the National Academy of Sciences of the United States of America, 101*(26), 9827–9832.

Ishai, A., Schmidt, C. F., & Boesiger, P. (2005). Face perception is mediated by a distributed cortical network. *Brain Research Bulletin, 67*(1–2), 87–93.

Ito, S. I. (1982). Prefrontal unit activity of macaque monkeys during auditory and visual reaction time tasks. *Brain Research, 247*, 39–47.

Ito, M., Tamura, H., Fujita, I., & Tanaka, K. (1995). Size and position invariance of neuronal responses in monkey inferotemporal cortex. *Journal of Neurophysiology, 73*(1), 218–226.

Jones, J. A., & Callan, D. E. (2003). Brain activity during audiovisual speech perception: An fMRI study of the McGurk effect. *Neuroreport, 14*(8), 1129–1133.

Kaas, J. H., & Hackett, T. A. (1998). Subdivisions of auditory cortex and levels of processing in primates. *Audiology & Neurotology, 3*, 73–85 [review] [66 refs].

Kesler-West, M. L., Andersen, A. H., Smith, C. D., Avison, M. J., Davis, C. E., Kryscio, R. J., et al. (2001). Neural substrates of facial emotion processing using fMRI. *Brain Research. Cognitive Brain Research, 11*(2), 213–226.

Kim, J. N., & Shadlen, M. N. (1999). Neural correlates of a decision in the dorsolateral prefrontal cortex of the macaque. *Nature Neuroscience, 2*(2), 176–185.

Klin, A., Jones, W., Schultz, R., Volkmar, F., & Cohen, D. (2002). Visual fixation patterns during viewing of naturalistic social situations as predictors of social competence in individuals with autism. *Archives of General Psychiatry, 59*(9), 809–816.

Kojima, S., & Goldman-Rakic, P. S. (1984). Functional analysis of spatially discriminative neurons in prefrontal cortex of rhesus monkey. *Brain Research, 291*, 229–240.

LoPresti, M. L., Schon, K., Tricarico, M. D., Swisher, J. D., Celone, K. A., & Stern, C. E. (2008). Working memory for social cues recruits orbitofrontal cortex and amygdala: A functional magnetic resonance imaging study of delayed matching to sample for emotional expressions. *The Journal of Neuroscience: The Official Journal of the Society for Neuroscience, 28*(14), 3718–3728.

Maestripieri, D., & Wallen, K. (1997). Affiliative and submissive communication in rhesus macaques. *Primates, 38*(2), 127–138.

Marinkovic, K., Trebon, P., Chauvel, P., & Halgren, E. (2000). Localised face processing by the human prefrontal cortex: Face-selective intracerebral potentials and post-lesion deficits. *Cognitive Neuropsychology, 17*(1), 187–199.

Marriott, B. M., & Salzen, E. A. (1978). Facial expressions in captive squirrel monkeys (*Saimiri sciureus*). *Folia Primatologica; International Journal of Primatology, 29*(1), 1–18.

McGurk, H., & MacDonald, J. (1976). Hearing lips and seeing voices. *Nature, 264*, 746–748.

Miller, E. K., & Cohen, J. D. (2001). An integrative theory of prefrontal cortex function. *Annual Review of Neuroscience, 24*, 167–202.

Miller, B. T., & D'Esposito, M. (2005). Searching for "the top" in top-down control. *Neuron, 48*(4), 535–538.

Miller, C., Diauro, A., Pistorio, A., Hendry, S. H., & Wang, X. (2010). Vocalization induced cFos expression in marmoset cortex. *Frontiers in Integrative Neuroscience, 4*, 128.

Miller, E. K., Erickson, C. A., & Desimone, R. (1996). Neural mechanisms of visual working memory in prefrontal cortex of the macaque. *Journal of Neuroscience, 16*, 5154–5167.

Miyashita, Y. (1993). Inferior temporal cortex: Where visual perception meets memory. *Annual Review of Neuroscience, 16*, 245–263.

Moeller, S., Freiwald, W. A., & Tsao, D. Y. (2008). Patches with links: A unified system for processing faces in the macaque temporal lobe. *Science (New York, NY), 320*(5881), 1355–1359.

Mohler, C. W., Goldberg, M. E., & Wurtz, R. H. (1973). Visual receptive fields of frontal eye field neurons. *Brain Research, 61*, 385–389.

Morel, A., Garraghty, P. E., & Kaas, J. H. (1993). Tonotopic organization, architectonic fields, and connections of auditory cortex in macaque monkeys. *Journal of Comparative Neurology, 335*, 437–459.

Nahm, F. K., Perret, A., Amaral, D. G., & Albright, T. D. (1997). How do monkeys look at faces? *Journal of Cognitive Neuroscience, 9*(5), 611.

Naumer, M. J., Doehrmann, O., Muller, N. G., Muckli, L., Kaiser, J., & Hein, G. (2009). Cortical plasticity of audio-visual object representations. *Cerebral Cortex (New York, NY: 1991), 19*(7), 1641–1653.

Newman, J. D., & Lindsley, D. F. (1976). Single unit analysis of auditory processing in squirrel monkey frontal cortex. *Experimental Brain Research, 25*(2), 169–181.

Niki, H., & Watanabe, M. (1976). Prefrontal unit activity and delayed response: Relation to cue location versus direction of response. *Brain Research, 105*(1), 79–88.

Nomura, M., Ohira, H., Haneda, K., Iidaka, T., Sadato, N., Okada, T., et al. (2004). Functional association of the amygdala and ventral prefrontal cortex during cognitive evaluation of facial expressions primed by masked angry faces: An event-related fMRI study. *NeuroImage, 21*(1), 352–363.

O'Scalaidhe, S. P., Wilson, F. A., & Goldman-Rakic, P. S. (1997). Areal segregation of face-processing neurons in prefrontal cortex. *Science, 278*, 1135–1138.

O'Scalaidhe, S. P. O., Wilson, F. A. W., & Goldman-Rakic, P. G. R. (1999). Face-selective neurons during passive viewing and working memory performance of rhesus monkeys: Evidence for intrinsic specialization of neuronal coding. *Cerebral Cortex, 9*, 459–475.

Ojanen, V., Mottonen, R., Pekkola, J., Jaaskelainen, I. P., Joensuu, R., Autti, T., et al. (2005). Processing of audiovisual speech in broca's area. *NeuroImage, 25*, 333–338.

Pandya, D. N., Hallett, M., & Kmukherjee, S. K. (1969). Intra- and interhemispheric connections of the neocortical auditory system in the rhesus monkey. *Brain Research, 14*, 49–65.

Pandya, D. N., & Kuypers, H. G. (1969). Cortico-cortical connections in the rhesus monkey. *Brain Research, 13*, 13–36.

Parr, L. A., Dove, T., & Hopkins, W. D. (1998). Why faces may be special: Evidence of the inversion effect in chimpanzees. *Journal of Cognitive Neuroscience, 10*(5), 615–622.

Parr, L. A., & Heintz, M. (2006). The perception of unfamiliar faces and houses by chimpanzees: Influence of rotation angle. *Perception, 35*(11), 1473–1483.

Parr, L. A., & Heintz, M. (2009). Facial expression recognition in rhesus monkeys, *Macaca mulatta. Animal Behaviour, 77*(6), 1507–1513.

Parr, L. A., Winslow, J. T., & Hopkins, W. D. (1999). Is the inversion effect in rhesus monkeys face specific? *Animal Cognition, 2*, 123–129.

Parron, C., & Fagot, J. (2007). Baboons (*Papio papio*) spontaneously process the first-order but not second-order configural properties of faces. *American Journal of Primatology, 70*(5), 415–422.

Partan, S. (2002). Single and multichannel signal composition: Facial expressions and vocalizations of rhesus macaques (*Macaca mulatta*). *Behaviour, 139*(8), 993–1027.

Perrett, D. I., Hietanen, J. K., Oram, M. W., & Benson, P. J. (1992). Organization and functions of cells responsive to faces in the temporal cortex. *Philosophical Transactions of the Royal Society of London. Series B, Biological Sciences, 335*(1273), 23–30.

Perrett, D. I., Mistlin, A. J., & Chitty, A. J. (1987). Visual neurones responsive to faces. *Trends in Neurosciences, 10*(9), 358–364.

Perrett, D. I., Rolls, E. T., & Caan, W. (1982). Visual neurones responsive to faces in the monkey temporal cortex. *Experimental Brain Research, 47*, 329–342.

Perrett, D. I., Smith, P. A., Potter, D. D., Mistlin, A. J., Head, A. S., Milner, A. D., et al. (1985). Visual cells in the temporal cortex sensitive to face view and gaze direction. *Proceedings of the Royal Society of London. Series B, Containing Papers of a Biological Character Royal Society (Great Britain), 223*(1232), 293–317.

Petrides, M., & Pandya, D. N. (1988). Association fiber pathways to the frontal cortex from the superior temporal region in the rhesus monkey. *Journal of Comparative Neurology, 273*, 52–66.

Petrides, M., & Pandya, D. N. (2002). Comparative cytoarchitectonic analysis of the human and the macaque ventrolateral prefrontal cortex and corticocortical connection patterns in the monkey. *The European Journal of Neuroscience, 16*(2), 291–310.

Pigarev, I. N., Rizzolatti, G., & Schandolara, C. (1979). Neurons responding to visual stimuli in the frontal lobe of macaque monkeys. *Neuroscience Letters, 12*, 207–212.

Poldrack, R. A., Wagner, A. D., Prull, M. W., Desmond, J. E., Glover, G. H., & Gabrieli, J. D. E. (1999). Functional specialization for semantic and phonological processing in the left inferior prefrontal cortex. *NeuroImage, 10*, 15–35.

Pourtois, G., Schwartz, S., Seghier, M. L., Lazeyras, F., & Vuilleumier, P. (2006). Neural systems for orienting attention to the location of threat signals: An event-related fMRI study. *NeuroImage, 31*(2), 920–933.

Quintana, J., & Fuster, J. M. (1992). Mnemonic and predictive functions of cortical neurons in a memory task. *Neuroreport, 3*(8), 721–724.

Quintana, J., Yajeya, J., & Fuster, J. M. (1988). Prefrontal representation of stimulus attributes during delay tasks. I. Unit activity in cross-temporal integration of sensory and sensory-motor information. *Brain Research, 474*, 211–221.

Rainer, G., Asaad, W. F., & Miller, E. K. (1998). Selective representation of relevant information by neurons in the primate prefrontal cortex. *Nature, 393*, 577–579.

Rao, S. C., Rainer, G., & Miller, E. K. (1997). Integration of what and where in the primate prefrontal cortex. *Science (New York, NY), 276*(5313), 821–824.

Rauschecker, J. P., Tian, B., & Hauser, M. (1995). Processing of complex sounds in the macaque nonprimary auditory cortex. *Science, 268*(5207), 111–114.

Redican, W. K. (1975). Facial expressions in nonhuman primates. In L. Rosenblum (Ed.), *Primate behavior* (pp. 103–194). New York, NY: Acedemic.

Rizzolatti, G., Scandolara, C., Matelli, M., & Gentilucci, M. (1981). Afferent properties of periarcuate neurons in macaque monkeys. II. Visual responses. *Behavioural Brain Research, 2*(2), 147–163.

Rolls, E. T. (1996). The orbitofrontal cortex. *Philosophical Transactions of the Royal Society of London. Series B, Biological Sciences, 351*(1346), 1433–1443 [discussion 1443–1444].

Rolls, E. T., Critchley, H. D., Browning, A. S., & Inoue, K. (2006). Face-selective and auditory neurons in the primate orbitofrontal cortex. *Experimental Brain Research. Experimentelle Hirnforschung. Experimentation Cerebrale, 170*(1), 74–87.

Rolls, E. T., & Tovee, M. J. (1995). Sparseness of the neuronal representation of stimuli in the primate temporal visual cortex. *Journal of Neurophysiology, 73*(2), 713–726.
Romanski, L. M. (2004). Domain specificity in the primate prefrontal cortex. *Cognitive, Affective, & Behavioural Neuroscience, 4*, 421–429.
Romanski, L. M. (2007). Representation and integration of auditory and visual stimuli in the primate ventral lateral prefrontal cortex. *Cerebral Cortex, 17*, i61–i69.
Romanski, L. M., & Averbeck, B. B. (2009). The primate cortical auditory system and neural representation of conspecific vocalizations. *Annual Review of Neuroscience, 32*, 315–346.
Romanski, L. M., Averbeck, B. B., & Diltz, M. (2005). Neural representation of vocalizations in the primate ventrolateral prefrontal cortex. *Journal of Neurophysiology, 93*(2), 734–747.
Romanski, L. M., Bates, J. F., & Goldman-Rakic, P. S. (1999). Auditory belt and parabelt projections to the prefrontal cortex in the rhesus monkey. *Journal of Comparative Neurology, 403*, 141–157.
Romanski, L. M., & Diehl, M. M. (2011). Neurons responsive to face-view in the primate ventrolateral prefrontal cortex. *Neuroscience, 189*, 223–235.
Romanski, L. M., & Goldman-Rakic, P. S. (2002). An auditory domain in primate prefrontal cortex. *Nature Neuroscience, 5*, 15–16.
Romanski, L. M., Tian, B., Fritz, J., Mishkin, M., Goldman-Rakic, P. S., & Rauschecker, J. P. (1999). Dual streams of auditory afferents target multiple domains in the primate prefrontal cortex. *Nature Neuroscience, 2*(12), 1131–1136.
Rosenfeld, S. A., & Van Hoesen, G. W. (1979). Face recognition in the rhesus monkey. *Neuropsychologia, 17*(5), 503–509.
Russ, B. E., Ackelson, A. L., Baker, A. E., & Cohen, Y. E. (2008). Coding of auditory-stimulus identity in the auditory non-spatial processing stream. *Journal of Neurophysiology, 99*(1), 87–95.
Sergerie, K., Lepage, M., & Armony, J. L. (2005). A face to remember: Emotional expression modulates prefrontal activity during memory formation. *NeuroImage, 24*(2), 580–585.
Stein, B. E., & Meredith, M. A. (1993). *The merging of the senses*. Cambridge, MA: MIT Press.
Stevens, A. A., Goldman-Rakic, P. S., Gore, J. C., Fulbright, R. K., & Wexler, B. E. (1998). Cortical dysfunction in schizophrenia during auditory word and tone working memory demonstrated by functional magnetic resonance imaging. *Archives of General Psychiatry, 55*(12), 1097–1103.
Stromswold, K., Caplan, D., Alpert, N., & Rauch, S. (1996). Location of syntactic comprehension by positron emission tomography. *Brain Language. 52*(3), 452–473.
Sugase, Y., Yamane, S., Ueno, S., & Kawano, K. (1999). Global and fine information coded by single neurons in the temporal visual cortex. *Nature, 400*(6747), 869–873.
Sugihara, T., Diltz, M. D., Averbeck, B. B., & Romanski, L. M. (2006). Integration of auditory and visual communication information in the primate ventrolateral prefrontal cortex. *The Journal of Neuroscience: The Official Journal of the Society for Neuroscience, 26*, 11138–11147.
Tanaka, K. (1996). Inferotemporal cortex and object vision. *Annual Review of Neuroscience, 19*, 109–139.
Tanaka, K., Saito, H., Fukada, Y., & Moriya, M. (1991). Coding visual images of objects in the inferotemporal cortex of the macaque monkey. *Journal of Neurophysiology, 66*(1), 170–189.
Tanila, H., Carlson, S., Linnankoski, I., & Kahila, H. (1993). Regional distribution of functions in dorsolateral prefrontal cortex of the monkey. *Behavioural Brain Research, 53*, 63–71.
Tanila, H., Carlson, S., Linnankoski, I., Lindroos, F., & Kahila, H. (1992). Functional properties of dorsolateral prefrontal cortical neurons in awake monkey. *Behavioral Brain Research, 47*, 169–180.
Thorpe, S. J., Rolls, E. T., & Maddison, S. (1983). The orbitofrontal cortex: Neuronal activity in the behaving monkey. *Experimental Brain Research, 49*, 93–115.
Tian, B., Reser, D., Durham, A., Kustov, A., & Rauschecker, J. P. (2001). Functional specialization in rhesus monkey auditory cortex. *Science, 292*, 290–293.
Tomonaga, M. (1999). Inversion effect in perception of human faces in a chimpanzee (*Pan troglodytes*). *Primates, 40*(3), 417–438. doi:10.1007/BF02557579.
Tomonaga, M. (2007). Visual search for orientation of faces by a chimpanzee (*Pan troglodytes*): Face-specific upright superiority and the role of facial configural properties. *Primates, 48*(1), 1–12. doi:10.1007/s10329-006-0011-4.

Tsao, D. Y., Schweers, N., Moeller, S., & Freiwald, W. A. (2008). Patches of face-selective cortex in the macaque frontal lobe. *Nature Neuroscience, 11*(8), 877–879.
Ungerleider, L. G., & Mishkin, M. (1982). Two cortical visual systems. In D. J. Ingle, M. A. Goodale, & R. J. W. Mansfield (Eds.), *Analysis of visual behavior* (pp. 549–586). Cambridge, MA: MIT Press.
Vaadia, E., Benson, D. A., Hienz, R. D., & Goldstein, M. H., Jr. (1986). Unit study of monkey frontal cortex: Active localization of auditory and of visual stimuli. *Journal of Neurophysiology, 56*, 934–952.
van Hooff, J. A. (1962). Facial expressions in higher primates. *Symposium of the Zoological Society of London, 8*, 97–125.
Vinette, C., Gosselin, F., & Schyns, P. G. (2004). Spatio-temporal dynamics of face recognition in a flash: It's in the eyes. *Cognitive Science, 28*(2), 289–301.
Vuilleumier, P., Armony, J. L., Driver, J., & Dolan, R. J. (2001). Effects of attention and emotion on face processing in the human brain. An event-related fMRI study. *Neuron, 30*(3), 829–841.
Watanabe, M. (1986). Prefrontal unit activity during delayed conditional Go/No-go discrimination in the monkey. I. Relation to the stimulus. *Brain Research, 382*(1), 1–14.
Watanabe, M. (1992). Frontal units of the monkey coding the associative significance of visual and auditory stimuli. *Experimental Brain Research, 89*, 233–247.
Webster, M. J., Bachevalier, J., & Ungerleider, L. G. (1994). Connections of inferior temporal areas TEO and TE with parietal and frontal cortex in macaque monkeys. *Cerebral Cortex, 4*, 470–483.
Weigel, R. M. (1979). The facial expressions of the brown capuchin monkey (*Cebus apella*). *Behaviour, 68*(3/4), 250–276.
Weiskrantz, L., & Mishkin, M. (1958). Effects of temporal and frontal cortical lesions on auditory discrimination in monkeys. *Brain; A Journal of Neurology, 80*, 406–414.
Werner, S., & Noppeney, U. (2010). Distinct functional contributions of primary sensory and association areas to audiovisual integration in object categorization. *The Journal of Neuroscience: The Official Journal of the Society for Neuroscience, 30*(7), 2662–2675.
Wilson, F. A., O'Scalaidhe, S. P., & Goldman-Rakic, P. S. (1993). Dissociation of object and spatial processing domains in primate prefrontal cortex. *Science, 260*, 1955–1958.
Xu, J., Gannon, P. J., Emmorey, K., Smith, J. F., & Braun, A. R. (2009). Symbolic gestures and spoken language are processed by a common neural system. *Proceedings of the National Academy of Sciences of the United States of America, 106*(49), 20664–20669.
Xu, G. Q., Lan, Y., Huang, D. F., Rao, D. Z., Pei, Z., Chen, L., et al. (2010). Visuospatial attention deficit in patients with local brain lesions. *Brain Research, 1322*, 153–159.
Yamane, S., Kaji, S., & Kawano, K. (1988). What facial features activate face neurons in the inferotemporal cortex of the monkey? *Experimental Brain Research. Experimentelle Hirnforschung. Experimentation Cerebrale, 73*(1), 209–214.
Zatorre, R. J., Meyer, E., Gjedde, A., & Evans, A. C. (1996). PET studies of phonetic processing of speech: Review, replication, and reanalysis. *Cerebral Cortex, 6*(1), 21–30.

Chapter 4
Intersensory Perception of Faces and Voices in Infants

Ross Flom

Abstract This chapter describes and provides behavioral and neurophysiological evidence articulating how the intersensory redundancy hypothesis addresses the question of how infants integrate faces and voices in perceiving other people. Infants' learning of the arbitrary relationship between faces and voices occurs in two tightly coupled steps. First, between 3 and 5 months of age infants attend to various amodal properties such as a common tempo, rhythm, and affective expressions that unite a particular face and voice. Second, around 6 months of age, when infants' attention is more flexible and they perceive amodal and modality-specific properties, infants perceive and remember various arbitrary features (i.e., the sound of particular voice and the visual appearance of a particular face) associated with a particular face–voice pairing.

Animals, including human infants, are adept at perceiving a world filled with a variety of objects, events, conspecifics, as well as the occasional enemy. The question of how we, along with most other organisms, are able to detect the perceptual relationships between various sources of stimuli in arriving at a unitary and veridical perception of the world has long perplexed philosophy, psychology, and more recently neuroscience. One of the first to address this question was William James (1890, p. 159) who, citing Royce (1881, p. 376), stated that

> A statue is an aggregation of particles of marble...For the spectator, however it is one; in itself it is an aggregate; just as, to the consciousness of an ant crawling over it, it may again appear a mere aggregate.

While James agreed, at least in this instance, with Royce that objects and events, in the presence of a conscious observer are perceived as a whole. James was somewhat perplexed, however, at how we come to integrate these different sensory inputs

To appear in P. Belin, S. Campanella, & T. Ethofer (Eds.) *Integrating Face and Voice in Person Perception.* Springer.

R. Flom (✉)
Department of Psychology, Brigham Young University, Provo, UT 84602, USA
e-mail: flom@byu.edu

P. Belin et al. (eds.), *Integrating Face and Voice in Person Perception*,
DOI 10.1007/978-1-4614-3585-3_4, © Springer Science+Business Media New York 2013

(what James described as "fusing") and is captured in one of James' (1890, p. 488) more famous quotations, "The baby, assailed by eyes, ears, nose, skin, and entrails at once, feels it all as one great blooming, buzzing confusion…" Thus from James' perspective initially we are bombarded with separate sensations and over the course of development we come to perceive a unitary world.

The question of how we arrive at a unitary perception of the world has been addressed for decades by many different theoretical perspectives and empirical approaches. Within the visual system, for example, many have examined how we associate different cues in arriving at a unitary visual perception of the world in a process known as perceptual binding. Some have argued that we solve the binding problem through top-down approaches such as attention (Treisman, 1996, 1998). Or because we posses separate cortical areas and pathways (Goodale & Milner, 1992; Ungerlieder & Mishkin, 1982; Zeki, 1991, 1993) we come to perceive the world as a unitary whole through higher order computational or neuro-cognitive maps (Shadlen & Movshon, 1999; Shafritz, Gore, & Marios, 2002). In contrast, a more bottom-up solution argues that a unitary perception arises through a lower-level process involving the "selective synchronization of dynamically formed neuronal groups" (Edelman, 1987, 1993; Gray, 1999; Seth, McKinstry, Edelman & Krichmar, 2004; Singer, 1999). While these two proposed solutions (i.e., higher order attention and neurological synchronization), as well as others, have done much to further our understanding of perceptual binding within a *single* sensory systems, i.e., intramodal perception, less is known about perceptual integration across *different* systems, i.e., intermodal perception. Moreover, nearly all of these approaches have examined how *adults* solve the binding problem and are silent on issues of development.

The fact that less is known about perceptual binding across sensory systems (hereafter referred to as intermodal or intersensory perception), including its development, is unfortunate because nearly all objects and events that we encounter, including other people, are experienced through and activate multiple sensory systems. For example, the approach of another person will stimulate and provide information to the visual system in terms of their appearance, the sound of their voice will activate the auditory system, and the approach and potential contact of another person may activate the olfactory as well as activate the tactile receptors. In this example, however, while we may perceive the modality-specific properties, we initially perceive the person. Thus just as it is important to understand processes of intramodal perception it is equally important to study how and when we come to perceive processes of intermodal perception because undoubtedly our perception of objects, events, conspecifics, etc., requires intermodal as well as intramodal perception.

In this chapter I will examine how young infants arrive at unitary perception of the world given the diversity of information across different sensory systems. Specifically, I will examine the development of intersensory perception of auditory and visual information within the context of faces and voices. I will also discuss and provide evidence that both attentional processes and neurophysiological synchrony are complementary processes—at least within the context of perceiving faces and voices. Finally, while much of literature in perception and perceptual development has focused on unimodal perception (perception of faces separate from perception of voices), we perceive and experience objects and events in a multisensory manner. Finally, it will be argued that research must coordinate unimodal with multimodal or intermodal research in order to reach a complete and veridical explanation of face–voice perception.

1 Developmental Approaches to Intermodal Perceptual Binding

Because objects and events provide a wealth of information, where we can only attend to a small amount of the available information, how do we arrive at a veridical perception of these objects and events? One historical developmental view proposed that between 6 months and 1 year of age infants learn to coordinate or associate information from one sense modality with information from another sense modality (Birch & Lefford, 1963; Piaget, 1952). This process of perceiving the perceptual correspondences across different sense modalities occurred as a result of infants' interactions and explorations of various objects and events, including other people. Thus early in development infants do in fact experience what James described as the "great blooming and buzzing confusion" of sensory information. The coordination or integration of sights and sounds is constructed through experience and interaction with objects, events and other people—thus experience builds or associates information across the different sense modalities.

In a departure from traditional views of perception, including the traditional or Piagetian explanation to the binding problem, J.J. Gibson (1966, 1979) argued that presence of different forms of sensory stimulation is not one that needs to be solved through association or integration. Instead Gibson (1966, 1979) argued that the senses are coordinated, from birth, and work together to perceive information that is common across the different modalities. In other words organisms, including young infants, begin not by "binding" information; rather they begin by directly perceiving that information which is common to more than one sense modality. In other words, perceiving information that is amodal or not bound to one sense modality.

On the one hand and according to Piaget, it is proposed that intermodal perception, including the coordinated perception of faces and voices, occurs as a result of associating information across different sense modalities. On the other hand, according to James and Eleanor Gibson (J. Gibson, 1966; E. Gibson, 1969), from birth infants perceive information that is common across sense modalities and perceptual development is not a process of *associating* different sensory stimulation rather perceptual development is a process *differentiating* different sensory stimulation (see Spector & Maurer, 2009 for a recent review). While research examining intermodal perception over the past 30-years largely supports for the latter, or perceptual differentiation perspective, the process of integrating faces and voices, however, provides evidence that both processes are necessary.

2 Nature of Information

Before we can describe how young infants learn to perceptually integrate faces and voices we must first describe the nature of information that is available to the sensory systems. Objects and events, including faces and voices provide two types of information; amodal and modality-specific information. Modality-specific information is defined as information that can be perceived or detected in only one sense modality. For example, the color and pattern printed on an object can be perceived only within the visual system. The pitch and timbre (i.e., complexity) of a person's voice is restricted to the auditory system and the cologne or perfume worn by another person can only be perceived through the olfactory system. Modality-specific information is thus restricted or tied to a specific sensory system and from early in development infants are adept at using information such as the appearance of a face or the sound of a voice in recognizing others.

In addition to modality-specific information objects and events also provide amodal information. Amodal information is defined as information that is "without modality" or according to Aristotle where "there should be a special sense organ to perceive common sensibles". In other words the information is common across two or more sense modalities and is not specific to any one sensory system. Amodal properties include, but are not limited to, tempo or rate, rhythm, shape and texture, and some "social" information, like affect or emotion, is also categorized as amodal. For example a speaker's happy or positive affective expression can be conveyed in their voice and facial expression where their communicated affect is not restricted to the auditory or the visual system. In addition to the communicated affect the visual movements of the speaker's lips and the onset/offset of their voice also share a common synchrony, tempo, and rhythm. Given that all objects and events provide a wealth of modality-specific and amodal information, it is important that we account for how, and under what conditions, we perceive amodal as well as modality-specific information.

3 Attending to Amodal and Modality-Specific Information

A recent explanation for how young infants come to perceive amodal and modality-specific information is what Bahrick and her colleagues have termed an Intersensory Redundancy Hypothesis (Bahrick & Lickliter, 2000, 2002; Bahrick, Lickliter, & Flom, 2004).

> Intersensory redundancy refers to a particular type of multimodal stimulation in which the same information is presented simultaneously and in a spatially coordinated manner to two or more sensory modalities. For the auditory–visual domain, it also entails the temporally synchronous alignment of the information in each modality. Only amodal information can be specified redundantly because by definition amodal information is information that can be conveyed by more than one sense modality. (Bahrick & Lickliter, 2002, p. 163).

According to this hypothesis infants' attention, and subsequent perceptual learning, is initially directed toward amodal properties of objects and events when redundant multimodal stimulation is available to the infant. One prediction of the intersensory redundancy hypothesis is

1. The perception and learning of amodal properties is facilitated in the context of redundant multimodal stimulation, i.e., within the context of intersensory redundancy and is attenuated in the context of unimodal stimulation. (Bahrick et al., 2004; Bahrick & Lickliter, 2002).

In other words, because amodal properties are not tied to a specific sense modality, and are often redundantly available in more than one sense modality, they capture or recruit infants' attention toward these properties. Thus redundant multimodal stimulation recruits infants' attention and promotes their initial learning and subsequent discrimination of these amodal properties.

As noted by Bahrick and Lickliter (2002; Bahrick et al., 2004) while many events provide infants with multimodal stimulation, in some situations redundant multimodal stimulation is not always present. Listening to an upset friend on the phone for example provides acoustic information specifying that the person is upset, but not in a redundant multimodal manner. In this example the amodal property of affect or emotion would be specified in a unimodal manner, i.e., intonation of voice. In such cases where amodal information is available, but not presented redundantly, intersensory redundancy does not exist and infants are equally likely to direct their attention to the amodal property that is specified unimodally or any other unimodal property that is also made available. The intersensory redundancy hypothesis therefore makes a second prediction.

2. Infants' perception and learning of modality-specific properties is initially enhanced when unimodal stimulation is available compared to the perception of the same modality-specific property in the context of multimodal stimulation (Bahrick & Lickliter, 2002; Bahrick et al., 2004).

These first two predictions made by the intersensory redundancy hypothesis are based on the interaction between the nature of information available for perceptual exploration, redundant multimodal stimulation and nonredundant unimodal stimulation, and the property of the object or event, amodal or modality specific. The first prediction, the attentional capture of amodal properties at the expense of unimodal properties in the context of redundant multimodal stimulation, is depicted in the top row of Fig. 4.1. The second prediction, modality-specific properties are more easily perceived in the context of unimodal stimulation, is depicted in the bottom row of Fig. 4.1.

Finally, the intersensory redundancy hypothesis also provides a developmental prediction.

3. Over the course of development, perceptual processing becomes more efficient where both amodal and modality-specific properties can be detected in either redundant multimodal stimulation or nonredundant unimodal stimulation (Bahrick et al., 2004, p. 101).

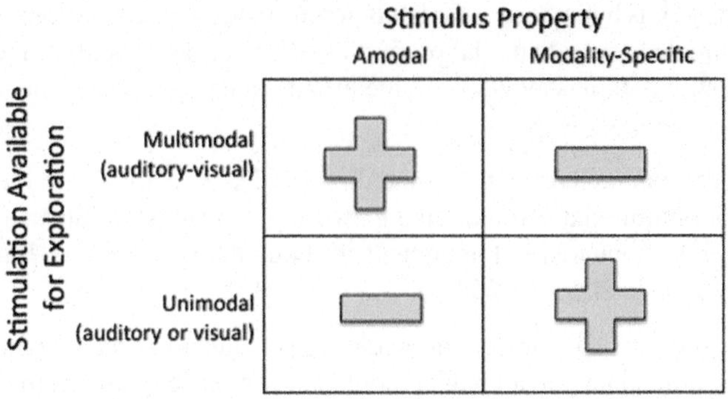

Fig. 4.1 Predictions of the intersensory redundancy hypothesis. Facilitation vs. attenuation of attention and perceptual processing for amodal vs. modality-specific properties of stimulation as a function of the type of stimulation (multimodal or unimodal) available for exploration. Reprinted from "Intersensory Redundancy Guides. Early Perceptual and Cognitive Development," by L.E. Bahrick and R. Lickliter, in R. Kail (Ed.), Advances in Child Development and Behavior, Vol. 30, p. 166, New York: Academic Press. Copyright 2002 by Academic Press. Reprinted with permission from Elsevier

4 Evidence Supporting the Intersensory Redundancy Hypothesis: Infants' Discrimination of Tempo and Rhythm

Over the past 10–12 years research has examined and found support for these three predictions offered by the intersensory redundancy hypothesis in human infants, a precocial avian species, and most recently children with autism (see Bahrick, 2010 for a review). Some of the earliest studies examining the intersensory redundancy hypothesis examined 3- to 5-month-olds' discrimination of amodal properties such as rhythm and tempo (Bahrick & Lickliter, 2000; Bahrick, Flom, & Lickliter, 2002). While others have examined infants' discrimination of tempo and rhythm (e.g., Allen, Walker, Symonds, & Marcell, 1977; Balaban & Dannemiller, 1992; Gardner & Karmel, 1995; Lewkowicz, 1988a, 1988b; Morrongiello & Trehub, 1987) the studies of Bahrick and colleagues were the first to examine whether infants' discrimination of these amodal properties is facilitated when infants are provided redundant bimodal stimulation compared to unimodal auditory or unimodal visual stimulation.

5 Infants' Discrimination of Rhythm and Tempo

In these early experiments infants were familiarized, i.e., habituated, to a plastic toy hammer moving up and down and striking a wooden surface accompanied by the appropriate impact sounds. In each experiment during habituation infants' were provided redundant and temporally synchronous bimodal stimulation (auditory–visual) or unimodal auditory or unimodal visual stimulation specifying the common rhythm or tempo of the event. Following habituation infants were shown the same event as habituation with the exception that the rhythm of the hammer or the tempo or rate of the hammer striking the surface changed in order to examine whether infants noticed the change in tempo or rhythm. The results of Bahrick and Lickliter (2000) revealed that 5-month-olds discriminate a change in rhythm when provided redundant and temporally synchronous bimodal auditory–visual stimulation. Five-month-olds, however, failed to discriminate a change in rhythm when provided unimodal auditory, unimodal visual stimulation, or temporally asynchronous bimodal stimulation (Bahrick & Lickliter, 2000). Likewise the results of Bahrick, Flom, and Lickliter (2002) reveal that 3-month-olds discriminate a change in tempo when provided redundant and temporally synchronous bimodal auditory–visual stimulation and not when provided unimodal auditory or unimodal visual stimulation.

These two experiments provide support for the first prediction of the intersensory redundancy hypothesis. Specifically, amodal properties such as rhythm and tempo are more easily perceived when conveyed in redundant multimodal stimulation than when conveyed in unimodal stimulation. More recently these researchers examined the third or the developmental prediction of the intersensory redundancy hypothesis. Namely, within the context of multimodal stimulation, infants' attention is initially captured by amodal properties, and over the course of development, infants perceive and discriminate changes in the same amodal property in multimodal or unimodal stimulation. In this experiment Bahrick and Lickliter (2004) examined older infants' (i.e., 5-month-olds compared with 3-month-olds) discrimination of tempo and 8-month-olds' compared with 5-month-olds' discrimination of rhythm. The results of this experiment showed that the slightly older infants are able to discriminate changes in tempo or rhythm when provided redundant bimodal or unimodal stimulation. Thus supporting the prediction that early in development, infants' discrimination of amodal properties occurs first within the context of multimodal stimulation and is later extended to unimodal stimulation.

6 Intersensory Redundancy: Infants' Discrimination of Affective Expressions

As described earlier most objects and events, including faces and voices, provide both amodal and modality-specific information. In a follow-up study we examined the developmental prediction of the intersensory redundancy hypothesis that infants' discrimination of affect, as conveyed in faces and voices, would first occur in multimodal stimulation and over the course of development discrimination of affect would extend to unimodal stimulation. We also examined within the context of unimodal stimulation whether infants' discrimination of affect occurred first for faces alone or voices alone (Flom & Bahrick, 2007).

In this experiment, like others, infants between 3 and 7 months were familiarized, using an infant-controlled habituation procedure, to an unfamiliar adult conveying

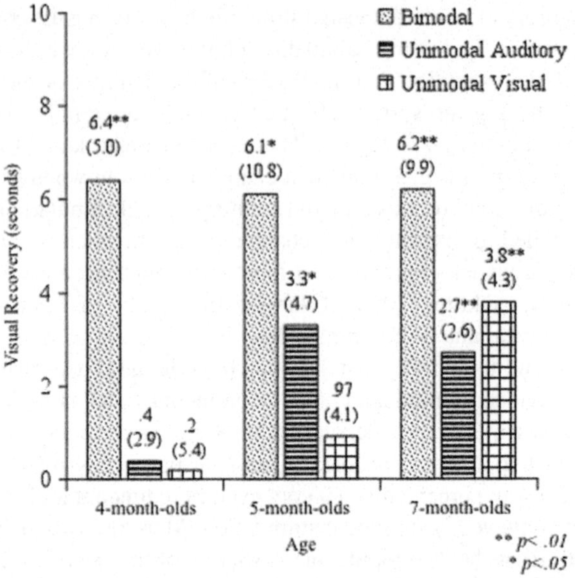

Fig. 4.2 Mean visual recovery (and standard deviations) as a function of condition (bimodal, unimodal auditory, unimodal visual) at 4, 5, and 7 months of age during the habituation phase. Visual recovery is the difference between infants' visual fixation during the test trials and visual fixation during the post-habituation trials and reflects infants' discrimination of affect. Reprinted from Flom, R., & Bahrick, L. (2007). The effects of multimodal stimulation on infants' discrimination of affect: An examination of the intersensory redundancy hypothesis. *Developmental Psychology, 43*, 238–252. Copyright 2007 by American Psychological Association. Reprinted with permission from APA

either a happy, sad, or angry affective expression bimodally (face–voice) or unimodally (face alone or voice alone). Results revealed that by 4 months of age infants reliably discriminated the different affective expressions when provided redundant bimodal stimulation. By 5 months of age infants reliably discriminated the different expressions when provided unimodal auditory information (voice alone) and by 7 months of age infants showed reliable discrimination when provided unimodal visual information (face alone). These results are shown in Fig. 4.2, and they demonstrate that infants' discrimination of affect emerges first in the context of redundant bimodal stimulation and is later extended to unimodal auditory and then unimodal visual stimulation.

Importantly, however, we also examined whether the fact that infants first showed discrimination of affect in the context of redundant bimodal stimulation is due to the fact that bimodal stimulation (face–voice) stimulates or activates two sensory systems or whether, as predicted by the intersensory redundancy hypothesis, that the temporal synchrony between the auditory and visual stimulation makes the amodal property of affect more perceptually salient. In this experiment we examined 4- and 5-month-olds' discrimination of affect when conveyed bimodally, but temporally misaligned such that the auditory information was delayed by 2 s (Flom & Bahrick, 2007). Results of this manipulation showed that 5-month-olds, but not 4-month-olds, reliably discriminated the change in affect when the face–voice was presented asynchronously. Thus 4-month-olds' discrimination of affect in the earlier condition was not due to providing more information, or activating two sense modalities, rather discrimination was a result of the temporally aligned auditory and visual stimulation, i.e., the intersensory redundancy.

While the above experiments provide evidence in support of Bahrick and Lickliter's (2000, 2002, 2004) intersensory redundancy hypothesis, they also provide a foundation for examining and understanding how infants perceive and integrate faces and voices into a unitary perception of a person. Specifically, this early sensitivity to tempo, rhythm, and a common or amodal affective expression is essential to infants' perception of unitary faces and voices. That is, within the context of faces and voices, we are able to unite a given face with a given voice in part based on various amodal properties such as the common tempo and rhythm of mouth and facial movements with the tempo and rhythm of audible speech as well as common affective expression.

7 Infants' Perception of Faces and Voices

As described above, infants, within the first 3 to 5 months of age, perceive changes in the amodal properties of rhythm, tempo, and affect. Slightly younger infants, however, are adept perceivers of unimodal faces and voices. Over the past 60-years a vast literature has accrued demonstrating that infants are excellent perceivers of faces (see Farah, Wilson, Drain, & Tanaka, 1998; Nelson, 2001; Pascallis & Kelly, 2009 for reviews). For example, newborns prefer faces compared to other visual stimuli (Fantz, 1963; Maurer & Barerra, 1981) and discriminate a static image of their mother's face from the face of an unfamiliar woman (Barrera & Maurer, 1981; Bushnell, 1982; Pascalis de Schonen, 1994). In addition, 2- to 4-day-old newborns discriminate dynamic images of their mother and a stranger's face (Field, Cohen, Garcia, & Greenberg, 1984; Sai & Bushnell, 1988). By 3 months of age infants discriminate between static images, as well as brief videos, of themselves and a peer (Bahrick, Moss, & Fadil, 1996). Finally, infants' face perception is likely shaped by their social experiences where younger, but not older, infants are adept perceivers of faces of other races and species (see Pascallis & Kelly, 2009 for a review).

In terms of voices, fetuses are able to hear during their last trimester (Querleu, Renard, Boutteville, & Crepin, 1989; Querleu, Renard, Versyp, Paris-Delrue, & Crepin, 1988) and show a preference for their mother's voice compared to a stranger's voice or the voice of their father (DeCasper & Fifer, 1980; DeCasper & Prescott, 1984). Four-day old infants discriminate between their "own" language and an unfamiliar language—but not between two unfamiliar languages (Mehler, Jusczyk, Lambertz, Halsted, Bertoncini, & Amiel-Tison, 1988; Moon, Cooper, & Fifer, 1993). Likewise 2-day-old infants, who prenatally heard their mother read a story once a day during their last trimester, showed a preference for their mother reading the familiar story compared to her reading of a novel story (DeCasper & Spence, 1986). It is also well known that newborns prefer infant-directed speech compared to adult-directed speech (Cooper & Aslin, 1990; Fernald, 1985; Pegg, Werker, & McLeod, 1992) where infants' preference for infant-directed speech is based on the affective properties of that speech rather than higher or more variable pitch (Singh, Morgan & Best, 2002). Thus from very early in development infants discriminate and recognize various faces and voices as well as various features of faces and voices.

8 Infants' Recognition of Amodal Properties Uniting Faces and Voices

While infants are remarkable at perceiving and discriminating changes in voices and faces, for the most part infants do not encounter faces or voices in isolation, rather infants are typically exposed to integrated or temporally and spatially collocated faces and voices and are perceived as a unitary whole. In addition, and as described above, infants are adept at perceiving and discriminating various amodal properties such as tempo, rhythm, and common affect that unite faces and voices. Along with infants' *discrimination* of amodal properties infants also *recognize* various amodal properties associated with their unitary perception of faces and voices. For example, by 2 months of age infants can recognize the synchrony between lip movements and the onset/offset of speech (Dodd, 1979), and by 4 months infants match the visual speech with the appropriate auditory information (Kuhl & Meltzoff, 1984). Like adults, infants are also susceptible to the McGurk effect where infants experience a unique or "oddball" speech sound when one auditory speech sound is artificially synchronized with different visible speech (i.e., lip movements) sound (Rosenblum, Schmuckler, & Johnson, 1997). Research also demonstrates that between 5 and 7 months of age infants are able to recognize, or match, human faces and voices on the basis of a common affective expression (Walker, 1982), can match canine vocalizations (barks) and posture on the basis of a common affective expression (Flom, Whipple & Hyde, 2009), and can match faces and voices on the basis of age and gender (Bahrick, Netto, & Hernandez-Reif, 1998; Walker-Andrews, Bahrick, Raglioni, & Diaz, 1991). Therefore just as infants are able to discriminate changes in various amodal properties uniting faces and voices they are also able to recognize these same properties when conveyed in different sense modalities.

The intersensory redundancy hypothesis was initially proposed to explain how infants arrive at unitary perception of their world as well as infants' perception of amodal and modality-specific properties including those properties that are necessary for infants to unite faces with voices (Bahrick et al., 2004). In addition to explaining infants' perceptual learning of amodal and modality-specific properties, the intersensory redundancy hypothesis has been used to explain infants' learning of arbitrary relationships (Bahrick, Hernandez-Reif & Flom, 2005).

9 Infants' Learning of Arbitrary Face–Voice Relations

As reviewed above there is ample evidence supporting the intersensory redundancy hypothesis. Specifically infants' attention, in the context of multimodal stimulation, is first directed toward amodal properties and over the course of development infants attend to amodal as well as modality-specific properties in the context of multimodal stimulation. Similarly it is also predicted, and evidence suggests, infants' attention and memory for modality-specific properties is initially facilitated within the context of unimodal stimulation and is later extended to redundant multimodal stimulation (Bahrick, Lickliter, & Flom, 2006; Flom & Bahrick, 2010).

The intersensory redundancy hypothesis is relevant to our understanding how infants learn arbitrary relationships—like the pairing of faces and voices—as these pairings involve both amodal and arbitrary relationships. In learning arbitrary face–voice relationships it is proposed that infants' initial attention will be directed toward amodal properties such as temporal and spatial synchrony, rhythm, and a common affect. This early focus on amodal properties will help infants learn which sights and sounds belong together and which do not thereby reducing the probability of learning, or forming, an incorrect audio–visual relationship. While a face–voice pairing is united through various amodal properties, each face–voice pairing also consists of arbitrary associations. For example, the sound (i.e., pitch, timbre, etc.) of particular voice and the visual appearance of a particular face (i.e., hair color, facial features, etc.) represent such an arbitrary relationship. According to the intersensory redundancy hypothesis infants' unitary perception of faces and voices therefore reflects an attentional and perceptual process of differentiation or increasing specificity beginning with amodal properties and later including arbitrary properties (Bahrick, 2001; Bahrick et al., 2005).

Until recently few studies examined whether, and under what conditions, infants are able to match face and voices. One of the first studies to examine this question found that 4-month-olds reliably match the faces and voices of their mother and father when the face–voice synchrony was matched (Spelke & Owsley, 1979). Because Spelke and Owsley (1979) used faces and voices of males and females (i.e., the infant's parents) it is not clear whether these infants matched the face and voice on the basis of their arbitrary face–voice relationship or whether they used gender in making the match? That is males tend to have, relative to females, deeper more resonate voices and tend to have larger more pronounced facial features compared to females. Moreover, Walker-Andrews et al. (1991) demonstrate that 4-month-olds match faces and voices on the basis of gender when the faces and voices are unfamiliar to the infant. In a more recent study researchers examined 3-month-olds' learning of arbitrary face–voice pairings (Brookes, Slater, Quinn, Lewkowicz, Hayes, & Brown, 2001). In this experiment infants were habituated to two temporally synchronous face–voice pairings. Following habituation infants were presented with a mismatched face and voice. For some of the infants the male face of habituation was now presented, in synchrony, with the voice of a female (and vice-versa). For others the mismatched face–voice pairing involved pairing the face of habituation with a different voice, but of the same gender. The results of this study were asymmetrical as infants showed reliable discrimination when the face–voice pairing was switched to a voice of a different gender and minimal discrimination when the novel face was of the same gender (Brookes et al., 2001). These studies are important as they lay a foundation for examining infants' learning of arbitrary face–voice relationships; however, in the case of Spelke and Owsley (1979) and Walker-Andrews et al. (1991) infants could have used stimulus/subject gender in matching faces and voices and in the example of Brookes et al. (2001) infants could use gender or temporal synchrony.

Recently we examined 2-, 4-, and 6-month-olds' learning of face–voice relations in two interrelated experiments (Bahrick et al., 2005). Moreover we wanted to examine the prediction made by the intersensory redundancy hypothesis that infants would initially focus on amodal properties uniting a face and voice and over the course of development infants would then focus on the arbitrary features associated with the face–voice pairing. In the first experiment we familiarized (i.e., habituated) infants to two synchronized male or two female face–voice pairings. It is worth noting in this experiment, unlike previous experiments, infants were exposed two same gender pairings to avoid the potential confound of infants learning/matching of faces–voices on the basis of gender. Therefore half of the infants for example were habituated to Jeff's face and voice and Matt's face and voice and the other half were habituated to Margie's face and voice and Shirley's face and voice. Following habituation infants received two test trials that depicted a change in the face–voice pairing. Unlike Brookes et al. (2001) and Spelke and Owsley (1979) during the test trials infants received a change in face–voice pairings for the same gender, e.g., during the test trials Matt's face would be paired with Jeff's voice. Results revealed that 4- and 6-month-olds, but not the 2-month-olds, detected a change in the face–voice pairing. Thus during the first phase of this experiment 4- and 6-month-olds, but not 2-month-olds, learned the arbitrary face–voice pairings.

Following infants' habituation to the two face–voice pairings, and subsequent discrimination or test trials (i.e., phase 1), infants were given a 10-min break and we then examined 4- and 6-month-olds' memory for the face–voice pairing (i.e., phase 2). During this intermodal matching phase infants were shown, on side-by-side monitors, the two faces they viewed during habituation. For half of the 12 trials infants heard the voice of one face and on the other half of the trials infants heard the voice that when went with the other faces. The dependent variable was the proportion of time infants looked to the matching face. The results of the intermodal matching phase revealed that only the 6-month-olds looked preferentially longer to the correct face (Bahrick et al., 2005).

The results of phase one of this experiment reveal that 4- and 6-month-olds, but not 2-month-olds, learned and noticed a change in a face–voice pairing. Results of phase two revealed that following a 10-min delay only the 6-month-olds remembered which voice was paired with which face. One question raised by this first experiment is whether 2-month-olds failed to learn the arbitrary face–voice pairings or whether 2-month-olds simply could not discriminate the voices used in this experiment. That is unlike previous studies, each infant in the current study was only exposed to faces and voices of one gender thus a follow-up experiment was conducted to examine 2-month-olds' unimodal discrimination of the faces and voices. In this second experiment we examined and found evidence that 2-month-olds do discriminate the voices used in the first experiment. Thus the poor performance of the 2-month-olds is not due to their inability to discriminate the individual voices (Bahrick et al., 2005).

In general the results of this study are important as they are among the first to examine the developmental origins of young infants' learning of arbitrary face–voice relations (e.g., Brookes et al., 2001). The results are also congruent with the perceptual differentiation or increasing specificity view of perceptual learning as described earlier (Bahrick, 2001; Gibson, 1969). Finally, the results are consistent with and support predictions made by the intersensory redundancy hypothesis.

According to the first prediction of the intersensory redundancy hypothesis younger infants' attention should be captured by amodal properties such as temporal synchrony, tempo, and rhythm of speech, as the faces and voices were conveyed in multimodal stimulation. Thus it is not surprising that 2-month-olds did not reliably discriminate a change in the face–voice relationship. This conclusion is likely warranted because the results of a follow-up experiment revealed that 2-month-olds do notice a change in the face or voice when the events were presented in a unimodal visual (face change) or unimodal auditory (voice change) context. According to the second prediction, infants' attention should become more flexible over the course of development and thus infants should be able to attend to amodal as well as arbitrary relationships. Our results support this prediction as well because 4-month-olds were able to attend to the amodal properties as well as the arbitrary face–voice pairing. Finally, only the 6-month-olds noticed a change in the arbitrary face–voice relationship and remembered this relationship following a 10-min delay.

10 Neurophysiological Foundations of Intersensory Perception

The intersensory redundancy hypothesis was generated as a way to explain how infants arrive at unitary perception of object and events—including faces and voices—when provided an ever changing array of multimodal and unimodal stimulation (Bahrick & Lickliter, 2000, 2002, 2004; Bahrick et al., 2004). As just described there is ample behavioral support for each of the predictions made by the intersensory redundancy hypothesis across a variety of behavioral studies with human infants and more recent evidence is now available with a precocial avian species (e.g., Lickliter, Bahrick & Honeycutt, 2004; Lickliter, Bahrick, & Markham, 2006). Until recently, however, little was known about the neurophysiological foundations for infants' intersensory perception—including infants' perception of faces and voices. In addition, much of what is known regarding the neurophysiological bases of intersensory perception is based on indirect evidence from single cell recordings within nonhuman animals and EEG and fMRI evidence from human adults (e.g., Calvert, Hansen, Iversen, & Brammer, 2001; Giard & Peronnet, 1999; Puce, 2012; Wallace & Stein, 1997).

Recordings from single cells within the deep layers of a cat's superior colliculus, for example, reveal an exponentially greater response to synchronous bimodal audio–visual stimulus than to unimodal auditory, unimodal visual, or to the summation of these two unimodal responses (Wallace & Stein, 1997; Wallace, Wilkinson, & Stein, 1996). Likewise, when adults are asked to identify an object based on bimodal auditory–visual stimulation, unimodal auditory, or unimodal visual stimulation the amplitude of their EEG/ERP is larger to the bimodal stimulation than to the sum of the unimodal auditory and unimodal visual responses (Giard & Peronnet, 1999; Santangelo, Van der Lubbe, Olivetti Berlardinelli & Postma, 2008). It has also been shown that the increased neurophysiological response to bimodal stimulation is evident in auditory and visual regions of the cortex as well as nonsensory regions of the right frontotemporal regions (Fort, Delpuech, Pernier & Giard, 2002a, 2002b) demonstrating that many areas of the cortex are responsive to multimodal stimulation. Given this rapidly growing literature with adult humans, as well nonhumans, we have begun to examine human infants neurophysiological response to redundant multimodal stimulation as well as infants response to temporal synchrony in faces and voices (Hyde, Jones, Flom, & Porter, 2011; Hyde, Jones, Porter, & Flom, 2010).

In our first experiment we examined whether we could replicate earlier work with human adults demonstrating neurophysiological enhancement to bimodal stimulation compared to unimodal stimulation. In this experiment we examined 3-month-old infants' and adults' neurophysiological responses to bimodal stimulation (i.e., a large colored circle paired with a "bong" sound and a small circle paired with a "ping" sound), unimodal auditory stimulation (i.e., the lower pitched "bong" and higher pitched "ping" sounds), and unimodal visual stimulation (i.e., the large and small colored circles). Each event was presented for thirty 1,000 ms trials. Data were filtered with 30-Hz low pass filter and segmented into −200 ms pre-stimulus to 800 ms epochs. We used a subtraction technique to compare the auditory response during the bimodal A–V stimulation to the unimodal auditory response ((AV-V) − A) (see Besle, Fort, Delpuech, & Giard, 2004; Giard & Peronnet, 1999 for similar analyses with adults). The results are summarized in Fig. 4.3 below.

The results of the 3-month-olds revealed a significant effect of stimulation over frontotemporal sites with bimodal stimulation eliciting more negative potentials (N450) compared to unimodal auditory stimulation (see Panel b of Fig. 4.3). In general, the results for the adults were similar to the results of the 3-month-olds. Adults in the bimodal condition elicited a larger negative amplitude compared to the unimodal conditions — albeit somewhat earlier (N2) — see Panel d of Fig. 4.3. Unlike the 3-month-olds, however, the response of the adults revealed a hemispheric difference where the left frontotemporal grouping showed more negative amplitude waveform compared to the right frontotemporal grouping (see Panels a and c of Fig. 4.3). Still, at both ages the results revealed increased auditory processing in the context of bimodal stimulation. Moreover, the results of Hyde et al. (2010) are consistent with previous adult ERP work comparing EEG/ERP responses to bimodal and unimodal stimulation (e.g., Fort et al., 2002a, 2002b; Giard & Peronnet, 1999) as our results demonstrate that increased neurological processing associated with bimodal stimulation is present by 3 months of age.

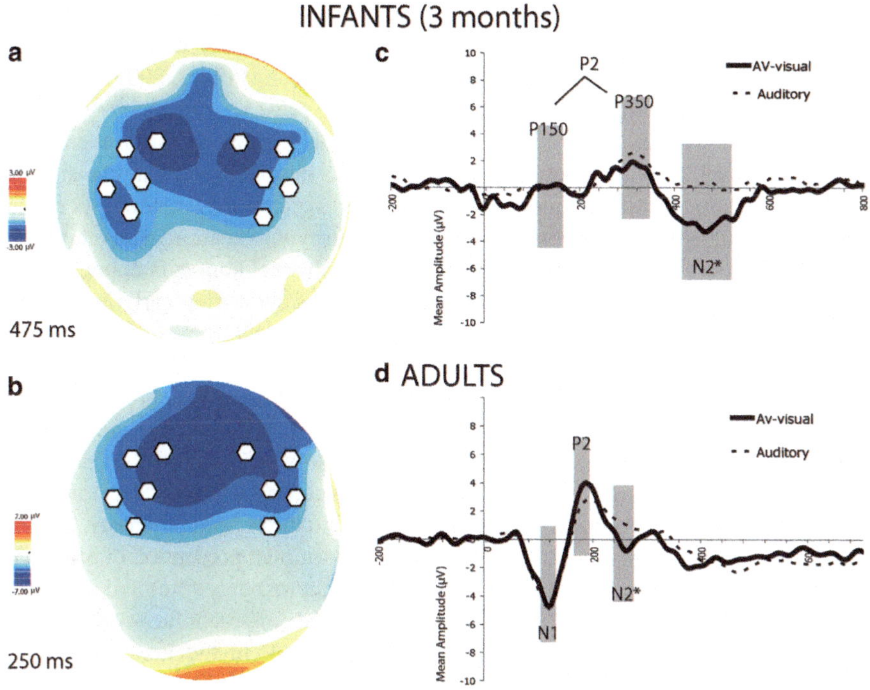

Fig. 4.3 Summary of ERP results for infants and adults. (**a**) Difference map of the (AV-visual) minus Auditory for infants at 475 ms *White* items represent electrode sites used to calculate the average waveforms. (**b**) Average waveforms for the bimodal (AV-visual) and unimodal conditions over frontotemporal scalp locations from −200 to 800 ms for infants. The *gray boxes* highlight components of interest. The *asterisk* indicates a significant difference between conditions. (**c**) Difference map of the (AV-visual)−Auditory for adults at 250 ms *White* items represent electrode sites used to calculate the average waveforms. (**d**) Average waveforms for the bimodal (AV-visual) and unimodal conditions over frontotemporal scalp locations from −200 to 800 ms for adults. The *gray boxes* highlight components of interest. The *asterisk* indicates a significant difference between conditions. Reprinted from Hyde, D. C., Jones, B. L., Porter, C. L., & Flom, R. (2010). Visual stimulation enhances auditory processing in 3-month-old infants and adults. *Developmental Psychobiology, 52*(2), 181–189. Reprinted with permission from John Wiley and Sons copyright 2010

As discussed earlier one of the central features of the intersensory redundancy hypothesis—including infants learning about faces and voices—is the role of temporal synchrony. While there is a wealth of behavioral evidence showing infants' sensitivity to temporal synchrony (see Bahrick, 2000; 2001; Bahrick & Lickliter, 2002; Lewkowicz, 2000, 2010 for reviews), and some evidence is available examining adults' neurophysiological response to temporal synchrony, there is little to no evidence examining infants' neurophysiological response to temporal synchrony in faces and voices. The last study I will describe examines 5-month-olds' neurophysiological response to temporal synchrony and asynchrony in faces and voices (Hyde et al., 2011).

Research examining the neural signatures of face–voice synchrony in adults indicates integration occurs during the early stages of sensory processing (Braida, 1991;

Green, 1998). For example, when presented with synchronous audio–visual speech adults' early auditory N1-P2 components are attenuated compared to unimodal auditory speech but not to asynchronous audio–visual speech (Pilling, 2009). In other words, with adults, auditory attenuation occurs in the presence of synchronous, but not asynchronous bimodal speech. The fact that adults show an early attenuated auditory neurophysiological response to synchronous bimodal speech has been used to address arguments that the auditory attenuation in synchronous bimodal speech is based on a shift in attention between visual and auditory information or an inhibition of auditory processing (Besle et al., 2004; van Wassenhove, Grant & Poeppel, 2005). Given that face–voice integration is hypothesized to occur during early in sensory processing in adults we examined whether face–voice integration similarly occurs during early sensory processing in human infants (Hyde et al., 2011).

Because young infants are behaviorally sensitive to temporal synchrony (Kuhl & Meltzoff, 1984; Lewkowicz, 2000) it was hypothesized that audio–visual integration will occur during early sensory processing (Bristow, et al., 2009). In addition, it was also predicted that changes in face–voice synchrony will also affect infants' neurophysiological processes associated with attention and memory. More specifically, research examining 7-month-olds' cross-modal perception of affective expressions reveals an attenuated attentional orienting response (Nc) and a greater positive slow wave response (PSW) to synchronous and congruent face–voice affect pairings (e.g., Grossmann, Striano, & Friederici, 2006). In the first experiment (Experiment 1) we presented 5-month-olds with 30 trials where the onset and offset of a static face and voice saying "hi" were in perfect temporal synchrony. In the asynchronous condition infants were presented with 30 trials where the voice occurred 400 ms before the appearance of the face. In the synchronous and asynchronous conditions each trial lasted for 1,000 ms.

The results of this first experiment are summarized in Fig. 4.4. Like adults (e.g., Besle et al., 2004; Pilling, 2009), our results revealed that 5-month-olds' auditory–visual integration occurred during early sensory processing (auditory P2) and continued during later attentional processing (see Panels b and d in Fig. 4.4). However, the results of the 5-month-olds reveal a larger response for the auditory component during the *synchronous* condition whereas adults tend to reveal a similar early response for the *asynchronous* condition (see Panel B). One possible explanation for why adults, but not infants, show attenuation to synchrony may reflect the fact that infants are highly sensitive to temporal synchrony, and as predicted by the intersensory redundancy hypothesis, may be biased to attend to temporal synchrony when it is present (Bahrick et al., 2004). In this first experiment infants also showed a large slow negative component (Nc) in both the synchronous and asynchronous conditions and this response is associated with the onset of the visual stimulus in both conditions. Interestingly, however, the asynchronous condition resulted in a larger amplitude (i.e., greater negativity) than the synchronous condition. Infants' later and significantly greater Nc response to asynchrony may reflect the fact that after integration occurs infants may find the asynchronous condition more novel or interesting (Reynolds & Richards, 2005). Thus initially (250 ms) 5-month-olds show a larger response to temporal synchrony, potentially reflecting early sensory integration, and

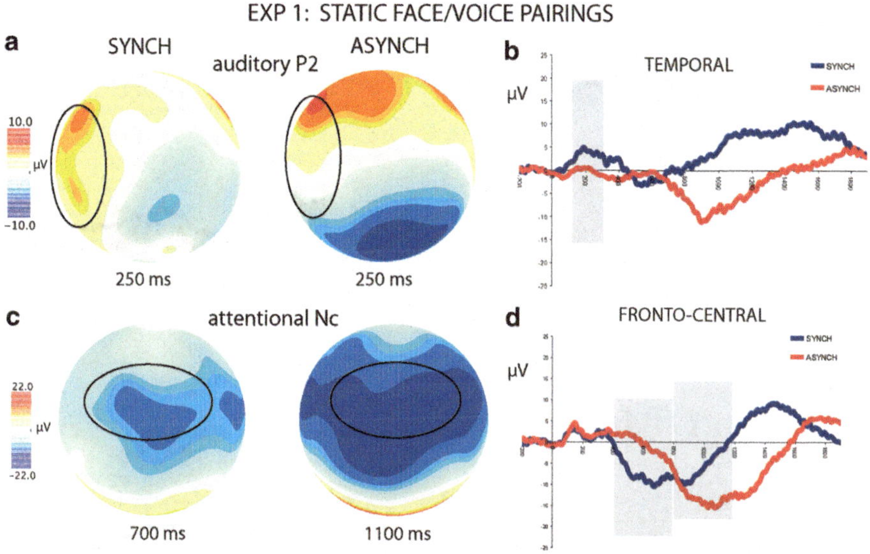

Fig. 4.4 Summary of Experiment 1 results. (**a**) Grand average scalp topography at 250 ms over left temporal sites for each experimental condition. (**b**) Average waveform from –200 to 1,900 ms averaged over left temporal sites characterizing auditory processing. The *shaded* region represents the statistical comparison of experimental conditions for the auditory *P2* (**c**). Grand average scalp topography at 700 and 1,100 ms characterizing the *Nc* for the synchronous and asynchronous conditions over frontocentral sites. (**d**) Average waveform from –200 to 1,900 ms averaged over frontocentral sites. The *shaded* region represents the comparison of experimental conditions for the *Nc*. Reprinted from Hyde, D.C., Jones, B.L., Flom, R. & Porter, C.L. (2011). Neural signatures of face–voice synchrony in 5-month-old human infants. *Developmental Psychobiology*. Reprinted with permission from John Wiley and Sons copyright 2011

later (1,100 ms) show a larger response to temporal asynchrony, potentially reflecting an attentional shift toward novelty (Reynolds & Richards, 2005).

Because temporal synchrony in this first experiment reflected a synchronous onset and offset of static stimuli, compared to more naturalistic and dynamic stimuli, we conducted a second experiment (Experiment 2) using dynamic stimuli (Hyde et al., 2011). In this second experiment 5-month-olds saw dynamic faces saying, "oh hi baby", in perfect synchrony (synchronous condition). Infants also saw dynamic faces mouthing different words (you're such a beautiful baby) but hearing "oh hi baby" (asynchronous condition). The onset/offset of the events was the same for the synchronous/asynchronous conditions. What differed between the conditions was whether what was visually articulated matched what was heard. In general, the result of this second experiment replicated Experiment 1 of Hyde et al. (2011) as infants showed an early response to synchrony, again hypothesized to reflect early integration, and a later attentional or Nc response to asynchrony. The results are summarized in Fig. 4.5 below.

In addition, 5-month-olds in the second experiment showed a positive slow wave (PSW) for the synchronous events compared to the asynchronous events. The fact

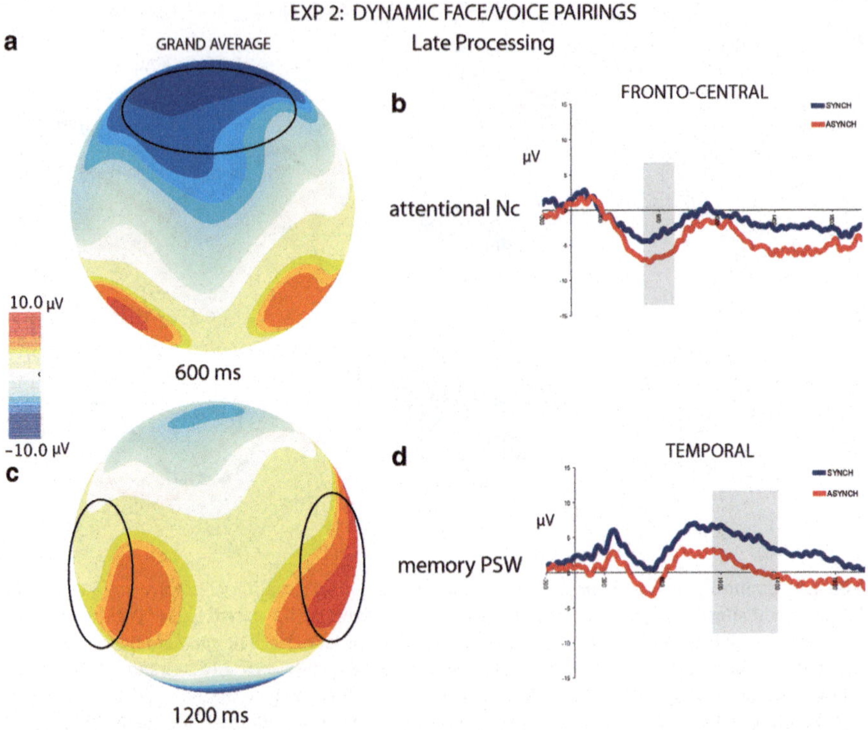

Fig. 4.5 Summary of Experiment 2 late processing results. (**a**) Grand average scalp topography at 600 ms over frontocentral sites (average of all experimental conditions). (**b**) Average waveform from −200 to 2,000 ms averaged over frontocentral sites. The *shaded* region represents the comparison of experimental conditions for the *Nc*. (**c**) Grand average scalp topography at 1,200 ms characterizing the positive slow wave (*PSW*) over temporal sites. (**d**) Average waveform from −200 to 2,000 ms averaged over left and right temporal sites. The *shaded* region represents the comparison of experimental conditions for the *PSW*. Reprinted from Hyde, D.C., Jones, B.L., Flom, R. & Porter, C.L. (2011). Neural signatures of face–voice synchrony in 5-month-old human infants. *Developmental Psychobiology*. Reprinted with permission from John Wiley and Sons copyright 2011

that we observed this PSW, or memory component, in the second experiment but not the first may reflect the more familiar and/or more ecologically valid stimuli used in Experiment 2 of Hyde et al. (2011).

The results of Hyde et al. (2010) are important as they demonstrate that 3-month-olds, like adults, show an enhanced response to bimodal compared to unimodal stimulation. The results of Hyde et al. (2011) are important as they demonstrate 5-month-olds show an early neurological response associated with synchronous bimodal speech where adults show early auditory attenuation to synchronous bimodal speech. Taken together, early in development infants show an enhanced neurological response to two properties (i.e., redundancy and temporally synchrony) hypothesized to affect infants' learning about faces and voices and these neurophysiological results converge with behavioral evidence associated with the intersensory redundancy hypothesis.

11 Summary and Future Directions

In answering the initial question "How do infants learn which face and voice go together and arrive at a unitary and veridical perception of people?" I have described and provided evidence articulating how the intersensory redundancy hypothesis addresses this question. Infants' learning of the arbitrary relationship between faces and voices occurs in two tightly coupled steps. First, between 3 and 5 months of age infants attend to various amodal properties such as a common tempo, rhythm, and affective expressions that unite a particular face and voice. Second, around 6 months of age, when infants' attention is more flexible and they perceive amodal and modality-specific properties, infants now perceive and remember various arbitrary features (i.e., the sound of particular voice and the visual appearance of a particular face) associated with a particular face–voice pairing.

From a neurophysiological perspective substantially less is known in terms of how infants (as well as adults) arrive at unitary perception of other people. We do know, however, that infants like adults, show an enhanced neurological response toward bimodal compared to unimodal speech (Hyde et al., 2010). In addition, infants, like adults, show early neurophysiological markers of sensory integration of faces and voices. Infants, unlike adults, however, show an early processing bias toward temporal synchrony and show later attentional and memory components associated with temporal asynchrony of faces and voices (Hyde et al., 2011).

Undoubtedly these neurophysiological and behavioral processes are interrelated and concurrently promote early learning about the unitary nature of other people. Future research is needed that examines the developmental convergence of these behavioral and neurophysiological processes. It is less than clear if infants' early neurological development precedes their behavioral integration (i.e., intersensory perception), or if early behavioral integration guides later neurological development? More probable, however, behavioral and neurophysiological processes associated with face–voice integration develop in a mutually interacting and dynamic manner.

As this volume attests, many behavioral and neurophysiological questions surrounding person perception have only begun to be addressed. For instance, additional research is needed examining how infants (and adults) use affect or emotion within the context of person perception (e.g., Flom & Bahrick, 2007; Flom & Pick, 2005; Grossman, 2012; Grossmann et al., 2006). Research is also needed in identifying how the different cortical and subcortical structures associated with face–voice perception interact, how processes associated with face–voice perception may go awry, and whether the processes associated with face–voice integration in humans is similar to related processes in other species. Finally, from a developmental perspective, additional research is needed that continues to examine the behavioral and neurophysiological development of face–voice perception within multimodal as well as unimodal contexts.

It seems William James' assertion that "infants feel it as one great blooming and buzzing confusion" is overstated. Understanding how infants come to perceive which face and voice belong together (and which do not) is certainly important. It is important to our understanding of perceptual development and it is relevant to our understanding of early cognitive and social development. It seems, however, from very early in development infants behaviorally, and neurologically, integrate and attend to information across different sense modalities and most importantly infants use this convergent information to learn about different objects, events, and of course people.

Acknowledgments Sir Isaac Newton is credited with the expression "If I have seen further, it is by standing on the shoulders of giants." The research and thoughts contained within this chapter reflect a portion of the knowledge I acquired as a post-doc working and conversing with Lorraine Bahrick, her husband and colleague Robert Lickliter, and countless conversations with David Lewkowicz. In many respects each of these individuals could rightly claim authorship. Without their guidance and collaboration this chapter would not be possible and I am indebted to each. I am equally indebted to the students and the parents of the many infants who participated in experiments described herein.

References

Allen TW, Walker K, Symonds L, Marcell M (1977) Intrasensory and intersensory perception of temporal sequences during infancy. Developmental Psychology 13:225–229

Bahrick LE (2000) Increasing specificity in the development of intermodal perception. In: Muir D, Slater A (eds) Infant development: The essential readings. Blackwell, Maklen, MA, pp 117–136

Bahrick LE (2001) Increasing specificity in perceptual development: Infants' detection of nested levels of multimodal stimulation. Journal of Experimental Child Psychology 79:253–270

Bahrick LE (2010) Intermodal perception and selective attention to intersensory redundancy: Implications for typical social development and autism. In: Bremner G, Wachs TD (eds) Blackwell handbook of infant development, 2nd edn. Blackwell, London

Bahrick LE, Flom R, Lickliter R (2002) Intersensory redundancy facilitates discrimination of tempo in 3-month-old infants. Developmental Psychobiology 41:352–363

Bahrick LE, Hernandez-Reif M, Flom R (2005) The development of infant learning about specific face-voice relations. Developmental Psychology 41:541–552

Bahrick LE, Lickliter R (2000) Intersensory redundancy guides attentional selectivity and perceptual learning in infancy. Developmental Psychology 36:190–201

Bahrick LE, Lickliter R (2002) Intersensory redundancy guides early perceptual and cognitive development. In: Kail R (ed) Advances in child development and behavior, vol 30. Academic, New York, pp 153–187

Bahrick LE, Lickliter R (2004) Infants' perception of rhythm and tempo in unimodal and multimodal stimulation: A developmental test of the intersensory redundancy hypothesis. Cognitive, Affective, & Behavioral Neuroscience 4:137–147

Bahrick LE, Lickliter R, Flom R (2004) Intersensory redundancy guides the development of selective attention, perception, and cognition in infancy. Current Directions in Psychological Science 13:99–102

Bahrick LE, Lickliter R, Flom R (2006) Up versus down: The role of intersensory redundancy in infants' sensitivity to object orientation and motion. Infancy 9:73–96

Bahrick LE, Moss L, Fadil C (1996) The development of visual self-recognition in infancy. Ecological Psychology 8:189–208

Bahrick LE, Netto DS, Hernandez-Reif M (1998) Intermodal perception of adult child faces and voices by infants. Child Development 69:1263–1275

Balaban MT, Dannemiller JL (1992) Age differences in responses to temporally modulated patterns at 6 and 12 weeks. Infant Behavior & Development 15:359–375

Barrera M, Maurer D (1981) Recognition of mother's photographed face by the three month old infant. Child Development 52:714–716

Besle J, Fort A, Delpuech C, Giard M (2004) Bimodal speech: Early suppressive visual effects in human auditory cortex. The European Journal of Neuroscience 20(8):2225–2234

Birch, H., & Lefford, A. (1963). Intersensory development in children. *Monographs of the Society for Research in Child Development, 28(5, Serial No. 89).*

Braida LD (1991) Crossmodal integration in the identification of consonant segments. The Quarterly Journal of Experimental Psychology 43(3):647–677

Bristow D, Dehaene-Lambertz G, Mattout J, Soares C, Gliga T, Mangin J (2009) Hearing faces: How the infant brain matches the face it sees with the speech it hears. Journal of Cognitive Neuroscience 21(5):905–921

Brookes H, Slater A, Quinn PC, Lewkowicz DJ, Hayes R, Brown E (2001) Three-month-old infants learn arbitrary auditory-visual pairings between voices and faces. Infant and Child Development 10:75–82

Bushnell IWR (1982) Discrimination of faces by young infants. Journal of Experimental Psychology 33:298–308

Calvert GA, Hansen PC, Iversen SD, Brammer MJ (2001) Detection of audio-visual integration sites in humans by application of electrophysiological criteria to the BOLD effect. NeuroImage 14(2):427–438

Cooper RP, Aslin RN (1990) Preference for infant-directed speech in the first month after birth. Child Development 61:1584–1595

DeCasper AJ, Fifer W (1980) Of human bonding: Newborns prefer their mothers' voices. Science 208:1174–1176

DeCasper AJ, Prescott PA (1984) Human newborns' perception of male voices: Preference, discrimination, and reinforcing value. Developmental Psychobiology 5:481–491

DeCasper AJ, Spence M (1986) Prenatal maternal speech influences newborns' perception of speech sounds. Infant Behavior & Development 9:133–150

Dodd B (1979) Lip reading in infants: Attention to speech presented in-and-out of synchrony. Cognitive Psychology 11:478–484

Edelman GM (1987) Neural Darwinism: The theory of neuronal group selection. Basic Books, New York, NY

Edelman GM (1993) Neural Darwinism: Selection and reentrant signaling in higher brain function. Neuron 10:115–125

Fantz R (1963) Pattern vision in newborn infants. Science 140(3564):296–297

Farah MJ, Wilson KD, Drain M, Tanaka JN (1998) What is "special" about face perception. Psychological Review 105(3):482–498

Fernald A (1985) Four-month-old infants prefer to listen to motherese. Infant Behavior & Development 8:181–195

Field TM, Cohen D, Garcia R, Greenberg R (1984) Mother-stranger face discrimination by the newborn infant. Infant Behavior & Development 7:19–25

Flom R, Bahrick L (2007) The effects of multimodal stimulation on infants' discrimination of affect: An examination of the intersensory redundancy hypothesis. Developmental Psychology 43:238–252

Flom R, Bahrick LE (2010) The effects of intersensory redundancy on attention and memory: Infants' long-term memory for orientation in audiovisual events. Developmental Psychology 46:428–436

Flom R, Pick AD (2005) Experimenter affective expression and gaze following in 7-month-olds. Infancy 7:207–218

Flom R, Whipple H, Hyde D (2009) Infants' intermodal perception of canine (Canis familairis) facial expressions and vocalizations. Developmental Psychology 45:1143–1151

Fort A, Delpuech C, Pernier J, Giard MH (2002a) Dynamics of cortico-subcorical crossmodal operations involved in audio-visual object detection in humans. Cerebral Cortex 12:1031–1039

Fort A, Delpuech C, Pernier J, Giard MH (2002b) Early auditory-visual interactions in human cortex during nonredundant target identification. Cognitive Brain Research 14:20–30

Gardner JM, Karmel BZ (1995) Development of arousal/attention preference interactions in early infancy. Developmental Psychology 31:473–482

Giard MH, Peronnet F (1999) Auditory-visual integration during multimodal object recognition in humans: A behavioral and electrophysiological study. Journal of Cognitive Neuroscience 11: 473–490

Gibson JJ (1966) The senses considered as perceptual systems. Houghton Mifflin, Boston

Gibson EJ (1969) Principles of perceptual learning and development. Appleton Century Crofts, New York, NY

Gibson JJ (1979) The ecological approach to visual perception. Houghton Mifflin, Boston

Goodale MA, Milner AD (1992) Separate visual pathways for perception and action. Trends in Neurosciences 15:20–25

Gray CM (1999) The temporal correlation hypothesis of visual feature integration: Still alive and well. Neuron 24:31–47

Green KP (1998) The use of auditory and visual information during phonetic processing: Implications for theories of speech perception. In: Campbell R, Dodd B, Burnham D (eds) Hearing by eye II. Psychology Press, East Sussex, UK

Grossman T (2012) The early development and brain bases of emotion perception in face and voice. In: Belin P, Campanella S, Ethofer T (eds) Integrating face and voice in person perception. Springer, Berlin

Grossmann T, Striano T, Friederici AD (2006) Crossmodal integration of emotional information from face and voice in the infant brain. Developmental Science 9(3):309–315

Hyde DC, Jones BL, Flom R, Porter CL (2011) Neural signatures of face-voice synchrony in 5-month-old human infants. Developmental Psychobiology 53:359–370

Hyde DC, Jones BL, Porter CL, Flom R (2010) Visual stimulation enhances auditory processing in 3-month-old infants and adults. Developmental Psychobiology 52(2):181–189

James W (1890) The principles of psychology. Dover Publications, New York, 1950 (vol. 1, pp. 159, 488)

Kuhl PK, Meltzoff AN (1984) The intermodal representation of speech in infants. Infant Behavior & Development 7:361–381

Lewkowicz DJ (1988a) Sensory dominance in infants: 1. Six-month-old infants' response to auditory-visual compounds. Developmental Psychology 24:155–171

Lewkowicz DJ (1988b) Sensory dominance in infants: 2. Ten-month-old infants' response to auditory-visual compounds. Developmental Psychology 24:172–182

Lewkowicz DJ (2000) The development of intersensory temporal perception: An epigenetic systems/limitations view. Psychological Bulletin 126:281–308

Lewkowicz D (2010) Infant perception of audio-visual speech synchrony. Developmental Psychology 46(1):66–77

Lickliter R, Bahrick LE, Honeycutt H (2004) Intersensory redundancy enhances memory in bobwhite quail embryos. Infancy 5:253–269

Lickliter R, Bahrick LE, Markham RG (2006) Intersensory redundancy educates selective attention in bobwhite quail embryos. Developmental Science 9:604–615

Maurer D, Barerra M (1981) Infants' perception of natural and distorted arrangements of a schematic face. Child Development 52:196–202

Mehler J, Jusczyk P, Lambertz G, Halsted N, Bertoncini J, Amiel-Tison C (1988) A precursor of language acquisition in young infants. Cognition 29:143–178

Moon C, Cooper RP, Fifer WP (1993) Two-day-olds prefer their native language. Infant Behavior & Development 16(4):495–500

Morrongiello BA, Trehub SE (1987) Age-related changes in auditory temporal perception. Journal of Experimental Child Psychology 44:413–426

Nelson CA (2001) The development and neural bases of face recognition. Infant and Child Development 10:3–18

Pascalis O, de Schonen S (1994) Recognition in 3- to 4-day-old human neonates. Neuroreport 5:1721–1724

Pascallis O, Kelly DJ (2009) The origins of face processing in humans: Phylogeny and ontogeny. Perspectives on Psychological Science 4:200–209

Pegg JE, Werker JF, McLeod PJ (1992) Preference for infant-directed over adult- directed speech: Evidence from 7-week-old infants. Infant Behavior & Development 15(3):325–345
Piaget J (1952) The origins of intelligence in children. International Universities Press, New York
Pilling M (2009) Auditory event-related potentials (ERPs) in audiovisual speech perception. Journal of Speech, Language, and Hearing Research 52(4):1073–1081
Puce A (2012) Neurophysiological correlates of face and voice integration. In: Belin P, Campanella S, Ethofer T (eds) Integrating face and voice in person perception. Springer, Berlin
Querleu D, Renard X, Boutteville C, Crepin G (1989) Hearing by the human fetus? Seminars in Perinatology 13:409–420
Querleu D, Renard X, Versyp F, Paris-Delrue L, Crepin G (1988) Fetal hearing. European Journal of Obstetrics, Gynecology, and Reproductive Biology 29:191–212
Reynolds GD, Richards JE (2005) Familiarization, attention, and recognition memory in infancy: An event-related potential and cortical source localization study. Developmental Psychology 41:598–615
Rosenblum LD, Schmuckler MA, Johnson JA (1997) The McGurk effect in infants. Perception & Psychophysics 59:347–357
Royce J (1881) "Mind-stuff" and reality. Mind 6(23):365–377
Sai F, Bushnell IWR (1988) The perception of faces in different poses by 1-month-olds. The British Journal of Developmental Psychology 6:35–41
Santangelo V, Van der Lubbe RHJ, Olivetti Berlardinelli M, Postma A (2008) Multisensory integration affects ERP components elicited by exogenous cues. Experimental Brain Research 185:269–277
Seth AK, McKinstry JL, Edelman GM, Krichmar JL (2004) Visual binding through reentrant connectivity and dynamic synchronization in a brain-based device. Cerebral Cortex 14(11):1185–1199
Shadlen MN, Movshon JA (1999) Synchrony unbound: A critical evaluation of the temporal binding hypothesis. Neuron 26:703–714
Shafritz KM, Gore JC, Marios R (2002) The role of parietal cortex in feature binding. Proceedings of the National Academy of Sciences 99:10917–10922
Singer W (1999) Neuronal synchrony: A versatile code for the definition of relations? Neuron 24: 49–65
Singh L, Morgan JL, Best CT (2002) Infants' listening preferences: Baby talk or happy talk? Infancy 3:365–394
Spector F, Maurer DM (2009) Synesthesia: A new approach to understanding the development of perception. Developmental Psychology 45:175–189
Spelke ES, Owsley CJ (1979) Intermodal exploration and knowledge in infancy. Infant Behavior & Development 2:13–27
Treisman A (1996) The binding problem. Current Opinion in Neurobiology 6:171–178
Treisman A (1998) Feature binding, attention and object perception. Philosophical transactions of the Royal Society of London Series B, Biological sciences 353:1295–1306
Ungerlieder LG, Mishkin M (1982) Two cortical visual systems. MIT Press, Cambridge, MA
van Wassenhove V, Grant KW, Poeppel D (2005) Visual speech speeds up the neural processing of auditory speech. Proceedings of the National Academy of Sciences of the United States of America 102:1181–1186
Walker AS (1982) Intermodal perception of expressive behaviors by human infants. Journal of Experimental Child Psychology 13:514–535
Walker-Andrews AS, Bahrick LE, Raglioni SS, Diaz I (1991) Infants' bimodal perception of gender. Ecological Psychology 3:55–75
Wallace MT, Stein BE (1997) Development of multisensory neurons and multisensory integration in cat superior colliculus. Journal of Neuroscience 17:2429–2444
Wallace MT, Wilkinson LK, Stein BE (1996) Representation and integration of multiple sensory inputs in primate superior colliculus. Journal of Neurophysiology 76(2):1246–1266
Zeki SM (1991) Cerebral akinetopsia (visual motion blindness). Brain 114:811–824
Zeki SM (1993) A vision of the brain. Blackwell, Oxford, UK

Chapter 5
The Early Development of Processing Emotions in Face and Voice

Tobias Grossman

Abstract Processing facial and vocal emotional expressions is a critical aspect of person perception. How this ability develops during infancy and what brain processes underpin infants' perception of emotion in face and voice are the questions dealt with in this chapter. I present a set of new electrophysiological studies that provide insights into the brain processes underlying infants' developing abilities. Evidence from unimodal (face or voice) and multimodal (face and voice) processing of emotion is considered. The reviewed infant data suggest that (1) early in development, emotion enhances the sensory processing of faces and voices, (2) infants' ability to allocate increased attentional resources to negative emotional information develops earlier in the vocal domain than in the facial domain, (3) at least by the age of 7 months, infants reliably integrate and recognize emotional information across face and voice. Futhermore, I present some recent work suggesting that already in infancy genetic variation in neurotransmitter systems is associated with individual differences in facial and vocal emotion processing. Finally, I propose new directions for research in this area.

1 Introduction

Infants develop in a world filled with other people, including parents, siblings, other family members, friends, and strangers. Relating socially to others not only has profound effects on what they feel, think, and do, but is also essential for their healthy development and for optimal functioning throughout life. Therefore, to develop an understanding of other people is one of the most fundamental tasks infants face in learning about the world.

T. Grossman (✉)
Centre for Brain and Cognitive Development, Birkbeck, University of London, Malet Street, Bloomsbury, London, WC1E 7HX, UK
e-mail: t.grossmann@bbk.ac.uk

P. Belin et al. (eds.), *Integrating Face and Voice in Person Perception*,
DOI 10.1007/978-1-4614-3585-3_5, © Springer Science+Business Media New York 2013

Interacting with others by reading their emotional expressions is an essential skill for humans. Reading emotional expressions during interpersonal interactions permits us to detect another person's emotional state or reactions, and can provide cues on how to respond appropriately in different social situations. It has been suggested that it may be adaptive for humans to recognize emotional expressions early in development (Darwin, 1872; Nelson, 1987). Thus, although the development of emotion perception expands beyond infancy (e.g., Russell, 1980, 1983; Russell & Bullock, 1986), developmental psychologists have focused on the question of how the perception of emotion develops during the first year of life.

At birth, the infant enters the world well prepared to rapidly develop competencies related to the perception of emotions by extracting relevant information from other's face and voice. Even though neonates' ability to discern fine visual detail is limited (Banks, 1980; Banks & Ginsburg, 1985), they do look preferentially at visual stimuli that are patterned, high-contrast, or moving (Walker-Andrews, 1997). Newborns look longer at face-like stimuli and track them farther than non-face-like stimuli (Goren, Sarty, & Wu, 1975; Johnson, Dziurawiec, Ellis & Morton, 1991). Not only do newborns look preferentially to faces in general, but also gaze longer at their mother's face specifically, even after very brief exposure to it (Bushnell, 2001; Bushnell, Sai, & Mullin, 1989; Field, Cohen, Garcia, & Greenberg, 1984).

Neonates are already sensitive to auditory information such as frequency, intensity, and temporal structure, and they prefer human voices to similar nonsocial auditory stimuli (Ecklund-Flores & Turkewitz, 1996; Hutt, Hutt, Leonard, von Bermuth, & Muntjewerff, 1968). Newborns also prefer their mother's voice over the voice of another newborn's mother (DeCasper & Fifer, 1980). It has been argued that newborns may prefer particular voices because of prenatal experience (Turkewitz, Birch, & Cooper, 1972). Furthermore, 1-month-old infants are able to make fine discriminations among different human speech sounds (Eimas, Siqueland, Jusczyk, & Vigorito, 1971).

Together, this suggests that from very early on, infants are highly attentive to social stimuli such as faces and voices, and they detect information that later may allow for the discrimination and recognition of emotional expressions. What is detected by the infant changes rapidly with the development of the perceptual systems. Thus, the development of emotion perception depends on the interplay of the maturation of perceptual systems, and the developing psychological capacities related to discriminating and recognizing emotional information. In the visual domain, for example, a newborn can only discern a blurry face and distinguish the hairline, eyes, nose, and mouth (Banks & Ginsburg, 1985). Therefore, it seems unlikely that newborns discriminate facial expressions on anything other than feature information. Then, by 6 months of age, visual acuity has improved substantially (Gwiazda, Bauer, & Held, 1989), and contrast sensitivity is sufficient to detect most static facial expression contrasts (Hainline & Abramov, 1992). Now infants can detect additional details (e.g., laugh lines) and relational information (e.g., distance between eyebrows and eye) that characterize particular facial expressions. This exemplifies how the postnatal maturation of the sensory systems can constrain the development of processing emotional information at least in the visual domain.

The next sections aim at describing infants' developing abilities in perceiving others' emotional information unimodally from face (Sect. 2.1), voice (Sect. 2.2), and multimodally from both face and voice (Sect. 2.3) by reviewing electrophysiological studies in these areas.

2 Event-Related Potential Studies of Infants' Perception of Emotion

How infants' perception of emotional expressions develops has been studied extensively using behavioral methods (see Chap. 4). However, we only poorly understand what the brain processes are that underlie infants' behaviorally exhibited capacities. The major objective of this section is to explore how the infant brain processes emotional information by reviewing some of our own work on this topic.

The focus of this work was a systematic examination of the electrophysiological bases of infants' perception of others' facial and vocal emotional expressions. Therefore, we conducted a series of three event-related potential (ERP) studies in which infants' perception of facial emotional information (Grossmann, Striano, & Friederici, 2007) and vocal emotional information (Grossmann, Striano, & Friederici, 2005) were examined unimodally, and then, in a third study (Grossmann, Striano, & Friederici, 2006), the integration of emotional information from face and voice was investigated.

The main focus of the work was on infants 7 months of age and older. This age group was selected for multiple reasons. First, at this age, infants' visual acuity has improved substantially (Gwiazda et al., 1989), and contrast sensitivity is sufficient to detect most static facial expression contrasts (Hainline & Abramov, 1992) so that they can perceive additional details (e.g., laugh lines) and relational information (e.g., distance between eyebrows and eye) that characterizes particular facial expressions. Second, by 7 months, infants are able to detect common emotion across face and voice (see previous section). Thus, in order to be able to examine and compare the underlying brain processes, one age group was chosen at which all three aspects (facial, vocal, and crossmodal information processing abilities) necessary for emotion perception are developed (see previous sections).

2.1 ERP Correlates of Emotion Processing in the Face

We measured ERPs in 7- and 12-month-old infants to examine the development of processing happy and angry facial expressions (Grossmann et al., 2007). In 7-month-olds we observed a larger negativity with a maximum at anterior (frontal and central) electrodes in response to happy faces when compare to angry faces. However, in 12-month-olds no ERP differences between emotions were measured at anterior electrodes. In this group of older infants a larger negativity to angry faces when compared to happy faces was observed at posterior (occipital) electrodes. Although the ERP data indicate that infants of both ages are able to discriminate between the facial expressions, the difference in topography (anterior: 7-month-olds; posterior: 12-month-olds) suggests that different brain systems are involved in processing of the same stimuli depending on the age of the infant. More specifically, at 12 months, enhanced negativity to an angry face at occipital sites might indicate greater sensitivity to angry faces during sensory processing in the visual cortices. In support of this interpretation, ERP findings show that angry compared to happy and neutral facial expressions elicit a larger early posterior negativity at occipital sites in adults (Schupp et al., 2004). This negative ERP component is thought to indicate facilitated sensory processing of emotional cues and appears uniformly also for other experimental designs and stimulus materials (Schupp, Junghöfer, Weike, & Hamm, 2003). Furthermore, a recent fMRI study with adults revealed increased activation of occipital regions for angry versus other facial expressions (Kesler-West et al., 2001). The finding of an adult-like electrophysiological response in 12-month-old infants is also in accordance with recent theoretical accounts that predict an increased sensory specificity through cortical specialization during development (Grossmann & Johnson, 2007; Johnson, 2001). This account of postnatal human brain development proposes that cortical areas will gain increasing functional specialization by selective loss of synapses and neurons, which might be partly determined by extrinsic (experiential) factors.

One possible developmental account is that although infants can discriminate between both facial expressions at 7 months of age and younger (see Barrera & Maurer, 1981), they still have not had sufficent exposure to angry faces to learn the signal value (threat) that an angry expression conveys (Campos et al., 2000). With increased exposure to angry faces towards the end of the first year, infants begin to detect the angry face as a signal of threat that signifies potential negative consequences. In support of this interpretation, following the onset of self-produced locomotion around 10 months of age (Illingworth, 1983), the frequency and quality of emotional communications from the adult to the infant changes. Specifically, self-produced locomotion increases the number of opportunities for caregivers to regulate infant's explorations facially and vocally. Indeed, mothers of locomotor as compared to prelocomotor infants reported a sharp increase in their expression of anger toward their infants (Campos et al., 2000; Campos, Kermoian, & Zumbahlen, 1992).

One potential avenue for future research could therefore be to assess interindividual differences in facial expression processing as a function of locomotion or affective experience. Along these lines, processing of happy and angry faces has been studied as a function of particular experiences such as physical abuse (Pollak, Cicchetti, Klorman, & Brumaghim, 1997; Pollak, Klorman, Thatcher, & Cicchetti, 2001), and maternal personality (de Haan, Belsky, Reid, Volein, & Johnson, 2004). In general, these various studies suggest that experiential factors influence the ways that infants and children process facial expressions. It is important to note that the reverse may also be true, i.e., that neural development occurring at the end of the first year (Diamond, 1991, 2000) may impact infant behavior and subsequently infants' experiences with others.

In a behavioral experiment of the study (Grossmann et al., 2007), we examined 7- and 12-month-olds' looking behavior in a visual-paired comparison task in which the two facial expressions were presented side-by-side and looking time to the expressions was measured. Contrary to the ERP data in which we found differences in the processing between ages, the looking time data revealed that both 7- and 12-month-old infants looked significantly longer at happy than angry facial expressions. It is possible that 7-month-olds simply showed a visual preference for the familiar happy face whereas 12-month-olds, who showed an adult-like brain response, avoided looking at the angry face because they perceived it as threatening and therefore preferred to look at the happy face instead. This scenario would result in longer looking to the happy expression at both ages. On a more general note, the phenomenon that different neurocognitive processes can result in similar overt behavior underlines the importance of a cognitive neuroscience approach to the study of development. Behavioral looking methods alone would have suggested that there is no development between 7 and 12 months because the looking preferences did not differ but with ERP methods we were able to show that the neural processing differs between 7 and 12 months.

2.2 *ERP Correlates of Emotion Processing in the Voice*

We examined 7-month-old infants' processing of emotional speech using ERP measures (Grossmann et al., 2005). We had infants listen to words with neutral, happy, and angry prosody in order to investigate whether and how ERP correlates differ between (a) neutral and emotionally charged prosody (happy and angry), and (b) positive emotion (happy) and negative emotion (angry).

We found that words with an angry prosody elicited a more negative response in infants' ERPs than did words with happy or neutral prosody. This effect was elicited over frontocentral sites and reached its peak amplitude around 450 ms. The negative shift observed in the current study resembles previous ERP work with 4-month-old infants, in which the mother's voice was compared to unfamiliar voices (Purhonen, Kilpeläinen-Lees, Valkonen-Korhonen, Karhu, & Lehtonen, 2004). In that study, 4-month-olds' ERPs revealed a negative shift in response to the mother's voice, while in the current study, angry prosody elicited a negative shift in 7-month-old infants' ERPs. In several infant ERP studies on visual processing it has been suggested that a larger amplitude of a negative component (Nc) indicates increased allocation of attention (de Haan, Johnson, & Halit, 2003). Based on this view, Purhonen et al. (2004) argued that the 4-month-olds in their study allocated more attention to process their own mother's voice compared to unfamiliar voices. Hence, we suggest that the 7-month-old infants in our study allocated more attentional resources to the angry than to the happy or neutral voice.

Furthermore, we found that words spoken with angry and happy prosody elicited a positive slow wave in infants' ERPs, whereas ERPs to words with neutral prosody returned to baseline. This effect was observed over temporal electrodes at a latency from 500 to 1,000 ms. It has been argued that infants' slow waves reflect more diffuse activation of neural systems (de Haan & Nelson, 1997). It is thus possible that the observed positive slow wave to happy and angry prosody indexes dispersed activation in auditory (temporal) brain structures to affectively loaded stimuli that is not evoked by neutral voices. This suggests an enhanced sensory processing only of the affectively loaded auditory stimuli.

Concordant with this interpretation is evidence from fMRI work in adults showing that emotionally charged words undergo more extensive processing than words with neutral prosody (Mitchell, Elliott, Barry, Cruttenden, & Woddruff, 2003). For example, relative to neutral prosody, angry prosody evoked enhanced activity in adults' associative auditory cortex, namely, in the middle portion of the superior temporal sulcus (Grandjean et al., 2005). Similarly, an enhancement in the processing of faces was reported in the right midfusiform gyrus for fearful relative to neutral faces (Vuilleumier, Armony, Driver, & Dolan, 2001). Therefore, it has been proposed that enhanced sensory responses to emotional facial and vocal stimuli might be a fundamental neural mechanism. It is possible that this mechanism might also account for the observed positive slow wave to happy and angry prosody over temporal sites in the 7-month-olds, indicating an enhanced sensory processing of the emotional stimuli. This enhanced processing, which we have demonstrated on an electrophysiological level, could be linked to the behavioral finding that vocal affect facilitates infants' spoken word recognition (Singh, Morgan, & White, 2004), suggesting a method by which emotional information in the speech signal might help infants develop language comprehension capacities.

2.3 ERP Correlates of Emotion Processing in Face and Voice

The ERP measure has been found to be sensitive to infants' crossmodal (haptic to visual) recognition of objects (Nelson, Henschel, & Collins, 1993), and has proven to be a valuable tool in assessing these underlying mechanisms of infants' processing of unimodal emotional information conveyed by the face (Grossmann et al., 2007; Nelson & de Haan, 1996) and by the voice (Grossmann et al., 2005). To extend this work into the domain of multisensory processing we investigated the electrophysiological processes underlying crossmodal integration of emotion in 7-month-old infants (Grossmann et al., 2006). As infants watched a static facial expression (happy or angry), they heard a word spoken in a tone of voice that was either emotionally congruent or incongruent with the facial expression. The ERP data revealed that the amplitude of a negative component and a subsequently elicited positive component in infants' ERPs varied as a function of crossmodal emotional congruity. We found that words spoken with a tone of voice that was emotionally incongruent to the facial expression elicited a larger negative component in infants' ERPs than did emotionally congruent words. Conversely, the amplitude of the positive component was larger to emotionally congruent words than to incongruent words. These findings provide electrophysiological evidence that 7-month-olds recognize common affect across modalities, which is in line with previous behavioral work (Soken & Pick, 1992; Walker, 1982; Walker-Andrews, 1986).

Extending behavioral findings, the ERP data from Grossmann et al. (2006) reveals insights into the time course and characteristics of the processes underlying the integration of emotional information across the senses in the infant brain. Numerous ERP studies in adults have investigated old–new effects in recognition memory tests with a variety of stimuli (for a review, see Rugg & Coles, 1995). The uniform finding across studies is that old (familiar) items evoke more positive-going ERPs than do new (unfamiliar) items. This general old–new effect comprises the modulation of two ERP components: an early negativity (early N400), which consistently shows an attenuated amplitude to old items, and a late positive component or complex (LPC), which shows an enhanced amplitude to old items.

Old (familiar) items have also been found to elicit an attenuated N400 and an enhanced LPC in children's ERPs when compared to new (unfamiliar) items (Friedman, 1991; Friedman, Putnam, & Sutton, 1989; Friedman, Putnam, Ritter, Hamberger, & Berman, 1992; Coch, Maron, Wolf, & Holcomb, 2002). Furthermore, similar effects have been observed in infants' ERPs (Nelson, Thomas, de Haan & Wewerka, 1998), where old (familiar) items elicited a more positive-going brain response with an attenuated early negative component (Nc) and an enhanced late positive component (Pc). Given the similarities in response properties, latency, and topography of these components across ages (infancy to adulthood), it is plausible to assume that the adult and child N400 corresponds with the infant Nc and that the adult and child LPC corresponds with the infant Pc. Thus, a coherent picture begins to emerge about the developmental continuity of recognition memory effects in the ERP.

In Grossmann et al. (2006), emotionally congruent face–voice pairs elicited similar ERP effects as recognized items in previous memory studies with infants, children, and adults. This suggests that 7-month-old infants recognize common affect in face and voice. Since the face–voice pairs presented to the infants were novel to them, the ERP data not only indicate that these infants recognized common affect, but, moreover, that they applied their knowledge about emotions in face and voice to draw inferences about what might be appropriate emotional face–voice associations when encountering novel bimodal events. Multimodal audiovisual events usually make two kinds of information available: amodal and modality specific information (for a detailed discussion of amodal and modality-specific information processing in infancy see Bahrick, Lickliter, & Flom, 2004). An example of amodal information is that the movements of the lips and the timing of speech share temporal synchrony, rhythm, and tempo, and have common intensity shifts. Since we used static facial expressions, there was no such amodal information available to the infants. Thus, infants could not simply determine that a face and voice belonged together by detecting amodal audiovisual relations; instead, they had to draw inferences based on their prior knowledge.

Another finding from this study was that the amplitude of infants' Nc not only differed between congruent and incongruent face–voice pairs but also between two incongruent conditions. Namely, when a happy face was presented with an angry voice, the Nc was more negative in its amplitude than when an angry face was presented with a happy voice. As mentioned earlier, we know that prior to the onset of crawling (around 10 months), infants have only little exposure to others' expression of anger, whereas happy emotional expressions are ubiquitous in infants' everyday social interactions (Campos et al., 1992, 2000). Based on this observation, it can be assumed that a happy face is more familiar than an angry face (see also Vaish, Grossmann, & Woodward, 2008). It is thus possible that the presentation of the more familiar happy face triggered a stronger expectation about the appropriate emotional prosody, causing an especially strong expectancy violation and a larger Nc when the angry voice was presented. This suggests a sensitivity of infants' Nc to familiarity-based processes, confirming previous research on infants' Nc (see Csibra, Kushnerenko, & Grossmann, 2008).

3 Discussion of ERP Findings

Together, the presented ERP findings indicate that infants' perception of emotional expressions in the face and voice elicited both sensory-specific and sensory-unspecific (general) effects in infants' ERPs. The ERP data revealed two sensory-specific effects: (1) a negative component observed over occipital sites to angry faces in 12-month-old infants (Grossmann et al., 2007) and (2) a positive slow wave elicited over temporal sites by emotionally loaded words in 7-month-old infants (Grossmann et al., 2005). These effects are likely to be sensory-specific because their observed scalp topography suggests that the visual (occipital) and the auditory (temporal) sensory processing were specifically affected. Specifically, an enhanced negativity to an angry face at occipital sites in 12-month-olds as shown in Grossmann et al. (2007) might indicate greater sensitivity to angry faces during sensory processing in the visual cortices. Moreover, the observed positive slow wave to happy and angry prosody might reflect sensory-specific processes in auditory (temporal) brain structures to affectively loaded stimuli that is not evoked by neutral voices. Based on fMRI work with adults, researchers have proposed that enhanced sensory responses to emotional facial and vocal stimuli might be a fundamental mechanism by which the brain highlights emotionally loaded information (e.g., Grandjean et al., 2005; Mitchell et al., 2003; Vuilleumier et al., 2001). The reviewed ERP data suggest the early emergence and effectiveness of this mechanism, since infants' enhanced sensory processing of emotional stimuli was demonstrated on an electrophysiological level for vocal and facial cues. More generally, the mechanism of an emotion-induced more elaborate sensory processing of stimuli could also help infants' developing cognitive skills. It is possible that this enhanced processing contributes to the facilitating effects emotion has on infants' learning in different domains (see Malatesta & Haviland, 1982; Singh et al., 2004).

In addition to the sensory-specific ERP effects emotional expressions also elicited sensory-unspecific (general) effects in infants' ERPs. By 7 months of age, happy faces evoked a negative component that was larger than that evoked by angry faces (Grossmann et al., 2007). However, at the same age, angry voices elicited a negative shift that was not observed in response to neutral and happy voices (Grossmann et al., 2005). Both, the negative component in face processing (de Haan et al., 2003) and the negative shift in voice processing (Purhonen et al., 2004) are thought to reflect the allocation of attentional resources. Based on this view, a larger amplitude of these components indexes increased allocation of attention. This suggests that 7-month-olds on the one hand allocate more attention to the processing of happy faces, but on the other hand they devote more attentional resources to the processing of angry voice.

These findings thus suggest that infants' ability to show a heightened attentional sensitivity to negative emotional information develops earlier in the vocal domain. Interestingly, it has been suggested that the advantage of the auditory sensory modality might result from the fact that the auditory system in mammals develops much earlier than the visual system (Gottlieb, 1971). The emergence of the different sensory systems begins early in gestation and is sequential, which leads to different amounts and types of sensory experience. The sequential nature of the sensory development is thought to have substantial impact on the development of intersensory function such that the early-developing sensory modalities become functionally differentiated without the competing influence of the later-developing ones, whereas the later-developing ones have to compete with the earlier-developing ones (Turkewitz & Devenny, 1993). The neonate comes into the world with a set of sensory systems that already have had differential sensory experience and that are, therefore, not functionally equivalent. After birth the sensory systems continue to interact with each other, as they did prenatally, but now they do so in a radically different postnatal setting characterized by a new and rich multimodal array of information.

Lewkowicz (1988a, 1988b) has designed studies on sensory dominance, and put forward a theory of early auditory dominance. In these studies infants 6 and 10 months of age were habituated to flashing checkerboards (visual information) accompanied by beeps (auditory information). The younger infants dishabituated only to audiovisual or auditory changes. At 10 months infants also dishabituated to visual changes, but overall infants at both ages were more sensitive to the auditory change than to the visual change. However, these data are limited to socially irrelevant stimuli. Based on evidence indicating that when infants are not yet showing consistent differential responsiveness to positive and negative facial expressions, they are responding differentially to positive and negative vocal expressions (Fernald, 1992), it has been proposed that in early development, emotional information in the voice is more powerful than in the face (see, Vaish et al., 2008; Vaish & Striano, 2004). The ERP data presented here suggests that infants' ability to show a heightened sensitivity to negative emotional information develops earlier in the vocal domain when compared to the visual domain, are consistent with this view.

Another finding in the present studies was that 7-month-olds can recognize common emotion in face and voice (Grossmann et al., 2006). This finding seems surprising, given that infants at the same age do not recognize anger by looking only at a facial display (Grossmann et al., 2007). A developmental sequence has been proposed in which infants learn to discriminate and recognize emotional expressions based on multimodal, then vocal, and finally, as visual acuity improves, facial cues

(Walker-Andrews, 1997). This notion that emotion discrimination and recognition occurs earlier in multimodal contexts is supported by the current ERP findings when multimodal context (Grossmann et al., 2006) is compared to unimodal facial context (Grossmann et al., 2007). It has been shown that the perception of multimodally specified events appears to be generally more efficient, because multimodal cues confer a significant advantage over unimodal cues both in perception and discriminative learning across a variety of species (Rowe, 1999). Moreover, the availability of multimodal information in the current study might have had advantageous multiplicative effects (Stein, Meredith & Wallace, 1993) on infants' perceptual abilities that cannot be anticipated by simply adding their performances in the unimodal contexts. In other words, although 7-month-olds failed to exhibit the ability to detect facial anger unimodally, they discovered commonalities across face and voice that allowed them to recognize the congruent emotion.

It has been suggested that through the detection of intermodal invariants in multimodal contexts, infants also discover the meaning of emotional expression (Walker-Andrews, 1997). However, note that although infants might first recognize the affective expressions of others as a unified multimodal event, and only later begin to recognize the same emotional information unimodally, this does not necessarily mean that infants also discover the meaning of emotional expressions through this process. Infants' ability to match facial and vocal expressions of emotion might merely be based on learning to associate a certain facial expression with the vocal expression that consistently accompanies it. This ability can be expressed by the infant without the appreciation of the meaning of the emotional expression. Thus, although the presented ERP data (Grossmann et al., 2006) suggest that 7-month-olds detect common affect in the face–voice pairs presented, it cannot be concluded that they also discover the meaning of these emotional expressions.

It is important to note that many mother–infant studies using live interaction suggest that infants recognize the emotional expressions of their own caregivers and respond to them meaningfully as early as 2–3 months (Cohn & Ellmore, 1988; Malatesta & Haviland, 1982). For example, 10-week-old infants respond differentially and contingently to their mothers' live presentation of happy, sad, and angry emotional expressions (Haviland & Lelwica, 1987). When mothers expressed happiness, infants expressed more joy and interest. When mothers presented sad expressions, infants expressed less joy, and they also showed increased mouthing behavior and gaze aversion. To maternal expressions of anger infants responded with increased anger and their movement appeared to freeze. This and other studies suggest that infants as young as 3 months of age have a wide repertoire of emotional expressions and respond effectively to their mothers' emotional expression.

As opposed to studies using live interactions, findings from the reviewed ERP studies and other experimental investigations examining infants' recognition of emotional expression suggest that only at around 7 months of age do infants match facial and vocal expressions of the same emotion (Soken & Pick, 1992; Walker-Andrews, 1986) or recognize that different examples of the same emotion belong to the same category (Ludemann & Nelson, 1988). This apparent age difference might be due to several differences between live interaction and experimental studies.

First, in most experimental studies, the emotional expressions presented are restricted to either the visual or the auditory domain, whereas in interaction studies, infants are provided with multimodal presentation of the emotion in face, voice, gesture, and touch. Second, in most experimental studies the emotional expression is displayed by an unfamiliar actress, whereas in interaction studies they are typically displayed by infants' mothers. Indeed, 3-month-old infants are better at discriminating among facial expressions when the expressions are portrayed by their own mother than by a stranger (Barrera & Maurer, 1981). Furthermore, Kahana-Kalman and Walker-Andrews (2001) found that infants presented with familiar faces and voice were able to recognize common affect across modalities at 3.5 months of age, whereas infants tested with unfamiliar faces and voices did not recognize common affect across modalities until 7 months of age (Walker-Andrews, 1986). Kahana-Kalman and Walker-Andrews propose that maternal emotional responses are not only more familiar to a young infant, but also more informative with respect to ensuing actions. Infants may have been more motivated to attend to the emotional expressions of their mothers because these may foreshadow more specific outcomes to them. For example, maternal smiles are likely to be followed by positive caretaking interactions, whereas maternal negative expressions may frequently be followed by experiences were the infant is left alone.

Based on these findings, which underline the prominent role of maternal expressions of emotion for infants' developing understanding of others' emotion, what are the implications for the presented ERP studies? In general, it can be stated that the findings described in these studies are limited to perception of emotional expressions displayed by strangers, and previous work seems to suggest that the abilities observed here might well be observable at an earlier age when investigated with maternal expressions. Therefore, for future studies it seems promising to examine the role of experience and familiarity on the electrophysiological correlates of infants' perception of emotion by using maternal stimuli.

The presented ERP work has provided insights into how the human brain processes emotional information very early in development. The systematic investigation of the electrophysiological correlates of perceiving facial, vocal, and multimodal emotional cues provided empirical data on the brain mechanisms guiding infants' emerging understanding of emotional expressions.

4 Genetic Factors Associated with Individual Differences in Emotion Processing in Face and Voice

An important further question is whether and how genetic variation might influence infants' brain responses to facial expressions and thus contribute to individual differences in emotional sensitivity and temperament. Addressing this question of specific genetic pathways that contribute to social behavior is critical to our understanding of how such differences confer vulnerability to psychiatric diseases (Meyer-Lindenberg & Weinberger, 2006). In addition, studying emotion processing in infancy provides the opportunity to examine gene effects at a time in development when genetic association might be more robustly demonstrated because effects of postnatal experience are still relatively small (Ebstein, 2006).

In adults, variations in specific genes acting on neurotransmitter systems have been found to impact emotion processing. Specifically, a number of genetic neuroimaging studies have shown effects of Catechol-*O*-methyltransferase (*COMT*) and Serotonintransporter (*SLC6A4/5-HTTLPR*) genotypes on the processing of emotional stimuli in general and of facial expressions in particular (for reviews, see Canli & Lesch, 2007; Heinz & Smolka, 2006).

COMT is an important enzyme involved in the elimination of dopamine (DA) in the prefrontal cortex (Goldberg & Weinberger, 2004). A functional polymorphism in the *COMT* gene (val158met) accounts for a significant difference in enzyme activity: while the high-active val allele is presumed to be associated with lower concentration of synaptic DA, the low-active met allele is thought to result in higher concentrations of DA (Chen et al., 2004; Heinz & Smolka, 2006). At the cognitive level, the met allele is associated with improved working memory and executive functioning (Goldberg & Weinberger, 2004). This better performance in executive functions and working memory is reflected in a more focal response in prefrontal cortex as measured with functional magnetic resonance imaging (fMRI), indexing more efficient neural processing (Egan et al., 2001). The met allele, moreover, is associated with an increased sensitivity to emotionally unpleasant stimuli. More specifically, in an fMRI study with adults, the met allele was associated with increased activity in limbic and prefrontal brain regions in response to fearful and angry facial expressions (Drabant et al., 2006). This increased neural sensitivity associated with the met allele was not found in response to positive stimuli, suggesting that it is specific to negative stimuli (Herrmann et al., 2009; Smolka et al., 2005).

Serotonin (5-HT) plays a major role in emotion regulation and social behavior. A functional polymorphism (*5-HTTLPR*) in the regulatory regions of the serotonin transporter gene has a short (s) and a long (l) allele (14- and 16-repeat alleles, respectively) that alter promoter activity: the s variant produces significantly less serotonin transporter mRNA and protein than the l variant, resulting in higher concentrations of serotonin in the synaptic cleft (Canli & Lesch, 2007). Individuals carrying the s allele appear to have increased anxious temperament, resulting in an elevated risk to develop depression (Lesch et al., 1996). On the neural level, healthy nondepressed adults carrying the s allele showed an increased amygdala response to threatening stimuli such as fearful faces (Hariri et al., 2002). Furthermore, structural analyses revealed reduced gray matter in s allele carriers in anterior cingulate and amygdala, and during the processing of fearful faces, these regions showed less functional coupling in carriers of the s allele (Pezawas et al., 2005).

Taken together, in adults, both the met allele of the *COMT* gene and the s allele of the *5-HTTLPR* gene appear to be associated with an increased sensitivity to negative, specifically fearful, expressions. Although both polymorphisms affect neural processes in the limbic system, the *COMT* variation is thought to be more specifically implicated in affecting prefrontal brain processes (Goldberg & Weinberger, 2004; Heinz & Smolka, 2006). Event-related brain potential (ERP) studies that allow for the precise investigation of the timing of neural processes have shown that, in adults, variation in

COMT and *5-HTTLPR* genotype affect the brain processing of emotional stimuli at early stages at occipital electrodes, starting approximately 200 ms after stimulus onset (Herrmann et al., 2006, 2009). Furthermore, in a recent study with adolescent twins, individual differences in ERP responses to emotional facial expressions have been found to be highly heritable (Anokhin, Golosheykin, & Heath, 2010).

In a recent study (Grossmann et al., 2011), we thus assessed the effects of *COMT* and *5-HTTLPR* genotypes on the brain processing of facial expressions (fearful and happy) in 7-month-old infants using ERPs. The analysis of genotype effects was focused on the Negative central (Nc) component in infants' ERPs, and the preceding so-called Positivity before (Pb). Both components have been shown to be similarly modulated by facial expressions in infancy (Nelson & de Haan, 1996). The Nc, is generated in the prefrontal cortex, occurs from approximately 300–600 ms, has its maximum at central electrodes, and is thought to reflect the allocation of attention to a stimulus, with a greater amplitude indexing increased allocation of attention (Richards, 2002). In 7-month-olds, fearful faces when compared to happy faces elicited a more negative-going waveform consisting of a decreased Pb and an enhanced Nc, indicating increased attention allocation to fearful expressions (Nelson & de Haan, 1996). Moreover, Peltola and colleagues (2009) found that 7-month-olds showed an enhanced Nc to fearful faces whereas 5-month-olds did not, suggesting that an enhanced sensitivity to fearful faces emerges between 5 and 7 months of age. Such an enhanced attention to fearful faces is also found in adults and is thought to be a fundamental mechanism to prioritize the processing of evolutionarily significant stimuli (Vuilleumier, 2006). Furthermore, in order to see whether the observed effects were specific to emotional face processing rather than related to general face processing, we analyzed the face-sensitive infant N170 as a function of genetic variation at the two loci. Finally, we examined effects of genotype on infant temperament as measured by the *Infant Behavior Questionnaire-R* (Garstein & Rothbart, 2003). On the basis of the adult work discussed above, it would be predicted that both polymorphisms affect the processing of fearful expressions. However, it is also possible that these genetic polymorphisms might be associated with different effects in infancy than in adulthood, since effects of genetic variation observed in adulthood may be an outcome of developmental processes that have distinct origins and manifestations in infancy (Gottlieb, 2006; Karmiloff-Smith, 1998).

The results of this study (Grossmann et al., 2011) revealed that variation in these genes is differentially associated with how infants process facial expressions of emotion. Specifically, variation at the *COMT* locus is associated with the processing of fearful facial expressions, whereas variation at the *5-HTTLPR* locus is associated with the processing of happy facial expressions. These differences were also reflected in the distinct topography of the ERP effects, suggesting the involvement of distinct brain processes: *COMT* variation was associated with centroparietal processing of fearful faces, whereas *5-HTTLPR* was associated with frontotemporal processing of happy faces. These genetic associations were specific to the processing of emotional faces as no such effects were observed for the processing of neutral facial expressions. This pattern suggests that, early in postnatal development, variations of these genes affect distinct brain systems involved in the processing of positive versus negative facial expressions.

In line with findings from adults, *COMT* variation was associated with processing negative (fearful) emotions in infants (Drabant et al., 2006). More specifically, the carriers of the met allele showed an enhanced negativity to fearful expressions at central and parietal electrodes, indicating increased attentional sensitivity to these expressions, whereas infants with the val/val genotype responded with an increased positivity to fearful expressions, suggesting that this genotype processes fearful expressions less sensitively. This finding might have important implications for clinical disorders insofar as work with patients with schizophrenia has found that these patients are impaired in the recognition of fearful faces, and there is evidence to suggest that schizophrenia is more common among individuals with the val/val *COMT* genotype (Egan et al., 2001; Harrison & Weinberger, 2005; Morris, Weickert, & Loughland, 2009). The increased attentional sensitivity to fearful faces associated with the met allele has also been reported in neuroimaging studies with adults (Drabant et al., 2006), thus suggesting developmental continuity in the influence of *COMT* on the processing of facial expressions.

In contrast to what has been shown in adults (where variation in *5-HTTLPR* like variation in *COMT* is associated with the processing of negative [fearful] affect), the current infant data revealed that *5-HTTLPR* variation is associated with the processing of positive (happy) affect. Specifically, carriers of the l allele showed a negativity in response to happy expressions at frontal and temporal electrodes, whereas infants with the s/s genotype showed a positivity in response to happy expressions, suggesting that s/s genotype infants process happy expressions differently and might be less sensitive to positive affect. Thus, our findings suggest that there are differences as to how *5-HTTLPR* variants influence emotion processing in the human brain depending on age. It is important to note that fMRI work comparing children (average age of 11 years) and adults has revealed that adults but not children show increased amygdala activity to fearful faces when compared to neutral faces (Thomas et al., 2001). This late development of amygdala sensitivity to fearful faces reported in the fMRI work might help explain the difference between the current findings with infants and the adult work. That is, if older children do not show specific amygdala sensitivity to fear, it seems unlikely that infants' processing of fearful faces will be influenced by a gene that affects amygdala sensitivity only in adults. Furthermore, it should be noted that we measured ERPs from the scalp, and these potentials might not be sensitive to amygdala activity.

Nonetheless, one intriguing developmental hypothesis derived from the current findings is that early in postnatal development, variation in *5-HTTLPR* may critically alter the processing of positive emotion, which later in development has effects on how adults respond to negative emotions. One mechanism that has been proposed is that infants have been responding sensitively to positive emotions from birth, which has established a positive default (or background) mode against which negative emotions stand out (see Vaish et al., 2008). It is possible that less sensitive responding to positive emotion in early development due to a specific genotype impairs the way in which positive affect becomes the background mode. According to this scenario, hypersensitivity in the processing of negative affect in adults could thus partly be a consequence of a reduced or impaired acquisition of positive affective stability during infancy and childhood (see Sprangler, Johann, Ronai, & Zimmermann, 2009). Future research investigating this hypothesis across the life span is needed to understand the impact of *5-HTTLPR* on the developmental trajectory of emotional sensitivity.

Further support for distinct influences of *COMT* and *5-HTTLPR* on emotional processes in infancy comes from our analysis of infant temperament as judged by their parents. The results showed that while *COMT* variation is associated with reported recovery from distress, *5-HTTLPR* variation was associated with reported smiling and laughter and duration of orienting. It is interesting to note that infants with the short/short genotype of *5-HTTLPR*, who were judged as smiling and laughing significantly less than infants with the other *5-HTTLPR* genotypes, also showed a different brain response to watching others' happy facial expressions. This may point to a link between infants' own experience of positive affect and processing positive affect from facial expressions in others, raising the possibility that so-called mirroring or simulation mechanisms could be influenced by temperament and genotype. The finding that COMT was associated with infants' recovery from distress is in line with work implicating this gene in prefrontal control and regulatory brain mechanisms (Goldberg & Weinberger, 2004; Heinz & Smolka, 2006). Surprisingly, the met allele appeared to be associated with better emotion regulation (recovery from distress) in infants, which seems to contradict findings with adults indicating that the met allele might be linked to anxiety and difficulties in emotion regulation (Heinz & Smolka, 2006). However, the met allele has also been linked to better cognitive control, a notion that is also supported by behavioral work with children and infants (Diamond, Briand, Fossella, & Gehlbach, 2004; Holmboe et al., 2010). Thus, better recovery from distress associated with the met allele as found in our infant sample might relate to generally improved control processes across cognitive and emotional domains, at least at this young age.

With respect to the timing of the brain processes that were found to be affected by variation in *COMT* and *5-HTTLPR*, our ERP analysis revealed that both genes are associated with infants' brain responses as early as 200 ms after face onset. The timing of these effects is in line with the adult ERP work (Herrmann et al., 2006, 2009). However, in the adult ERP work, both genotypes were associated with posterior brain processes at occipital sites, whereas there were no associations with occipital sites in the current infant ERP data, suggesting that there might be a change during development in the topography of the effects. However, we cannot further interpret these topographic differences between infants and adults because in the adult work the analysis of genetic effects was focused only on posterior (occipital) sites and no data for other regions were presented (Herrmann et al., 2006, 2009). This is problematic because, in adults, ERP effects can be obtained at frontal and central electrodes in response to fearful faces (see, e.g., Eimer & Holmes, 2002).

Taking such a genetic imaging approach has been shown to be of great value for our understanding of individual differences in adults, and studying the association of genetic variation with brain responses as intermediate phenotypes, or so-called endophenotypes, has been argued to be a more powerful approach than studying gene effects on behavior (or personality traits) (Goldberg & Weinberger, 2004).

Applying this approach to infants in the current study has revealed novel insights by adding a developmental component to the complex picture of how genetic variation may affect human emotion. The finding that, in infancy, *COMT* and *5-HTTLPR* variation are associated with emotion processing in distinct ways raises interesting hypotheses about how genetic variation may bias certain brain mechanisms and thereby give rise to early individual differences that ultimately contribute to complex phenotypes such as temperament and personality. This might be a promising novel approach to the study of early emotional development, but it is only a first step. To gain a fuller understanding of the relationship between genetic variation, brain and emotion in development, we will need to examine genetic influences longitudinally in a larger sample of infants.

We followed up on these findings by extending this approach to study how genetic variation at the same loci affects the processing of emotional information in the voice in 7-month-old infants. Strikingly, in this line of new work (Grossmann, Hughes, Stoneking, and Friederici, in preparation) we were able to replicate our ERP findings using facial expressions of emotion by showing that *COMT* was associated with variation in processing negative (angry) affect in the voice, whereas *5-HTTLPR* was associated with variation in processing positive (happy) affect in the voice. This finding is an important extension of the prior work and it clearly suggests that the patterns of genetic association can be observed across face and voice, pointing to a robust effect of the way in which genetic variation is linked to specific differences in emotion processing.

5 Directions in the Study of Early Emotion Processing from Face and Voice

It is my hope that this chapter might stimulate future work that extends these findings on four levels. First, it seems worthwhile to test infants and children at other ages to further examine the roles maturation and experience play in the development of processing emotional information. Specifically, one important issue that has not been addressed is the question of how infants' own production of facial and vocal expressions relates and possibly shapes their understanding and processing of emotional expressions in others. This question seems partiularly pertinent because there is much debate about the role of the so-called human mirror neuron system' (Rizzolatti & Craighero, 2004) in the understanding of emotion and action in adults. According to the mirror neuron system view, infants are not expected to show an understanding of other people's actions or emotions before they can perform the action or express the emotion themselves. Indeed, there is some first evidence from studies on action understanding to support this hypothesis (Falck-Ytter, Gredebäck & von Hofsten, 2006; Sommerville, Woodward, & Needham, 2005), but whether this also holds for emotional facial and vocal expressions remains to be seen. Second, this line of electrophysiological work should be extended to other emotions in order to understand the emotion-specificity of the found effects (see Kobiella,

Grossmann, Striano, & Reid, 2008). Third, it is crucial to identify the neural sources that are involved in infants' processing of emotional facial and vocal information by using methods that can localize brain activity in the infant. One method that permits spatial localization of brain activation by measuring hemodynamic responses is near-infrared spectroscopy (NIRS) (see Lloyd-Fox, Blasi, & Elwell, 2010 for a review of this method and its use with infants). Other neuroimaging techniques that are well established in adults are limited in their use with infants because of methodological concerns. NIRS is better suited for infant research because it can accommodate a good degree of movement from the infants, enabling them to sit upright on their parent's lap and behave relatively freely while watching or listening to certain stimuli. In addition, unlike PET and fMRI, NIRS systems are portable. Finally, despite its inferior spatial resolution, NIRS, like fMRI, measures localized patterns of hemodynamic responses, thus allowing for a comparison of infant NIRS data with adult fMRI data (see Strangman, Culver, Thompson, & Boas, 2002). In a recent study using NIRS (Grossmann, Oberecker, Koch, & Friederici, 2010), we were able to show that 7-month-olds but not 4-month-olds showed increased responses in left and right superior temporal cortex to the human voice when compared to nonvocal sounds, suggesting that voice sensitive brain systems emerge between 4 and 7 months of age. Moreover, hearing emotional prosody, specifically angry prosody resulted in increased responses in a region identified as voice-sensitive in 7-month-old infants. This demonstrates the power of this method in identifying the neural sources of the voice processing in early development and should encourage future studies mapping the brain basis of processing emotional information across face and voice (see Grossmann, 2008 for limitations in using NIRS to study face processing). Fourth, once we know how the typically developing brain processes emotional information it might be possible to examine how this differs from the processing in atypically developing infants. By this comparison we might gain knowledge about atypical brain indices that, in conjunction with other measures, can contribute to an early diagnosis of the specific deficit (see Elsabbagh & Johnson, 2010). An early diagnosis allows early intervention and may therefore help the affected children and families.

Acknowledgments The work on this chapter was supported by a *Sir Henry Wellcome Fellowship* awarded by the Wellcome Trust (082659/Z/07/Z). I would like to thank Amrisha Vaish for comments

References

Anokhin AP, Golosheykin S, Heath AC (2010) Heritability of individual differences in cortical processing of facial affect. Behavior Genetics 40:178–185

Bahrick LE, Lickliter R, Flom R (2004) Intersensory redundancy guides infants' selective attention, perceptual and cognitive development. Current Directions in Psychological Science 13:99–102

Banks MS (1980) The development of visual accommodation during early infancy. Child Development 51:646–666

Banks MS, Ginsburg AP (1985) Infant visual preferences, a review and new theoretical treatment. In: Reese HW (ed) Advances in child development and behavior. Academic, New York, pp 207–246
Barrera ME, Maurer D (1981) The perception of facial expressions by three-month-olds. Child Development 5:203–206
Bushnell I (2001) Mother's face recognition in newborn infants: Learning and memory. Infant and Child Development 10:67–74
Bushnell IWR, Sai F, Mullin JT (1989) Neonatal recognition of the mother's face. The British Journal of Developmental Psychology 7:3–15
Campos JJ, Anderson DI, Barbu-Roth MA, Hubbard EM, Hertenstein MJ, Witherington D (2000) Travel broadens the mind. Infancy 1:149–219
Campos JJ, Kermoian R, Zumbahlen MR (1992) Socioemotional transformations in the family system following infant crawling onset. In: Eisenberg N, Fabes RA (eds) Emotion and its regulation in early development. Jossey-Bass, San Francisco, pp 25–40
Canli T, Lesch KP (2007) Long story short: The serotonin transporter in emotion regulation and social cognition. Nature Neuroscience 10:1103–1109
Chen J, Lipska BK, Halim N, Ma QD, Matsumoto M, Melhem S et al (2004) Functional analysis of genetic variation in catechol-O-methyltransferase (COMT): Effects on mRNA, protein, and enzyme activity in postmortem human brain. American Journal of Human Genetics 75: 807–821
Coch D, Maron L, Wolf M, Holcomb PJ (2002) Word and picture processing in children: An event-related potential study. Developmental Neuropsychology 22:373–406
Cohn JF, Ellmore M (1988) Effect of contingent changes in mothers' affective expression on the organization of behavior in 3-month-old infants. Infant Behavior & Development 11:493–505
Csibra G, Kushnerenko E, Grossmann T (2008) Electrophysiological methods in studying infant cognitive development. In: Nelson CA, Luciana M (eds) Handbook of developmental cognitive neuroscience, 2nd edn. MIT Press, Cambridge, pp 247–262
Darwin C (1872) The expression of emotions in man and animals. John Murray, London
de Haan M, Belsky J, Reid V, Volein A, Johnson MH (2004) Maternal personality and infants' neural and visual responsitivity to facial expressions of emotion. Journal of Child Psychology and Psychiatry 45:1209–1218
de Haan M, Johnson MH, Halit H (2003) Development of face-sensitive event-related potentials during infancy: A review. International Journal of Psychophysiology 51:45–58
de Haan M, Nelson CA (1997) Recognition of the mother's face by six-month-old infants: A neurobehavioral study. Child Development 68:187–210
DeCasper AJ, Fifer WP (1980) Of human bonding: Newborns prefer their mothers' voices. Science 280:1174–1176
Diamond A (1991) Frontal lobe involvement in cognitive changes during the first year of life. In: Gibson KR, Peterson AC (eds) Brain maturation and cognitive development: Comparative and cross-cultural perspectives. Aldine de Gruyter, New York, pp 127–180
Diamond A (2000) Close interrelation of motor development and cognitive development and of the cerebellum and prefrontal cortex. Child Development 71:44–56
Diamond A, Briand L, Fossella J, Gehlbach L (2004) Genetic and neurochemical modulation of prefrontal cognitive functions in children. The American Journal of Psychiatry 161:125–132
Drabant EM, Hariri AR, Meyer-Lindenberg A, Munoz KE, Mattay VS, Kolachana BS et al (2006) Catechol-O-methyltransferase val158met genotype and neural mechanisms related to affective arousal and regulation. Archives of General Psychiatry 63:1396–1406
Ebstein RP (2006) The molecular genetic architecture of human personality: Beyond self-report questionnaire. Molecular Psychiatry 11:427–445
Ecklund-Flores L, Turkewitz G (1996) Asymmetric headturning to speech and nonspeech in human newborns. Developmental Psychobiology 29:205–217
Egan MF, Goldberg TE, Kolachana BS, Callicott JH, Mazzanti CM, Straub RE et al (2001) Effect of COMT val$^{108/158}$met genotype on frontal lobe function and risk for schizophrenia. Proceedings of the National Academy of Sciences of the United States of America 98:6917–6922

Eimas PD, Siqueland ER, Jusczyk PW, Vigorito J (1971) Speech perception in infants. Science 220:21–23

Eimer M, Holmes A (2002) An ERP study on the time course of emotional face processing. Neuroreport 13:427–431

Elsabbagh M, Johnson MH (2010) Getting answers from babies about autism. Trends in Cognitive Science 14:81–87

Falck-Ytter T, Gredebäck G, von Hofsten C (2006) Infants predict other people's action goals. Nature Neuroscience 9:878–879

Fernald A (1992) Human maternal vocalizations to infants as biologically relevant signals: An evolutionary perspective. In: Barkow JH, Cosmides L, Tooby J (eds) The adapted mind: Evolutionary psychology and the generation of culture. Oxford University Press, Oxford, UK, pp 391–428

Field TM, Cohen D, Garcia R, Greenberg R (1984) Mother-stranger face discrimination by the newborn. Infant Behavior & Development 7:19–25

Friedman D (1991) The endogenous scalp-recorded brain potentials and their relation to cognitive development. In: Jennings JR, Coles MGH (eds) Handbook of cognitive psychophysiology: Central and autonomic nervous system approaches. John Wiley, New York, pp 621–656

Friedman D, Putnam L, Ritter W, Hamberger M, Berman S (1992) A developmental study of picture matching in children, adolescents, and young-adults: A replication and extension. Psychophysiology 29:593–610

Friedman D, Putnam L, Sutton S (1989) Cognitive brain potentials in children, young adults and senior citizens: Homologous components and changes in scalp distribution. Developmental Neuropsychology 5:33–60

Garstein MA, Rothbart MK (2003) Studying infant temperament via the revised infant behavior questionnaire. Infant Behavior & Development 26:64–86

Goldberg TE, Weinberger DR (2004) Genes and the parsing of cognitive processes. Trends in Cognitive Sciences 8:325–335

Goren CC, Sarty M, Wu P (1975) Visual following and pattern discrimination of face-like stimuli by newborn infants. Pediatrics 56:544–549

Gottlieb G (1971) Ontogenesis of sensory function in birds and mammals. In: Tobach E, Aronson LR, Shaw E (eds) The biopsychology of development. Academic, New York, pp 67–128

Gottlieb G (2006) Probabilistic epigenesis. Developmental Science 10:1–11

Grandjean D, Sander D, Pourtois G, Schwartz S, Seghier ML, Scherer KR et al (2005) The voices of wrath: Brain responses to angry prosody in meaningless speech. Nature Neuroscience 8:145–146

Grossmann T (2008) Shedding light on infant brain function: The use of near-infrared spectroscopy (NIRS) in the study of face perception. Acta Paediatrica 97:1156–1158

Grossmann T, Hughes DA, Stoneking M, & Friederici AD (in preparation) Individual differences in emotional voice processing in infancy: Insights from studying genetic variation in neurotransmitter systems

Grossmann T, Johnson MH (2007) The development of the social brain in infancy. The European Journal of Neuroscience 25:909–919

Grossmann T, Johnson MH, Vaish A, Hughes D, Quinque D, Stoneking M et al (2011) Genetic and neural dissociation of individual responses to emotional expressions in human infants. Developmental Cognitive Neuroscience 1:57–66

Grossmann T, Oberecker R, Koch SP, Friederici AD (2010) Developmental origins of voice processing in the human brain. Neuron 65:852–858

Grossmann T, Striano T, Friederici AD (2005) Infants' electric brain responses to emotional prosody. Neuroreport 16:1825–1828

Grossmann T, Striano T, Friederici AD (2006) Crossmodal integration of emotional information from face and voice in the infant brain. Developmental Science 9:309–315

Grossmann T, Striano T, Friederici AD (2007) Developmental changes in infants' processing of happy and angry facial expressions: A neurobehavioral study. Brain and Cognition 64:30–41

Gwiazda J, Bauer J, Held R (1989) From visual acuity to hyperacuity: A 10-year update. Canadian Journal of Psychology 43:109–120

Hainline L, Abramov I (1992) Assessing visual development: Is infant vision good enough? In: Rovee-Collier C, Lipsitt LP (eds) Advances in infancy research. Ablex, Norwood, NJ, pp 30–102
Hariri AR, Mattay VS, Tessitore A, Kolachana B, Fera F, Goldman D et al (2002) Serotonin transporter genetic variation and the response of the human amygdala. Science 297:400–403
Harrison PJ, Weinberger DR (2005) Schizophrenia genes, gene expression, and neuropathology: On the matter of their convergence. Molecular Psychiatry 10:40–68
Haviland JM, Lelwica (1987) The induced affect response: 10-week-old infants' responses to three emotion expression. Developmental Psychology 23:97–104
Heinz A, Smolka MN (2006) The effects of catechol O-methyltransferase on brain activity elicited by affective stimuli and cognitive tasks. Reviews in the Neurosciences 17:359–367
Herrmann MJ, Huter T, Müller F, Mühlberger A, Pauli P, Reif A et al (2006) Additive effects of serotonin transporter and tryptophan hydroxylase-2 gene variation on emotional processing. Cerebral Cortex 17:1160–1163
Herrmann MJ, Würflein H, Schreppel T, Koehler S, Mühlberger A, Reif A et al (2009) Catechol-O-methyltransferase val$^{108/158}$met genotype affects neural correlates of aversive stimuli processing. Cognitive, Affective, & Behavioral Neuroscience 9:168–172
Holmboe K, Nemoda Z, Fearon RMP, Csibra G, Sasvari-Szekely M, Johnson MH (2010) Polymorphisms in dopamine system genes associated with individual differences in attention in infancy. Developmental Psychology 46:404–416
Hutt SJ, Hutt C, Leonard HG, von Bermuth H, Muntjewerff WF (1968) Auditory responsitivity in the human neonate. Nature 218:888–890
Illingworth RS (1983) The development of the infant and young child: Normal and abnormal. Churchill Livingstone, New York
Johnson MH (2001) Functional brain development in humans. Nature Reviews Neuroscience 2:475–483
Johnson MH, Dziurawiec S, Ellis HD, Morton J (1991) Newborns' Preferential tracking of face-like stimuli and its subsequent decline. Cognition 40:1–21
Kahana-Kalman R, Walker-Andrews AS (2001) The role of person familiarity in young infants' perception of emotional expression. Child Development 72:352–369
Karmiloff-Smith A (1998) Development itself is the key to understanding developmental disorders. Trends in Cognitive Sciences 2:389–398
Kesler-West ML, Andersen AH, Smith CD, Avison MJ, Davis CE, Kryscio RJ et al (2001) Neural substrates of facial emotion processing using fMRI. Cognitive Brain Research 11:213–226
Kobiella A, Grossmann T, Striano T, Reid VM (2008) The discrimination of angry and fearful facial expressions in 7-month-old infants: An event-related potential study. Cognition & Emotion 22:134–146
Lesch KP, Bengel D, Heils A, Sabol SZ, Greenberg BD, Petri S et al (1996) Association of anxiety-related traits with a polymorphism in the serotonin transporter gene regulatory region. Science 274:1527–1531
Lewkowicz DJ (1988a) Sensory dominance in infants 1: Six-month-old infants' response to auditory-visual compounds. Developmental Psychology 24:155–171
Lewkowicz DJ (1988b) Sensory dominance in infants 2: Ten-month-old infants' response to auditory-visual compounds. Developmental Psychology 24:172–182
Lloyd-Fox S, Blasi A, Elwell CE (2010) Illuminating the developing brain: The past, present and future of functional near-infrared spectroscopy. Neuroscience and Biobehavioral Reviews 34:269–284
Ludemann PM, Nelson CA (1988) The categorical representation of facial expressions by 7-month-old infants. Developmental Psychology 24:492–501
Malatesta CZ, Haviland JM (1982) Learning display rules: The socialization of emotion expression in infancy. Child Development 53:991–1003
Meyer-Lindenberg A, Weinberger DR (2006) Intermediate phenotypes and genetic mechanisms of psychiatric disorder. Nature Reviews Neuroscience 7:818–827

Mitchell RLC, Elliott R, Barry M, Cruttenden A, Woodruff PWR (2003) The neural response to emotional prosody, as revealed by functional magnetic resonance imaging. Neuropsychologia 41:1410–1421

Morris RW, Weickert CS, Loughland CM (2009) Emotional face processing in schizophrenia. Current Opinion in Psychiatry 22:140–146

Nelson CA (1987) The recognition of facial expressions in the first year of life: Mechanisms of development. Child Development 56:58–61

Nelson CA, de Haan M (1996) Neural correlates of infants' visual responsiveness to facial expression of emotion. Developmental Psychobiology 29:577–595

Nelson CA, Henschel M, Collins PF (1993) Neural correlates of crossmodal recognition memory by 8-month-old human infants. Developmental Psychology 29:411–420

Nelson CA, Thomas KM, de Haan M, Wewerka S (1998) Delayed recognition memory in infants and adults as revealed by event-related potentials. International Journal of Psychophysiology 29:145–165

Peltola MJ, Leppänen JM, Mäki S, Hietanen JK (2009) Emergence of enhanced attention to fearful faces between 5 and 7 months of age. Social Cognitive and Affective Neuroscience 4:134–142

Pezawas L, Meyer-Lindenberg A, Drabant E, Verchinski BA, Munoz KE, Kolachana BS et al (2005) 5-HTTLPR polymorphism impacts human cingulate-amygdala interactions: A genetic susceptibility mechanism for depression. Nature Neuroscience 8:828–834

Pollak SD, Cicchetti D, Klorman R, Brumaghim J (1997) Cognitive brain event-related potentials and emotion processing in maltreated children. Child Development 68:773–787

Pollak SD, Klorman R, Thatcher JE, Cicchetti D (2001) P3b reflects maltreated children's reactions to facial displays of emotion. Psychophysiology 38:267–274

Purhonen M, Kilpeläinen-Lees R, Valkonen-Korhonen M, Karhu J, Lehtonen J (2004) Cerebral processing of mother's voice compared to unfamiliar voice in 4-month-old infants. International Journal of Psychophysiology 52:257–266

Richards JE (2002) The development of visual attention and the brain. In: de Haan M, Johnson MH (eds) The cognitive neuroscience of development. Psychology Press, Hove, UK, pp 73–98

Rizzolatti G, Craighero L (2004) The mirror-neuron system. Annual Review of Neuroscience 27:169–192

Rowe C (1999) Receiver psychology and the evolution of multicomponent signals. Animal Behaviour 58:921–931

Rugg MD, Coles MGH (1995) Electrophysiology of mind: Event-related brain potentials and cognition. Oxford University Press, Oxford

Russell J (1980) A circumplex model of affect. Journal of Personality and Social Psychology 39: 1161–1178

Russell J (1983) Dimensions underlying children's emotion-concepts. Developmental Psychology 19:795–804

Russell J, Bullock M (1986) Fuzzy concepts and the perception of emotion in facial expressions. Social Cognition 4:309–341

Schupp HT, Junghöfer M, Oehmann A, Weike AI, Stockburger J, Hamm AO (2004) The facilitated processing of threatening faces: An ERP analysis. Emotion 4:189–200

Schupp HT, Junghöfer M, Weike AI, Hamm AO (2003) Attention and emotion: An ERP analysis of facilitated stimulus processing. Neuroreport 14:1107–1110

Singh L, Morgan JL, White KS (2004) Preference and processing: The role of speech affect in early spoken word recognition. Journal of Memory and Language 51:173–189

Smolka MN, Schumann G, Wrase J, Grusser SM, Flor H, Mann K et al (2005) Catechol-O-methyltransferase val158met genotype affects processing of emotional stimuli in the amygdala and prefrontal cortex. Journal of Neuroscience 25:836–842

Soken, NH, Pick AD (1992) Intermodal perception of happy and angry expressive behaviors by seven-month-old infants. Child Development 63:787–795

Sommerville JA, Woodward AL, Needham AN (2005) Action experience alters 3-month-old infants' perception of other's actions. Cognition 96:1–11

Sprangler G, Johann M, Ronai Z, Zimmermann P (2009) Genetic and environmental influence on attachment disorganization. Journal of Child Psychology and Psychiatry 50:952–961

Stein BE, Meredith MA, Wallace MT (1993) Development and neural basis of multisensory integration. In: Lewkowicz DJ, Lickliter R (eds) The development of intersensory perception. Erlbaum, Hillsdale, NJ, pp 81–106

Strangman G, Culver JP, Thompson JH, Boas DA (2002) A quantitative comparison of simultaneous BOLD fMRI and NIRS recordings during functional brain activation. NeuroImage 17: 719–731

Thomas KM, Drevets WC, Whalen PJ, Eccard CH, Dahl RE, Ryan ND et al (2001) Amygdala response to facial expressions in children and adults. Biological Psychiatry 49:309–316

Turkewitz G, Birch HG, Cooper KK (1972) Responsiveness to simple and complex auditory stimuli in the human newborn. Developmental Psychobiology 5:7–19

Turkewitz G, Devenny DA (1993) Developmental time and timing. Erlbaum, Hillsdale, NJ

Vaish A, Grossmann T, Woodward A (2008) Not all emotions are created equal: The negativity bias in social-emotional development. Psychological Bulletin 134:383–403

Vaish A, Striano T (2004) Is visual reference necessary? Vocal versus facial cues in social referencing. Developmental Science 7:261–269

Vuilleumier P (2006) How brains beware: Neural mechanisms of emotional attention. Trends in Cognitive Sciences 9:585–594

Vuilleumier P, Armony JL, Driver J, Dolan RJ (2001) Effects of attention and emotion on face processing in the human brain: An event-related fMRI study. Neuron 30:829–841

Walker AS (1982) Intermodal perception of expressive behaviors by human infants. Journal of Experimental Child Psychology 33:514–535

Walker-Andrews AS (1986) Intermodal perception of expressive behaviors: Relation of eye and voice? Developmental Psychology 22:373–377

Walker-Andrews AS (1997) Infants' perception of expressive behaviors: Differentiation of multimodal information. Psychological Bulletin 121:1–20

Part II
Identity Information

Chapter 6
Audiovisual Integration in Speaker Identification

Stefan R. Schweinberger

Abstract Audiovisual integration (AVI) is well-known during speech perception, but evidence for AVI in speaker identification has been less clear. This chapter reviews evidence for face–voice integration in speaker identification. Links between perceptual representations mediating face and voice identification, tentatively suggested by behavioral evidence more than a decade ago, have been recently supported by neuroimaging data indicating tight functional connectivity between the fusiform face and temporal voice areas. Research that recombined dynamic facial and vocal identities with precise synchrony provided strong evidence for AVI in identifying personally familiar (but not unfamiliar) speakers. Electrophysiological data demonstrate AVI at multiple neuronal levels and suggest that perceiving time-synchronized speaking faces triggers early (~50–80 ms) audiovisual processing, although audiovisual speaker identity is only computed ~200 ms later.

1 Introduction

Just like speech perception or the recognition of others' emotional state, the recognition and identification of people is an extremely important everyday ability for human social functioning. Yet, it can be argued that whereas audiovisual integration (AVI) is now acknowledged to be important for the perception of speech and emotion, a role of audiovisual processing for speaker recognition is considered by very few researchers, or such a role is thought to be minor at best (Bruce & Young, 2011). One reason for this may be that person recognition is often thought to depend heavily on the face, with only a minor contribution of the voice (e.g., Bruce & Young, 1986; Walker, Bruce, & O'Malley, 1995). In fact, this is not unlike the situation in

S.R. Schweinberger (✉)
Department of General Psychology and Cognitive Neuroscience,
DFG Research Unit Person Perception, Friedrich-Schiller-University of Jena,
Am Steiger 3/1, 07743 Jena, Germany
e-mail: stefan.schweinberger@uni-jena.de

P. Belin et al. (eds.), *Integrating Face and Voice in Person Perception*,
DOI 10.1007/978-1-4614-3585-3_6, © Springer Science+Business Media New York 2013

speech perception at the time before the seminal paper by McGurk and Macdonald (1976) was published: speech perception had been typically regarded as a purely auditory process, before McGurk and Macdonald demonstrated a surprising and involuntary contribution of visual information from the articulating face.

Of course we are able to recognize speech, or person identity, from one modality alone—otherwise, we would not be able to talk to a person over the telephone, or to recognize a celebrity depicted in a magazine. But despite the fact that the vast majority of research on person recognition has been conducted on faces, speaker recognition from the voice has been studied experimentally for more than 50 years (Bricker & Pruzansky, 1966; Pollack, Pickett, & Sumby, 1954). A while ago now, it has been suggested that the processes involved in voice recognition are organized in a broadly similar manner to the ones involved in face recognition (Ellis, Jones, & Mosdell, 1997; Schweinberger, Herholz, & Sommer, 1997; for a review, see Belin, Fecteau, & Bedard, 2004). Neuropsychological evidence has demonstrated that deficits can occur in voice recognition (VanLancker & Kreiman, 1987), and some such impairments can be surprisingly selective (Garrido et al., 2009; Neuner & Schweinberger, 2000). However, a major increase in research interest in voice recognition seems to have occurred only after the first identification of voice-selective areas in the human temporal cortex (Belin, Zatorre, Lafaille, Ahad, & Pike, 2000).

Of course, the investigation of person recognition via unimodal stimuli (e.g., static portraits of faces, or voice samples) is an easy and convenient procedure. Unfortunately, its ecological validity is rather limited. Think of a typical communication situation: here we experience other people as dynamic and multimodal stimuli, and the auditory signal from the voice is closely linked in space and time to the dynamic visual signal from the speaking face. Regularities in face–voice co-occurrence exist in speech (e.g., which dynamic mouth shapes, or "visemes," correspond with which acoustic phonemes), but they also exist in person identification (i.e., which facial identity corresponds to which voice quality). Because of the systematic nature of this correspondence, one might expect that our brain should be able to efficiently process these multimodal signals of speaker identity. The purpose of this chapter is to evaluate the degree to which this is the case.

The traditional view of the sensory brain holds that multimodal integration occurs only after extensive unimodal processing in unisensory brain areas. Similarly, traditional models of person perception (those in the tradition of Bruce & Young, 1986) hold that face–voice integration, if it occurs at all, is limited to late postperceptual processing stages. For instance, the influential interactive activation and competition (IAC) model (Burton, Bruce, & Johnston, 1990) assumes that information from the face and the voice first converges at a postperceptual stage of the so-called person identity node (PIN), at which the access to a person's identity and semantic information becomes available. By contrast, no face–voice integration was assumed at earlier stages of perceptual processing of faces and voices.

These views have recently become more controversial. First, newer data are interpreted to the effect that multisensory integration occurs very early, and at many cortical levels, including those traditionally thought of as unisensory areas (Ghazanfar & Schroeder, 2006). Second, as I hope will become clearer in this chapter, there is new and convincing evidence for perceptual integration of faces and voices in speaker recognition before the PIN level.

Audiovisual integration of faces and voices is best established in speech perception. For instance, seeing the speaking face while listening to someone provides a considerable facilitation in auditory speech comprehension, broadly equivalent to improving the signal-to-noise ratio by 10–15 dB (Summerfield, MacLeod, McGrath, & Brooke, 1989). In the McGurk effect, when subjects are presented with an auditory syllable (e.g. /ba/), synchronized with a face articulating a different visual syllable (e.g., /ga/), they report hearing yet a different syllable (e.g., /da/ or /tha/). AVI in speech perception is often thought to occur at a precategorical level of early perceptual processing. For example, neurophysiological recordings suggest that the brain detects an incongruence between auditory and visual speech within the first 100–200 ms (Saint-Amour, De Sanctis, Molholm, Ritter, & Foxe, 2007; van Wassenhove, Grant, & Poeppel, 2005), and that the effect is relatively independent of voluntary control (Green, Kuhl, Meltzoff, & Stevens, 1991; van Wassenhove et al., 2005; but see Soto-Faraco & Alsius, 2009, for qualifications).

Audiovisual integration is much less well investigated for the perception of paralinguistic information. It may also be noted that even fewer studies used dynamic bimodal stimuli. Many studies on crossmodal recognition of emotional expression in particular combined voices with *static* pictures of faces (e.g., de Gelder & Vroomen, 2000; Hagan et al., 2009). An obvious limitation of this design is that it does not provide the brain with audiovisual stimuli exhibiting the temporal correspondence of a face and a voice as it occurs in the natural environment. In contrast, and in line with substantial evidence (e.g., Calvert, Brammer, & Iversen, 1998; Welch & Warren, 1980), I consider a relatively precise temporal correspondence of matching *dynamic* visual and auditory events as a major determinant for AVI to occur. That being said, audiovisual synchrony apparently does not need to be absolutely perfect in order to elicit the McGurk effect, and current evidence suggests a short temporal window of AVI lasting up to a few hundred milliseconds (Munhall, Gribble, Sacco, & Ward, 1996). The integration window has been suggested to show a degree of flexibility to respond to small regular asynchronies in recent exposure (Navarra et al., 2005), and successful attempts were made to model the integration window (Colonius, Diederich, & Steenken, 2009).

2 Early Indirect Evidence for Face–Voice Integration in Speaker Recognition

While the most influential models of face perception were quite explicit in claiming that information from the face and the voice is only combined in postperceptual processing stages, and that “a face recognition unit will respond when any view of the appropriate person’s *face* is seen, but will not respond at all to his or her voice or name” (Bruce & Young, 1986, p. 312), some older evidence tentatively suggested a link between the perceptual representations that mediated the recognition of faces and voices. First, it has been found that the simultaneous presentation of a face facilitates voice recognition (Legge, Grossmann, & Pieper, 1984). Second, an encounter with a famous person's face (but not with that person's name) caused a degree of long-term repetition priming for the later recognition of that famous person's voice (Schweinberger, Herholz, & Stief, 1997). Because long-term repetition priming effects usually exhibit strong domain-specificity, this finding suggests that famous faces (but not names) might have been able to activate a perceptual representation of the respective celebrity's voice.[1] Third, it has been shown that audiovisual speech facilitates voice learning, relative to a purely auditory learning condition (Sheffert & Olson, 2004).

3 Neural Mechanisms

The neural systems for face perception and voice perception and recognition, as investigated with fMRI studies, have been described elsewhere in detail (Belin, Bestelmeyer, Latinus, & Watson, 2011; Haxby, Hoffman, & Gobbini, 2000; Natu & O'Toole, 2011). Although it remains possible in principle that audiovisual face–voice integration involves specific neural convergence zones, or "integration" areas, a more parsimonious assumption is that audiovisual face–voice integration is mediated by direct connections between face-selective and voice-selective areas. In a recent review, Campanella and Belin (2007) advocate a model according to which at least three different mechanisms of face–voice integration can be distinguished which mediate crossmodal speech perception, emotion recognition, and speaker identification, respectively. It is important to note that although each of these integrative mechanisms is thought to be mediated by different neural systems, each operates at a perceptual, presemantic stage of analysis of the stimulus. From that point of view, AVI in speaker identification could be expected to be mediated via an interaction between the bilateral or right fusiform face area (FFA) and a bilateral or right temporal voice area (TVA).

Although using auditory stimulation only, the results from an fMRI study by von Kriegstein, Kleinschmidt, Sterzer, and Giraud (2005) were broadly in line with that hypothesis, in that familiar voices on their own were able to activate the FFA. Here, it is important to note that another region in posterior cingulate cortex including retrosplenial cortex has been implicated as a neural site for processing person familiarity (i.e., the PIN stage in the model by Burton et al., 1990), independent of stimulus modality (Shah et al., 2001; but also see Sugiura, Shah, Zilles, & Fink, 2005). In the study by von Kriegstein et al. (2005), functional connectivity analysis showed that the TVA and the

[1] Like many other studies, it needs to be noted that this experiment used static faces. On the one hand, the study is therefore subject to the limitations mentioned earlier; on the other hand, this might be further evidence that even static faces can elicit some crossmodal effects (Joassin, Maurage, Bruyer, Crommelinck, & Campanella, 2004; Joassin et al., 2011).

FFA were tightly coupled. However, as fMRI does not allow to determine the precise time course of neural activation, and because TVA activation was also tightly coupled with retrosplenial activation, it may be difficult to reject the possibility that postperceptual processing at the PIN stage at least partly contributed to these findings.

4 Face–Voice Integration from Corresponding and Noncorresponding Identities

In a recent series of experiments, my colleagues and I sought direct experimental evidence for face–voice integration in speaker identification. To this end, we developed a paradigm that would allow us to combine dynamic faces and voices from real speakers, with precise time synchrony. We decided to take what we considered to be the most naturalistic approach while still exerting a precise control over the stimuli. One of our hypotheses when we started this project was that face–voice integration for speaker identity would likely depend on prior perceptual experience with the systematic correspondence of a specific face and a voice, and would thus depend strongly on speaker familiarity. Working in a Psychology Department teaching a large number of new undergraduates each year, we decided to use the professors of our student participants as personally familiar speakers; an equal number of matched professors from other departments were used as unfamiliar speakers. An additional advantage of this was that it was straightforward to both define and homogenize the degree of audiovisual exposure our undergraduates had with a speaker. Our criterion for inclusion typically was that a participant had attended at least one course for one full term (equivalent to 12–14 90-min sessions of lecturing).

Since earlier research had indicated that continuous speech in the region of 1.5–2 s is required to recognize a familiar voice with reasonable accuracy (Schweinberger, Herholz, & Sommer, 1997), we decided to record full-sentence stimuli approximating, or slightly exceeding, that duration. We then edited the video and audio signals towards a complete time-standardization defined by the population average, such that each face could be recombined with each voice with precise synchrony (see Fig. 6.1 for a schematic example; some stimulus examples are available at http://www2.uni-jena.de/svw/Allgpsy1/avi.htm). Although time-standardization was of course indispensable for our paradigm, it should be kept in mind that this largely eliminated temporal aspects of speech as a source of speaker identification. This is clearly a limitation in principle, but one we considered as acceptable. This was because the temporal aspects of speech may be of relatively minor importance for speaker identification when compared to voice quality (Bricker & Pruzansky, 1966), and because even substantial alterations (33 %) in speaking rate seems to have an only moderate, though significant, effect on voice identification (VanLancker, Kreiman, & Wickens, 1985). It may be noted that none of the listeners reported anything unusual in the voice samples of these familiar speakers.

In our first experiment, we asked undergraduate participants to classify a professor's *voice* as familiar or unfamiliar. The voice was either presented alone (voice-only

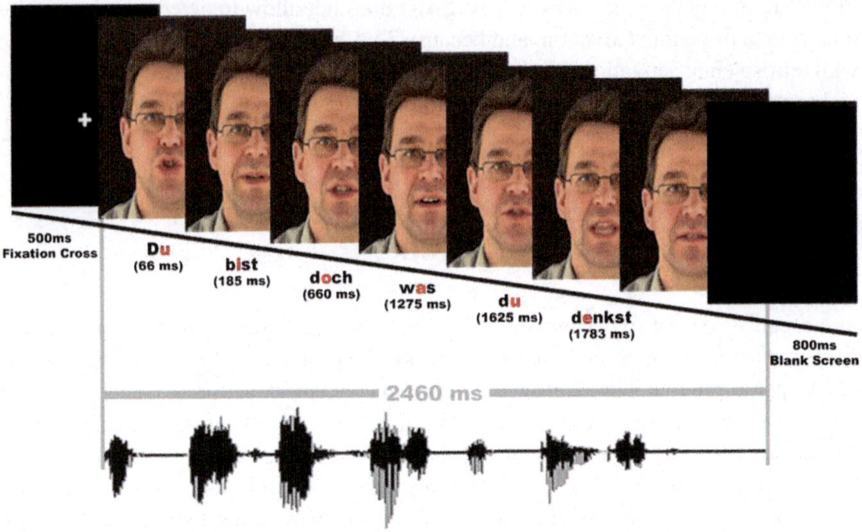

Fig. 6.1 A schematic example of a typical dynamic audiovisual trial, with vowel onset (VO) timings in parentheses. Figure reproduced from Schweinberger, Kloth, and Robertson (2011), Cortex. Reprinted with permission from Elsevier (license no. 2798351198825)

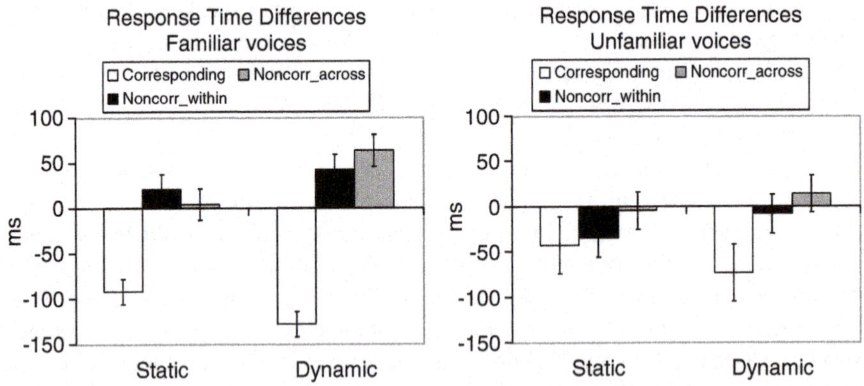

Fig. 6.2 *Left*: Reaction time differences relative to the voice only baseline condition (in ms) for voice recognition responses when familiar voices were combined with corresponding faces, with different noncorresponding faces of the same familiarity level (i.e., familiar faces, *NonCorr within*), or with different noncorresponding faces of the opposite familiarity level (i.e., unfamiliar faces, *NonCorr_across*) as the voice, separate for static and dynamic presentations of the face. Negative values indicate RT benefits, positive values indicate costs. *Top right*: Same for unfamiliar voices. Data from Schweinberger et al. (2007), The Quarterly Journal of Experimental Psychology. Reprinted with permission from Taylor & Francis (license no. 2798370863042)

baseline) or in combination with static or dynamic facial videos. Moreover, a face could show either a corresponding or a noncorresponding identity to the voice. Figure 6.2 shows results from our first experiment (Schweinberger, Robertson, & Kaufmann, 2007), plotted as response time (RT) benefits and costs relative to a voice-only baseline. In short, a corresponding facial identity facilitated RT, and this facilitation was significantly larger for a dynamic and synchronized face than for a static face. Moreover, a noncorresponding face (of a different identity to the voice) interfered with RT in voice recognition only when the face was presented dynamically. It therefore seems that observers could ignore noncorresponding static faces, but were unable to ignore synchronously articulating faces when recognizing voices. Finally, this pattern was much reduced or eliminated in the case of unfamiliar speakers. Thus the provisional conclusion by Schweinberger et al. (2007) was that AVI effects in speaker identification strongly depended on familiarity with a speaker, enabling the brain to "know" which facial identity corresponds to which voice quality. We also concluded that the AVI effects observed depended on the specific temporal correspondence of the facial and vocal signals, and not just on the combination of a voice and a face, or the fact that the face was moving per se. That latter conclusion was somewhat tentative, and was based on the fact that although a static corresponding face also caused some RT benefit to voice recognition (albeit smaller than a dynamically synchronized face did), a static noncorresponding face did not cause any RT cost. It therefore remained possible that the identification of a static face could still be used as a semantic "cue" to voice recognition in that experiment, particularly in view of evidence that person identification from the face is much faster compared to the voice (Hanley, Smith, & Hadfield, 1998; Schweinberger, Herholz, & Stief, 1997). We also considered that there is very little evidence that motion is a potent cue per se for face recognition, unless the face is shown strongly degraded (Lander & Chuang, 2005). Nevertheless, it was clearly desirable to study the role of audiovisual asynchrony for AVI in speaker identification in more detail.

Of relevance for that purpose, the role of asynchrony for AVI in speech perception has been studied by measuring the McGurk effect while manipulating audiovisual asynchrony in small steps from auditory-lead to auditory-lag, relative to the facial articulation (Munhall et al., 1996; van Wassenhove, Grant, & Poeppel, 2007). With a good degree of overlap, these studies suggested that (1) the time window for most efficient AVI in speech perception lasts between about 200 and 300 ms, and that (2) best integration does not actually occur at perfect synchrony, but rather when the voice slightly lags behind the facial movement. As an aside, that finding might be speculated to reflect an adaptation of the perceptual system to different velocities of light and sound in the physical world (Schweinberger, 1996), considering that the resolution of the visual system permits us to perceive broad articulatory mouth movements over distances of 30–50 m. From those studies, the window of integration was estimated to last approximately from −50 to +200 ms (Fig. 6.3).

The comparison between synchronized and static faces (Schweinberger et al., 2007) was clearly unsatisfactory for assessing the role of asynchrony in more detail. Accordingly, Robertson and Schweinberger (2010) performed an experiment on speaker recognition, in which they used similar stimuli as Schweinberger et al. (2007) while systematically manipulating audiovisual asynchrony as in the abovementioned studies on the McGurk effect. The most important finding from this study was that the benefit in terms of shorter RTs elicited by a corresponding face depended strongly on synchrony, and recognition was fastest at a small auditory lag of

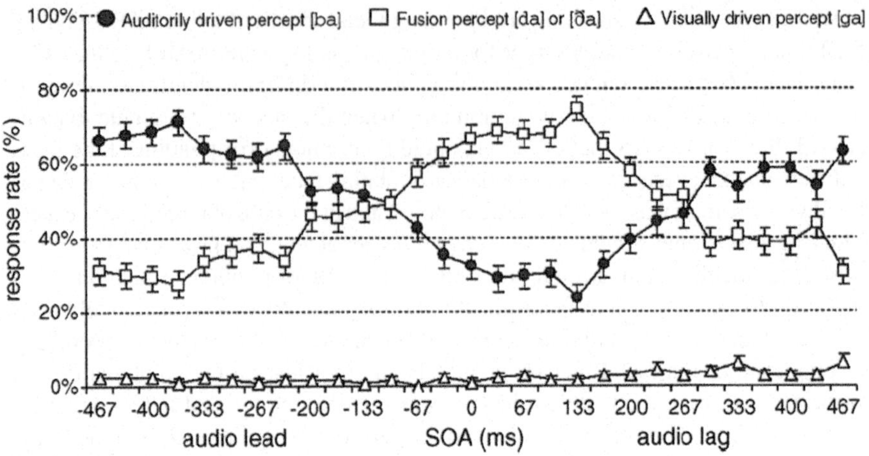

Fig. 6.3 Response categories as a function of audiovisual asynchrony (in ms) in the McGurk illusion for an auditory /ba/ and visual /ga/ pair. Note that integration in terms of fusion responses (/da/ or /ða/) are maximal at a slight auditory lag, and auditorily driven responses are reduced in parallel. Data from van Wassenhove et al. (2007), Neuropsychologia. Reprinted with permission from Elsevier (license no. 2798370283436)

+100 ms, with an approximate window of integration between −100 and +300 ms. Importantly, no such effects were seen for unfamiliar speakers (see Fig. 6.4). Somewhat unexpectedly, no costs from noncorresponding speakers were seen in RT in this study, although such costs were reported in response accuracy, at the same asynchrony around +100 ms auditory lag. Overall then, the study by Robertson and Schweinberger (2010) provides initial evidence about the role of audiovisual asynchrony. As in speech perception, largest AVI effects tend to occur at a small auditory lag. Thus, the time window of AVI in person recognition seems qualitatively similar to what has been reported for AVI in speech perception. Quantitatively, it appears that the window of integration for person recognition may be somewhat wider (i.e., in the region of 400 ms). As a limitation, a direct comparison between AVI time windows in speech perception vs. speaker identification remains to be done.

More recently, we have measured event-related brain potentials (ERPs) in the audiovisual speaker identification paradigm (Schweinberger, Kloth, & Robertson, 2011). Like in earlier related studies on the McGurk effect (Saint-Amour et al., 2007; Sams et al., 1991; van Wassenhove et al., 2005) we reasoned that if AVI took place at early perceptual processing, then ERPs with their very high time resolution would be much more informative than performance measures or functional magnetic resonance imaging (fMRI) data (with their comparatively poor time resolution). It has also been suggested that bimodal stimulation speeds up neural processing in speech recognition (van Wassenhove et al., 2005), so it would be interesting to see whether similar findings would be obtained in speaker identification. However, an ERP experiment required a few modifications to our design. First, we had to add a second unimodal condition (face only). This was because it is often argued that

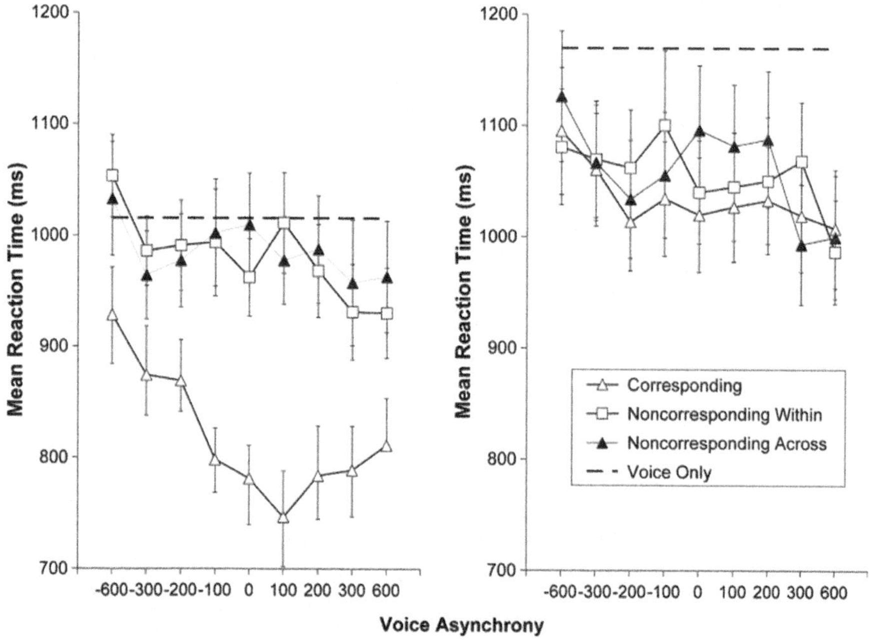

Fig. 6.4 *Left*: Mean RTs in ms for familiar voices when combined with faces of corresponding (matching) or noncorresponding identities. Asynchronies range from 600 ms auditory lead (−600) to 600 ms auditory lag (600). *Right*: Same for unfamiliar voices. The unimodal voice-only baseline is represented by a *dotted line*. Data from Robertson and Schweinberger (2010), The Quarterly Journal of Experimental Psychology. Reprinted with permission from Taylor & Francis (license no. 2798371106587)

AVI is demonstrated when a response to bimodal stimulation is larger than the sum of the responses to the same stimuli when presented in either modality alone—the criterion of superadditivity (Hagan et al., 2009; Stein & Stanford, 2008), and because we wanted to be in a position to test for that possibility. Second, we changed our speaker familiarity task to a speaker identification task, which enabled us to use familiar speakers only, and thus to maximize the signal-to-noise ratio which depends on the number of trials available for ERP averaging.

The analysis of performance in this ERP study revealed the usual pattern of RT benefits (relative to a voice only baseline) from a corresponding face, and costs from a noncorresponding face. The most important finding from ERPs was that face–voice integration seemed to involve not just one, but several mechanisms at different points in time: first, at 50–80 ms after stimulus onset, AV presentation elicited an earlier frontocentral negativity compared to the added unimodal responses (Fig. 6.5a). This suggests that audiovisual presentation speeded up neural responses. It has been suggested recently that anticipatory visual motion may be crucial for the modulation of early ERPs (Stekelenburg & Vroomen, 2007), and so it may be important to point out that the stimuli presented by Schweinberger, Kloth, and Robertson (2011) and Schweinberger, Walther, Zäske, and Kovacs (2011) did not

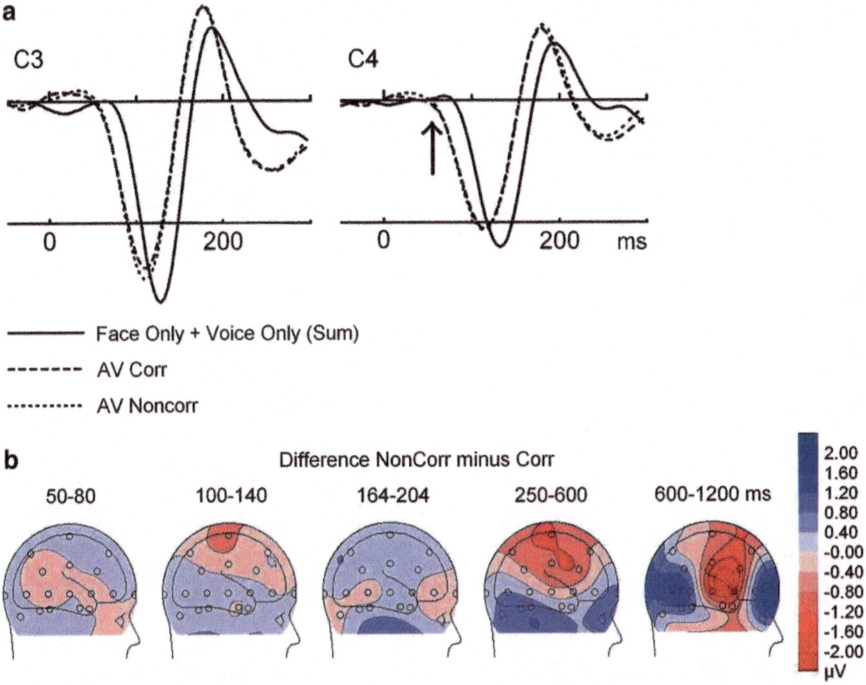

Fig. 6.5 (**a**) *Top*: Independent of speaker correspondence, a frontocentral negativity to audiovisual stimuli emerges around 50–80 ms (*arrow*), substantially earlier when compared to the algebraically summed ERPs to the same individual unimodal stimuli. (**b**) *Bottom*: Scalp voltage maps of the correspondence effect difference between AV noncorresponding minus AV corresponding condition. Reliable correspondence effects do not emerge before 250 ms. Around 250 ms, a central negativity is seen for noncorresponding pairs, and this negativity increases and shifts to a right frontotemporal maximum between 600 and 1,200 ms. Figure modified from Schweinberger, Kloth, and Robertson (2011). Reprinted with permission from Elsevier (license no. 2798351198825)

contain anticipatory visual motion, and that visual and auditory stimulus onset was identical. Second, we observed a surprising enhancement of the face-sensitive N170 component (164–204 ms) in the AV conditions relative to the face only condition. Importantly, these two earlier effects were completely independent of face–voice identity correspondence. Third, the earliest point in time when the neural response seen in ERPs picked up face–voice identity correspondence was around 250 ms. This happens to be similar in latency to the face-sensitive N250r component, which may be the earliest consistent ERP correlate of familiar face recognition (Schweinberger, Pickering, Jentzsch, Burton, & Kaufmann, 2002; see Schweinberger, 2011, for a recent review). This latency also broadly corresponds to the earliest ERP modulations of voice identity processing independent of speech content (Schweinberger, Walther, Zäske, & Kovacs, 2011). Between 250 and 600 ms, noncorresponding face–voice combinations elicited a larger central negativity than corresponding face–voice combinations, and this effect was enhanced and shifted to a right frontotemporal maximum in the subsequent time segment between 600 and 1,200 ms (Fig. 6.5b).

Our findings therefore partially confirm, and substantially extend, recent suggestions of a face–voice integrative mechanism between around 150 and 200 ms (Charest et al., 2009). Importantly, the first effects of speaker correspondence occurred only after ~250 ms, such that we concluded that AVI in speaker identification may require more time compared to AVI in speech perception, where congruence effects are seen substantially earlier, after 100–200 ms (Sams et al., 1991; van Wassenhove et al., 2005).[2] The lateralization of correspondence effects observed by Schweinberger, Kloth, and Robertson (2011) and Schweinberger, Walther, Zäske, and Kovacs (2011) tentatively suggests a greater role of the right hemisphere for AVI in person identification, and which of course would be in keeping with the assumption that the right FFA and the right TVA play a greater role in face and voice recognition respectively, compared to their left hemisphere counterparts.

5 Crossmodal Adaptation

Although of more indirect relevance to the question of audiovisual face–voice integration, I briefly consider recent research in the field of high-level perceptual adaptation. While adaptation to simple stimulus attributes such as color or motion has long been known to elicit contrastive aftereffects, the demonstration of similar aftereffects in high-level face perception is a relatively recent discovery, and one that arguably was boosted by the availability of sophisticated techniques of image manipulation such as image morphing, a technique first published around 20 years ago (Benson & Perrett, 1991). Systematic contrastive aftereffects were subsequently demonstrated for a range of facial signals including gender, expression, eye gaze, or even identity (Leopold, O'Toole, Vetter, & Blanz, 2001). Similarly suitable software for auditory morphing has been developed more recently (Kawahara & Matsui, 2003), and so the first study using this technique to show systematic effects of high-level adaptation to voice quality was only published in 2008 (Schweinberger et al., 2008). In that study, adaptation to unfamiliar female voices was shown to cause a subsequent voice to be perceived as more male, and vice versa. No such aftereffects could be elicited by adaptation to sinusoidal tones which were matched to the fundamental frequencies of female or male voices. Crucially, adaptation to silent videos of female or male speakers also failed to elicit any aftereffects to the perception of gender in subsequent voices.

[2] It could be speculated whether differences in timing might have been a consequence of the use of temporally extended sentence stimuli in Schweinberger, Kloth, and Robertson (2011) and Schweinberger, Walther, Zäske, and Kovacs (2011). However, in as yet unpublished research, we have now repeated the same experiment using brief syllabic stimuli similar to those used in the McGurk-paradigm, and replicated the crucial results, in terms of an early frontocentral negativity around 50–80 ms to bimodal stimuli, and an onset of speaker identity correspondence effects around 250 ms.

This absence of a crossmodal adaptation effect on voice perception is mirrored by findings from a few other studies that also failed to find crossmodal adaptation effects on the perception of facial emotion or gender (Fox & Barton, 2007; Kovács et al., 2006), and by findings demonstrating an insensitivity of the McGurk effect to cross-gender audiovisual speaker combinations (Green et al., 1991). On the one hand, it therefore seems possible that information about (unfamiliar) speaker gender may not be integrated across modalities; on the other hand, a number of other findings could be taken to suggest that information about familiar speakers is integrated across modalities (Schweinberger et al., 2007; von Kriegstein et al., 2005).

The hypothesis of crossmodal adaptation for familiar speakers was tested more specifically by Zäske, Schweinberger, and Kawahara (2010), in a study on voice identity adaptation to familiar voices. Experiment 1 of this study demonstrated that adaptation to speaker A's voice biased the perception of identity-ambiguous voice morphs between speakers A and B towards speaker B, and vice versa. Importantly, similar (though reduced) aftereffects in voice identity perception were also seen in Experiment 2, when adaptors were videos of speakers' silently articulating faces. Judging from earlier neuroimaging work (Belin & Zatorre, 2003), we tentatively assumed the right anterior TVA to be involved in voice identity adaptation. The novel finding from Zäske et al. (2010) was to establish a visual influence on voice adaptation, and although the precise mechanisms of this crossmodal influence remain to be determined, the ERP topography of the audiovisual correspondence effect observed by Schweinberger, Kloth, and Robertson (2011) and Schweinberger, Walther, Zäske, and Kovacs (2011); cf. Fig. 6.5b) is at least consistent with the idea that facial identity in speaking faces is able to modulate processing in the right TVA.

6 Conclusion and Outlook

Taken together, I would like to emphasize that a research program on audiovisual face–voice integration in speaker identification is both technically demanding, and in its early stages empirically. Nevertheless, some progress has already been made, such that we can conclude that (1) AVI is an important factor in the recognition of people, (2) AVI strongly depends on familiarity with a speaker, and (3) AVI shows sensitivity to temporal synchronization of the facial and vocal articulation, such that best integration occurs with a very slight auditory lag. Across neurophysiological studies, differences in timing, topography, and lateralization of effects also suggest that different mechanisms of face–voice integration are probably invoked for speaker vs. speech recognition.

Compared to around 15 years ago, we have made enormous progress in understanding the human neurocognitive system for face perception, and its various components in particular that subserve the processing of facial identity, emotion, facial speech, or gender. With the present research focus on the neurocognitive system for voice perception, we should soon be able to understand in more detail the components that subserve the auditory processing of speech, emotional prosody, voice identity, or gender, and promising progress has already been made here (Andics et al., 2010; Belin et al., 2011; Formisano, De Martino, Bonte, & Goebel, 2008). There can be little doubt that such knowledge will also facilitate the investigation and understanding of the mechanisms that mediate audiovisual face–voice integration for different social signals such as speech content, emotional expression, or speaker identity. Finally, since amazing technical progress in stimulus manipulation now enables us to present well-controlled and yet highly realistic dynamic audiovisual stimuli, researchers should take advantage of these methods to obtain the best and most ecologically valid results.

Acknowledgments The author's research is supported by grants from the Deutsche Forschungsgemeinschaft (Grants Schw 511/6-2 and Schw511/10-1) in the context of the DFG Research Unit Person Perception (FOR1097). I am very grateful to Romi Zäske for helpful comments on an earlier draft of this chapter.

References

Andics, A., McQueen, J. M., Petersson, K. M., Gal, V., Rudas, G., & Vidnyanszky, Z. (2010). Neural mechanisms for voice recognition. *NeuroImage, 52*, 1528–1540.

Belin, P., Bestelmeyer, P. E. G., Latinus, M., & Watson, R. (2011). Understanding voice perception. *British Journal of Psychology, 102*, 711–725.

Belin, P., Fecteau, S., & Bedard, C. (2004). Thinking the voice: Neural correlates of voice perception. *Trends in Cognitive Sciences, 8*, 129–135.

Belin, P., & Zatorre, R. J. (2003). Adaptation to speaker's voice in right anterior temporal lobe. *NeuroReport, 14*, 2105–2109.

Belin, P., Zatorre, R. J., Lafaille, P., Ahad, P., & Pike, B. (2000). Voice-selective areas in human auditory cortex. *Nature, 403*, 309–312.

Benson, P. J., & Perrett, D. I. (1991). Perception and recognition of photographic quality facial caricatures: Implications for the recognition of natural images. *European Journal of Cognitive Psychology, 3*, 105–135.

Bricker, P. D., & Pruzansky, S. (1966). Effects of stimulus content and duration on talker identification. *Journal of the Acoustical Society of America, 40*, 1441–1449.

Bruce, V., & Young, A. (1986). Understanding face recognition. *British Journal of Psychology, 77*, 305–327.

Bruce, V., & Young, A. (2011). *Face perception*. Hove, UK: Psychology Press.

Burton, A. M., Bruce, V., & Johnston, R. A. (1990). Understanding face recognition with an interactive activation model. *British Journal of Psychology, 81*, 361–380.

Calvert, G. A., Brammer, M. J., & Iversen, S. D. (1998). Crossmodal identification. *Trends in Cognitive Sciences, 2*, 247–253.

Campanella, S., & Belin, P. (2007). Integrating face and voice in person perception. *Trends in Cognitive Sciences, 11*, 535–543.

Charest, I., Pernet, C. R., Rousselet, G. A., Quinones, I., Latinus, M., Fillion-Bilodeau, S., et al. (2009). Electrophysiological evidence for an early processing of human voices. *BMC Neuroscience, 10*(127), 1–11.

Colonius, H., Diederich, A., & Steenken, R. (2009). Time-Window-of-Integration (TWIN) model for saccadic reaction time: Effect of auditory masker level on visual-auditory spatial interaction in elevation. *Brain Topography, 21*, 177–184.

de Gelder, B., & Vroomen, J. (2000). The perception of emotions by ear and by eye. *Cognition & Emotion, 14*, 289–311.

Ellis, H. D., Jones, D. M., & Mosdell, N. (1997). Intra- and inter-modal repetition priming of familiar faces and voices. *British Journal of Psychology, 88*, 143–156.

Formisano, E., De Martino, F., Bonte, M., & Goebel, R. (2008). "Who" Is Saying "What"? Brain-based decoding of human voice and speech. *Science, 322*, 970–973.

Fox, C. J., & Barton, J. J. S. (2007). What is adapted in face adaptation? The neural representations of expression in the human visual system. *Brain Research, 1127*, 80–89.

Garrido, L., Eisner, F., McGettigan, C., Stewart, L., Sauter, D., Hanley, J. R., et al. (2009). Developmental phonagnosia: A selective deficit of vocal identity recognition. *Neuropsychologia, 47*, 123–131.

Ghazanfar, A. A., & Schroeder, C. E. (2006). Is neocortex essentially multisensory? *Trends in Cognitive Sciences, 10*, 278–285.

Green, K. P., Kuhl, P. K., Meltzoff, A. N., & Stevens, E. B. (1991). Integration speech information across talkers, gender, and sensory modality: Female faces and male voices in the McGurk effect. *Perception & Psychophysics, 50*, 524–536.

Hagan, C. C., Woods, W., Johnson, S., Calder, A. J., Green, G. G. R., & Young, A. W. (2009). MEG demonstrates a supra-additive response to facial and vocal emotion in the right superior temporal sulcus. *Proceedings of the National Academy of Sciences of the United States of America, 106*, 20010–20015.

Hanley, J. R., Smith, S. T., & Hadfield, J. (1998). I recognise you but I can't place you: An investigation of familiar-only experiences during tests of voice and face recognition. *Quarterly Journal of Experimental Psychology, 51A*, 179–195.

Haxby, J. V., Hoffman, E. A., & Gobbini, M. I. (2000). The distributed human neural system for face perception. *Trends in Cognitive Sciences, 4*, 223–233.

Joassin, F., Maurage, P., Bruyer, R., Crommelinck, M., & Campanella, S. (2004). When audition alters vision: An event-related potential study of the cross-modal interactions between faces and voices. *Neuroscience Letters, 369*, 132–137.

Joassin, F., Pesenti, M., Maurage, P., Verreckt, E., Bruyer, R., & Campanella, S. (2011). Cross-modal interactions between human faces and voices involved in person recognition. *Cortex, 47*, 367–376.

Kawahara, H., & Matsui, H. (2003). Auditory morphing based on an elastic perceptual distance metric in an interference-free time-frequency representation. *IEEE Proceedings of ICASSP, 1*, 256–259.

Kovács, G., Zimmer, M., Banko, E., Harza, I., Antal, A., & Vidnyanszky, Z. (2006). Electrophysiological correlates of visual adaptation to faces and body parts in humans. *Cerebral Cortex, 16*, 742–753.

Lander, K., & Chuang, L. (2005). Why are moving faces easier to recognize? *Visual Cognition, 12*, 429–442.

Legge, G. E., Grossmann, C., & Pieper, C. M. (1984). Learning unfamiliar voices. *Journal of Experimental Psychology: Learning, Memory, and Cognition, 10*, 298–303.

Leopold, D. A., O'Toole, A. J., Vetter, T., & Blanz, V. (2001). Prototype-referenced shape encoding revealed by high-level aftereffects. *Nature Neuroscience, 4*, 89–94.

McGurk, H., & MacDonald, J. (1976). Hearing lips and seeing voices. *Nature, 264*, 746–748.

Munhall, K. G., Gribble, P., Sacco, L., & Ward, M. (1996). Temporal constraints on the McGurk effect. *Perception & Psychophysics, 58*, 351–362.

Natu, V., & O'Toole, A. J. (2011). The neural processing of familiar and unfamiliar faces: A review and synopsis. *British Journal of Psychology, 102*, 726–747.

Navarra, J., Vatakis, A., Zampini, M., Soto-Faraco, S., Humphreys, W., & Spence, C. (2005). Exposure to asynchronous audiovisual speech extends the temporal window for audiovisual integration. *Cognitive Brain Research, 25*, 499–507.

Neuner, F., & Schweinberger, S. R. (2000). Neuropsychological impairments in the recognition of faces, voices, and personal names. *Brain and Cognition, 44*, 342–366.

Pollack, I., Pickett, J. M., & Sumby, W. H. (1954). On the identification of speakers by voice. *Journal of the Acoustical Society of America, 26*, 403–406.

Robertson, D. M. C., & Schweinberger, S. R. (2010). The role of audiovisual asynchrony in person recognition. *Quarterly Journal of Experimental Psychology, 63*, 23–30.

Saint-Amour, D., De Sanctis, P., Molholm, S., Ritter, W., & Foxe, J. J. (2007). Seeing voices: High-density electrical mapping and source-analysis of the multisensory mismatch negativity evoked during the McGurk illusion. *Neuropsychologia, 45*, 587–597.

Sams, M., Aulanko, R., Hämalainen, M., Hari, R., Lounasmaa, O. V., Lu, S.-T., et al. (1991). Seeing speech: Visual information from lip movements modifies activity in the human auditory cortex. *Neuroscience Letters, 127*, 141–145.

Schweinberger, S. R. (1996). Recognizing people by faces, names, and voices: Psychophysiological and neuropsychological investigations. University of Konstanz: Habilitation Thesis.

Schweinberger, S. R. (2011). Neurophysiological correlates of face recognition. In A. J. Calder, G. Rhodes, M. H. Johnson, & J. V. Haxby (Eds.), *The handbook of face perception* (pp. 345–366). Oxford: Oxford University Press.

Schweinberger, S. R., Casper, C., Hauthal, N., Kaufmann, J. M., Kawahara, H., Kloth, N., et al. (2008). Auditory adaptation in voice perception. *Current Biology, 18*, 684–688.

Schweinberger, S. R., Herholz, A., & Sommer, W. (1997). Recognizing famous voices: Influence of stimulus duration and different types of retrieval cues. *Journal of Speech, Language, and Hearing Research, 40*, 453–463.

Schweinberger, S. R., Herholz, A., & Stief, V. (1997). Auditory long-term memory: Repetition priming of voice recognition. *Quarterly Journal of Experimental Psychology, 50A*, 498–517.

Schweinberger, S. R., Kloth, N., & Robertson, D. M. C. (2011). Hearing facial identities: Brain correlates of face-voice integration in person identification. *Cortex, 47*, 1026–1037.

Schweinberger, S. R., Pickering, E. C., Jentzsch, I., Burton, A. M., & Kaufmann, J. M. (2002). Event-related brain potential evidence for a response of inferior temporal cortex to familiar face repetitions. *Cognitive Brain Research, 14*, 398–409.

Schweinberger, S. R., Robertson, D., & Kaufmann, J. M. (2007). Hearing facial identities. *Quarterly Journal of Experimental Psychology, 60*, 1446–1456.

Schweinberger, S. R., Walther, C., Zäske, R., & Kovacs, G. (2011). Neural correlates of adaptation to voice identity. *British Journal of Psychology, 102*, 748–764.

Shah, N. J., Marshall, J. C., Zafiris, O., Schwab, A., Zilles, K., Markowitsch, H. J., et al. (2001). The neural correlates of person familiarity. A functional magnetic resonance imaging study with clinical implications. *Brain, 124*, 804–815.

Sheffert, S. M., & Olson, E. (2004). Audiovisual speech facilitates voice learning. *Perception & Psychophysics, 66*, 352–362.

Soto-Faraco, S., & Alsius, A. (2009). Deconstructing the McGurk–MacDonald illusion. *Journal of Experimental Psychology: Human Perception and Performance, 35*, 580–587.

Stein, B. E., & Stanford, T. R. (2008). Multisensory integration: Current issues from the perspective of the single neuron. *Nature Reviews Neuroscience, 9*, 255–266.

Stekelenburg, J. J., & Vroomen, J. (2007). Neural correlates of multisensory integration of ecologically valid audiovisual events. *Journal of Cognitive Neuroscience, 19*, 1964–1973.

Sugiura, M., Shah, N. J., Zilles, K., & Fink, G. R. (2005). Cortical representations of personally familiar objects and places: Functional organization of the human posterior cingulate cortex. *Journal of Cognitive Neuroscience, 17*, 183–198.

Summerfield, Q., MacLeod, A., McGrath, M., & Brooke, M. (1989). Lips, teeth, and the benefits of lipreading. In A. W. Young & H. D. Ellis (Eds.), *Handbook of research on face processing* (pp. 223–233). Amsterdam: North-Holland.

van Wassenhove, V., Grant, K. W., & Poeppel, D. (2005). Visual speech speeds up the neural processing of auditory speech. *Proceedings of the National Academy of Sciences of the United States of America, 102*, 1181–1186.

van Wassenhove, V., Grant, K. W., & Poeppel, D. (2007). Temporal window of integration in auditory-visual speech perception. *Neuropsychologia, 45*, 598–607.

VanLancker, D., & Kreiman, J. (1987). Voice discrimination and recognition are separate abilities. *Neuropsychologia, 25*, 829–834.

VanLancker, D., Kreiman, J., & Wickens, T. D. (1985). Familiar voice recognition: Patterns and parameters. Part II: Recognition of rate-altered voices. *Journal of Phonetics, 13*, 39–52.

von Kriegstein, K., Kleinschmidt, A., Sterzer, P., & Giraud, A. L. (2005). Interaction of face and voice areas during speaker recognition. *Journal of Cognitive Neuroscience, 17*, 367–376.
Walker, S., Bruce, V., & O'Malley, C. (1995). Facial identity and facial speech processing: Familiar faces and voices in the McGurk effect. *Perception & Psychophysics, 57*, 1124–1133.
Welch, R. B., & Warren, D. H. (1980). Immediate perceptual response to intersensory discrepancy. *Psychological Bulletin, 88*, 638–667.
Zäske, R., Schweinberger, S. R., & Kawahara, H. (2010). Voice aftereffects of adaptation to speaker identity. *Hearing Research, 268*, 38–45.

Chapter 7
Audiovisual Integration of Face–Voice Gender Studied Using "Morphed Videos"

Rebecca Watson, Ian Charest, Julien Rouger, Christoph Casper, Marianne Latinus, and Pascal Belin

Abstract Both the face and the voice provide us with not only linguistic information but also a wealth of paralinguistic information, including gender cues. However, the way in which we integrate these two sources in our perception of gender has remained largely unexplored. In the following study, we used a bimodal perception paradigm in which varying degrees of incongruence were created between facial and vocal information within audiovisual stimuli. We found that in general participants were able to combine both sources of information, with gender of the face being influenced by that of the voice and vice versa. However, in conditions that directed attention to either modality, we observed that participants were unable to ignore the gender of the

R. Watson (✉) • M. Latinus
Voice Neurocognition Laboratory, Institute of Neuroscience and Psychology,
College of Medical, Veterinary and Life Sciences, University of Glasgow, Glasgow, UK
e-mail: r.watson@psy.gla.ac.uk

I. Charest
MRC Cognition and Brain Sciences Unit, Cambridge, UK

J. Rouger
Brain Innovation BrainVoyager, Maastricht, The Netherlands

C. Casper
Department of Business Administration and Human Resource Management,
University of Cologne, Cologne, Germany

P. Belin
Voice Neurocognition Laboratory, Institute of Neuroscience and Psychology,
College of Medical, Veterinary and Life Sciences, University of Glasgow, Glasgow, UK

International Laboratories for Brain, Music and Sound (BRAMS), Université de
Montréal & McGill University, Montreal, Quebec, Canada
e-mail: pascal.belin@glasgow.ac.uk

P. Belin et al. (eds.), *Integrating Face and Voice in Person Perception*,
DOI 10.1007/978-1-4614-3585-3_7, © Springer Science+Business Media New York 2013

voice, even when instructed to. Overall, our results point to a larger role of the voice in gender perception, when more controlled visual stimuli are used.

1 Introduction

In addition to communicating linguistic information, faces and voices are both rich in information on a person's biological characteristics, including unique identity and gender. The ability to not only recognise this information but also integrate these into a unified percept is a crucial part of social interaction. Despite our natural, bimodal perception of paralinguistic information such as this, the overwhelming amount of literature on identity and gender recognition has concentrated on facial cues (reviewed in Haxby, Hoffman, & Gobbini, 2000). Fewer studies have looked at the perception of non-linguistic vocal identity information and, until recently, studies examining the combination of the both were scarce. However, particularly in the past few years, there has been an increase in studies within this area—particular those using neuroimaging techniques such as fMRI. These latest advances have formed a solid groundwork into further investigation of the integration of audiovisual, paralinguistic signals.

Yet, despite increasing research within this area, the majority of studies on face–voice integration have used stimuli that are impoverished and distant from the normal ecological situation (i.e. static photographs of faces coupled with audio recordings of voices). This is always going to provide a somewhat unrealistic experience for the participant, as in real life we almost constantly see a dynamic presentation of audio and visual information, synchronised in time: that is, we see moving faces with their respective voices simultaneously and integrate them into one. Articulatory movements of the face are especially related to speech perception, due to physical changes in the face occurring during vocal production (Munhall et al., 2006). In speech perception, the dynamic presentation of faces and speech in approximate time synchrony appears to be crucial. A clear illustration of this point is shown when testing the McGurk effect, where clips of faces in movement, but not still photograph, influence speech perception (Campanella & Belin, 2007; McGurk & MacDonald, 1976). Evidence also suggests that audiovisual integration effects for dynamic information are greater than those that exist for static stimuli. For example, Schweinberger, Robertson, and Kaufmann (2007) observed that naturalistic, dynamic faces caused a far more pronounced audiovisual integration effect than static ones; and Kamachi, Hill, Lander, and Vatikiotis-Bateson (2003) found that the integration effect they discovered was *dependent* on articulatory facial movements. A number of neuroimaging studies have also reported that cerebral regions understood to be involved in the processing of facial emotion (e.g. the pSTS and the amygdala) appear to show a stronger response to dynamic, as opposed to static, emotive expressions (e.g. Haxby et al., 2000; Kilts, Egan, Gideon, Ely, & Hoffman, 2003).

These findings present strong reason for using stimuli that approximate real-life conditions as much as possible—specifically, matched, and time-synchronised, dynamic faces and voices. Our studies aim to use morphing software (both video (facial) and auditory) in order to generate a number of stimuli continua, which can then be used to investigate audiovisual integration effects. The ecological validity of our stimuli is superior to those used in the majority of other studies in this area: we use realistic, dynamic stimuli, obtained from video-recordings, with the morphing software allowing us to preserve the naturalistic qualities of the stimuli.

A main aim of our work is to create realistic, ecological dynamic face-voice movies for use in studies of face–voice integration. We have used state-of-the-art facial and vocal morphing techniques in order to parametrically vary gender and affective information in the face and voice. Such techniques allow us to examine not only relatively crude pairings of face–voice information (e.g. completely congruent, completely incongruent) but also more fine-grained combinations of varying gender information. We then have examined and contrasted responses to different presentations of audiovisual and unimodal information, in order to address specific questions of integrative processes, for example:

1. Does pairing different faces with a voice (AV presentation) significantly alter unimodal categorisation of gender, and vice versa?
2. How do (if at all) gender categorisation ratings differ between audiovisual stimuli with different congruence between the face and the voice?
3. Overall, do people place more emphasis on the face or voice when categorising gender and emotion?
4. Does the integration effect differ in response to a dynamic face, as compared to a static one?

The aim of the following experiment was to investigate the crossmodal audiovisual interactions during gender processing with dynamic faces and voices, in a more ecological approach of face–voice integration processes. We used an experimental paradigm similar to other behavioural studies of paralinguistic audiovisual processing (e.g. de Gelder & Vroomen, 2000; Joassin, Maurage, & Campanella, 2011; Joassin, Pesenti et al., 2011), in which both faces and voices were presented synchronously in an audiovisual condition (AV), and also separately (A and V) within unimodal conditions. This allowed us to compare responses between bimodal and single-mode conditions. Additionally, our stimuli are unique in that both faces and voices were morphed between gender parametrically and independently. Morphed stimuli have been previously used in audiovisual studies: however, typically these studies have morphed static photos of faces, and then paired these with distinct, original voice clips. Our study develops upon this, by using morphed audiovisual movies, composed of different face *and* voice morph pairings. This has allowed us to obtain a more detailed picture of audiovisual interactions between face and voice gender.

2 Methods

2.1 *Subjects*

Twenty-two English-speaking subjects (3 non-native speakers; 12 females; all right handed; mean age = 22 years) participated in the study. The ethical committee from the University of Glasgow approved the study. All volunteers provided informed written consent before, and received payment for, participation.

2.2 *Stimuli*

2.2.1 Video Recording and Editing

Ten participants (five males and five females, selected to match in age) were video-recorded saying the word "had" multiple times. Participants were instructed to speak the word with standardised timing. All participants were native speakers of the English language. The males were clean-shaven, and the females wore no make up. None had any distinctive facial markings or piercings. This ensured that morphs of the faces would not contain any cues which related to the gender of either individual. Recordings took place in the television studio at the Learning and Teaching Centre, Glasgow University, and participants were paid at the rate of £6/h. The participants were shot under standard studio lighting conditions (standard tungsten light), and sat 235 cm away from the camera, directly facing it. Videos were recorded with 25 frames/s (40 ms/frame) using a Panasonic DVC Pro AJD 610 camera, fitted with a Fujiform A17 × 7.8 BERM-M28 lens, and transferred and edited using Adobe Premier Elements. Within the video recording, vocalisations were recorded with 16-bit resolution at a sampling frequency of 44,100 Hz. Videos were edited so that every pronunciation of the words by all male and female formed a separate clip. One clip from each volunteer was selected for use. Each of the clips was then separated into their visual and audio components.

2.2.2 Face Morphing

In all clips, seven important landmarks in terms of facial movements, and the frames at which they occurred, were identified. These landmarks were the first movement of the chin, first opening of lips, maximum opening of the mouth, first movement of the lips inwards, point at which the teeth met, closing of the lips, and the last movement of the chin. The average frames for these landmarks were then calculated, so the occurrence of these landmarks matched in all clips. Editing consisted of inserting or deleting video frames during fairly motionless periods. Due to the speakers pronouncing the word with standardised timing, little editing was necessary. The editing produced ten adjusted clips, each 36 frames (40-ms) long. These were then used to create a "composite" male and female face frames (i.e. an average of the five female faces and five male faces). Morphing software "Psychomorph" (Tiddeman & Perrett, 2001) was used to generate the average morphs, with 36 frames for both the average male and average female face. Morphing software "Videomorph" was then used to create a morphed continuum of faces, which extended from 10% female to 90% female, morphed in 10% steps. Each of these morphed faces was therefore a ten-face composite (five male faces, five female). All frames were all converted to greyscale, matched for luminance, and an oval mask fitted around each face to conceal the artefacts such as the hair, which could act as a gender cue. New videos were then created using these masked frames. In order to create the static, control video we lengthened the 18th frame of each of the videos, to last 36 frames. We used the 18th frame as this when the mouth was at its maximum opening.

2.2.3 Auditory Morphing

Auditory stimuli were edited using Adobe Audition 2.0. Stimuli were first normalised for mean amplitude.

In order to generate the auditory components to the "morph-videos" a similar procedure was used. As with the visual stimuli "landmarks" were identified which occur at some specific points during vocal production. In total five temporal landmarks in the word production (beginning of the production, beginning and end of the voicing of "HA(-d)", as well as the plosive "(ha-)D" and the end of the production). In addition, nine landmarks were used in the frequencies, three anchor points for each of the three formants. Landmarks were placed at the beginning and end of each formant, as well as on the formant shift, the points where each formant lowered in amplitude. All these landmarks were set in the MATLAB-based morphing algorithm STRAIGHT (Kawahara & Matsui, 2003), and then used to generate first an average male and average female voice, and second a morph continuum between the two average voices equivalent for that for face. This resulted in nine different voices, with varying amounts of gender information within them. See audiovisual material for an example.

2.2.4 Audiovisual Video Production

One hundred and sixty-two audiovisual videos were produced by pairing static and dynamic face videos with the morphed voices (18 faces videos (nine face morphs, still and articulating) matched to each one of the nine voices). This provided a variety of congruous and incongruous stimuli. The audiovisual videos were then cut from the 10th frame to the 30th frame. This was in order that in the dynamic videos, the participant saw mainly lip movement, as opposed to this in addition to periods of a static face at the beginning and end of the clip, where the lips were closed. The videos started at the frame before movement of the lips occurred. It should be noted that in our original videos, the onset of the faces preceded the onset of the audio speech. Indeed, in terms of natural vocalisation, the first facial movements precede vocalisation. Therefore, the onset of visual articulation did not correspond with the first frame of speech production. Instead, the vocalisation (defined by the first burst of the "a" of "had") began 120 ms after visual onset, and 80 ms after the first movement of the lips. This auditory delay was matched in the static videos. Videos were then cut again, so that each began at the 10th frame, and finished at the 30th frame. This was to ensure that participants saw the face and heard the paired voice for almost an identical amount of time, whilst still preserving the naturalistic quality of the video.

2.3 *Design and Procedure*

All videos were presented at 720 × 576 pixels, using Matlab 2007b (MATHWORKS Inc., Natick, MA) and the Psychophysics Toolbox (PTB3) extensions running on a PC. The auditory stimuli were presented in mono, via Beyerdynamic DT 770 headphones at approximately 70 dB. Participants saw and heard all stimuli in a soundproof booth. Instructions were given to the participants before each condition. All participants undertook all five of the following conditions.

2.3.1 Audiovisual (No Direction to Modality)

Participants were instructed to watch the screen and listen to the presented voices, and asked to indicate their gender decision via a two choice button press. Before the experiment began a fixation cross appeared on the screen for 2 s. A total of 162 AV stimuli were presented ten times each. These were presented in randomised order in ten blocks, over two sessions, consisting of 162 trials each. Breaks were given between each block. Of these 162 AV stimuli, there were 18 completely congruent stimuli. If the participant indicated their response *during* the movie presentation, the next movie was presented 1 s after the end of that movie presentation. If the participant indicated their response after the movie presentation, the next movie was played 1 s after their response. If the participant responded more than 2,000 ms after the start of the movie presentation, their response was not counted.

2.3.2 Audiovisual (Attend to Face)

In this condition the same stimuli were used as in condition 1, and again participants were told to listen to the voice and watch the screen, but the instructions emphasised that their task was to judge the gender of the face, and ignore the voice. A randomised order was used again, and timings were the same as in condition one.

2.3.3 Audiovisual (Attend to Voice)

Again, the same stimuli were used as in condition 1 and 2; participants were told to listen to the voice and watch the screen, but the instructions were such that their task was to judge the gender of the voice and ignore the face.

2.3.4 Audio Only

In this condition, participants heard a series of voices alone. They were instructed to listen to each voice, and make a decision on gender based on the voice they had just heard. Again they indicated their decision via a button press. The nine voice morphs were presented ten times each, in randomised order in two blocks consisting of 45 trials each. There was a break between blocks.

2.3.5 Video Only

Participants saw all face videos, uncoupled with a voice. They were instructed to watch the screen and indicate their decision regarding gender in the same way as before. The 18 faces—nine static and nine dynamic—were presented ten times each, in randomised order in two consecutive blocks consisting of 90 trials each. There was a break between each block.

Participants always completed all three of the audiovisual conditions before the two unimodal conditions. Conditions 2 and 3 were counterbalanced between participants, as were conditions 4 and 5.

The data from one male subject was excluded, due to anomalous gender categorisation.

3 Results

3.1 Dynamic Versus Static

We initially submitted results from each audiovisual condition to three separate three factor ANOVAs with Movement (dynamic or static), Voice (voice morph 1 (90% female)–9 (90% male)) and Face (face morph 1 (90% female)–9 (90% male)) as within-subjects factors for each (2×9×9 ANOVA); and results from the unimodal face condition to an ANOVA with Movement and Face as within factors (2×9 ANOVA). In all ANOVAs, the effect of movement was not significant (face alone $F(1,19)=1.45$, $p=0.244$, audiovisual $=F(1,19)=0.282$, $p=0.601$, audiovisual (attend to face) $F(1,20)=1.24$, $p=0.279$, audiovisual (attend to voice) $F(1,19)=0.141$, $p=0.712$) and we thus averaged our results from the static and dynamic conditions. The results described below are those from the average of these two conditions. It should be noted that for each factor, in each of the following described ANOVAs, degrees of freedom were corrected using Greenhouse–Geisser estimates of sphericity (epsilon <0.75).

3.2 Unimodal Versus Audiovisual

We initially performed simple analyses in which one of the two modalities only took the two extreme values, for the sake of comparison with previous literature.

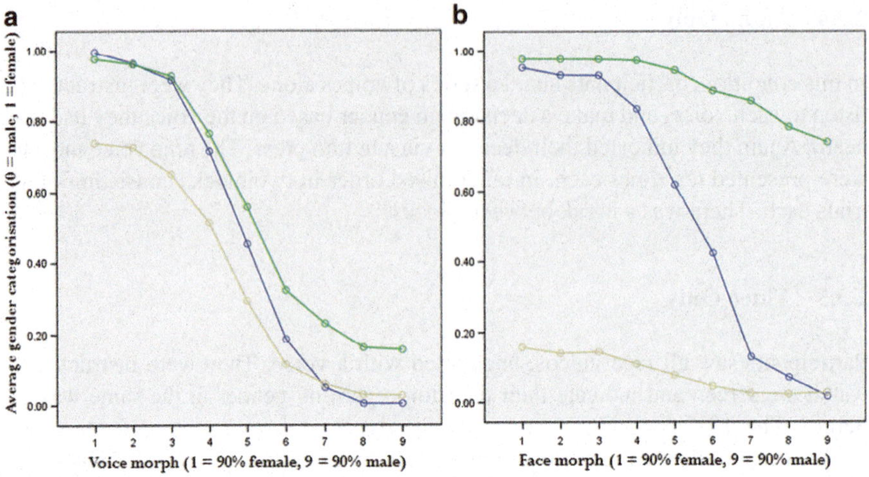

Fig. 7.1 Unimodal vs. audiovisual gender categorisation. (**a**) Green=90% female face, Beige=90% male face; (**b**) green=90% female voice, Beige=90% male voice

Face categorisation. Here we compared categorization values for the nine face morphs across three conditions: unimodal (no pairing with voice), pairing with male voice, pairing with female voice (Fig. 7.1a). Data was submitted to an ANOVA with Face (1–9) and Voice (none, 90% female and 90% male) as within-subjects factors. There was a main effect of Face ($F(1.45,28.5)=65.4$, $p<10^{-4}$), indicating different gender ratings for the different faces and reflecting correct gender categorization of the face gender continuum. The main effect of Voice was also highly significant, however ($F(1.23,26.5)=197$, $p<10^{-4}$), indicating an influence of the voice on the face gender judgments. As can be seen in Fig. 7.1a, overall, when a female voice was paired with any given face, categorisation ratings on average increased (with the exception of the congruent condition); and when a male voice was paired with a face, categorisation ratings lowered (again, with the exception of the congruent pairing). There was also a significant Face × Voice interaction ($F(3.27,65.4)=66.5$, $p<10^{-4}$).

Voice categorization. Symmetrically, we compared categorization values for the nine voice morphs across three conditions: unimodal (no pairing with face), pairing with male face, pairing with female face (Fig. 7.1b). Data was submitted to an ANOVA with Voice (1–9) and Face (none, 90% female and 90% male) as within-subjects factors. There was a main effect of Voice ($F(2.45,49.4)=159$, $p<10^{-4}$) again showing correct categorization of the voice gender continuum. The main effect of Face was also highly significant, indicating an influence of the visual modality on voice gender ratings ($F(1.15,23.0)=8.39$, $p=0.006$). As shown in Fig. 7.1b, the pairing of faces with vocal information caused categorisation shifts as in Fig. 7.1a. There was also a significant interaction between Voice and Face ($F(4.33,86.7)=4.64$, $p=0.001$).

However, comparison of Fig. 7.1a, b highlights that, although in both cases there was a significant main effect of both modes of information, vocal information caused

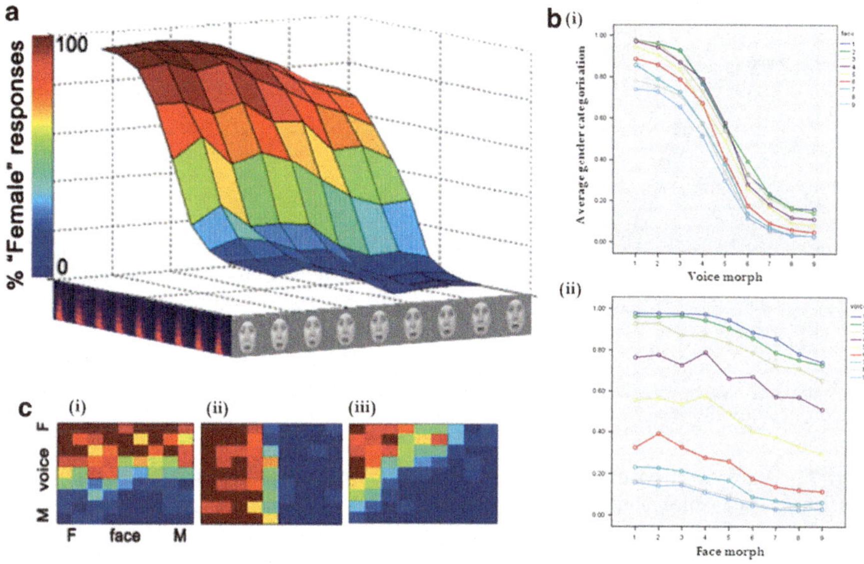

Fig. 7.2 Audiovisual condition (uncontrolled attention). (**a**) 3D plot of categorisation responses; (**b**) 2D plots of categorisation responses. 1 = 90% female information, 9 = 90% male information, in both face and voice; (**c**) individual categorisation strategies (i) Voice information, (ii) Face information, (iii) both sources

a more pronounced shift in ratings when compared to the unimodal face condition, than facial information did when compared to the unimodal voice condition.

3.3 *Audiovisual Condition: Uncontrolled Attention*

Here we compared ratings obtained in the different audiovisual conditions when subjects were free to attend to any modality. Figure 7.2a shows a 3D plot of the average ratings for the 9 × 9 morph steps in the audiovisual condition. Here it can be seen that although both face and voice morph caused shifts in categorisation ratings (indicated by change in colour) these changes were not symmetrical between the two modes—voice shows a stronger visible effect. Data was submitted to an ANOVA with Face (1–9) and Voice (1–9) as within-subject factors. The main effect of voice was significant ($F(1.80,35.9) = 126$, $p < 10^{-4}$), as well as that of the Face ($F(1.11,22.1) = 8.23$, $p = 0.007$), indicating that both face and voice gender affected overall gender ratings. The Voice × Face interaction was also significant ($F(9.4,188) = 2.27$, $p = 0.018$), indicating that the effect of one modality depended on values in the other modality. The effect of voice was larger overall, highlighting that subjects were weighting the auditory modality more when making the gender judgment (Fig. 7.2b). Figure 7.2c suggests, however, that not all subjects show this effect; indeed some rarer individuals weighted the face modality more than the voice, or presented a mixed strategy.

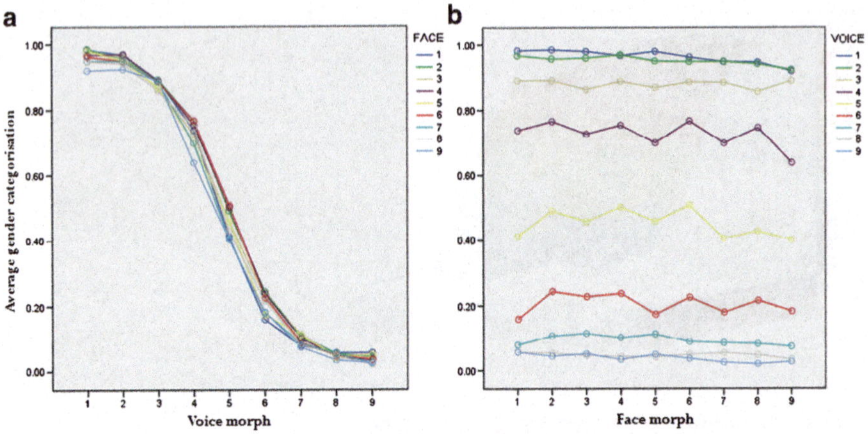

Fig. 7.3 Audiovisual condition (attention to voice). Average gender categorisation: 0 = male, 1 = female. Face–Voice morph: 1 = female, 9 = male

3.4 Audiovisual Condition: Attention to Voice

Data was submitted to an ANOVA with Face (1–9) and Voice (1–9) as within subjects' factors.

The main effect of voice was significant ($F(1.86,35.4) = 295, p < 10^{-4}$) as expected, indicating adequate categorization of the voice gender continuum. However, there was no significant effect of face gender, indicating a lack of influence of the visual modality on gender perception when attention was attracted to the voice ($F(2.10,39.9) = 2.81$, $p = 0.07$). This can be observed in Fig. 7.3: the little visible difference between the curves in Fig. 7.3a and the lack of slope of the curves in Fig. 7.3b indicate the non-significant effect of face information. There was no significant interaction between factors.

3.5 Audiovisual Condition: Attention to Face

Here participants were presented with a face–voice stimulus, but instructed to rate gender based only upon the face. Data was submitted to an ANOVA with Face (1–9) and Voice (1–9) as within subjects' factors. The effect of Face was, as expected, highly significant ($F(2.11,42.3) = 205$, $p < 10^{-4}$). However, the effect of voice was also highly significant, however ($F(1.97,39.3) = 16.6$, $p < 10^{-4}$), indicating a strong influence of the voice gender on face gender categorization even under instructions to ignore the voice. The influence of the voice can be seen in Fig. 7.4: particularly, its notable effect on perceived face gender in the central, androgynous portion of the face continuum (Fig. 7.4b).The Voice × Face interaction was also significant ($F(12.6,252) = 1.88$, $p = 0.034$).

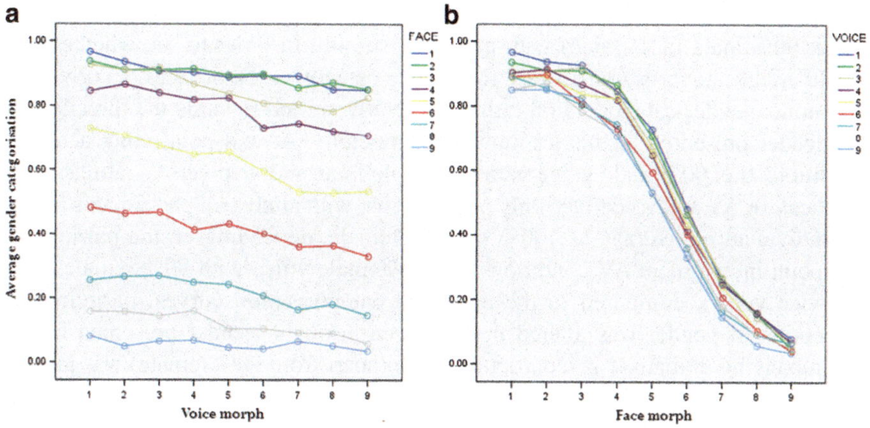

Fig. 7.4 Audiovisual condition (attention to face). Average gender categorisation: 0 = male, 1 = female. Face–Voice morph: 1 = female, 9 = male

4 Discussion

The main objective of the present study was to explore the combination of information from facial and vocal cues in the recognition of gender. Overall, the experiment showed that both face and voice gender influenced overall gender ratings, highlighted by different average gender ratings under audiovisual presentation, as compared to unimodal; and that overall, the effect of voice gender in this experiment was stronger than that of face gender. This was confirmed by the results of the audiovisual conditions which controlled for attention: attending to voices resulted in the previous influence of face on gender categorisation disappearing, whereas attending to face showed an influence of both modalities.

First, we investigated the effect of facial movement in the categorisation of gender. Our results showed that participants' categorisation of gender was not dependent on whether the video contained an articulating or static face. This is in contrast to some other behavioural results (e.g. Kamachi et al., 2003; Munhall, Gribble, Sacco, & Ward, 1996; Schweinberger et al., 2007) which suggest that dynamic faces may result in a differing audiovisual effect compared to that of static faces. However, it should also be taken into account that our investigation differs from them in that it explores gender recognition, and not speech, or familiar person perception. The lack of difference between dynamic and static conditions in our study may be due to the fact that a dynamic image did not offer any more gender information that a static one—or at least, no more information crucial for making a decision on gender. A static image appears to immediately present us with all the facial information necessary to make such a categorisation. Specifically, crucial gender discriminators such as facial features (Brown & Perrett, 1993) and the natural configuration of the face (Bruce et al., 1993) were available in both types of video.

Second, we paired the end-point face and voice morphs (i.e. 90% female face–voice, 90% male face–voice) with unimodal stimuli, in order to see whether this would affect categorisation ratings. Relative to categorisation of faces (F) or voices (V) alone, gender categorisation ratings generally shifted towards the direction of the gender presented in the accompanying modality. At end-point congruent AV conditions (i.e. 90% male voice with 90% male face, and vice versa), ratings were identical for AV and voice/face only presentation, with almost all participants rating the individual on average as 100% male or female, depending on the pairing. At end-point incongruent AV conditions (i.e. 90% male voice with 90% female face, and vice versa), compared to the unimodal categorisation curves, participant's perception of gender was altered in the direction of the gender presented in the accompanying domain (e.g. a particular voice (apart from 90% female) was judged as more female if a female face was shown alongside it). This alteration in ratings was not only apparent in cases of complete incongruence, but also when there was an intermediate difference in face–voice gender information (for example, an "androgynous", or 50% male–female face–voice). Although an audiovisual pairing of gender information significantly shifted categorisation ratings in both ANOVAs (audiovisual vs. voices alone, audiovisual vs. faces alone), the audiovisual effect was significantly stronger when these conditions were contrasted with the unimodal face condition: that is, there was a greater difference in ratings between audiovisual pairings and faces alone, as compared to audiovisual pairings and voices alone, thus highlighting a stronger influence of voice in this experiment.

We then compared ratings within the audiovisual (uncontrolled attention) condition only. This analysis involved comparisons between all pairings of all face and voice morphs. Again, we observed a main effect of both face and voice, indicating that participants could combine data from the two sources to arrive at a unique judgement on gender. However, as in the previous analysis, the main effect of voice was greater, indicating that participants, on average, used vocal information more when categorising gender. Overall, categorisation responses shifted proportionally in accordance with the amount of gender information in the face and voice, indicating a somewhat additive effect of our parametric manipulations of gender information (Fig. 7.2b(i and ii)).

Finally, we investigated the role of attention. Our reason for doing so was based on a result of previous studies (de Gelder & Vroomen, 2000; Vroomen, Driver, & de Gelder, 2001), which both found that participants were unable to ignore information presented in another mode, thus indicating a possibly automatic, mandatory integration of inputs at a pre-perceptual level. In these conditions, participants were still presented with audiovisual stimuli, but were instructed to ignore either the face or voice, and make their judgements purely on the basis of what they heard or saw in the other mode. Participants were able to ignore the face when instructed, indicated by no significant main effect of face—which had been observed in the audiovisual condition with uncontrolled attention. However, in contrast, participants were unable to ignore the vocal information. Although the effect of voice was visibly smaller than in the uncontrolled attention condition, there were still significant shifts in categorisation depending on the degree of gender information contained within the voice of the audiovisual stimulus. This result underlines the strong effect of voice observed the previous analyses, particularly in contrast to that of the face.

Our study advances on previous work, in multiple ways. We have made an effort to develop the ecological validity of stimuli—specifically, by creating articulating faces with time-matched voices. Our inclusion of static portraits enabled us to directly compare, for the first time, whether there was a significant difference between categorisation of dynamic and still faces. Although we found there was no change in gender ratings between these two types, we still feel that our effort to raise the environmental validity of our stimuli was nonetheless important. Our study also utilised morphing techniques in order to create parametric manipulations of both face and voice gender morph. Using morphing techniques allowed us to also create ambiguous face–voice pairs, manipulate face–voice congruence in a more fine-grained manner, and to test, using controlled experimental manipulation, the respective influence of faces and voices in the crossmodal processing of gender. In contrast to some previous studies on face–voice identity integration, we found that—although both sources of information were used in categorisation of gender—auditory information dominated over visual information with regard to gender categorisation. A number of reasons could account for this result. First, gender perception can be viewed as a low-level processing task, as compared to familiar/unfamiliar person recognition which requires a higher level processing due to accessing of identity information. The level of processing may play an important role in the way in which faces and voices interact, and the relative influence of each. Second, with regard to gender, voices are arguably more dimorphic than faces. For example, the fundamental frequency (f0), which determines the perceived pitch of a voice, is almost always significantly higher for females—typically by one octave (Linke, 1973). Future research using stimuli allowing a greater independent control of relevant social dimensions in faces and voices while preserving naturalness and audiovisual synchrony such as the present "morphed video" stimuli will allow addressing this important issue.

References

Brown, E., & Perrett, D. I. (1993). What gives a face its gender? *Perception, 22*(7), 829–840.

Bruce, V., Burton, A. M., Hanna, E., Healey, P., Mason, O., Coombes, A., Fright, R., & Linney, A. (1993). Sex discrimination: How do we tell the difference between male and female faces? *Perception, 22*(2), 131–152.

Campanella, S., & Belin, P. (2007). Integrating face and voice in person perception. *Trends in Cognitive Sciences, 11*(12), 535–543.

de Gelder, B., & Vroomen, J. (2000). The perception of emotions by ear and by eye. *Cognition and Emotion, 14*(3), 289–311.

Haxby, J. V., Hoffman, E. A., & Gobbini, M. I. (2000). The distributed human neural system for face perception. *Trends in Cognitive Sciences, 4*(6), 223–233.

Joassin, F., Maurage, P., & Campanella, S. (2011). The neural network sustaining the crossmodal processing of human gender from faces and voices: An fMRI study. *Neuroimage, 54*(2), 1654–1661.

Joassin, F., Pesenti, M., Maurage, P., Verreckt, E., Bruyer, R., & Campanella, S. (2011). Cross-modal interactions between human faces and voices involved in person recognition. *Cortex, 47*(3), 367–376.

Kamachi, M., Hill, H., Lander, K., & Vatikiotis-Bateson, E. (2003). "Putting the face to the voice": Matching identity across modality. *Current Biology, 13*(19), 1709–1714.

Kawahara, H. (2003). Exemplar-based voice quality analysis and control using a high quality auditory morphing procedure based on straight. In: VoQual 03: Voice Quality: Functions, Analysis and Synthesis. Geneva (Switzerland): ISCA Tutorial and Research Workshop.

Kilts, C. D., Egan, G., Gideon, D. A., Ely, T. D., & Hoffman, J. M. (2003). Dissociable neural pathways are involved in the recognition of emotion in static and dynamic facial expressions. *Neuroimage, 18*, 156–168.

Linke, C. E. (1973). A study of pitch characteristics of female voices and their relationship to vocal effectiveness. *Folia Phoniatrica, 25*, 173–185.

McGurk, H., & MacDonald, J. (1976). Hearing lips and seeing voices. *Nature, 64*(5588), 746–748.

Munhall, K. G., Gribble, P., Sacco, L., & Ward, M. (1996). Temporal constraints on the McGurk effect. *Perception and Psychophysics, 58*(3), 351–362.

Schweinberger, S. R., Robertson, D., & Kaufmann, J. M. (2007). Hearing facial identities. *The Quarterly Journal of Experimental Psychology, 60*(10), 1446–1456.

Tiddeman, B., & Perrett, D. (2001). *Moving facial image transformations based on static 2D prototypes*. Paper presented at the 9th International conference in Central Europe on Computer Graphics, Visualization and Computer Vision 2001 (WSCG 2001), Plzen, Czech Republic.

Vroomen, J., Driver, J., & de Gelder, B. (2001). Is cross-modal integration of emotional expressions independent of attentional resources? *Cognitive, Affective and Behavioural Neurosciences, 1*(4), 382–387.

Chapter 8
Cross-Modal Integration of Identity and Gender Information Through Faces and Voices Involves a Similar Cortical Network

Salvatore Campanella and Frédéric Joassin

Abstract We investigate the cerebral cross-modal interactions between human faces and voices involved during gender and identity categorization in two separate functional magnetic resonance imaging (fMRI) studies. In each of these experiments, participants were scanned in four runs that contained three conditions consisting in the presentation of faces, voices, or congruent face–voice pairs. The task consisted in categorizing each trial (visual, auditory, or associations) according to its gender or identity. The subtraction between the bimodal condition and the sum of the unimodal ones, as well as psychophysiological interaction analyses (PPI), were performed. Main results suggest that the cross-modal auditory–visual categorization of human gender and identity is sustained by a network of highly similar cerebral regions. This network included several regions such as the unimodal visual and auditory regions processing the perceived faces and voices and inter-connected via a subcortical relay located in the striatum, the left superior parietal gyrus, part of a larger parieto-motor network dispatching the attentional resources to the visual and auditory modalities, and the right inferior frontal gyrus sustaining the integration of the semantically congruent information into a coherent multimodal representation. Therefore, we suggest that cross-modal processing of human stimuli requires the activation of a network of cortical regions, including both unimodal visual and auditory regions and supramodal parietal and frontal regions involved in the integration of both faces and voices and in the cross-modal attentional processes.

S. Campanella (✉)
Laboratory of Psychological Medicine, Free University of Brussels, Brussels, Belgium,

CHU Brugmann, Psychiatry Department (EEG), The Belgian Fund for Scientific Research (FNRS), Brussels, Belgium
e-mail: salvatore.campanella@chu-brugmann.be; salvatore.campanella@ulb.ac.be

F. Joassin
Clinique de la mémoire, CHU Ambroise Paré, Mons, Belgium
e-mail: frederic.joassin@hap.be

P. Belin et al. (eds.), *Integrating Face and Voice in Person Perception*,
DOI 10.1007/978-1-4614-3585-3_8,

1 Introduction

In daily life, our social interactions are guided by our ability to integrate distinct sensory inputs into a coherent multimodal representation of our interlocutors. For instance, we are able to integrate the auditory information of what is said and the visual information of who is saying it, so that we can attribute a particular speech to a particular person (Kerlin, Shahin, & Miller, 2010) and thus take part to a conversation. Numerous studies have examined the cerebral correlates of these kinds of "cross-modal" auditory–visual speech perception (e.g., Calvert, Campbell, & Brammer, 2000). Nevertheless, cross-modal interactions occur not only during speech perception but also during the memory processes allowing the identification of familiar people (Campanella & Belin, 2007). Indeed, integration of information from face and voice plays a central role in our social interactions as, for instance, both faces and voices are rich in information on a person's identity. Therefore, many studies have been devoted to the investigation of specific neural correlates of identity processing from faces or voices.

On the one hand, a large body of neuroimaging investigations has demonstrated that unimodal face identity processing is mediated by a distributed, hierarchical network of cerebral areas, including the fusiform face area (FFA) of the middle fusiform gyrus (e.g., Kanwisher, McDermott, & Chun, 1997), the occipital face area (OFA) in the inferior occipital cortex (e.g., Gauthier, Skudlarski, Gore, & Anderson, 2000), and an area located in the posterior part of the superior temporal sulcus (pSTS) (e.g., Puce, Allison, Gore, & McCarthy, 1995). These three bilateral areas, which show a strong right hemisphere advantage, are thought to form the core system for normal face perception (see Haxby et al., 2000 for a review). Moreover, electrophysiological studies showed that a difference in event-related potentials (ERPs) related to face identity can be observed in the latency of the face-selective N170 response recorded on occipito-temporal sites (e.g., Campanella et al., 2000). The attempt to clarify the functional neuroanatomy of face perception has been also largely constrained by the cases of neurological patients (brain-damaged) or congenital patients (individuals experiencing problems throughout their lives in the absence of neurological damage) suffering from prosopagnosia—the inability to individualize faces following brain damage (Bodamer, 1947). By fruitfully combining functional imaging and neuropsychology, current studies on prosopagnosic patients are still trying to refine the functional organization of the cortical areas mediating face processing in the human brain (e.g., Steeves et al., 2009).

On the other hand, processing identity information is also possible from the voice alone, and this important ability has been shown to rely on anterior temporal lobe regions of the right hemisphere, particularly along the right anterior STS close to the temporal voice areas (TVAs) (Belin, Zatorre, Lafaille, Ahad, & Pike, 2000). Electrophysiological data suggest that familiarity effects on voice processing first occur at about 200 ms after voice onset (e.g., Beauchemin et al., 2006). Here also the investigation of brain-damaged or congenital patients presenting disorders of familiar voice recognition (phonagnosia, e.g., Van Lancker & Canter, 1982; Garrido et al., 2009) has been of the greatest interest.

Overall, there is nowadays clear evidence about the segregated neural mechanisms implied in the separate processing of identity on the basis of faces and voices. However, normal adult humans are able and take advantage from combining identity information from face and voice. Therefore, the question arises as to how the brain manages to create a single coherent representation of a person on the basis of these different attributes, processed by distinct cortical regions.

2 Cross-Modal Interactions Between Faces and Voices Involved in Identity Processing

Human social interactions are shaped by our ability to identify individuals, a process to which face and voice recognition contributes both separately and jointly. Indeed, for instance, Sheffert and Olson (2004) have shown that the learning of voice identities was easier when the voices to learn were associated with a face. Schweinberger, Robertson, and Kaufmann (2007) showed that voice recognition was easier when simultaneously presented with an associated face, whereas it was hampered when presented with a face that did not share the same identity. This demonstrates that listeners cannot ignore a face as soon as it is presented in time synchrony with a voice. With this in mind, several questions may arise, such as: How is identity information from face and voice combined in the brain? For person identification, does the association of unimodal information require a supramodal stage of cortical processing involved in representing semantic information about the identity of each known person? Or can "unimodal" face and voice processing neural systems interact directly without a relay through supramodal regions?

These questions summarized the two main hypotheses that have emerged to explain the cross-modal cerebral integration process. The first one postulates direct links between the unimodal regions processing the distinct sensory stimuli (Von Kriegstein, Kleinschmidt, Sterzer, & Giraud, 2005; Von Kriegstein & Giraud, 2006). For instance, the authors showed that the right FFA had an enhanced connectivity with the right STS during speaker recognition, suggesting that multimodal person recognition does not necessarily engage supramodal cortical substrates but can result from the direct sharing of information between the unimodal auditory and visual regions (Von Kriegstein & Giraud, 2006). One possible neural mechanism for such direct links between unimodal regions could be the synchronization of the oscillatory activities of assemblies of neurons, especially in the gamma-band frequency range (30 Hz, for a review, see Senkowski, Schneider, Foxe, & Engel, 2008). On the other hand, the alternative hypothesis proposes that the cross-modal integration of faces and voices relies on the activation of a neural network including supramodal convergence regions (Driver & Spence, 2000; Bushara et al., 2003). This hypothesis was supported by a study of ours (Joassin, Maurage, Bruyer, Crommelinck, & Campanella, 2004). ERPs were measured during an identification task in which participants were exposed either to the simultaneous presentation of previously learned faces and voices or to their separate presentation. The comparison of the responses evoked during the bimodal condition with those observed during the unimodal condition revealed: (1) a first central positive/posterior negative wave at about 110 ms, which is best explained by a pair of dipoles originating in the associative visual cortex; (2) a central negative/posterior positive wave at about 170 ms, which is best explained by a pair of dipoles localized in the associative auditory cortex; and (3) a central positive wave at about 270 ms, which is best explained by a network of cortical regions including the fusiform gyrus, the associative auditory cortex, and the superior frontal gyrus and superior colliculi, two multimodal convergence regions. These results constitute the first direct evidence for additional cerebral processes at different latencies when combining face and voice information for person identification, possibly corresponding to first integrative responses in "unimodal" sensory cortices (1, 2) and a supramodal stage of integration (3). Accordingly, Bernstein, Auer, Wagner, and Ponton (2008), using ERPs, observed a specific cerebral activity of the left angular gyrus during audiovisual speech perception, suggesting that this region plays a role in the multimodal integration of visual and auditory speech perception.

To investigate this issue, Joassin, Pesenti et al. (2011) measured brain activity using fMRI while 14 participants were recognizing previously learned static faces, voices, and voice–static face associations. During the fMRI session, three different conditions were presented: faces (F), voices (V), and voice–face associations (VF, see Fig. 8.1). Only two of the four identities were included in each run (for instance, the identities "Detiez" and "Goffin" in the first run, "Detiez" and "Gillet" in the second run, and so on) and these were varied across runs. Participants were informed of the two identities used in each run by a written instruction ("Detiez or Goffin?") appearing on the screen before the beginning of each run. The task consisted of categorizing each trial (face, voice, or association) according to its identity (i.e., its name) by pressing one of two keys on a stimpad with two fingers of the right hand (left button for the first identity and right button for the second identity). Each volunteer participated in six runs each consisting of six experimental blocks of 32 s (2 blocks per condition), interleaved with 16-s fixation periods (white cross on black background). The order of the various conditions within the run was pseudo-randomly balanced across runs and subjects. Each experimental block comprised 12 trials. Each trial was composed of a fixation cross (300 ms), followed by the stimulus for 700 ms and an empty interval of 1,500 ms. The importance of both speed and accuracy was emphasized.

Using a subtraction method between bimodal and unimodal conditions (VF − (F + V)), we observed that voice–face associations activated both unimodal visual and auditory areas, and specific multimodal regions located in the left angular gyrus and the right hippocampus (see Fig. 8.2). Moreover, a functional connectivity analysis confirmed the connectivity of the right hippocampus with the unimodal areas. Taken together, the present results demonstrate that cross-modal person recognition relies on the activation of a cerebral network including unimodal face and voice areas along with multimodal regions such as the left angular gyrus, involved in cross-modal attentional processing, and the hippocampus, sustaining the forming and retrieval of auditory–visual representations of people in memory. They also

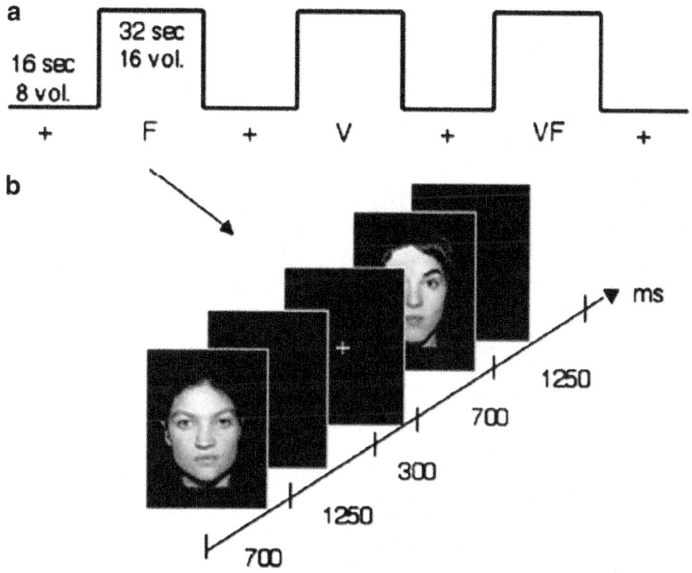

Fig. 8.1 (**a**) fMRI design: each run consisted in three alternances of a 16-s fixation period (*white cross on black background*) and a 32-s activation period. Each activation period corresponded to a different condition, presented in a pseudorandom order. (**b**) Examples of behavioral task: participants were presented with 12 trials in each condition. Each trial comprised a fixation cross for 300 ms, a stimulus—faces (F), voices (V), or face–voice associations (VF)—for 700 ms and a black intertrial interval for 1,500 ms

support a dynamic vision of cross-modal interactions in which heteromodal areas are not simply the final stage of a hierarchical unimodal-to-multimodal processing model, but rather, they may work in parallel and influence each other.

Nevertheless, the results of our previous experiments raised several questions, notably about the specificity of the neural network involved in the multimodal recognition of familiar people. The classical cognitive models of face identification have postulated that recognition, i.e., the access to the biographical information and the name of a familiar person, is independent from the processing of the other facial features such as the ethnicity, the age, or the gender (Bruce & Young, 1986; Burton, Bruce, & Johnston, 1990). In the next paragraph, we will focus on the gender dimension.

3 Cross-Modal Interactions Between Faces and Voices Involved in Gender Processing

Several recent studies have challenged this idea and proposed that gender and identity are processed by a single route. Ganel and Goshen-Gottstein (2002) showed that participants could not selectively attend to either sex or identity without being

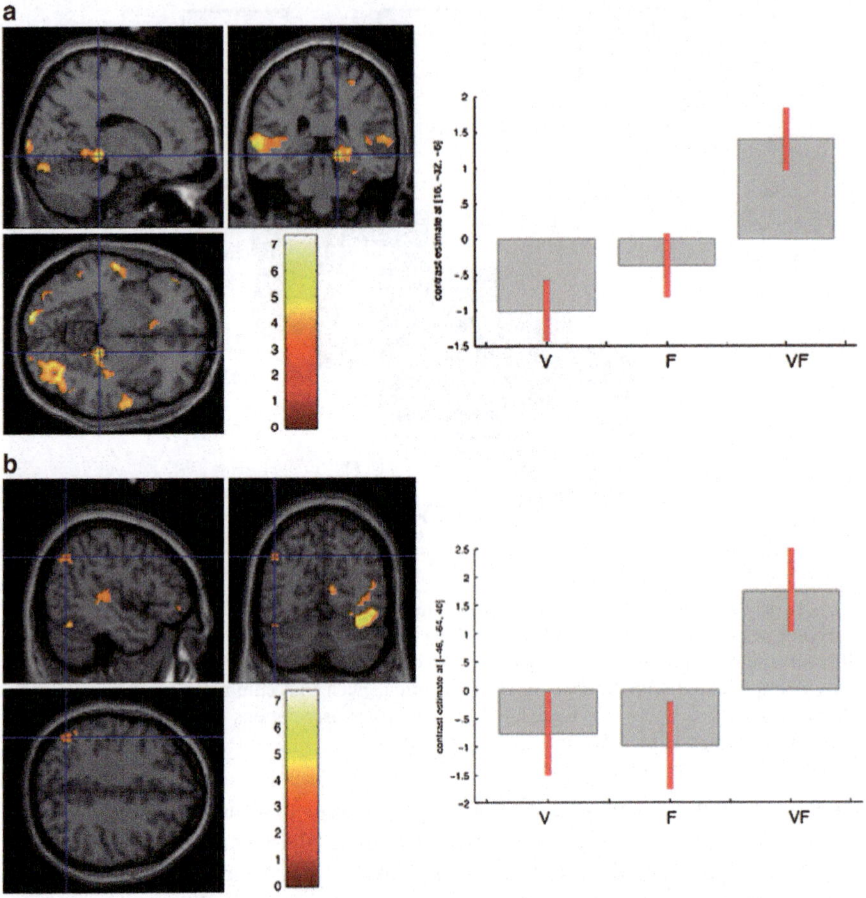

Fig. 8.2 (**a**) Brain sections of the contrast [VF − (V + F)] centered on the right hippocampus (*left side*). Activation changes for each condition in the right hippocampus (*right side*). $p < 0.05$ corrected for multiple comparisons at cluster size. (**b**) Brain sections of the contrast [VF − (V + F)] centered on the left angular gyrus (*left side*). Activation changes for each condition in the left angular gyrus (*right side*). $p < 0.05$ corrected for multiple comparisons at cluster size. *V* voices, *F* faces, *VF* face–voice associations

influenced by the other feature, suggesting that both informations are processed by a single route. Moreover, Smith, Grabowecky, and Suzuki (2007) have recently shown that auditory and visual information interact during face gender processing. In their experiment, participants had to categorize androgynous faces according to their gender. These faces were coupled with pure tones in the male or female fundamental-speaking-frequency range. They found that faces were judged as male faces when coupled with a pure male tone while they were judged as female ones when coupled with a pure female tone. The aim of the present experiment was thus to investigate the cross-modal audiovisual interactions during gender processing

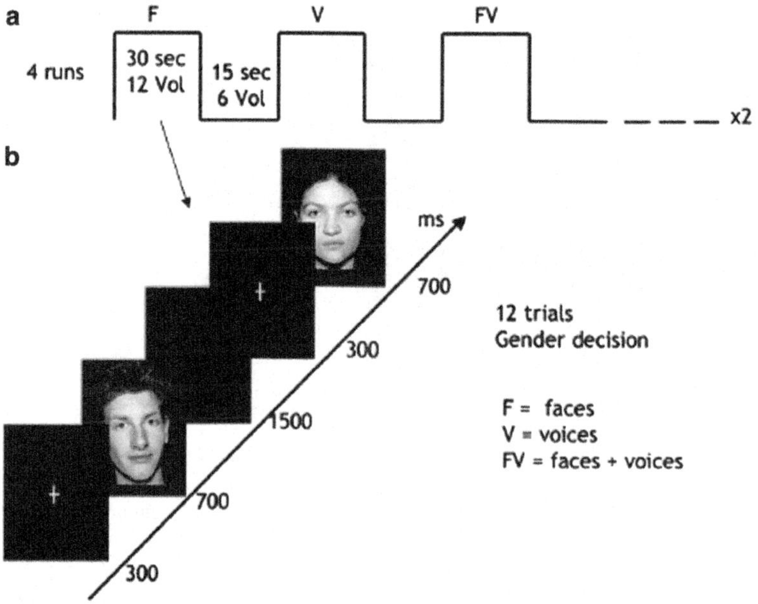

Fig. 8.3 (**a**) fMRI design: each run consisted in six alternances of a 15-s fixation period (*white cross on black background*) and a 30-s activation period. Each activation period corresponded to a different condition (F, V, FV), presented twice in a pseudorandom order. Participants were presented with 12 trials in each condition. Each trial comprised a fixation cross for 300 ms, a stimulus—faces (F), voices (V), or face–voice associations (FV)—for 700 ms and a black inter-trial interval for 1,500 ms

with real faces and voices, in a more ecological approach of face–voice integration processes. We used an experimental paradigm similar to those used in our previous studies (Campanella et al., 2001; Joassin, Campanella et al., 2004; Joassin, Maurage et al., 2004; Joassin, Meert, Campanella, & Bruyer, 2007; Joassin, Pesenti et al., 2011), enabling the direct comparison between a bimodal condition (FV) in which both faces and voices were presented synchronously and two unimodal conditions in which faces and voices were presented separately (F and V). This paradigm allowed us to perform the main contrast [FV − (F + V)] in order to isolate the specific activations elicited by the integration of faces and voices during gender categorization (see Fig. 8.3 for illustration).

This method uses a super-additive criterion to detect these specific activations, requiring multisensory responses larger than the sum of the unisensory responses (Calvert 2001; Beauchamp, 2005). This criterion has often been considered as overly strict in the sense that it can introduce type II errors (false negative), due to the fact that, in a single voxel, the activity of super- and sub-additive neurons is measured (Laurienti et al., 2005). Nevertheless, as the activations observed in our previous experiments have been obtained by this way (Campanella et al., 2001; Joassin, Campanella et al., 2004; Joassin, Maurage et al., 2004; Joassin et al., 2007; Joassin, Pesenti et al., 2011), we decided to continue to apply the same super-additive

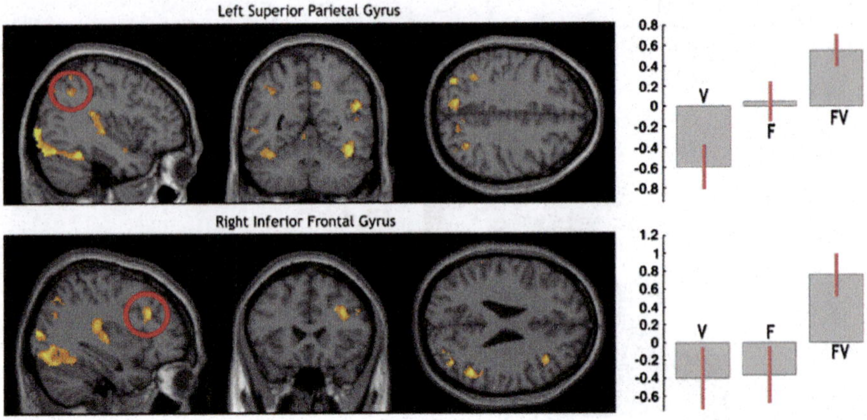

Fig. 8.4 *Left side*: brain sections of the contrast [FV – (V + F)] centered on the left superior parietal gyrus (*upper part*) and the right inferior frontal gyrus (*lower part*). *Right side*: activation changes for each condition in the left superior parietal gyrus (*upper part*) and the right inferior frontal gyrus (*lower part*). $p < .05$ corrected for multiple comparisons at cluster size. *V* voices, *F* faces, *FV* face–voice associations

criterion. In the same way, we used static faces identical to those used in our previous experiments (Joassin, Maurage et al., 2004; Joassin, Pesenti et al., 2011) in order to keep the same general methods and to be able to compare the results of these distinct experiments between each other. We predicted that if gender and identification processing share a single cognitive route, audiovisual gender categorization should activate the same cerebral network than the recognition of face–voice associations, i.e., a network of cerebral regions composed of the unimodal face and voice areas and supramodal integration regions including left parietal and prefrontal regions.

Results showed that judging the sex of human faces activated the bilateral fusiform gyri, the right inferior frontal gyrus, the left calcarine sulcus, the left thalamus, the left and right inferior parietal gyri and the left putamen. Judging the sex of human voices activated the left and right superior temporal gyri, the right inferior frontal gyrus, and the bilateral regions of the cerebellum. Judging the sex of face–voice associations activated the left and right superior and middle temporal gyri, including the left supramarginal and angular gyri, the right inferior occipital gyrus, the left putamen, the left precuneus, and the right inferior parietal gyrus. The main contrasts of this experiment consisted in subtracting the cerebral activities elicited by the gender categorization of unimodal visual and auditory stimuli from the cerebral activities elicited by the gender categorization of audiovisual stimuli, in order to isolate the specific activations involved in the integration of visual and auditory information during gender processing. The contrast [FV – (F + V)] revealed an extensive activation of the visual and auditory regions including, respectively, the right calcarine sulcus and the left fusiform and middle occipital gyri, and the left and right superior temporal gyri (see Fig. 8.4). We also observed specific integrative activations in the left superior parietal gyrus including the angular gyrus and the right inferior frontal gyrus. Moreover, psychophysiological interaction (PPI) analyses were performed to examine the functional connectivity of the cerebral regions observed in the subtraction. It showed that the left inferior parietal gyrus had an enhanced connectivity with the right fusiform gyrus, the left supplementary motor area, and the right cerebellum. The right STS had an enhanced connectivity with the left auditory STS, the left visual fusiform gyrus, and the left and right putamen. The right calcarine sulcus had an enhanced connectivity with the right STS and the left putamen. Finally, the right inferior frontal gyrus had an enhanced connectivity with the right supramarginal gyrus, the left inferior occipital gyrus, and the left and right STS (Joassin, Maurage et al., 2011).

These results showed that the cross-modal processing of faces and voices was sustained by a neural network composed not only of the unimodal visual and auditory regions but also of two regions, the left superior parietal cortex and the right inferior frontal gyrus, whose activations was specific to the bimodal condition. The PPI analysis centered on the left parietal cortex showed that this region had an enhanced connectivity with the cerebellum and the supplementary motor area. This cerebello-parieto-motor network is important in the cross-modal control of attention (Bushara et al., 1999; Driver & Spence, 2000; Shomstein & Yantis, 2004), and it could sustain the integration of faces and voices by allowing an optimal dispatching of the attentional resources between the visual and the auditory modalities. The PPI analyses also showed that the unimodal visual and auditory regions were interconnected, but had also an enhanced connectivity with several other cerebral regions. At first, the left putamen, as a part of the striatum, is known to play the role of a subcortical integration relay allowing to access and regulate multimodal information by means of dopaminergic channels (Haruno & Kawato, 2006). Secondly, we observed that the unimodal regions were also connected to the right inferior frontal gyrus (Brodmann area 44). This region was activated in both unimodal conditions and is known to receive inputs from face (Rolls, 2000; Leube et al., 2001) and voice sensitive areas (Hesling et al., 2005; Rama and Courtney, 2005). This activation was also interpreted as reflecting a sensibility of this region to semantic congruency and also an involvement in the learning of novel visuo-auditory associations, as suggested by Gonzalo et al. (2000). Supporting this interpretation, McNamara et al. (2008) showed that the right BA44 was activated by the learning of new associations between an arbitrary sound and a gesture. Further experiments, investigating the encoding of such face–voice associations would be helpful to better understand the role of the frontal regions in the cross-modal processing of human stimuli.

It is important to note that this network is highly similar to the network of cerebral regions observed in our previous experiment testing the cross-modal recognition of face–voice associations (Joassin, Pesenti et al., 2011) (see Fig. 8.5). In this experiment, the subtraction between unimodal and bimodal conditions also revealed an activation of the unimodal visual and auditory regions and of the left angular gyrus. It seems thus that the involvement of this network does not depend on the level of processing of faces and voices or the task to perform, but is rather specific to the human stimuli.

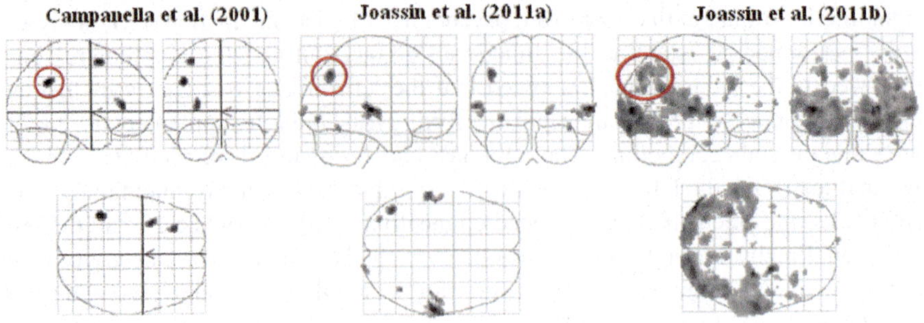

Fig. 8.5 Illustration of the activations of the left parietal gyrus observed by Campanella et al. (2001), Joassin, Pesenti et al. (2011), and Joassin, Maurage et al. (2011)

4 Conclusions

In conclusion, the auditory–visual integration of human faces and voices during the multimodal processing of identity and gender was associated with the activation of a specific network of cortical and subcortical regions. This network included several regions devoted to the different cognitive processing implied in face and voice categorization task—the unimodal visual and auditory regions processing the perceived faces and voices and inter-connected via a subcortical relay located in the striatum, the left superior parietal gyrus, part of a larger parieto-motor network dispatching the attentional resources to the visual and auditory modalities, and the right inferior frontal gyrus sustaining the integration of the semantically congruent information into a coherent multimodal representation. The similarity between the results observed in our two studies supports the hypothesis that the integration of human faces and voices is sustained by a network of cerebral regions activated independently of the task to perform or the cognitive level of processing (gender processing or recognition).

These results raise several new questions that further experiments will help to answer, notably about the possible specificity of the observed network for the processing of the human stimuli relative to other kinds of visuo-auditory associations or the explicit/controlled versus implicit/automatic aspects in the integration of highly ecological social stimuli such as the human faces and voices. Moreover, if the study of the cross-modal processes between sensory modalities is particularly important for a better understanding of the neural networks operating in the healthy brain, it is also important to better understand the neuro-functional impairments in several psychopathological and developmental disorders. For instance, exploring the multimodal integration in the growing brain is particularly important, as it seems that difficulties in information integration may lead to some developmental disorders, such as autism (Melillo & Leisman, 2009) or the pervasive developmental disorders (PDD, Magnée, de Gelder, van Engeland, & Kemner, 2008). Moreover, it appears that the impairments in the recognition of the emotions might be due to distinct neuro-functional impairments such as a deficit of the connectivity between several brain regions in autism (the amygdala and the associative temporal and prefrontal gryi, Monk et al., 2010) or a hypoactivation of the visual regions in other pathological conditions such as schizophrenia (Seiferth et al., 2009). Accordingly, we have recently shown that chronic alcoholism is associated with a specific impairment of the visuo-auditory recognition of emotions (Maurage, Campanella, Philippot, Pham, & Joassin, 2007) and that it is linked to a hypoactivation of the prefrontal regions (Maurage et al., 2008). These studies exploring the impairments of the multimodal processing of faces and voices in psychopathology are particularly important to develop, as they may lead to multimodal therapy that would include a combination of somatosensory, cognitive, behavioral, and biochemical interventions (Melillo and Leisman, 2009). Chapters of the section IMPAIRMENT of the present book will address these questions thoroughly.

References

Beauchamp MS (2005) Statistical criteria in fMRI studies of multisensory integration. Neuroinformatics 3:93–113

Beauchemin M et al (2006) Electrophysiological markers of voice familiarity. European Journal of Neuroscience 23:3081–3086

Belin P, Zatorre RJ, Lafaille P, Ahad P, Pike B (2000) Voice-selective areas in human auditory cortex. Nature 403:309–312

Bernstein LE, Auer ET Jr, Wagner M, Ponton CW (2008) Spatiotemporal dynamics of audiovisual speech processing. Neuroimage 39:423–435

Bodamer J (1947) Die Prosop-Agnosia (Die Agnosie des Physionomeerkennens). Archives fur Psychiatrie und Nervenkrankenheiten 179:6–33

Bruce V, Young A (1986) Understanding face recognition. British Journal of Psychology 77(3):305–327

Burton AM, Bruce V, Johnston RA (1990) Understanding face recognition with an interactive model. British Journal of Psychology 81:361–380

Bushara KO, Hanakawa T, Immish I, Toma K, Kansaku K, Hallett M (2003) Neural correlates of cross-modal binding. Nature Neuroscience 6(2):190–195

Bushara KO, Weeks RA, Ishii K, Catalan MJ, Tian B, Rauschecker JP et al (1999) Modality-specific frontal and parietal areas for auditory and visual spatial localization in humans. Nature Neuroscience 2:759–766

Calvert GA (2001) Crossmodal processing in the human brain: Insights from functional neuroimaging studies. Cerebral Cortex 11:1110–1123

Calvert GA, Campbell R, Brammer MJ (2000) Evidence from functional magnetic resonance imaging of crossmodal binding in human heteromodal cortex. Current Biology 10:649–657

Campanella S, Belin P (2007) Integrating face and voice in person perception. Trends in Cognitive Sciences 11(12):535–543

Campanella S, Hanoteau C, Depy D, Rossion B, Bruyer R, Crommelinck M (2000) Right N170 modulation in a face discrimination task: An account for categorical perception of familiar faces. Psychophysiology 37:796–806

Campanella S, Joassin F, Rossion B, De Volder AG, Bruyer R, Crommelinck M (2001) Associations of the distinct visual representations of faces and names: A PET activation study. Neuroimage 14:873–882

Driver J, Spence C (2000) Multisensory perception: Beyond modularity and convergence. Current Biology 10:731–735

Ganel T, Goshen-Gottstein Y (2002) Perceptual integrity of sex and identity of faces: Further evidence for the single-route hypothesis. Journal of Experimental Psychology Human Perception and Performance 28:854–867

Garrido L, Eisner F, McGettigan C, Stewart L, Sauter D, Hanley JR, Schweinberger SR, Warren JD, Duchaine B (2009) Developmental phonagnosia: A selective deficit of vocal identity recognition. Neuropsychologia 47(1):123–131

Gauthier I, Skudlarski P, Gore JC, Anderson AW (2000) Expertise for cars and birds recruits brain areas involved in face recognition. Nature Neuroscience 3:191–197

Gonzalo D, Shallice T, Dolan R (2000) Time-dependent changes in learning audiovisual associations: A single-trial fMRI study. Neuroimage 11:243–255

Haruno M, Kawato M (2006) Heterarchical reinforcement-learning model for integration of multiple cortico-striatal loops: fMRI examination in stimulus–action–reward association learning. Neural Networks 19:1242–1254

Haxby JV et al (2000) The distributed human neural system for face perception. Trends in Cognitive Sciences 4:223–233

Hesling I, Clément S, Bordessoules M, Allard M (2005) Cerebral mechanisms of prosodic integration: Evidence from connected speech. Neuroimage 24:937–947

Joassin F, Campanella S, Debatisse D, Guérit JM, Bruyer R, Crommelinck M (2004a) The electrophysiological correlates sustaining the retrieval of face–name associations: An ERP study. Psychophysiology 41:625–635

Joassin F, Maurage P, Bruyer R, Crommelinck M, Campanella S (2004b) When audition alters vision: An event-related potential study of the cross-modal interactions between faces and voices. Neuroscience Letters 369:132–137

Joassin F, Maurage P, Campanella S (2011a) The neural network sustaining the crossmodal processing of human gender from faces and voices: An fMRI study. Neuroimage 54(2):1654–1661

Joassin F, Meert G, Campanella S, Bruyer R (2007) The associative processes involved in faces–proper names vs. objects–common names binding: A comparative ERP study. Biological Psychology 75(3):286–299

Joassin F, Pesenti M, Maurage P, Verreckt E, Bruyer R, Campanella S (2011b) Cross-modal interactions between human faces and voices involved in person recognition. Cortex 47:367–376

Kanwisher N, McDermott J, Chun MM (1997) The fusiform face area: A module in human extrastriate cortex specialized for face perception. The Journal of Neuroscience 9:462–475

Kerlin JR, Shahin AJ, Miller LM (2010) Attentional grain control of ongoing cortical speech representations in a "cocktail party". The Journal of Neuroscience 30(2):620–628

Laurienti PJ, Perrault TJ, Stanford TR, Wallace MT, Stein BE (2005) On the use of superadditivity as a metric for characterizing multisensory integration in functional neuroimaging studies. Experimental Brain Research 166:289–297

Leube DT, Erb M, Grodd W, Bartels M, Kircher TTJ (2001) Differential activation in parahippocampal and prefrontal cortex during word and face encoding tasks. Neuroreport 12(12):2773–2777

Magnée M, de Gelder B, van Engeland H, Kemner C (2008) Atypical processing of fearful face–voice pairs in Pervasive Developmental Disorder: An ERP study. Clinical Neurophysiology 119:2004–2010

Maurage P, Campanella S, Philippot P, Pham T, Joassin F (2007) The crossmodal facilitation effect is disrupted in alcoholism: A study with emotional stimuli. Alcohol and Alcoholism 42: 552–559

Maurage P, Philippot P, Joassin F, Alonso Prieto E, Palmero Soler E, Zanow F, Campanella S (2008) The auditory–visual integration of anger is disrupted in alcoholism: An ERP study. Journal of Psychiatry and Neuroscience 33(2):111–122

McNamara A, Buccino G, Menz MM, Gläsher J, Wolbers T, Baumgärtner A, Binkofski F (2008) Neural dynamics of learning sound–action associations. PLoS One 3(12):1–10

Melillo R, Leisman G (2009) Autistic spectrum disorders as functional disconnection syndrome. Reviews in the Neurosciences 20(2):111–131

Monk C, Weng SJ, Wiggins J, Kurapati N, Louro H, Carrasco M, Maslowsky J, Risi S, Lord C (2010) Neural circuitry of emotional face processing in autism spectrum disorders. Journal of Psychiatry and Neuroscience 35(2):105–114

Puce A, Allison T, Gore JC, McCarthy G (1995) Face-sensitive regions in human extrastriate cortex studied by functional MRI. Journal of Neurophysiology 74(3):1192–1199

Rama P, Courtney SM (2005) Functional topography of working memory for face or voice identity. NeuroImage 24:224–234

Rolls ET (2000) The orbitofrontal cortex and reward. Cerebral Cortex 10:284–294

Schweinberger SR, Robertson D, Kaufmann JM (2007) Hearing facial identities. The Quarterly Journal of Experimental Psychology 60(10):1446–1456

Seiferth N, Pauly K, Kellermann T, Shah N, Ott G, Herpertz-Dahlmann B, Kircher T, Schneider F, Habel U (2009) Neuronal correlates of facial emotion discrimination in early onset schizophrenia. Neuropsychopharmacology 34:477–487

Senkowski D, Schneider TR, Foxe JJ, Engel AK (2008) Crossmodal binding through neural coherence: Implications for multisensory processing. Trends in Cognitive Sciences 31(8):401–409

Sheffert SM, Olson E (2004) Audiovisual speech facilitates voice learning. Perception & Psychophysics 66(2):352–362

Shomstein S, Yantis S (2004) Control of attention shifts between vision and audition in human cortex. The Journal of Neuroscience 24(47):10702–10706

Smith EL, Grabowecky M, Suzuki S (2007) Auditory–visual crossmodal integration in perception of face gender. Current Biology 17:1680–1685

Steeves J, Dricot L, Goltz HC, Sorger B, Peters J, Milner AD, Goodale MA, Goebel R, Rossion B (2009) Abnormal face identity coding in the middle fusiform gyrus of two brain-damaged prosopagnosic patients. Neuropsychologia 47(12):2584–2592

Van Lancker DR, Canter GJ (1982) Impairment of voice and face recognition in patients with hemispheric damage. Brain and Cognition 1(2):185–195

Von Kriegstein K, Giraud AL (2006) Implicit multisensory associations influence voice recognition. PLoS Biology 4:e326

Von Kriegstein K, Kleinschmidt A, Sterzer P, Giraud AL (2005) Interaction of face and voice areas during speaker recognition. Journal of Cognitive Neuroscience 17(3):367–376

Chapter 9
Neurophysiological Correlates of Face and Voice Integration

Aina Puce

1 Introduction

On meeting another individual for the first time during the course of a conversation we learn a lot unique and idiosyncratic facts about that person. If that meeting is socially or professionally significant, in the future, on crossing paths with that individual again, we can easily remember their details and the circumstances of the meeting. The person's age, gender, ethnic or racial background, place of birth, current place of residence, professional and personal interests are all important pieces of data for building up a mental picture of that individual which we notice without going to too much effort. We might also notice the tone of their voice, the prosody with which they speak, and a foreign accent, if present. Interestingly, all of those details can be readily recalled on either seeing their face or just hearing their voice. Many of the chapters in this book are devoted to the importance of the face and voice in the formed percept we have of another individual—assigning an individual their own unique identity. Other chapters focus on how animals decode these important conspecific details. The questions asked in this chapter pertain less to brain mechanisms active in identifying specific individuals and their characteristics, but focus on issues relating to how non-verbal face and voice cues are integrated by the human brain. Early behavioral studies have noted how important *non-verbal behaviors* are for the interpretation of the actions of others, in terms of presenting important information relating to the social interaction (Campbell & Rushton, 1978; Mehrabian & Ferris, 1967). Yet, this remains a poorly studied area in social neuroscience and is a major focus for our laboratory.

How does the human brain respond to cues sent by the human face and voice relative to audiovisual stimuli involving other animals or inanimate objects? How

A. Puce, Ph.D. (✉)
Department of Psychological and Brain Sciences, Indiana University,
1101 E 10th St., Bloomington, IN 47401, USA
e-mail: ainapuce@indiana.edu

P. Belin et al. (eds.), *Integrating Face and Voice in Person Perception*,
DOI 10.1007/978-1-4614-3585-3_9,

does information from the senses become integrated in the human brain when processing dynamic human face and voice cues? These two questions form the main focus of this chapter and two studies with noninvasive electrophysiological techniques in human subjects are described here.

2 Human Brains Generate Distinct Neural Signatures to Viewing Changes in Human Faces and Concurrent Vocalizations

In everyday life we usually see changes in facial expressions concurrently with heard vocalizations, which are most typically verbal utterances. Given this experience it could be argued that it is difficult to think of a face as an isolated unisensory entity, and the same could be said for the voice. Given our vast experience with the coupled human face–human voice stimulus it is not surprising that some might argue that the human brain might harbor specialized neural mechanisms for processing this very important compound multisensory stimulus type. To investigate this question we ran an electrophysiological study where neural responses to a human face–voice pairing were evaluated and compared to audiovisual pairings of a nonhuman primate and an inanimate control stimulus. By using audiovisual stimulus pairings that were congruent or incongruent we were able to investigate if human face–voice pairings could generate unique and distinct neural signatures in the human brain. I describe the experiment briefly, as the details of the data acquisition and analysis can be accessed elsewhere (Puce, Epling, Thompson, & Carrick, 2007).

Three types of overall luminance and contrast matched grayscale visual stimuli were used: a human male clean-shaven face, a monkey face, and an image of a house with two windows and front door positioned to spatially be similar to the eyes and mouths on the human and monkey faces. The three auditory stimuli consisted of a human burp, a monkey screech, and a creaking door sound. The harmonic-to-noise ratios (Lewis, Talkington, Puce, Engel, & Frum, 2011) on the auditory stimuli were altered to match as closely to one another as possible, but still render the auditory stimuli to be discriminated as distinct recognizable entities. All experimental trials had an audiovisual stimulus pairing which was randomly presented to be congruous (e.g., human face/burp, monkey face/screech, house/creaking door sound) or incongruous (e.g., human face/creaking door, monkey face/burp, house/screech, etc.). Subjects made a two-button forced choice response to indicate if a given stimulus pairing was congruous or incongruous. The structure of each experimental trial ensured that neural responses occurring to an initial visual stimulus had died away prior to the audiovisual stimulus of interest being presented. The audiovisual stimulus consisted of a sound onset and a concurrent change in the visual display (e.g., mouth or front door opening) so that it appeared that the sound was generated by the visual stimulus transition. When the sound was completed (after 400 ms) the visual display returned back to its initial baseline position (mouth or front door closed). The electroencephalogram (EEG) was recorded continuously during the experiment.

Averaged event-related potentials (ERPs) were generated to each stimulus condition after rejecting incorrect behavioral trials and trials with EEG artifacts for each subject, and also for the group as a whole. Repeated measures ANOVAs were performed on the peak amplitudes and latencies of prominent ERP components as a function of stimulus condition.

Elicited ERPs consisted of a stereotypical set of components consisting of an auditory negative potential with a central vertex maximum (N140), a visual negative potential that was maximal at bilateral posterior temporal sites (N170), and later ERPs. One of the later ERPs was a positivity seen over the posterior scalp at around 400 ms post-stimulus (P400). Figure 9.1a displays the characteristic scalp topography of these responses for a congruous audiovisual stimulus pairing. Notably, when the three congruous audiovisual conditions were contrasted (Fig. 9.1b), the N140 to the human and monkey audiovisual stimulus pairings was significantly larger relative to the inanimate (house) stimulus pair. One interpretation for these findings might be that our brains are more sensitive to these primate (animate) audiovisual stimulus categories. While there is always the possibility that a low-level auditory stimulus difference might be at the root of this amplitude difference, this is unlikely for two reasons. First, the stimuli were matched as best as possible to each other in terms of their harmonic-to-noise properties. Notwithstanding this, however, their fundamental auditory qualities will necessarily be different at some level, since despite our auditory adjustments they remained recognizable as distinct auditory objects. Second, no significant differences in N140 amplitudes were noted when these same auditory stimuli were paired with either the monkey face or the house visual stimuli (data not shown). This would suggest that the amplitude differences seen in N140 in the context of the human and nonhuman primate faces are modulated by the multisensory stimulus context and not by differences in the low-level auditory characteristics of the stimuli.

In the incongruous audiovisual stimulus pairings the only significant ERP differences were noted only when the *visual* human face stimulus was presented. This time a significantly smaller *auditory* N140 amplitude was observed when the human face was paired with an incongruous auditory stimulus, relative to the congruous condition (Fig. 9.1c, Cz ERP waveforms). These data indicate that in addition to showing sensitivity to primate (or animate) audiovisual stimuli for congruous stimulation (discussed above), there is a clear preference by the human brain for a *human* auditory and *human* visual stimulus combination. This preference occurs at an early point on time (N140) when the information is thought to be in the primary auditory cortex (Eggermont & Ponton, 2002; Giard & Peronnet, 1999). Again, it is unlikely that auditory low-level stimulus differences were the cause of this effect, as the audiovisual incongruous pairings with the monkey face and the image of the house did not show these differences (data not shown).

The only other observed differences in ERPs between congruous and incongruous conditions were in P400 amplitude, which appeared to be driven by the extent of the incongruity. Significantly larger P400s were observed over the posterior temporal scalp to the animate–inanimate audiovisual stimulus pair of the human face and house sound (Fig. 9.1c). The pairing of human face and monkey sound did not elicit late ERP effects that were different to the congruous condition (data not shown).

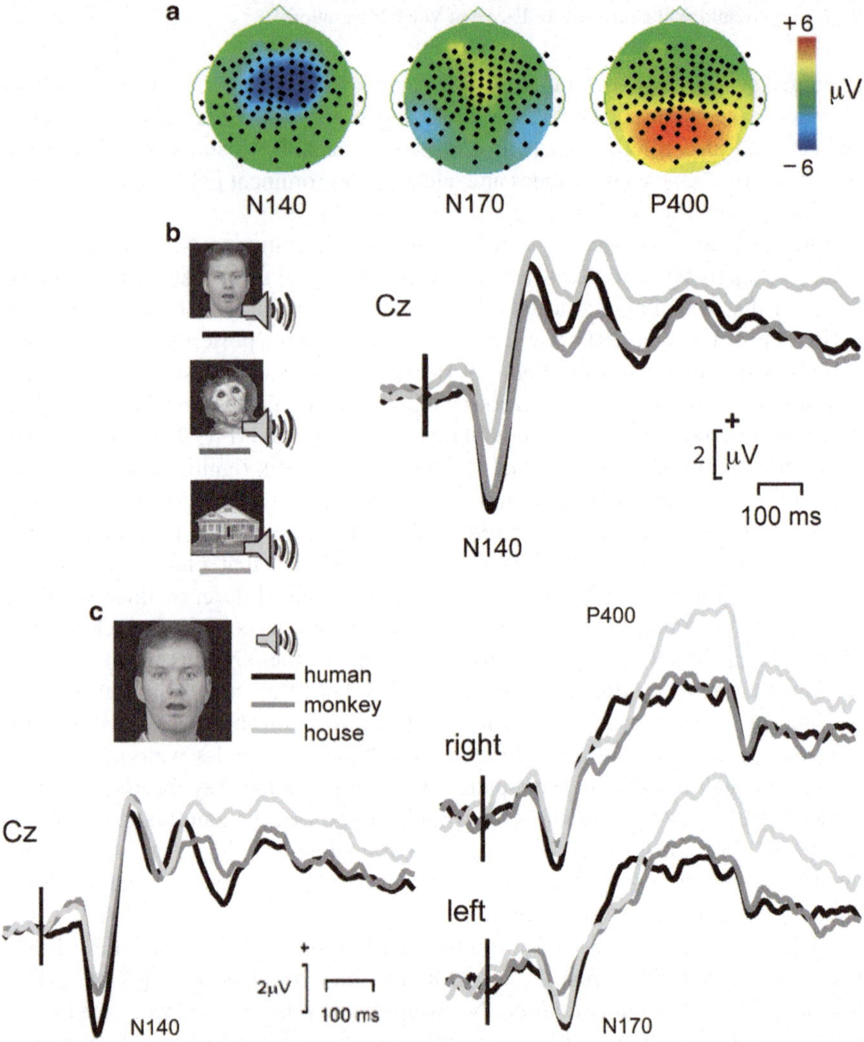

Fig. 9.1 Group data display typical ERPs and scalp topography elicited to congruous and incongruous audiovisual stimulus pairings. (**a**) Scalp topographic voltage maps shown at the peak of three different ERP components. N140 is a negative potential occurring over the central scalp (vertex or Cz), N170 a negative potential that occurs over the bilateral posterior occipitotemporal scalp, and P400 is a later positivity seen over the posterior scalp. The *color calibration bar* indicates voltage ranges of ±6 μV. These maps show the data from the congruous monkey face/monkey screech condition, however, all three congruous audiovisual stimulus pairings elicit similar scalp topographies. (**b**) N140 at the vertex (Cz) to the three congruous audiovisual stimulus pairings shows a significant difference between the responses to the face stimuli (human or monkey) relative to the house stimulus. No difference is seen in amplitude of response between the human and monkey stimuli. *Horizontal* and *vertical calibration bars* depict time (in ms) and response amplitude (in μV). (**c**) Vertex N140s (*left panel*), and P400s from the left and right posterior temporal scalp (*right panel*) elicited to the human face when paired with congruous and incongruous auditory stimulation. N140 was significantly decreased for both incongruous auditory stimuli relative to the matched stimulus. P400 was significantly larger for the human face and creaking door pairing relative to either the other incongruous condition (human face/monkey screech) or the congruous stimulus pairing

The multisensory incongruity response involving the human face stimulus and non-voice auditory pairing described here is intriguing, relative to the previous ERP literature dealing with incongruity. The first human incongruity potential, the N400, was described in 1980 and was elicited by sentences with anomalous endings and was named the "semantic incongruity" (Kutas & Hillyard, 1980). However, other types of incongruities were also studied, and relevant to the current data a so-called physical incongruity was reported (McCallum, Farmer, & Pocock, 1984). In this instance, if a sentence spoken by a speaker was suddenly completed by another speaker, the experienced auditory inconsistency produced a late ERP that was positive in polarity and which typically occurred at around 400 ms post-stimulus. Because this potential was elicited as a function of a physical stimulus change (different speaker's voice) and was not associated with a semantic violation, McCallum and colleagues designated it as a "physical incongruity" potential. The incongruity literature has always focused mainly on incongruities elicited in the *same* sensory stimulus modality. Here we are eliciting a multisensory incongruity P400 that occurs only to incongruous stimulus pairings with a *human* face. The nature of our multisensory incongruity could be regarded as physical, so in this sense we identify our multisensory incongruity P400 as being a variant of McCallum's original visual physical incongruity.

One somewhat unexpected finding in our multisensory incongruity experiment was no net difference in N170 amplitudes or latencies with our experimental manipulation. Our own work with unisensory manipulations of dynamic faces (Puce & Perrett, 2003; Puce, Smith, & Allison, 2000) has shown modulations in N170 amplitude and latency as a function of type of facial movement that do not appear to be influenced by social context when facial movement type is kept constant (Carrick, Thompson, Epling, & Puce, 2007; Puce et al., 2007). It is also well known that different visual stimulus categories such as static faces and objects elicit significantly different N170s, with face stimuli eliciting the largest N170s (Bentin, Allison, Puce, Perez, & McCarthy, 1996; Itier & Batty, 2009; Rossion & Jacques, 2008). It is interesting to speculate that since the auditory component of the audiovisual stimulus pair is probably processed first (based on the latencies of the sensory ERPs in each sensory modality), the audiovisual context in this experiment may have modulated N170 amplitude to produce no effective amplitude difference across dynamic mouth opening and dynamic door opening stimuli—stimulus categories which might be expected to produce amplitude differences when presented in the visual modality in isolation. The presence of the auditory information may have served to potentially enhance processing of the visual stimulus – consistent with the expected augmentation provided by multisensory stimulation. Better stimulus registration would optimize subsequent processing, which is typiclly signaled by later potentials such as P400. Typically, it is the later ERPs that are affected by (social or cognitive) context when subjects are asked to make judgments about face stimuli (Carrick et al., 2007; Puce et al., 2007; Sabbagh, Moulson, & Harkness, 2004). Typically, N170 experiments are usually performed in a unisensory paradigm, so it is hard to know what the underlying mechanism might be for behavior of the N170 elicited in a multisensory experimental manipulation. Our data underscore the need for subsequent experiments using faces or voices to utilize a multisensory context.

In sum, and in answer to our first question, the processing of the human face and human voice is augmented when compared to other animate and inanimate audiovisual pairings. Sensory ERP components did not show a unique morphology or scalp topography; however, their amplitudes were affected by multisensory context. Auditory N140 appears sensitive to the human face–human voice pairing—the amplitude of this response is significantly larger to that of the other congruous pairings, and in addition to some sensitivity in the auditory modality to an animate stimulus, the observed multisensory effect is clearly influenced by the presence of the visual stimulus—which set a clear visual context at the beginning of each trial. N170 also appears to be influenced by the presence of the auditory stimulus, and it is likely that auditory neural activity which occurs earlier than visual activity may modulate this visual processing.

3 How Is Information from Dynamic Human Faces and Vocalizations Integrated in the Human Brain?

In order to try and get at aspects of N170 amplitude differences across sensory stimulation type described above, one way to tackle this question could be to perform a classic unisensory versus multisensory experimental manipulation. In the experiment described below, unisensory only, auditory only, and audiovisual combined stimulation were performed to examine not only how visual N170 but also auditory N140 and subsequent ERPs might be modulated by a multisensory context.

We chose to work with an avatar face as a dynamic stimulus, since it could be easily controlled on a frame-by-frame basis. Our previous studies had used the apparent or real facial movements (e.g., eye aversions, mouth opening) that were very simple relative to the capability of humans to generate facial movements or expressions involving multiple parts of the face (Puce & Perrett, 2003; Puce et al., 2000; Wheaton, Pipingas, Silberstein, & Puce, 2001). Hence, we chose to use a more complex dynamic face stimulus, but which would not be typically associated with vocalizations. Non-affective, non-verbal facial movements are generated when we sneeze, cough, or yawn, for example. These types of facial movements are always accompanied by non-verbal vocalizations also, and hence for our purposes are an ideal audiovisual stimulus to use in this instance. We also avoided asking subjects to make overt social or affective evaluations of the stimuli, as we were interested in how the neurophysiological response to a complex facial motion would be modulated by the presence of a congruous auditory stimulus (as seen in daily life). Also, to ensure that subjects divided their attention equally between the auditory and visual modalities, two types of target stimuli, a visual and an auditory stimulus, were used. Finally, so that we could examine both the temporal dynamics and neuroanatomical loci of multisensory versus unisensory processing, one group of subjects were studied with noninvasive EEG and another group completed an fMRI study. Complete details of the tasks, stimuli, and data acquisition and analyses have been described elsewhere (Brefczynski-Lewis, Lowitszch, Parsons, Lemieux, & Puce, 2009). The results of the fMRI study will not be discussed here.

The visual stimulus consisted of a computer generated female avatar face displaying common facial movements depicting common actions which included yawns, sighs, coughs, sneezes, and burps. Matching audio stimuli were created by a female voice, so that when the concurrent audiovisual stimulus was presented, subjects reported that the stimulus was very realistic—so much so that some subjects noted that they had a strong desire to turn away when the avatar sneezed at them! Two target stimuli were created: a visual only target where the avatar blinked and no associated sound was presented, and a auditory only target where the avatar said "mmm" but did not change her facial configuration. Stimuli were presented in blocks of visual only (V), auditory only (A), and combined audiovisual (AV) stimulation. Subjects were instructed to be on the alert for either target stimulus that could occur randomly throughout the experiment. Peak amplitudes and latencies of ERP components were calculated from averaged EEG data as a function of stimulus condition.

Data from multisensory experiments typically produce complex effects, where the addition of another sensory modality can augment or diminish behavioral and neural responses elicited in the other sensory modality. We examined our data for such multisensory interaction effects as superadditivity and underadditivity, in line with previous ERP and fMRI studies (Calvert, 2001; Giard & Peronnet, 1999; Stein, Stanford, Ramachandran, Perrault, & Rowland, 2009). We defined our assessment criteria for ERP components (also applicable to fMRI activation) as follows:

1. Superadditive: where audiovisual (AV) > auditory alone (A) + visual alone (V)
2. Subadditive: AV > V or AV > A, where V > 0 or A > 0
3. Underadditive: AV < A + V, and AV < A, or AV < V
4. Common: AV = A, or AV = V, or AV = A = V

It should be noted that since our data were originally published, a more recent and thorough classification for defining effects of multisensory interactions has been proposed for the field (Stein et al., 2009).

ERP waveform morphology to the dynamic avatar stimulus was similar to that described in the previous study using an apparent motion task with a grayscale human (and also monkey) face. Clear vertex-centered auditory N140s and bilateral occipitotemporal visual N170s were produced to each respective unisensory condition in addition to the audiovisual condition. The two sensory ERP components were more likely to show the effects of underadditivity—visual N170 was significantly diminished in the audiovisual condition relative to visual only stimulation, particularly in the right hemisphere. Auditory N140 showed a trend for underadditivity in the audiovisual condition relative to unisensory stimulation. Hence, the presence of an auditory stimulus clearly influences the visual neurophysiological response. In the other experiment described earlier in this chapter, a similar conclusion was drawn, but in that instance concurrent audiovisual stimulation was always present.

Overall, the dataset consisted of a complex series of ERP components in response to the dynamic avatar-vocalization stimuli (depicted schematically in Fig. 9.2). Most effects were seen mainly in the posterior scalp, but also featured very late components that were seen across the frontal scalp. Bilateral temporoparietal electrodes

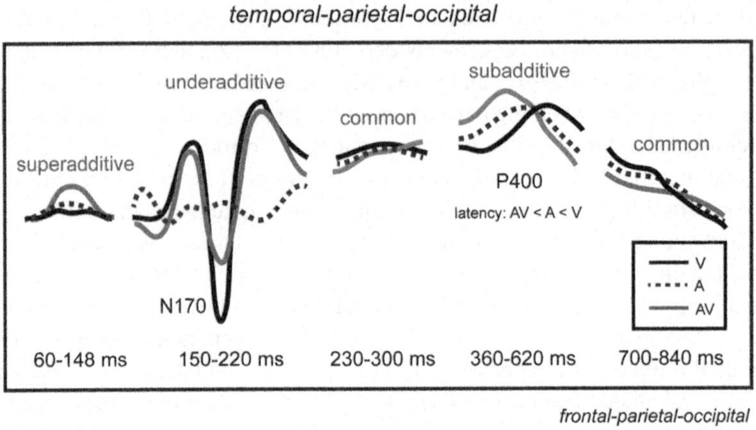

Fig. 9.2 Schematic representation of main ERP findings in a face–voice audiovisual integration study. Varied multisensory interaction effects occurred in ERP data as a function of post-stimulus time interval. Superadditivity and underadditivity were noted in the earliest neurophysiological responses, at 60–148 ms post-stimulus and 150–220 ms, respectively. More complex subadditive effects were noted after these time intervals. Common activation was observed during 200–300 ms post-stimulus and between 360 and 620 ms post-stimulus where the audiovisual condition (AV) produced a P400 potential that was larger than those to either respective unisensory condition (i.e., A or V), but where audiovisual ERP amplitudes could be larger or smaller relative to the sum of the amplitudes of the respective unisensory conditions (i.e., A+V). Finally, common activation was observed, as ERP waveforms appear to return to baseline during 700–840 ms post-stimulus. All the abovementioned effects were seen in selected electrode clusters overlying the posterior scalp. The late potentials (700–840 ms) were also prominent across the frontal scalp. Legend: AV=audiovisual, A=auditory, V=visual

showed common activation in the 230–300 ms post-stimulus latency range. The most prominent activity was that of a late positive component with a very broad scalp distribution that was vertex-centered and persisted for several hundred milliseconds. The peak latency for this potential was earliest for the audiovisual condition (392 ms) and auditory condition (400 ms), with the visual condition producing a very late peak response at around 492 ms. The sustained positivity, or P400, exhibited subadditivity, based on our criteria (see above).

Late ERP components that have typically associated with evaluative or cognitive processing of stimuli can exhibit complex effects such as subadditivity (Stein et al., 2009), which can produce larger amplitudes in the audiovisual condition (relative to each respective unisensory conditions), and audiovisual responses which are not necessarily larger than the sum of the respective unisensory conditions. While it is tempting to equate this late positive potential with the P400 discussed in the previous section, the scalp distribution of the multisensory incongruity P400 was considerably posterior to the response being described here, and unlike the incongruity potential, the peak latency of the avatar elicited P400 also shifted as a function of stimulus condition. Based on the current data, it is not clear what this neurophysiological response may index. It is unrelated to target effects, as ERPs to the target stimuli were not included in the averages. Further work using complex face-voice animations is needed, where explicit evaluation of these stimuli might be able to shed some light on the functional significance of this late neurophysiological response.

The dynamic avatar and vocalizations also produced several electrophysiological responses not previously observed in our studies. First, a unique early bilateral occipitotemporal ERP was observed in the audiovisual condition that preceded either the vertex-centered N140 or occipitotemporal N170—in the range 60–148 ms post-stimulus. Interestingly, this response was not present in either unisensory condition. We categorized this very early effect as superadditive (Stein et al., 2009). Early multisensory responses in this latency range have been described, albeit for audiotactile stimulation (Foxe et al., 2000).

Sustained activity was seen up to 700–840 ms post-stimulus that was most evident in the posterior scalp for all conditions and was also present in the frontal scalp. These very late neurophysiological responses did not form clear defined peaks of activity, relative to the other ERPs that have been discussed here. If anything, it appeared that this sustained neurophysiological activity was a gradually return back to baseline (zero voltage). Future studies using both single trial time-frequency analysis and ERP averaging will be needed to investigate these responses further. For example, it may be that the early response in the 60–148 ms range most likely might manifest as a transient burst of EEG activity in a discrete frequency range, whereas the prolonged sustained activity which persists out to 840 ms could be due to either sustained postsynaptic potential activity or induced changes in EEG rhythms potentially in beta or gamma frequency bands (Engel & Fries, 2010; Engel, Fries, & Singer, 2001; Herrmann, Frund, & Lenz, 2010; Mazaheri & Jensen, 2010; Young & Eggermont, 2009).

In sum, and in answer to the second question posed in this chapter, it is clear that multisensory processing for dynamic non-verbal face–voice information begins early in the post-stimulus period between 60 and 150 ms. Initial face–voice processing produces a superadditive response, setting the stage for underadditivity in subsequent sensory ERPs, and a complex set of multisensory interactions in subsequent later ERPs that are visible up to around 840 ms post-stimulus. Much of the activity occurs in the posterior temporal scalp, however, with increasing time post-stimulus the presence of more anterior activity can be seen.

4 Relevance of the Current Findings to Existing Literature and Implications for Future Studies

The ERP data generated in the second study in particular clearly show that in the post-stimulus time interval a very complex set of neurophysiological transitions and interactions can occur when auditory and visual stimulation is combined (summarized in Fig. 9.2). Therefore, over the entire post-stimulus time interval there was a clear effect of one stimulus modality on another in the human neural response. Because of the design of this study, we could not evaluate the effects of multisensory stimulation on behavior, as subjects performed a unisensory target detection task designed to discourage attention being explicitly focused on one particular sensory modality. Having said that, our data nevertheless underscore the power of the combined face–voice stimulus: in this instance all stimuli were clearly discriminable and the observed neural effects were clearly seen particularly in ERPs traditionally regarded as "auditory" or "visual" neural responses. Recently, however, more and more studies in humans and animals are beginning to challenge the notion of "unisensory" cortices and underscore that information from other stimulus modalities can influence early processing of unisensory stimuli (Foxe et al., 2000; Ghazanfar & Schroeder, 2006; Kayser, Petkov, Augath, & Logothetis, 2007). These types of effects have caused some controversy and debate about cortical function and putative hierarchical processing of sensory information in the cerebral cortex. Yet, it is well known that these multisensory interactions can occur at a *subcortical level*—a robust finding that was demonstrated many years ago (Meredith & Stein, 1986)!

Typically, the real benefits of a multisensory stimulus are conferred when one or the other unisensory stimulus is not clearly discriminable, or when the task to be performed is demanding or challenging. Indeed, it is known that when the intensity of one stimulus is very low, the addition of the other sensory modality will produce gains in the multisensory response that dwarf the augmentation of the response that occurs when the stimulus is readily discriminable. This is known as principle of inverse effectiveness (Meredith & Stein, 1986; Stein et al., 2009). The enhancement of an early audiovisual response in our ERP data in the 60–148 ms latency range where no clear unisensory response was present in either modality is intriguing and cannot be attributed to this effect as our stimuli were clearly discriminable. Hence, the functional significance of our early superadditive ERP response is unknown. However, further studies exploring supra- and subthreshold face–voice stimuli would be interesting to pursue in order to characterize these effects further and understand their functional significance. Traditionally, human visual ERP studies do not readily describe early visual responses, so it would be interesting to see if a clear response evolves in the presence of a poorly discriminable visual stimulus when it is accompanied by concurrent auditory stimulation in this early latency range. It is interesting to speculate that this type of effect might have been missed in the past as investigators have always strived to use high-quality perceptual visual stimuli to study "visual cortex": if earlier studies had performed this type of near- or subthreshold multisensory stimulation more often, perhaps concepts such as "primary sensory cortex" might have been defined using different criteria.

Recent noninvasive human studies using fMRI have shown how reliable the principle of inverse effectiveness is for different types of cross-modal stimulation (Kim & James, 2010; Stevenson & James, 2009; Stevenson, Kim, & James, 2009). fMRI studies also have the capability of potentially identifying the functional neuroanatomical loci for these multisensory interaction effects, however, one must always consider the sluggishness of the hemodynamic response. An additional important consideration is that multisensory interactions can evolve differently over the post-stimulus time epoch, as shown by the ERP data described in this chapter (see Fig. 9.2). Hence, there may be an issue with how an overall hemodynamic subadditive effect might be interpreted, given that the neurophysiological response to the same stimulation shows multiple types of multisensory interaction effects over the same time interval. Having said that, the neurophysiological response may well represent the sum of neural activity that may have a number of generators and itself may not be a unitary phenomenon.

Despite these caveats, fMRI studies have clearly demonstrated that the cortex of the superior temporal region is important for multisensory processing, e.g., (Nath & Beauchamp, 2011a; Stevenson, Geoghegan, & James, 2007; Stevenson & James, 2009; Wright, Pelphrey, Allison, McKeown, & McCarthy, 2003). A recent study also advances the idea of changes in functional connectivity between different brain regions when stimulus discriminability becomes an issue—a finding that is relevant for the previous discussion on inverse effectiveness. When the discriminability of either the visual or auditory stimulus is low, there appears to be an increase in functional connectivity between the superior temporal sulcus (STS) and the respective "primary sensory cortex" in which the sensory stimulus is the more discriminable (Nath & Beauchamp, 2011a). This is an interesting new line of research that will no doubt be extremely informative for the study of these complex multisensory interactive mechanisms.

Our non-verbal face–voice integration ERP data also need to be considered in the light of the face–voice interaction literature on interpreting visual speech, which many investigators have argued might form a special category of multisensory integration. A number of chapters in this volume deal with this phenomenon and hence will not be discussed here. However, phenomena such as the McGurk effect have been used as evidence for specialization for human face and voice (McGurk & MacDonald, 1976), where the experienced audiovisual stimulus produces a completely different percept relative to each individual unisensory stimulus. That there is a window of time of around 180 ms where the audiovisual stimuli can be jittered is known not only from behavioral studies (Munhall, Gribble, Sacco, & Ward, 1996) but also was suggested from the results of a very early magnetoencephalography (MEG) study examining the neural correlates of this effect (Sams et al., 1991). Sams and colleagues (1991) showed audiovisual speech stimuli to subjects in a heroic experiment for its time, given the status of the video technology of the day. They elicited the McGurk effect behaviorally and recorded from a limited MEG sensor array over the left temporal cortex (state-of-the-art MEG technology at that time used limited arrays of sensors positioned over the putative cortical regions of interest.). Their experimental design had concordant and discordant audiovisual phonemes, with the discordant stimuli producing a McGurk-like percept. Averaged MEG waveforms recorded from left temporal cortex showed differences across stimulus categories in the post-stimulus latency range beginning around 170 ms post-stimulus that persisted until the end of their recording epoch of around 400 ms.

More recently, invasive recordings in epilepsy surgery patients viewing dynamic lip images and hearing associated vocalizations have verified that the locus of the McGurk integration effect occurs in the superior temporal cortex, but can elicit clear neurophysiological signatures when the lip movement consists of *non-verbal* gunning movements also (Reale et al., 2007). A very recent fMRI study on the McGurk effect shows that the subject's likelihood of experiencing the McGurk effect percept correlates with increased neural activity in the left STS (Nath & Beauchamp, 2011b). Importantly, transcranial magnetic stimulation of the activated fMRI regions disrupts the McGurk effect when stimulation was delivered 100 ms before and up to 100 ms after the onset of the McGurk eliciting stimulus (Beauchamp, Nath, & Pasalar, 2010).

Overall, neurophysiological studies of non-verbal face–voice integration in human subjects have not been a popular area of study in cognitive and social neuroscience. There appears to be only one other human neurophysiological study that has attempted to investigate non-verbal aspects of face–voice integration. Hagan and colleagues (2009) conducted an MEG experiment where audiovisual, auditory, and visual only presentations of fearful and neutral face and voice stimuli were presented in a total of six conditions. Evoked MEG power was compared across conditions across different frequency ranges. The beta frequency range showed the strongest effects for audiovisual integration, although superadditive effects were demonstrated in a range of frequencies spanning 3–80 Hz at post-stimulus latencies of up to 250 ms. Source modeling of this activity for the fearful face condition identified multiple sources including anterior/posterior STS, parietal cortex, and anterior cingulate cortex (Hagan et al., 2009).

Examining the neurophysiological data from both the point of view of visualizing evoked (ERP) and induced (rhythmic oscillations) activity is important for obtaining the complete neurophysiological picture (Herrmann, Munk, & Engel, 2004). An excellent case in point is a recent audiovisual integration study that examined the effects of varying stimulus onset asynchrony (SOA) of an audiovisual stimulus pairing consisting of simple white noise bursts and flashes (Naue et al., 2011). Unisensory and multisensory stimulation conditions were compared. Three types of oscillatory responses (theta, beta, and gamma EEG frequency bands) were observed in the experiment. However, in the combined audiovisual stimulation condition, only the activity in the theta band (peak at around 6 Hz, latency of 50–200 ms) was observed to show subadditive enhancement associated with multisensory integration. Additionally, there was an interesting interaction effect where beta activity (peak at around 29 Hz, latency of 50–100) was enhanced for some SOAs and not others. For auditory stimulation only, beta activity was distributed over the posterior scalp and not centered over the vertex as would be expected if the activity was localized to auditory cortex. Source modeling confirmed that this activity occurred in the visual cortex. Hence auditory stimulation clearly modulates ongoing EEG activity in visual cortex (even in the absence of a visual stimulus) at latencies of 50–100 ms post-stimulus. Therefore, when a visual stimulus is presented concurrently with an auditory stimulus, there will be a modulation of the activity elicited to the visual stimulus as a function of incoming activity from auditory cortex (as seen in the two experiments described in this chapter). The authors suggest that the phenomenon of phase resetting of these cortical rhythms produces the resulting multisensory interaction effect which will be stronger at some SOAs than others (Naue et al., 2011). Indeed, the profile of ongoing EEG activity which is present at the time of the delivery of a *unisensory* visual stimulus can affect the subject's ability to detect the stimulus (Busch, Dubois, & VanRullen, 2009). These data underscore the importance of investigating both evoked *and* induced oscillatory EEG (and MEG) activity for any study of perception and cognition. Additionally, it would also be important to including the analysis of activity that occurs in the immediate pre-stimulus period. Multisensory stimulation would be expected to produce complex profiles of evoked and induced activity and it would be great to see future studies of face and voice integration progress in this direction. Additionally, future investigations using multi-modal concurrent EEG and fMRI might also be useful for source modeling purposes (Herrmann & Debener, 2008).

5 Conclusions

Neurophysiological studies have increased in complexity since their inception in the mid-twentieth century. Today multichannel EEG and MEG recordings of not only evoked but also induced activity are performed with relative ease, and using dynamic multisensory stimulation that would have been unthinkable 20 years ago. High-field fMRI studies can now be performed concurrently with neural and other measures. Transcranial magnetic stimulation offers another assessment tool for probing brain-behavior relationships. Potentially, however, there are challenges for understanding the observed differences between types of neuroimaging datasets, e.g., fMRI and EEG/MEG, due to the different sensitivity in time scales for each method. Running experiments using parallel techniques offer a valuable way forward to begin to understand some of these differences. This approach, unfortunately, is time-consuming and may require a series of experiments with which complex sets of findings across neuroimaging methods can be understood and synthesized.

For the emerging field of social neuroscience, there are currently very many knowledge gaps with respect to the processing of *non-verbal* face–voice cues by the human brain. Yet, our daily social interactions rely on this important additional information to give another individual's spoken word context and also credence. Little is known about how the brain's response changes to poorly discriminable face–voice stimuli. Similarly, how non-verbal face–voice cues affect how our brains process the incoming communications from other individuals is also poorly understood. Other important questions not addressed here relate to potential gender differences in how these types of stimuli are evaluated and how culture might influence these processes. Another big question to address would be how these neural responses evolve and change across the lifespan: from infancy, childhood, adolescence, adulthood, and finally senescence. From the existing literature it is clear that human face–voice stimuli elicit complex, but distinctive, neural responses in the adult human brain. The literature suggests that these neural responses are robust and amenable to many different forms of experimental manipulation, including the use of virtual environments—a potential way forward to create the naturalistic context that we experience in everyday life.

References

Beauchamp, M. S., Nath, A. R., & Pasalar, S. (2010). fMRI-Guided transcranial magnetic stimulation reveals that the superior temporal sulcus is a cortical locus of the McGurk effect. *Journal of Neuroscience, 30*(7), 2414–2417.

Bentin, S., Allison, T., Puce, A., Perez, A., & McCarthy, G. (1996). Electrophysiological studies of face perception in humans. *Journal of Cognitive Neuroscience, 8*, 551–565.

Brefczynski-Lewis, J., Lowitszch, S., Parsons, M., Lemieux, S., & Puce, A. (2009). Audiovisual non-verbal dynamic faces elicit converging fMRI and ERP responses. *Brain Topography, 21*(3–4), 193–206.

Busch, N. A., Dubois, J., & VanRullen, R. (2009). The phase of ongoing EEG oscillations predicts visual perception. *Journal of Neuroscience, 29*(24), 7869–7876.

Calvert, G. A. (2001). Crossmodal processing in the human brain: Insights from functional neuroimaging studies. *Cerebral Cortex, 11*(12), 1110–1123.

Campbell, A., & Rushton, J. P. (1978). Bodily communication and personality. *The British Journal of Social and Clinical Psychology, 17*(1), 31–36.

Carrick, O. K., Thompson, J. C., Epling, J. A., & Puce, A. (2007). It's all in the eyes: Neural responses to socially significant gaze shifts. *Neuroreport, 18*(8), 763–766.

Eggermont, J. J., & Ponton, C. W. (2002). The neurophysiology of auditory perception: From single units to evoked potentials. *Audiology & Neuro-Otology, 7*(2), 71–99.

Engel, A. K., & Fries, P. (2010). Beta-band oscillations–signalling the status quo? *Current Opinion in Neurobiology, 20*(2), 156–165.

Engel, A. K., Fries, P., & Singer, W. (2001). Dynamic predictions: Oscillations and synchrony in top-down processing. *Nature Reviews. Neuroscience, 2*(10), 704–716.

Foxe, J. J., Morocz, I. A., Murray, M. M., Higgins, B. A., Javitt, D. C., & Schroeder, C. E. (2000). Multisensory auditory-somatosensory interactions in early cortical processing revealed by high-density electrical mapping. *Brain Research. Cognitive Brain Research, 10*(1–2), 77–83.

Ghazanfar, A. A., & Schroeder, C. E. (2006). Is neocortex essentially multisensory? *Trends in Cognitive Sciences, 10*(6), 278–285.

Giard, M. H., & Peronnet, F. (1999). Auditory-visual integration during multimodal object recognition in humans: A behavioral and electrophysiological study. *Journal of Cognitive Neuroscience, 11*(5), 473–490.

Hagan, C. C., Woods, W., Johnson, S., Calder, A. J., Green, G. G., & Young, A. W. (2009). MEG demonstrates a supra-additive response to facial and vocal emotion in the right superior temporal sulcus. *Proceedings of the National Academy of Sciences of the United States of America, 106*(47), 20010–20015.

Herrmann, C. S., & Debener, S. (2008). Simultaneous recording of EEG and BOLD responses: A historical perspective. *International Journal of Psychophysiology, 67*(3), 161–168.

Herrmann, C. S., Frund, I., & Lenz, D. (2010). Human gamma-band activity: A review on cognitive and behavioral correlates and network models. *Neuroscience and Biobehavioral Reviews, 34*(7), 981–992.

Herrmann, C. S., Munk, M. H., & Engel, A. K. (2004). Cognitive functions of gamma-band activity: Memory match and utilization. *Trends in Cognitive Sciences, 8*(8), 347–355.

Itier, R. J., & Batty, M. (2009). Neural bases of eye and gaze processing: The core of social cognition. *Neuroscience and Biobehavioral Reviews, 33*(6), 843–863.

Kayser, C., Petkov, C. I., Augath, M., & Logothetis, N. K. (2007). Functional imaging reveals visual modulation of specific fields in auditory cortex. *Journal of Neuroscience, 27*(8), 1824–1835.

Kim, S., & James, T. W. (2010). Enhanced effectiveness in visuo-haptic object-selective brain regions with increasing stimulus salience. *Human Brain Mapping, 31*(5), 678–693.

Kutas, M., & Hillyard, S. A. (1980). Reading senseless sentences: Brain potentials reflect semantic incongruity. *Science, 207*(4427), 203–205.

Lewis, J. W., Talkington, W. J., Puce, A., Engel, L. R., & Frum, C. (2011). Cortical networks representing object categories and high-level attributes of familiar real-world action sounds. *Journal of Cognitive Neuroscience, 23*(8), 2079–2101.

Mazaheri, A., & Jensen, O. (2010). Rhythmic pulsing: Linking ongoing brain activity with evoked responses. *Frontiers in Human Neuroscience, 4*, 177.

McCallum, W. C., Farmer, S. F., & Pocock, P. V. (1984). The effects of physical and semantic incongruities on auditory event-related potentials. *Electroencephalography and Clinical Neurophysiology, 59*(6), 477–488.

McGurk, H., & MacDonald, J. (1976). Hearing lips and seeing voices. *Nature, 264*(5588), 746–748.

Mehrabian, A., & Ferris, S. R. (1967). Inference of attitudes from nonverbal communication in two channels. *Journal of Consulting Psychology, 31*(3), 248–252.

Meredith, M. A., & Stein, B. E. (1986). Visual, auditory, and somatosensory convergence on cells in superior colliculus results in multisensory integration. *Journal of Neurophysiology, 56*(3), 640–662.

Munhall, K. G., Gribble, P., Sacco, L., & Ward, M. (1996). Temporal constraints on the McGurk effect. *Perception & Psychophysics, 58*(3), 351–362.

Nath, A. R., & Beauchamp, M. S. (2011a). Dynamic changes in superior temporal sulcus connectivity during perception of noisy audiovisual speech. *Journal of Neuroscience, 31*(5), 1704–1714.

Nath, A. R., & Beauchamp, M. S. (2011b). A neural basis for interindividual differences in the McGurk effect, a multisensory speech illusion. *NeuroImage, 59*(1), 781–787.

Naue, N., Rach, S., Struber, D., Huster, R. J., Zaehle, T., Korner, U., et al. (2011). Auditory event-related response in visual cortex modulates subsequent visual responses in humans. *Journal of Neuroscience, 31*(21), 7729–7736.

Puce, A., Epling, J. A., Thompson, J. C., & Carrick, O. K. (2007). Neural responses elicited to face motion and vocalization pairings. *Neuropsychologia, 45*(1), 93–106.

Puce, A., & Perrett, D. (2003). Electrophysiology and brain imaging of biological motion. *Philosophical Transactions of the Royal Society of London. Series B, Biological Sciences, 358*(1431), 435–445.

Puce, A., Smith, A., & Allison, T. (2000). ERPs evoked by viewing facial movements. *Cognitive neuropsychology, 17*, 221–239.

Reale, R. A., Calvert, G. A., Thesen, T., Jenison, R. L., Kawasaki, H., Oya, H., et al. (2007). Auditory-visual processing represented in the human superior temporal gyrus. *Neuroscience, 145*(1), 162–184.

Rossion, B., & Jacques, C. (2008). Does physical interstimulus variance account for early electrophysiological face sensitive responses in the human brain? Ten lessons on the N170. *NeuroImage, 39*(4), 1959–1979.

Sabbagh, M. A., Moulson, M. C., & Harkness, K. L. (2004). Neural correlates of mental state decoding in human adults: An event-related potential study. *Journal of Cognitive Neuroscience, 16*(3), 415–426.

Sams, M., Aulanko, R., Hamalainen, M., Hari, R., Lounasmaa, O. V., Lu, S. T., et al. (1991). Seeing speech: Visual information from lip movements modifies activity in the human auditory cortex. *Neuroscience Letters, 127*(1), 141–145.

Stein, B. E., Stanford, T. R., Ramachandran, R., Perrault, T. J., Jr., & Rowland, B. A. (2009). Challenges in quantifying multisensory integration: Alternative criteria, models, and inverse effectiveness. *Experimental Brain Research, 198*(2–3), 113–126.

Stevenson, R. A., Geoghegan, M. L., & James, T. W. (2007). Superadditive BOLD activation in superior temporal sulcus with threshold non-speech objects. *Experimental Brain Research, 179*(1), 85–95.

Stevenson, R. A., & James, T. W. (2009). Audiovisual integration in human superior temporal sulcus: Inverse effectiveness and the neural processing of speech and object recognition. *NeuroImage, 44*(3), 1210–1223.

Stevenson, R. A., Kim, S., & James, T. W. (2009). An additive-factors design to disambiguate neuronal and areal convergence: Measuring multisensory interactions between audio, visual, and haptic sensory streams using fMRI. *Experimental Brain Research, 198*(2–3), 183–194.

Wheaton, K. J., Pipingas, A., Silberstein, R. B., & Puce, A. (2001). Human neural responses elicited to observing the actions of others. *Visual Neuroscience, 18*(3), 401–406.

Wright, T. M., Pelphrey, K. A., Allison, T., McKeown, M. J., & McCarthy, G. (2003). Polysensory interactions along lateral temporal regions evoked by audiovisual speech. *Cerebral Cortex, 13*(10), 1034–1043.

Young, C. K., & Eggermont, J. J. (2009). Coupling of mesoscopic brain oscillations: Recent advances in analytical and theoretical perspectives. *Progress in Neurobiology, 89*(1), 61–78.

Part III
Affective Information

Chapter 10
Integration of Face and Voice During Emotion Perception: Is There Anything Gained for the Perceptual System Beyond Stimulus Modality Redundancy?

Gilles Pourtois and Monica Dhar

Abstract In this chapter, we review empirical data and theoretical models which have been put forward in the affective science literature to account for the perception of emotions, when this process is simultaneously accomplished by sight and hearing. The visual component is provided by the face configuration that undergoes some geometric changes, which in turn lead to different and discrete emotion facial expressions. The auditory component is provided by the voice and its changes in pitch, duration, and/or intensity leading to different affective tones of voice. Face–voice integration during emotion perception occurs when affective information conveyed by the two sensory modalities is integrated into a unified percept, or multisensory object. Although one may assume that the rapid and mandatory combination of multiple or complementary affective cues is adaptive (i.e., it likely reduces the effects of adverse factors like drifts or intrinsic noise), the central nervous system must however show some selectivity regarding which inputs from separate senses may eventually combine, as compared with merely redundant emotion signals. Indeed, not all spatial or temporal coincidences or co-occurrences lead to the perception of unified objects. Interestingly, results of behavioral studies confirm this conjecture, and indicate that the combination of emotional facial expressions with affective prosody leads to the creation of genuinely multisensory emotional objects, which show different properties compared to the combination of an emotional facial expression with another redundant or distracting emotional facial expression, or an emotion written word. Hence, the findings and models reviewed in this chapter suggest that some selectivity can be found in the way visual and auditory information is actually combined during emotion perception. The rapid and automatic pairing of an emotional face with an affective voice might present a naturalistic situation in

G. Pourtois (✉) • M. Dhar
Department of Experimental-Clinical and Health Psychology, Ghent University, Henri Dunantlaan 2, 9000 Ghent, Belgium
e-mail: gilles.pourtois@ugent.be; monica.dhar@ugent.be

P. Belin et al. (eds.), *Integrating Face and Voice in Person Perception*,
DOI 10.1007/978-1-4614-3585-3_10,

the sense that there is no need for mediation by higher-level cognitive, attentional or linguistic processes, which may be necessary for the efficient decoding of other stimulus categories or multisensory objects.

1 General Introduction

The perception of emotions in conspecifics stands out as one of the most important social skills in human cognition (Damasio, 1994; Darwin, 1871, 1872; Frijda, 1989). Yet, emotion perception is a complex phenomenon, as affective information is usually processed and conveyed concurrently by multiple sensory channels. Not only the muscles innervating the face swiftly change their configurations to convey or communicate a specific emotional expression, but also the tone of voice as well as the gait or body posture undergo compatible dynamic alterations to promote the efficient expression, communication, as well as decoding by conspecifics, of this emotion. Such emotional multisensory stimulations are the rule rather than the exception in natural environments (De Gelder & Bertelson, 2003), yet still very little is known about the actual brain mechanisms and cognitive processes underlying this remarkable perceptual ability, despite a clear recrudescence of empirical contributions in the field of Multisensory Integration during the last decade (Foxe & Molholm, 2009).

This scarcity is probably related to the fact that one of the dominant paradigms in emotion research is the cognitive approach (Fodor, 1983), which by definition seeks to decompose or break down complex mental functions (including emotion perception) into elementary or basic processes or principles, hence providing a strong analytical (as opposed to integrative) bias. Accordingly, emotion perception has mostly been studied in social or cognitive sciences by looking at mental processes or brain functions concerning a specific, isolated sensory modality (with a clear preference or advantage for vision, relative to the other senses), as opposed to systematic investigations exploring how multiple sensory cues are actually integrated during emotion perception. This observation does not imply that the cognitive approach makes the detailed study of multisensory integration almost impossible, but the modular and analytical perspective embraced by this dominant approach is sometimes hardly compatible with the fact that object-based multisensory perception is by essence a complex, permeable, and nonencapsulated phenomenon, which does not necessarily obey to laws of organization that have been put forward primarily to account for modality-specific processes, like visual computations during object recognition or semantic processing during speech perception for instance. Consistent with this conjecture, many studies have been designed to better characterize mechanisms of emotion perception when the affective information is conveyed predominantly by the face only (Ekman, 1992; Ekman & Friesen, 1976), or by the voice only (Banse & Scherer, 1996; Osullivan, Ekman, Friesen, & Scherer, 1985; Scherer, Banse, & Wallbott, 2001), but by comparison, few studies have been carried out to look at the nature and extent of perceptual effects when both channels (face and voice) are concurrently conveying important social or emotional signals, and they eventually interact with one another to yield a unified emotion percept (De Gelder & Vroomen, 2000; Campanella & Belin, 2007; Massaro & Egan, 1996). Likewise, studies looking at correlations between emotion face, affective voice and emotion word perception do exist and have been performed in the past (e.g., Borod et al., 1998, 2000), but these valuable studies suggesting the likely existence of amodal perceptual mechanisms during emotion perception do not directly address the question of how multiple sensory cues (e.g., face and voice) may be integrated during emotion perception, and what the resulting emotion percept may be. To address this complex question, another methodology and experimental approach, going beyond correlation methods, is required.

Presumably, an emotion is eventually perceived and experienced as such when these modality specific signals are combined and integrated together to yield a unified multimodal percept. What are the rules or constraints, if any, of this multisensory integration during emotion perception? Does multisensory integration of emotions represent another instance of stimulus redundancy (Marzi, Tassinari, Aglioti, & Lutzemberger, 1986; Miniussi, Girelli, & Marzi, 1998), or is there anything distinctive to this audiovisual integration process? The research presented in this chapter addresses these fundamental questions. Yet, the goal of this chapter is not to study emotions per se (see Brosch, Pourtois, & Sander, 2010; Vuilleumier & Pourtois, 2007 for recent reviews), but rather to shed light on mechanisms allowing the perceptual system to bind together affective information conveyed concurrently by multiple sensory channels. The focus is limited to visual and auditory information, and we do not consider other sensory inputs, like nonlinguistic signals (e.g., vocalizations, see Morris, Scott, & Dolan, 1999; Panksepp, 2005; Scott et al., 1997), emotional body language or gait (de Gelder, 2006), which also accompany expression of emotions.

The visual component is provided by the face configuration that undergoes some changes, which can eventually lead to different discrete emotional facial expressions (Mckelvie, 1995). Note however that unlike person identity information (Tanaka & Farah, 1993), it is still debated whether facial expression perception actually relies on the configuration of the face (as a whole), or instead the selective perceptual processing of some diagnostic faces parts, including the mouth and the eyes (see Adolphs et al., 2005; deGelder, Teunisse, & Benson, 1997; Smith, Cottrell, Gosselin, & Schyns, 2005). In any case, it can be argued that in some cases at least, affect-relevant information from the face is carried by the whole facial configuration (see deGelder et al., 1997), and this is the assumption adopted in our work. The auditory component is provided by the voice and its subtle changes in pitch, duration or intensity/loudness when articulating and producing speech sounds or fragments, which may lead to different discrete affective tones of voice. The respective contribution of variations in these psychoacoustical parameters has been measured in natural or simulated affective speech (Cummings & Clements, 1995; Lieberman & Michaels, 1962; Scherer, 1989; Williams & Stevens, 1972). Many prosodic features contribute to the expression of vocal emotions, and it seems evident that the acoustic correlates are subject to large interindividual differences (see Lieberman & Michaels, 1962). Despite the large interspeaker variability, there is some general consensus that if prosodic features are ranked in terms of their respective contribution, then gross changes in pitch do contribute most to the transmission of emotions, duration is intermediate whereas loudness seems to be least important (Frick, 1985; Murray & Arnott, 1993), even if the simultaneous processing and integration of these different parameters is probably required to efficiently decode the emotion from the voice. Noteworthy, the manipulation of affective speech prosody can be done independently of the semantic content conveyed in the message and this is why the prosodic channel can be considered to be a separate channel. Hence, the primary goal is to explore the nature of the relationship unifying a given emotional face expression and a concurrent (either compatible or not) affective tone of voice. We refer to this integration effect as "multisensory perception of emotion," following standard practice (see de Gelder, Vroomen, & Pourtois, 2004).

This chapter is divided into two main and consecutive sections. In the first part, we review some classical empirical evidence and dominant theoretical frameworks that have been put forward in the cognitive sciences literature to account for multisensory perception in general, and multisensory perception of emotion more specifically. In the second part, we present new (unpublished) behavioral empirical data addressing the selectivity of multisensory perception of emotion. Several experiments were carried out to assess if the integration of face (emotional expression) and voice (affective tone of voice) during emotion perception may be somehow specific, relative to other forms of stimulus redundancy (e.g., two emotional faces shown simultaneously, relative to a single emotional face). The results of these experiments somehow enable to better demarcate the constraints on bimodal sensory inputs which have to be met to eventually yield genuine behavioral effects of multisensory perception of emotion.

2 Object-Based Multisensory Perception

2.1 Introduction

We restrict our review to what is usually referred to as object-based multisensory perception (Lehmann & Murray, 2005). As it turns out, multisensory perception of emotion can be considered as an instance of multisensory object recognition, and is similar to many other cases of object perception where convergent information about the same object is presented through different sensory modalities. As a first approximation, the kinds of audiovisual objects that have been mostly studied appear to be of two categories: simple/arbitrary audiovisual pairings vs. complex/natural pairings (see Pourtois & de Gelder, 2002). A common example of simple audiovisual pairings is the combination of light flashes with tone bursts, or the combination of specific tone frequencies with simple geometric figures (Fuster, Bodner, & Kroger, 2000; Giard & Peronnet, 1999; Stein & Meredith, 1993; Talsma, Senkowski, Soto-Faraco, & Woldorff, 2010). Such pairings are obviously arbitrary and usually the subject is trained intensively to associate them, and later perceive them as paired in the context of the experiment. The situation is quite different with more complex audiovisual pairs consisting of speech sounds and lip movements, or facial emotional expressions and affective tones of voice (Campanella & Belin, 2007; De Gelder & Bertelson, 2003). These complex pairings are natural, as they do not require any training for the perceiver to treat these pairs as such in the laboratory. In fact, in the course of studying these pairings naturally associated (e.g., an emotional facial expression combined with an affective tone of voice), the experimenter may even create conditions allowing pulling them apart and dissociating them (see De Gelder & Vroomen, 2000; Mcgurk & Macdonald, 1976). This is often done in order to obtain incongruent pairs and compare them with the more natural situation of congruence. Natural and arbitrary pairs thus seems to pull the researchers in opposite directions, to some extent, and it is plausible to argue that the underlying multisensory integration processes may be different depending on whether the audiovisual pairs are natural, or rather arbitrary (see Pourtois & de Gelder, 2002 for evidence).

Object-based multisensory perception is widespread in daily environments. However, there are only a few multisensory objects that have been studied in depth so far in cognitive sciences. Space perception, language perception, and the perception of temporal events are three domains of human cognition where multisensory research has brought valuable insight. In the domain of space perception, many multisensory or crossmodal effects have been shown previously that all reflect our ability to integrate spatial information when this information is concurrently provided by the visual and auditory (or proprioceptive or tactile) modality (Driver & Spence, 1998a, 1998b, 2000). For example, the distance between spatially disparate auditory and visual stimuli tends to be underestimated with temporally coincident presentations, a phenomenon known as the ventriloquist effect/illusion (Bermant & Welch, 1976; Bertelson, 1999). Visual capture is another instance found in the spatial domain (Hay, Pick, & Ikeda, 1965). It involves a spatial localization situation in which the visual information is in conflict with that of another modality, namely, proprioceptive information, and perceived location is determined predominantly by visual information. Likewise, when speech sounds (syllables) are presented simultaneously with incongruent lip movements, subjects report a percept that belongs neither to the visual modality nor to the auditory one, but that represents either a fusion or combination between the two inputs (Mcgurk & Macdonald, 1976). These results indicate that the visual and auditory components of syllables do combine and this combination translates as a new speech percept. Natural speech perception therefore provides a compelling case of multisensory integration (Dodd & Campbell, 1987). A third compelling instance or illusion of object-based multisensory integration is found in the temporal domain and may be seen, to some degree, as a symmetric case to that observed with the ventriloquist illusion. Here a visual illusion is induced by sound (Shams, Kamitani, & Shimojo, 2000). When a single flash of light is accompanied by multiple auditory beeps, the single flash is perceived as multiple flashes. This phenomenon is partly consistent with previous behavioral results that showed that sound can alter the visually perceived direction of motion (Sekuler, Sekuler, & Lau, 1997). Altogether, these effects suggest that visual perception is malleable by signals from other sensory modalities, such as auditory perception is malleable by signals from other sensory modalities. More generally, the dominance of one modality over the other does not seem therefore to be fixed or absolute, but instead may depend upon the context in which crossmodal effects take place. For space perception, the visual modality dominates over the auditory, and this situation is reversed during the perception of discrete temporal events (for which the auditory domain takes the lead on visual cues).

Traditionally, two sets of constraints have been envisaged in the literature (Bertelson, 1999). The first, referred to as structural factors, primarily concerns the spatial and temporal properties of the sensory inputs. The other set, often discussed as cognitive factors, is related to a whole set of higher-level, semantic or attention-related factors, including the subject's knowledge of and familiarity with the multisensory situation (Talsma et al., 2010). Structural factors are the ones that have attracted by far the most attention from researchers in the field of multisensory integration (see Calvert, Spence, & Stein, 2004). By comparison, the role of cognitive factors is still underinvestigated, although more recent work has started to explore the links between selective attention brain mechanisms and multisensory integration brain processes (see Talsma et al., 2010). However, from a conceptual viewpoint, it seems plausible to argue that some additional cognitive or object-based constraints on multisensory perception actually take place, to prevent the organism to register many invalid and spurious incidences of multimodality, as solely defined based on the spatial and temporal coincidences of the visual and auditory inputs. Yet, there are only a few studies that have addressed this question, and tested to which extent object-based constraints may influence mechanisms of multisensory perception (see De Gelder & Bertelson, 2003; Pourtois & de Gelder, 2002).

Object-based multisensory perception is a complex issue, since beyond the spatial and temporal determinants of the input, the nature of the object to perceive may vary a lot from one condition (or encounter) to another. In this context, one may consider emotions just as one class of perceptual objects, besides other categories like speech (i.e., speech sounds presented simultaneously with lipreading information/ lip movements, see Calvert et al., 1997; Mcgurk & Macdonald, 1976) or space (i.e., although spatial localization is determined predominantly by visual cues, the presentation of concurrent spatial auditory or tactile cues strongly biases and influences visual spatial localization abilities, see Bertelson, 1999; Driver & Spence, 1998a; Stein & Meredith, 1993), as reviewed here above. Several objects or dimensions are actually susceptible to being perceived by multiple sensory channels at the same time, and therefore, a central (still unanswered) question concerns the existence of general principles that would govern multisensory perception. Structural factors, such as temporal and spatial coincidence (see Stein & Meredith, 1993), may be envisaged as such. On the other hand and contrary to this view, one might postulate that each domain or object of perception (e.g., emotion, speech, space) actually possesses its own organization principles and that the overlap between these domains is fairly limited. Presumably, multisensory perception of emotion most likely shares some invariance in the basic perceptual mechanisms of audiovisual integration with these other domains (speech and space perception), while some specificity may well be present, although this question still remains open.

2.2 *Multisensory Perception: Behavioral Effects and Cognitive Models*

In behavioral research on audiovisual integration, a few classical models have been proposed (Bertelson, 1999; De Gelder & Bertelson, 2003; Dodd & Campbell, 1987; Massaro, 1998; Miller, 1982, 1986). The behavioral measures on the basis of which audiovisual integration is inferred are predominantly accuracy and response latency. When participants respond better and faster to the bimodal (audiovisual) stimulus than to either the visual only or auditory only stimulus, there is evidence that the response is presumably based on multisensory integration (Giard & Peronnet, 1999; Talsma et al., 2010). However, this evidence is inevitably indirect, and other accounts, like for example a race model (Raab, 1962) that do not assume integration of the two separate modality inputs can in principle still explain the same pattern of behavioral results. Although multisensory integration intuitively refers to the notion that the brain combines different input modalities, it is actually a theoretical notion advanced in order to account for a wide range of (behavioral) observations showing that there are bidirectional interaction effects between different sensory modalities. Traditionally, faster RTs for bimodal stimulus pairs than unimodal stimuli are compatible with the Redundant Signal Effect (RSE, see Miller, 1982, 1986). If a RSE is obtained for (congruent) audiovisual stimulus pairs, it does not necessarily mean that audiovisual integration (or neural interaction) occurs (Miller, 1986), however. Firstly, RSEs are also obtained with redundant stimuli presented in the same modality. The RSE is therefore not specific to multisensory perception, and is also found in spatial summation experiments in which a redundant simple visual stimulus (e.g., the simultaneous and synchronous presentation of the same simple visual stimulus at two separate spatial positions, usually on each side in the visual field to allow callosal interhemispheric transfer) is detected faster than a nonredundant visual stimulus, an effect classically referred to as Redundant Target Effect (RTE, see de Gelder, Pourtois, van Raamsdonk, Vroomen, & Weiskrantz, 2001; Marzi et al., 1986; Miniussi et al., 1998; Murray, Foxe, Higgins, Javitt, & Schroeder, 2001; Savazzi & Marzi, 2002, 2004, 2008; Turatto, Mazza, Savazzi, & Marzi, 2004). Secondly, faster RTs for (congruent) bimodal stimulus pairs (relative to unimodal stimuli) could be explained by a horse race model that does not imply interaction between sensory modalities (Raab, 1962), as briefly explained here above. In this perspective, each stimulus of a pair independently competes for response initiation and the faster of the two mediates the response. Thus, "simple" probability (or statistical) summation could yield the RSE. Indeed, the likelihood of either of two stimuli yielding a fast RT is higher than that from one (unisensory) stimulus alone. On the other hand, RSE could also be explained by a coactivation model that implies that the two modalities are integrated together and interact prior to motor response initiation (Miller, 1982). In order to distinguish between these two opposite accounts (race model vs. coactivation model), Miller (1982) proposed to analyze RTs using cumulative probability functions and to test for what he called the inequality assumption. The inequality places an upper limit on the cumulative probability of RTs at a given latency for a stimulus pair. For any latency, *t*, the race model holds when the cumulative probability value is less than or equal to the sum of the cumulative probabilities from each of the single stimulus minus an expression of their joint probability. Hence, based on a formal analysis of RT distribution (and the violation of the inequality assumption, or not, see Miller, 1982, 1986), it is possible to establish whether a simple statistical facilitation/summation, or instead a coactivation (integration) between the two modalities (or sensory inputs) occurs during the processing of bimodal stimulus pairs (see Molholm et al., 2002 for an example).

Besides these important technical considerations related to the definition or qualification of multisensory perception effects, in fact, very few (computational) models have been developed in the literature to account for these multisensory behavioral effects. Notably, the Fuzzy Logical Model of Perception (FLMP, see Massaro, 1987, 1998) represents such a valuable attempt. The key assumption behind the model of Massaro is that sensory information is always processed the same way, whatever the domain of application. In this perspective, audiovisual integration is just one instance of perception besides other cases and the underlying mechanisms responsible for audiovisual perception are similar to the mechanisms involved in other domains of cognition or perception. To validate his model, Massaro has provided data on bimodal speech perception from children, elderly, hearing impaired or bimodal emotion perception that all fit the FLMP (see also Massaro & Egan, 1996). This is an apparent strength of this computational model: this model is able to describe a wide range of human performance patterns during audiovisual perception. However, a possible downside is that the FLMP remains only descriptive, because this model does not implement any preconception about the nature of the components it seeks to describe (see Burnham, 1999). In the FLMP, four sequential stages of processing are postulated. The first step is feature evaluation, which is assumed to be carried out independently and separately for each modality source. The second stage is an integration of the features available after the first stage. This is of course the stage of interest, with regard to mechanisms of multisensory integration. Integration is achieved through a multiplicative combination of the response strengths of components of information input. Then, the result of this integration is matched against a prototype stored in memory during the assessment stage. Finally, a response is selected based on the most consistent prototype, given the visual and auditory cues. The proposal of a first evaluation stage carried out separately for each modality source is debated, and does not agree with independent evidence from neuroimaging or neurophysiology work showing reliable crossmodal effects not only in multimodal or heteromodal brain regions (Damasio, 1989; Ethofer, Pourtois, & Wildgruber, 2006; Mesulam, 1998; Pourtois, de Gelder, Bol, & Crommelinck, 2005) but also (and already) in unisensory or modality-specific cortices (see Calvert, 2001; Calvert et al., 1999; Macaluso, Frith, & Driver, 2000). Moreover, this independence of the auditory and visual components during audiovisual perception has been called into question, at least for the case of speech perception. An alternative account is the possibility of intermodal cues (see Campbell, Dodd, & Burnham, 1998). Another controversial property of the first evaluation step is related to the nature of the representations that drive this process. Indeed, in this model (Massaro, 1998), the algorithm of perception tags each feature with a continuous value and this characteristic runs against several empirical data that showed a categorical perception function during speech perception (see Liberman, Harris, Hoffman, & Griffith, 1957). Despite these critiques or limitations, the FLMP undoubtedly provides one of the few valuable computational models aimed at describing the critical computations involved during the perception and later integration of visual and auditory cues when presented simultaneously (object-based multisensory integration).

2.3 Multisensory Perception Effects Revealed Using the Crossmodal Paradigm

Several methods have been used to disclose, at the behavioral level, evidence of object-based multisensory perception. A classical method that we have used in this work is referred to as the crossmodal paradigm (see Bertelson, 1999). This specific paradigm is actually part of a larger set of methods, used to indirectly measure the impact of stimulus processing in one sensory modality onto another. Other indirect methods include the use of aftereffects (see Held, 1965), intersensory fusions (Mcgurk & Macdonald, 1976) or staircases (Bertelson, 1999).

The crossmodal paradigm is reminiscent of older studies on intermodal discrepancy following prismatic adaptation (Hay et al., 1965), on audiovisual space perception (Bermant & Welch, 1976), and has been used in audiovisual speech studies (Driver, 1996; Massaro, 1987, 1998) and in crossmodal attention studies (Driver & Spence, 1998a). In this paradigm, the systematic influence of one modality on the other is assessed using a strict methodology that requires a narrowing of the subject's attentional resources to one modality only, during stimulus processing. Then, a relative "automatic" crossmodal bias effect from the unattended modality to the attended modality can be measured. Therefore, the impact of one modality on the other is measured indirectly in the crossmodal paradigm. This procedure offers a double methodological advantage. Firstly, it has been shown to be more sensitive than direct measures and this procedure is better suited than other methodologies to capture genuine perceptual, as opposed to postperceptual effects (see Bertelson, 1999). Secondly, it allows to manipulate the level or amount of congruence between the two modalities, unlike other contrasting methods capitalizing solely on the direct comparison between unimodal and multimodal stimulus conditions (see Giard & Peronnet, 1999; Molholm et al., 2002). Hence, the experimenter may use this powerful method and set up in the laboratory artificial conditions in which the level of congruence between the two sensory modalities systematically or parametrically varies (see De Gelder & Vroomen, 2000). This method enables quantification of the actual crossmodal impact from one modality onto the other (e.g., from vision to audition, or vice versa) during the perception of bimodal stimulus pairs, including the combination of an emotional facial expression with an affective tone of voice (see Pourtois et al., 2005; Pourtois, de Gelder, Vroomen, Rossion, & Crommelinck, 2000; Pourtois, Debatisse, Despland, & de Gelder, 2002), as reviewed in the next section.

3 Multisensory Perception of Emotion

3.1 Possible Functions of Multisensory Perception of Emotion

The fact that emotional information concurrently presented in different sensory modalities is integrated is likely to occur for reasons that go far beyond a simple back-up function allowing the system to overcome a given sensory loss and to rely on the spared/redundant modality to continue to operate. At least three distinct arguments or points can be evoked to support the functionality and need for integration during multisensory perception of emotion.

A first support for functionality comes from several older developmental studies (see Lewkowicz, 2000) that have clearly shown that very young infants look longer at face stimuli accompanied by voices (see Haith, Bergman, & Moore, 1977). Five- to seven-month-old infants also look longer at a face that carries the same expression as the voice than at a face carrying a different expression (Walker & Grolnick, 1983). These results suggest that the recognition of affective expressions may be first multimodal, before a differentiation occurs between the face and the voice (Walker-Andrews, 1997). There would be an ontogenetic priority in favor of multisensory perception. Furthermore, these results suggest a possible modular organization for audiovisual perception of emotion, which is not consistent with a simple back-up function.

The second element is that each sense (here the visual and the auditory channel) actually provides a qualitatively distinct subjective impression of the environment, including emotion perception. Although referring to the same event (e.g., an angry affective state), the emotion conveyed by ear (voice of wrath) and by eye (a furious facial expression) is not simply redundant, but both senses complement each other given the specificity and specialization of each sense (de Gelder, Vroomen, & Pourtois, 1999). This argument is therefore about the sensory specificity and complementary of multisensory perception of emotion.

A third defense for the importance of the function of multisensory perception of emotion is the optimization. Indeed, multisensory perception of emotion consists of enhancing detection and discrimination of emotions, as well as speed responsiveness to these highly relevant biological stimuli (Sander, Grafman, & Zalla, 2003). The fact that the perception of emotions is by nature multimodal and audiovisual, allows the perceptual system to disambiguate the actual functional meaning of the emotional input using a stable amodal or supramodal representation (see Borod et al., 2000; Farah, Wong, Monheit, & Morrow, 1989). There are large interindividual differences between human beings (as well as animals) in the ability to express and perceive different emotions. Moreover, humans have numerous ways to express and perceive the same emotion. As a consequence, the rapid and automatic combination of different channels of communication probably acts as an optimizer or catalyzer to rapidly perceive and efficiently recognize a given emotional state or object. From an evolutionary perspective (see also Damasio, 1994), integration of multiple affective inputs across different sensory modalities makes adaptive sense, given the enhanced biological significance of emotional stimuli. It also makes sense given the fact that combining different sources of information (face and voice) usually leads to more accurate and faster judgments, as well as more appropriate behaviors, as stressed in the second section of this chapter here below. However, this compensatory function may not be specific to multisensory perception of emotions, and could also explain other multisensory perceptual phenomena, like crossmodal spatial mechanisms at stake during the ventriloquist illusion for instance (Bertelson, 1999).

3.2 Behavioral Evidence for Multisensory Perception of Emotion

In their seminal study, de Gelder and Vroomen (2000) performed a series of elegant behavioral experiments looking at crossmodal effects from the voice to the face, and vice versa, during emotion perception (see also Massaro & Egan, 1996). These authors used an experimental situation in which varying degrees of (in)congruence were created between emotional facial expression and affective tone of voice. Two contrasting emotions (happy vs. sad) in the voice were manipulated. The same sentence (with a neutral semantic content) was uttered by a semiprofessional actor, either with a happy or sad tone of voice. Using a standard morphing technique (see Beale & Keil, 1995; deGelder et al., 1997; Etcoff & Magee, 1992), a visual continuum of varying emotional facial expression with 11 steps starting from one emotion at one extreme (happy) and going to another emotion (sad) at the other extreme was created. In the two first experiments, de Gelder and Vroomen (2000) combined the 11 faces with the two auditory conditions and compared these pairings to the condition where the morphed faces were presented alone (no accompanying sound). The task of the participant was to judge the emotion (Experiment 1) or to judge the emotion conveyed by the face (happy vs. sad) while ignoring the concurrent voice (Experiment 2). Results clearly showed that the identification of the emotion in the face was categorical (see deGelder et al., 1997; Beale & Keil, 1995; Etcoff & Magee, 1992), but more importantly was systematically biased in the direction of the simultaneously presented affective tone of voice. More specifically, this effect consisted in the fact that the likelihood to give a sad response when judging the emotional face (along the continuum) was reduced if the face was paired with a happy voice, regardless of the amount of sadness perceived in the face (i.e., general lateral shift of the psychometric response function). Moreover, RT results also showed that congruent bimodal stimulus pairs were judged faster than either incongruent stimulus pairs or single-modality stimuli (i.e., faces only).

Another question addressed in this study was whether this crossmodal bias effect during emotion perception could also be obtained from the face to the voice, or only from the voice to the face as reviewed here above. Were these crossmodal bias effects during emotion perception bidirectional and symmetric? In a third experiment, de Gelder and Vroomen (2000) directly addressed this question and created for this purpose a symmetric situation where the crossmodal impact from the face to the voice during emotion perception could be measured and assessed. They created a symmetric experiment to Experiment 2 in which a seven-steps voice continuum between two extreme emotions (fear vs. happy) was made up. Like was the case for the visual continuum used in the first two experiments, a vocal continuum was created using a computer-assisted auditory morphing procedure, by manipulating in a parametric fashion the physical distance between several features of the voices. This sophisticated procedure essentially works out on a modeling and subsequent parametric modulation of the fundamental frequency (F0) of the two original auditory fragments (fear voice and happy voice, see Vroomen, Collier, & Mozziconacci, 1993). Changing simultaneously and parametrically the duration, pitch range and pitch register of the two original utterances allowed to create several discrete steps along a vocal continuum, progressively going from one emotion (happy) to the other (fear). These seven voice fragments were then combined with two facial expressions (fearful vs. happy) to yield 14 stimulus pairs with varying levels of emotion (in) congruence. In this experiment, participants were instructed to judge the emotion conveyed by the voice, while ignoring (though attending to) the face information. Results showed a systematic bias of emotional voice identification by the concurrent facial expression, as well as a RT facilitation for congruent bimodal stimulus pairs, relative to incongruent bimodal pairs. Altogether, these results suggest bidirectional (from face to voice and vice versa) crossmodal bias effects during emotion perception.

Although certainly convincing, these behavioral results (De Gelder & Vroomen, 2000) are also compatible with other explanations that do not postulate any access to the emotional content of the face or the voice in order to trigger the crossmodal bias effect during emotion perception. For example, one may speculate that the crossmodal bias effect from the face to the voice during emotion perception described here above may not be specific to the affective content of the face, but instead may be obtained with any other visual stimuli that have an affective content. Hence, an important additional evidence would be to show that the actual (covert) processing of the affective information from the emotional face is crucial in order to obtain a reliable crossmodal bias effect (from the face to the voice) during emotion perception. This question was addressed in a different study and the results basically confirmed this hypothesis (De Gelder, Vroomen, & Bertelson, 1998). Presenting emotional faces upside down disrupts the perceptual processing of the emotional facial expression (deGelder et al., 1997; Mckelvie, 1995). Based on this face inversion effect for emotional facial expressions, de Gelder et al. (1998) surmised that the crossmodal bias effect from the face to the voice would be strongly attenuated with the presentation of inverted, relative to upright emotional facial expressions. Results of this study confirmed this prediction, as when subjects were asked to identify the emotional tone of voice in the presence of inverted emotional facial expressions, the crossmodal bias effect from the face to the voice basically disappeared, whereas it was still well present when these faces were presented upright (see also De Gelder & Vroomen, 2000). These behavioral results suggest that the crossmodal bias effect (from the face to the voice) during emotion perception is a function of the expression conveyed by the face. More generally, these findings add support to the notion that crossmodal affective biases are to some extent automatic and perceptual in nature, and they cannot be easily reduced to some postperceptual voluntary adjustments.

3.3 A Role for Attention in Multisensory Perception of Emotion?

These findings indicate that these crossmodal effects during emotion perception are likely to be perceptual, mandatory and automatic, as opposed to postperceptual (e.g., response bias) or influenced by attention, subjective beliefs or decision processes (see Bertelson, 1999). Indeed, the fact that this systematic crossmodal influence was observed even when participants were instructed to voluntarily ignore one of the two sensory modalities seems to indicate that multisensory integration of affective information takes place "automatically" to some extent, regardless of attentional factors. This property (i.e., independence from demands on attentional capacity) has long been one of the defining characteristics of "automatic" processes (see Kahneman, 1973; Schneider & Shiffrin, 1977, but see Moors & De Houwer, 2006 for a more recent and refined theoretical account of "automaticity"). To some degree, this observation is also consistent with the notion that the integration between an emotional facial expression and an affective tone of voice is occurring at a preattentive level (see also Driver, 1996), as demonstrated using other investigation techniques, like the recording of event-related brain potentials in healthy adult participants, which suggest early perceptual effects within modality-specific cortices during multisensory perception of emotion (see Pourtois et al. 2000, 2002). This audiovisual integration of emotional signals could take place during an early perceptual stage of stimulus processing, before (selective) attention comes into play (Talsma et al., 2010).

On the other hand, strict perceptual properties associated with multisensory perception of emotion may appear unlikely, given the fact that recognition of emotion stands as a particularly content-rich process, and seems more akin to higher-level cognition than perception. In this context, the (multisensory) perception of emotion should make a poor candidate for qualifying as a case of perception-based audiovisual integration. Yet, there is a wealth of recent empirical studies (including in brain-imaging) that have brought support to the notion that emotion perception and recognition (at least for some specific emotions like fear or anger) is a true perceptual process, or at least has a hard perceptual core (see Bocanegra & Zeelenberg, 2009a, 2009b; Calder, Lawrence, & Young, 2001; Phelps, Ling, & Carrasco, 2006; Pourtois, Grandjean, Sander, & Vuilleumier, 2004; Vuilleumier, 2005 see also Chap. 12). Hence, multisensory perception of emotion is likely to rely on genuine perceptual mechanisms, which allow a preattentive binding of affective information simultaneously conveyed by multiple sensory cues (visual and auditory). Indirect evidence obtained in specific brain-damaged patients also lent support to this conclusion (i.e., multisensory perception of emotion is a perceptual process). In one of these neuropsychology studies, two patients with selective striate cortex damages but unaware low-level residual visual abilities ("Blindsight," see Weiskrantz, 1986) were nevertheless shown to benefit partly from the presentation of an emotional visual stimulus (either a face or a scene) in their blind visual field during multisensory integration of emotion (i.e., this visual stimulus for which they therefore remained unaware had nonetheless a reliable crossmodal influence on the processing of a concurrent affective tone of voice, see de Gelder, Pourtois, & Weiskrantz, 2002). These results speak for multisensory integration mechanisms occurring without attention, possibly even without (visual) stimulus awareness (see also de Gelder, Pourtois, Vroomen, & Bachoud-Levi, 2000).

As suggested here above, the observation that crossmodal influences (from the face to the voice and vice versa) during emotion perception are observed although the participants were instructed to ignore one of the two modalities (i.e., crossmodal bias effect) can be taken as evidence, at least partly (see Schneider & Shiffrin, 1977) that multisensory perception of emotion is automatic and does not depend upon the availability of attentional resources. However, even if the instructions are to ignore a stimulus in one modality and attend to the concurrent emotional stimulus in the other modality, it may well be that it is actually difficult to ignore this former stimulus and modality (e.g., because of an intrinsic "sensory" dominance for example). Indeed, research on attention has clearly shown that irrelevant/unattended visual stimuli may be particularly hard to ignore under low-load conditions (Lavie, 1995, 2005). Hence, one may speculate that in the case of crossmodal influence from the face (which has to be ignored) to the voice, it may be hard to ignore the face, a highly biologically relevant visual stimulus, despite the instructions, due to the low-load nature of the experimental situation (i.e., the emotional facial expression was the only visual stimulus present and the task required an identification of the affective tone of voice, see De Gelder & Vroomen, 2000). Interestingly, this issue was actually addressed in a different study (Vroomen, Driver, & de Gelder, 2001). The authors used a dual-task paradigm asking participants either to add two digits presented visually together (i.e., high load) while judging the affective tone of voice (presented simultaneously with a congruent or incongruent emotional facial expression), or simply detect zeros in a rapidly presented sequence of digits shown visually (i.e., low load) while performing the same task. Moreover, in a third condition, they also gave a secondary auditory task to participants, consisting of deciding whether a tone was high or low (i.e., low load) while judging the emotion from the voice. This experimental design allowed the authors to test whether the crossmodal bias effect from the face to the voice was (load) attention dependent or not, i.e., would be reduced by either a secondary auditory task or by a secondary visual task, the latter which could be either easy or more difficult. Results showed that the crossmodal bias effect was actually independent of whether or not subjects performed a demanding additional (distracting) task. In all three cases, the visible static emotional face had a reliable impact on judgments of the heard emotional voices. Moreover, the systematic influence of the seen emotional facial expression on judgments of the emotional tone of the heard voice was not eliminated under conditions of high load (see Vroomen et al., 2001). These behavioral results therefore confirmed that multisensory perception of emotion is automatic to some extent, since it arises regardless of the attentional demands imposed by an additional task (see Lavie, 2005).

4 Multisensory Perception of Emotion vs. Emotion Stimulus Redundancy

4.1 Introduction

The behavioral evidence reviewed so far is consistent with the notion that the perceptual system integrates emotional information from the face and the voice, probably at an early/preattentive stage following stimulus onset (see Pourtois et al., 2000). Indeed, one may speculate that the ability to combine multiple inputs from different sensory sources is advantageous for an organism in the case of affect perception, as it appears to be the case for other forms of multisensory perception (e.g., speech or space perception). But reasonably, in the absence of any limits on which inputs make "good" pairings, such a theoretical advantage would be quickly lost, to some extent. Yet, still little is known about possible constraints on affective pairings and on the possible role that such constraints may play when turning to the case of multisensory perception of emotion. This question is therefore somehow related to the selectivity of the crossmodal bias effect during emotion perception, as reviewed here above (see also de Gelder et al., 2002). Two extremes, but equally plausible alternatives, may be suggested in this respect. Either the crossmodal bias between the voice and the face during emotion perception actually reflects the existence of a general mechanism for affect perception whereby the perceptual system (blindly) samples and merges all sources of affect information available at a given moment (time) and position (space) (Borod et al., 2000). Or alternatively, the crossmodal bias effect during emotion perception is narrowly restricted to the combination of an emotional facial expression with an effective tone of voice, and this specific audiovisual situation requires selective (perceptual) mechanisms (see Pourtois et al., 2005; Pourtois & de Gelder, 2002). For example, we do not know if task irrelevant stimuli presented in the periphery, like for example an additional emotional facial expression or a written emotion word, would not have a comparable influence on emotional ratings of a central emotional facial expression, just like the affective prosody in a concurrent voice does (see De Gelder & Vroomen, 2000). If the crossmodal bias effect would present some of the same characteristics as for example the interference effect observed in Stroop-like tasks (see MacLeod, 1991), it would slow-down rather than speed up the response to the central target emotional face stimulus. Hence, unconstrained integration (i.e., occurring regardless of the object or content) would probably expose the organism to vicarious influences away from the main task at hand.

In a series of behavioral studies, we actually addressed this question of selectivity and compared the crossmodal bias effect (from the voice to the face) during emotion perception to other cases of pairings or emotion stimulus redundancy (while keeping the sensory modality—vision—constant). Either a central emotion expression was paired with a congruent or incongruent affective tone of voice (audiovisual integration of emotion), or instead with another/distracting congruent or incongruent emotional facial expression (visual redundancy, see Marzi et al., 1986; Miniussi et al., 1998). Likewise, we also looked at the pairing of the central emotional facial expression with a distracting written emotion word, shown at the same unattended spatial location. We predicted that the crossmodal bias effect would be qualitatively different, relative to these two other instances of emotion redundancy ("intramodal" bias). More specifically, we surmised a facilitation for congruent bimodal face–voice parings during emotion perception (see De Gelder & Vroomen, 2000), whereas the presentation of an additional emotional facial expression or written emotion word would primarily slow down perceptual decision during incongruent pairings, consistent with an interference effect (see MacLeod, 1991). Such an asymmetric outcome would indicate that multisensory perception of emotion cannot simply be assimilated to a case of emotion stimulus redundancy, and that there is more to gain for the perceptual system in the simultaneous presentation of a face and a voice during emotion perception, than the mere juxtaposition of multiple/redundant emotional stimuli within the same sensory modality (here with a focus on vision and emotion face stimuli).

4.2 Methods

Thirty-one adult participants (mean age: 20) were instructed to discriminate the emotion expression (angry vs. sad) of a central target face. This central target face was presented either alone, or accompanied by an affective distracter. This distracter could be either auditory (an angry or sad tone of voice), or visual (the written name of an emotion word or another face). Following standard practice, (see Bertelson, 1999; Driver, 1996), we manipulated the emotional congruence between the target face and the affective information presented concurrently and to be ignored. We used a within-subject design, the same procedure and stimulus duration of the central target face across the different conditions. Notably, the effect of the auditory distracter on emotion face perception was studied in a separate block than the effect of the visual distracters (either a written emotion name or an emotional facial expression). Note that only emotions with a negative valence, i.e., angry and sad, were used, in order to avoid possible confounds in the interpretation between the role of (in)congruence between affective content of target and distracter, and actual valence of the emotion displayed (positive vs. negative).

The target face (5 cm width × 6.5 cm height) consisted of the static black and white photograph of one out of six actors, posing either a sad or an angry emotional facial expression (see Ekman & Friesen, 1976) and was briefly presented in the center of a 17-in. screen for 150 ms. Auditory stimuli were 12 different spoken words always with the same neutral content (/plane/) pronounced by semiprofessional actors, either with a sad or angry affective tone of voice (see Pourtois et al. 2005, 2000, 2002 for additional details regarding these previously validated auditory stimuli). Mean duration of the auditory fragments was 348 ms. Based on the emotion content of the face and the voice, congruent and incongruent audiovisual pairs were created. The spoken distracter was always presented at such a time that its offset coincided with that of the central face stimulus (duration of 150 ms).

Face distracters were identical to the targets. All combinations involved two pictures of the same actor, displaying the same emotion (thus twice the same picture) on congruent trials, or different emotions on incongruent trials. The emotional face distracter was presented in full synchrony with the target face, 5 cm above it (distance from the screen was 60 cm). Emotion written words were two adjectives (/ANGRY/vs./SAD/, in French) printed in Times police 24 (3 cm width × 1 cm height). Like for the distracting face, the distracting word was presented synchronously with the central emotional face, 5 cm above it. Congruent and incongruent trials were created based on the (mis)match between the emotion displayed by the central face and that conveyed by the written word briefly presented in the upper visual field.

The experimental session included control trials during which the central target face was presented alone (no distracter) and trials during which it was accompanied by a distracter, in random order. All trials started with the presentation of a fixation cross in the center of the screen for 250 ms, followed by a 600 ms blank screen, and then the presentation of the central target face for 150 ms. Such a short stimulus presentation for the emotional face presumably reduced peripheral eye explorations (e.g., vertical saccades back and forth between the two positions in the visual field). At the offset of the central target face, the screen went blank again for 1,200 ms (allowing registration of the actual response made by the participant), before the next trial started. Participants were instructed to perform a two-alternative forced choice task about the actual emotional expression (sad vs. angry) of the central emotional face, and to respond as accurately and as fast as possible by pressing one of two keys of a response box with their dominant hand. They were explicitly told to base their response only on the emotional target face, and to ignore either the auditory or visual distracter (either a face or a written word). The testing included two main blocks. In one block, control trials ($n=60$) were intermixed with audiovisual trials ($n=120$, 60 per level of congruence). In the other, control trials ($n=60$) were intermixed with visual-redundant trials ($n=240$, 60 per type of distracter and per level of congruence). The order of the two blocks was counterbalanced across participants. Within each block and across participants, trial order was randomized.

4.3 Results

For the audiovisual condition/block, a repeated measures analysis of variance (ANOVA) with two factors (Emotion of the central face and Trial type) was computed on mean error rates (mean error rate was 15%). This analysis disclosed a significant main effect of trial type ([$F(2,60)=6.15$, $p<0.005$]), with no significant modulation by emotion. Post hoc comparisons (based on paired t-tests) showed that congruent trials produced on average less discrimination errors than control [$t30=3.37$, $p<0.005$] and incongruent [$t30=2.52$, $p=0.014$] trials. The difference between control and incongruent trials was not significant [$t30<1$]. The statistical analysis carried out on mean RTs (for correct discriminations only) basically revealed a similar outcome, indicated by a clear RT facilitation for congruent audiovisual pairs, relative to either control trials (i.e., emotional face alone) or incongruent audiovisual trials (see Fig. 10.1a). This analysis revealed a significant effect of trial type [$F(2,60)=12.3$, $p<0.001$], with no reliable modulation by the emotion content of the face. Post hoc comparisons confirmed faster perceptual decisions for congruent audiovisual trials, relative to control trials [$t30=2.47$, $p=0.016$] or incongruent audiovisual trials [$t30=2.49$, $p=0.016$]. There was therefore no speed-accuracy trade-off, as participants responded faster and made less errors for congruent audiovisual trials compared to the other conditions (emotional face alone or emotional face combined with an incongruent affective tone of voice). More importantly, these behavioral results confirmed a systematic and significant crossmodal bias effect from the to be ignored voice to the attended face (see De Gelder & Vroomen, 2000), indicated by a facilitation of the ratings/perceptual judgments of the central emotional face when it was accompanied by a congruent affective tone of voice (Fig. 10.1a). Central to the present investigation is the question whether a similar facilitatory perceptual effect could be observed when the same central emotional face is no longer combined with an affective tone of voice, but instead with another "distracting" emotional face or a written emotion word.

Results obtained for the other block (visual-redundant trials) show a very different outcome (see Fig. 10.1b). First, the 2 (Emotion) × 2 (Congruence) × 2 (trial type: written word vs. face) ANOVA performed on mean error rates did not yield any significant effect. By comparison, the ANOVA performed on mean RTs disclosed a significant effect of Congruence [$F(1,30)=25.85$, $p<0.001$], as well as a significant interaction between Emotion and Congruence [$F(1,30)=9.78$, $p<0.005$]. However, this significant effect of congruence clearly indicated slower RTs with incongruent trials than either control or congruent trials (Fig. 10.1b), and this effect turned out to be larger for sad faces than angry faces. Moreover, this significant interference effect was found to be the same, regardless of the nature of the affective distracter, either an emotional face or a written emotion word (Fig. 10.1b).

4.4 Discussion

Based on previous results and findings (see De Gelder & Vroomen, 2000; Pourtois et al., 2005), we predicted that a gain in accuracy and response latencies (RT) would be observed when an emotional facial expression had to be judged as part of a multisensory emotion object (i.e., audiovisual pairing). Results of this study confirmed this prediction. However, when the exact same emotional target face stimuli were rated in the presence of a concurrent distracting emotional face or visual emotion word, there was also a reliable influence from the latter unattended visual stimulus on the ratings of the central face, but from a different nature (compare Fig. 10.1a, b). Unlike what we found for face–voice pairs during emotion perception, when emotional visual distracters are congruent with the central emotional face targets, they do not facilitate or enhance the perceptual processing of these targets. Instead, these visual emotional distracters have a negative impact on response latencies when

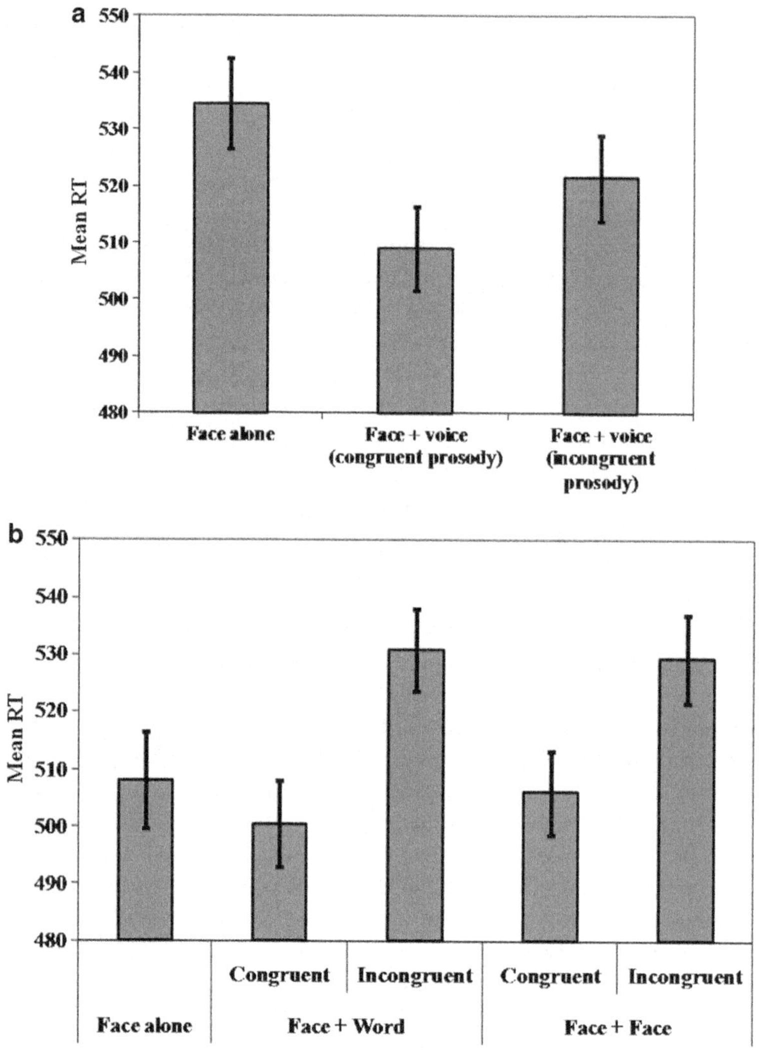

Fig. 10.1 (**a**) Mean RTs (±1 S.E.M) obtained during the block containing control trials (emotional face only) intermixed with either congruent or incongruent audiovisual trials (emotional face + affective tone of voice). Results show a facilitation of perceptual decisions for the central emotional face stimulus, when this face stimulus was paired with a congruent (though unattended) affective tone of voice. (**b**) Mean RTs (±1 S.E.M) obtained during the block containing control trials (emotional face only) intermixed with either congruent or incongruent within-modality redundant trials. The visual distracter was either an emotion written word, or another, secondary emotional facial expression (always shown at a fixed location in the upper visual field). Results show an interference effect created by the presentation of an incongruent (though unattended) affective distracter, as opposed to a benefit in perceptual processing when the exact same emotional face was paired with a congruent affective tone of voice

they carry an incongruent emotional meaning, relative to the central visual target. Taken together, these behavioral results somehow suggest a special status for face–voice combinations during emotion perception, relative to the mere redundancy of emotion information within the same (visual) sensory modality. As reviewed here above, it has been argued that facilitation or enhancement of responses/decisions to a target (i.e., perceptual "benefit") presented together with a congruent distracter may be indicative of perceptual integration between the two inputs (see Marzi et al., 1986; Massaro, 1998; Miller, 1982; Miniussi et al., 1998; Stein & Meredith, 1993). In contrast, a response "cost" associated with the presence of incongruent distracters inevitably evokes response competition (or response bias) phenomena, such as typically observed in the well-known Stroop effect (MacLeod, 1991).

More generally, these new behavioral results suggest a general mechanism for within-modality perception of affect (as the effect was the same for the unattended additional emotional face and the emotional written word), and are therefore consistent with the computational model of perception proposed by Massaro (1998). However, the new critical result is that in this condition (emotion stimulus redundancy within the same sensory modality), congruent trials are processed the same way as control trials (face alone, see Fig. 10.1b), and therefore this interference effect is substantially different from the response facilitation observed with audiovisual pairings during emotion perception (see Fig. 10.1a). In other words, emotion congruence across the concurrent inputs (face + face or face + word) does not lead to a gain in response latencies, but incongruence leads to a processing cost, whose origin is likely to be found at the level of the selection of the motor response (i.e., postperceptual effect) and which may be dependent upon the availability of attentional resources (see Talsma et al., 2010). Indeed, these findings may be compatible with a relatively late response competition view between visual stimuli, such as postulated previously in other interference situations (see MacLeod, 1991). These results therefore suggest a dissociation between the combinatory processes unifying an emotional face with a concurrent affective tone of voice (see Pourtois et al., 2005), and those underlying the perception of multiple or redundant visual emotional stimuli. Whereas the former may depend on perceptual, possibly even preattentive mechanisms (see also de Gelder et al., 2000, 2004), the latter may reflect another class of integration processes which do not rely so much on perceptual mechanisms, and which in turn may be influenced by higher order cognitive processes, including decision making and selective attention (Talsma et al., 2010).

A potential objection is that the response facilitation found with audiovisual pairs during emotion perception may be somehow a consequence of the auditory component in the pairing, rather than the affective congruence of the central emotion face stimulus with this emotional auditory distracter. For example, with distracters that are in the same modality as the central target, they may not produce any facilitatory effect due to an attentional bottleneck phenomenon (see Marois & Ivanoff, 2005; Pashler, 1994). This alternative interpretation is unlikely though, because prior behavioral studies have shown that the crossmodal bias effect during emotion perception was not altered by a secondary task, even if the latter actually required extra processing load in either the visual or auditory modality (see Driver, 1996;

Lavie, 2005; Vroomen et al., 2001). Hence, a putative limitation of processing resources does not seem to be a critical factor hampering multisensory perception of emotion (see also Pourtois et al. 2005, 2000).

5 General Conclusions

Sensory modalities are traditionally characterized by the type of physical stimulation they are more sensitive to, light for vision, sound for hearing, skin pressure for touch, molecules in the air for smell, etc. (Mesulam, 1998). This approach to the study of perception does not do justice to natural beginnings of perception. Indeed, in most of the cases, the organism is confronted with different sensory inputs often taking place at the same time and place, and the perceiver reports objects with multiple sensory attributes. Hence, in many natural situations, different senses receive more or less simultaneously correlated information about the same object or event. The sensory specificity does not correspond either to what usually happens at the other extreme of the perception process, namely, the perceiver's intuition that after perceiving or recognizing an object or event, or after remembering or imagining it, different sensory modalities are intimately linked with one another. In line with this notion of sensory specificity, there seems to be a sort of general consensus in the field about the assumption that information from primary and modality specific cortices is combined in heteromodal areas of the brain (see Calvert, 2001; Ethofer et al., 2006; Mesulam, 1998; Stein & Meredith, 1993), an integration process that eventually yields multisensory-determined objects.

In this chapter, we have reviewed evidence from behavioral studies and cognitive models showing that the perception of emotion can be qualified as an object-based multisensory phenomenon. Emotional facial expression and affective tone of voice do combine during emotion perception to eventually yield strong perceptual effects, which can be distinguished from the mere redundancy of emotional signals within a given sensory modality (here with a focus on the visual modality). This multisensory phenomenon is likely to be an early perceptual, maybe preattentive integration effect, which does not resemble behavioral manifestations of emotion stimulus redundancy, for which a clear-cut postperceptual cost, rather than a perceptual benefit was observed in our study. On the other hand, this observation does not contradict the notion that selective attention mechanisms can boost or alter multisensory perception effects (see Talsma et al., 2010), but under certain laboratory circumstances at least, the combination of an emotional facial expression with a concurrent affective tone of voice can take place irrespective of the fact that stimulus content in one modality is directly ignored (unattended), or attentional load is reliably increased (see Vroomen et al., 2001).

As we have argued throughout this chapter, this audiovisual integration during emotion perception rests on an argument of adaptiveness. The combination of complementary affective information conveyed by multiple channels is probably adaptive, because it can reduce the effects of potentially adverse factors like drifts or intrinsic noise on perceptual performance. But in the absence of any limitations, multisensory perception would be rather inefficient to deal with these natural variations. The new behavioral results presented in this chapter are in line with this assumption and previous studies (see De Gelder & Vroomen, 2000), as they show that the combination of an emotional facial expression with an affective tone of voice does not reflect a general (amodal) perceptual effect (see Borod et al., 2000; Massaro, 1998; Massaro & Egan, 1996), but instead it may serve a specific optimization function for the organism, aimed at binding selectively visual and nonlexical (psychoacoustic) auditory cues during emotion perception, as these two cues usually convey simultaneously critical emotional information about the actual mental state, and possible intentions or action tendencies of peers or conspecifics (Frijda, 1989). For this reason, is the combination of multisensory inputs during emotion perception probably a key perceptual process relying on specific cognitive processes and neural systems (see Pourtois et al., 2005), likely sharing similarities with other multisensory objects, including space perception (see Bertelson, 1999), even though this conjecture remains largely speculative at this stage and it would need some direct confirmation at the empirical level.

Acknowledgments Writing and elaboration of this chapter was made possible thanks to the financial support provided by the European Research Council (Starting Grant #200758) and Ghent University (BOF Grant #05Z01708) to G.P. The behavioral results presented in this chapter were already partly introduced and discussed in the unpublished thesis manuscript written up and submitted by Gilles Pourtois (Tilburg University, April 2002).

References

Adolphs, R., Gosselin, F., Buchanan, T. W., Tranel, D., Schyns, P., & Damasio, A. R. (2005). A mechanism for impaired fear recognition after amygdala damage. *Nature, 433*(7021), 68–72.

Banse, R., & Scherer, K. R. (1996). Acoustic profiles in vocal emotion expression. *Journal of Personality and Social Psychology, 70*(3), 614–636.

Beale, J. M., & Keil, F. C. (1995). Categorical effects in the perception of faces. *Cognition, 57*(3), 217–239.

Bermant, R. I., & Welch, R. B. (1976). Effect of degree of separation of visual-auditory stimulus and eye position upon spatial interaction of vision and audition. *Perceptual and Motor Skills, 43*(2), 487–493.

Bertelson, P. (1999). Ventriloquism: A case of crossmodal perceptual grouping. In G. Aschersleben, T. Bachman, & J. Musseler (Eds.), *Cognitive contributions to the perception of spatial and temporal events* (pp. 347–369). Amsterdam: Elsevier Science.

Bocanegra, B. R., & Zeelenberg, R. (2009a). Dissociating emotion-induced blindness and hypervision. *Emotion, 9*(6), 865–873.

Bocanegra, B. R., & Zeelenberg, R. (2009b). Emotion improves and impairs early vision. *Psychological Science, 20*(6), 707–713.

Borod, J. C., Cicero, B. A., Obler, L. K., Welkowitz, J., Erhan, H. M., Santschi, C., et al. (1998). Right hemisphere emotional perception: Evidence across multiple channels. *Neuropsychology, 12*(3), 446–458.

Borod, J. C., Pick, L. H., Hall, S., Sliwinski, M., Madigan, N., Obler, L. K., et al. (2000). Relationships among facial, prosodic, and lexical channels of emotional perceptual processing. *Cognition & Emotion, 14*(2), 193–211.

Brosch, T., Pourtois, G., & Sander, D. (2010). The perception and categorisation of emotional stimuli: A review. *Cognition & Emotion, 24*(3), 377–400.

Burnham, D. (1999). Perceiving talking faces: From speech perception to a behavioral principle. *Trends in Cognitive Sciences, 3*, 487–488. reviewed by D. Burnham.

Calder, A. J., Lawrence, A. D., & Young, A. W. (2001). Neuropsychology of fear and loathing. *Nature reviews, 2*(5), 352–363.

Calvert, G. A. (2001). Crossmodal processing in the human brain: Insights from functional neuroimaging studies. *Cerebral Cortex, 11*(12), 1110–1123.

Calvert, G. A., Brammer, M. J., Bullmore, E. T., Campbell, R., Iversen, S. D., & David, A. S. (1999). Response amplification in sensory-specific cortices during crossmodal binding. *Neuroreport, 10*(12), 2619–2623.

Calvert, G. A., Bullmore, E. T., Brammer, M. J., Campbell, R., Williams, S. C., McGuire, P. K., et al. (1997). Activation of auditory cortex during silent lipreading. *Science, 276*(5312), 593–596.

Calvert, G. A., Spence, C., & Stein, B. E. (2004). *The handbook of multisensory processes*. Cambridge: MIT Press.

Campanella, S., & Belin, P. (2007). Integrating face and voice in person perception. *Trends in Cognitive Sciences, 11*(12), 535–543.

Campbell, R., Dodd, B., & Burnham, D. (1998). *Hearing by eye II: Advances in the psychology of speechreading and audio-visual speech*. Hove, UK: Psychology Press.

Cummings, K. E., & Clements, M. A. (1995). Analysis of the glottal excitation of emotionally styled and stressed speech. *The Journal of the Acoustical Society of America, 98*, 88–98.

Damasio, A. R. (1989). Time-locked multiregional retroactivation: A system-level proposal for the neural substrates of recall and recognition. *Cognition, 33*, 25–62.

Damasio, A. R. (1994). *Descartes' error: Emotion, reason and the human brain*. New York: Putman Books.

Darwin, C. (1871). *The descent of man*. London: John Murray.

Darwin, C. (1872). *The expression of emotions in man and animals*. London: John Murray.

de Gelder, B. (2006). Towards the neurobiology of emotional body language. *Nature reviews, 7*(3), 242–249.

De Gelder, B., & Bertelson, P. (2003). Multisensory integration, perception and ecological validity. *Trends in Cognitive Sciences, 7*(10), 460–467.

de Gelder, B., Pourtois, G., van Raamsdonk, M., Vroomen, J., & Weiskrantz, L. (2001). Unseen stimuli modulate conscious visual experience: Evidence from inter-hemispheric summation. *Neuroreport, 12*(2), 385–391.

de Gelder, B., Pourtois, G., Vroomen, J., & Bachoud-Levi, A. C. (2000). Covert processing of faces in prosopagnosia is restricted to facial expressions: Evidence from cross-modal bias. *Brain and Cognition, 44*(3), 425–444.

de Gelder, B., Pourtois, G., & Weiskrantz, L. (2002). Fear recognition in the voice is modulated by unconsciously recognized facial expressions but not by unconsciously recognized affective pictures. *Proceedings of the National Academy of Sciences of the United States of America, 99*(6), 4121–4126.

De Gelder, B., & Vroomen, J. (2000). Perceiving emotions by ear and by eye. *Cognition & Emotion, 14*(289–311).

De Gelder, B., Vroomen, J., & Bertelson, P. (1998). Upright but not inverted faces modify the perception of emotion in the voice. *Current Psychology of Cognition, 17*, 1021–1031.

de Gelder, B., Vroomen, J., & Pourtois, G. (1999). Seeing cries and hearing smiles. Crossmodal perception of emotional expressions. In G. Aschersleben, T. Bachmann, & J. Müsseler (Eds.), *Cognitive contributions to the perception of spatial and temporal events* (pp. 425–438). Amsterdam: Elsevier.

de Gelder, B., Vroomen, J., & Pourtois, G. (2004). Multisensory perception of emotion, its time course, and its neural basis. In G. Calvert, C. Spence, & B. E. Stein (Eds.), *The handbook of multisensory processes* (pp. 581–597). Cambridge, MA: MIT Press.

deGelder, B., Teunisse, J. P., & Benson, P. J. (1997). Categorical perception of facial expressions: Categories and their internal structure. *Cognition & Emotion, 11*(1), 1–23.

Dodd, B., & Campbell, R. (1987). *Hearing by eye: The psychology of lip-reading*. Hillsdale, NJ: Lawrence Erlbaum Associates.
Driver, J. (1996). Enhancement of selective listening by illusory mislocation of speech sounds due to lip-reading. *Nature, 381*(6577), 66–68.
Driver, J., & Spence, C. (1998a). Cross-modal links in spatial attention. *Philosophical Transactions of the Royal Society of London, 353*(1373), 1319–1331.
Driver, J., & Spence, C. (1998b). Crossmodal attention. *Current Opinion in Neurobiology, 8*(2), 245–253.
Driver, J., & Spence, C. (2000). Multisensory perception: Beyond modularity and convergence. *Current Biology, 10*(20), R731–R735.
Ekman, P. (1992). An argument for basic emotions. *Cognition & Emotion, 6*, 169–200.
Ekman, P., & Friesen, W. V. (1976). *Pictures of facial affect*. Palo-Alto: Consulting Psychologists Press.
Etcoff, N. L., & Magee, J. J. (1992). Categorical perception of facial expressions. *Cognition, 44*(3), 227–240.
Ethofer, T., Pourtois, G., & Wildgruber, D. (2006). Investigating audiovisual integration of emotional signals in the human brain. *Progress in Brain Research, 156*, 345–361.
Farah, M. J., Wong, A. B., Monheit, M. A., & Morrow, L. A. (1989). Parietal lobe mechanisms of spatial attention – Modality-specific or supramodal. *Neuropsychologia, 27*(4), 461–470.
Fodor, J. (1983). *The modularity of mind*. Cambridge, MA: MIT Press.
Foxe, J. J., & Molholm, S. (2009). Ten years at the multisensory forum: Musings on the evolution of a field. *Brain Topography, 21*(3–4), 149–154.
Frick, R. W. (1985). Communicating emotion – The role of prosodic features. *Psychological Bulletin, 97*(3), 412–429.
Frijda, N. (1989). *The emotions*. Cambridge: Cambridge University Press.
Fuster, J. M., Bodner, M., & Kroger, J. K. (2000). Cross-modal and cross-temporal association in neurons of frontal cortex. *Nature, 405*(6784), 347–351.
Giard, M. H., & Peronnet, F. (1999). Auditory-visual integration during multimodal object recognition in humans: A behavioral and electrophysiological study. *Journal of Cognitive Neuroscience, 11*(5), 473–490.
Haith, M. M., Bergman, T., & Moore, M. J. (1977). Eye contact and face scanning in early infancy. *Science, 198*, 853–855.
Hay, J. C., Pick, H. L., & Ikeda, K. (1965). Visual capture produced by prism spectacles. *Psychonomic Science, 2*(8), 215–216.
Held, R. (1965). Plasticity in sensory-motor systems. *Scientific American, 213*, 84–94.
Kahneman, D. (1973). *Attention and effort*. Englewood Cliffs, NJ: Prentice Hall.
Lavie, N. (1995). Perceptual load as a necessary condition for selective attention. *Journal of Experimental Psychology. Human Perception and Performance, 21*(3), 451–468.
Lavie, N. (2005). Distracted and confused?: Selective attention under load. *Trends in Cognitive Sciences, 9*(2), 75–82.
Lehmann, S., & Murray, M. M. (2005). The role of multisensory memories in unisensory object discrimination. *Brain Research, 24*(2), 326–334.
Lewkowicz, D. J. (2000). The development of intersensory temporal perception: An epigenetic systems/limitations view. *Psychological Bulletin, 126*(2), 281–308.
Liberman, A. M., Harris, K. S., Hoffman, H. S., & Griffith, B. C. (1957). The discrimination of speech sounds within and across phoneme boundaries. *Journal of Experimental Psychology, 54*(5), 358–368.
Lieberman, P., & Michaels, S. B. (1962). Some aspects of fundamental frequency and envelope amplitude as related to emotional content of speech. *The Journal of the Acoustical Society of America, 34*, 922–927.
Macaluso, E., Frith, C. D., & Driver, J. (2000). Modulation of human visual cortex by crossmodal spatial attention. *Science, 289*(5482), 1206–1208.
MacLeod, C. M. (1991). Half a century of research on the Stroop effect: An integrative review. *Psychological Bulletin, 109*, 163–203.

Marois, R., & Ivanoff, J. (2005). Capacity limits of information processing in the brain. *Trends in Cognitive Sciences, 9*(6), 296–305.

Marzi, C. A., Tassinari, G., Aglioti, S., & Lutzemberger, L. (1986). Spatial summation across the vertical meridian in hemianopics: A test of blindsight. *Neuropsychologia, 24*(6), 749–758.

Massaro, D. W. (1987). *Speech perception by ear and by eye: A paradigm for psychological inquiry*. Hillsdale, NJ: Lawrence Erlbaum Associates.

Massaro, D. W. (1998). *Perceiving talking faces: From speech perception to a behavioral principle*. Cambridge: MIT Press.

Massaro, D. W., & Egan, P. B. (1996). Perceiving affect from the voice and the face. *Psychonomic Bulletin & Review, 3*, 215–221.

Mcgurk, H., & Macdonald, J. (1976). Hearing lips and seeing voices. *Nature, 264*(5588), 746–748.

Mckelvie, S. J. (1995). Emotional expression in upside-down faces – Evidence for configurational and componential processing. *The British Journal of Social Psychology, 34*, 325–334.

Mesulam, M. M. (1998). From sensation to cognition. *Brain, 121*, 1013–1052.

Miller, J. (1982). Divided attention – Evidence for co-activation with redundant signals. *Cognitive Psychology, 14*(2), 247–279.

Miller, J. (1986). Timecourse of coactivation in bimodal divided attention. *Perception & Psychophysics, 40*(5), 331–343.

Miniussi, C., Girelli, M., & Marzi, C. A. (1998). Neural site of the redundant target effect electrophysiological evidence. *Journal of Cognitive Neuroscience, 10*(2), 216–230.

Molholm, S., Ritter, W., Murray, M. M., Javitt, D. C., Schroeder, C. E., & Foxe, J. J. (2002). Multisensory auditory-visual interactions during early sensory processing in humans: A high-density electrical mapping study. *Brain Research, 14*(1), 115–128.

Moors, A., & De Houwer, J. (2006). Automaticity: A theoretical and conceptual analysis. *Psychological Bulletin, 132*, 297–326.

Morris, J. S., Scott, S. K., & Dolan, R. J. (1999). Saying it with feeling: Neural responses to emotional vocalizations. *Neuropsychologia, 37*(10), 1155–1163.

Murray, I. R., & Arnott, J. L. (1993). Toward the simulation of emotion in synthetic speech: A review of the literature on human vocal emotion. *The Journal of the Acoustical Society of America, 93*, 1097–1108.

Murray, M. M., Foxe, J. J., Higgins, B. A., Javitt, D. C., & Schroeder, C. E. (2001). Visuo-spatial neural response interactions in early cortical processing during a simple reaction time task: A high-density electrical mapping study. *Neuropsychologia, 39*(8), 828–844.

Osullivan, M., Ekman, P., Friesen, W., & Scherer, K. (1985). What you say and how you say it – The contribution of speech content and voice quality to judgments of others. *Journal of Personality and Social Psychology, 48*(1), 54–62.

Panksepp, J. (2005). Psychology. Beyond a joke: From animal laughter to human joy? *Science, 308*(5718), 62–63.

Pashler, H. (1994). Dual-task interference in simple tasks: Data and theory. *Psychological Bulletin, 116*(2), 220–244.

Phelps, E. A., Ling, S., & Carrasco, M. (2006). Emotion facilitates perception and potentiates the perceptual benefits of attention. *Psychological Science, 17*(4), 292–299.

Pourtois, G., & de Gelder, B. (2002). Semantic factors influence multisensory pairing: A transcranial magnetic stimulation study. *Neuroreport, 13*(12), 1567–1573.

Pourtois, G., de Gelder, B., Bol, A., & Crommelinck, M. (2005). Perception of facial expressions and voices and of their combination in the human brain. *Cortex; a journal devoted to the study of the nervous system and behavior, 41*(1), 49–59.

Pourtois, G., de Gelder, B., Vroomen, J., Rossion, B., & Crommelinck, M. (2000). The time-course of intermodal binding between seeing and hearing affective information. *Neuroreport, 11*(6), 1329–1333.

Pourtois, G., Debatisse, D., Despland, P. A., & de Gelder, B. (2002). Facial expressions modulate the time course of long latency auditory brain potentials. *Brain Research, 14*(1), 99–105.

Pourtois, G., Grandjean, D., Sander, D., & Vuilleumier, P. (2004). Electrophysiological correlates of rapid spatial orienting towards fearful faces. *Cerebral Cortex, 14*(6), 619–633.

Raab, D. H. (1962). Statistical facilitation of simple reaction times. *Transactions of the New York Academy of Sciences, 24*, 574–590.
Sander, D., Grafman, J., & Zalla, T. (2003). The human amygdala: An evolved system for relevance detection. *Reviews in the Neurosciences, 14*(4), 303–316.
Savazzi, S., & Marzi, C. A. (2002). Speeding up reaction time with invisible stimuli. *Current Biology, 12*(5), 403–407.
Savazzi, S., & Marzi, C. A. (2004). The superior colliculus subserves interhemispheric neural summation in both normals and patients with a total section or agenesis of the corpus callosum. *Neuropsychologia, 42*(12), 1608–1618.
Savazzi, S., & Marzi, C. A. (2008). Does the redundant signal effect occur at an early visual stage? *Experimental brain research. Experimentelle Hirnforschung. Expérimentation cérébrale, 184*(2), 275–281.
Scherer, K. (1989). Vocal measurement of emotion. In R. Plutchik & H. Kellerman (Eds.), *Emotion: Theory, research, and experience* (Vol. 4, pp. 233–259). San Diego, CA: Academic.
Scherer, K. R., Banse, R., & Wallbott, H. G. (2001). Emotion inferences from vocal expression correlate across languages and cultures. *Journal of Cross-Cultural Psychology, 32*(1), 76–92.
Schneider, W., & Shiffrin, R. M. (1977). Controlled and automatic human information-processing. 1. Detection, search, and attention. *Psychological Review, 84*(1), 1–66.
Scott, S. K., Young, A. W., Calder, A. J., Hellawell, D. J., Aggleton, J. P., & Johnson, M. (1997). Impaired auditory recognition of fear and anger following bilateral amygdala lesions. *Nature, 385*(6613), 254–257.
Sekuler, R., Sekuler, A. B., & Lau, R. (1997). Sound alters visual motion perception. *Nature, 385*(6614), 308.
Shams, L., Kamitani, Y., & Shimojo, S. (2000). Illusions. What you see is what you hear. *Nature, 408*(6814), 788.
Smith, M. L., Cottrell, G. W., Gosselin, F., & Schyns, P. G. (2005). Transmitting and decoding facial expressions. *Psychological Science, 16*(3), 184–189.
Stein, B. E., & Meredith, M. A. (1993). *The merging of the senses*. Cambridge: Bradford Books.
Talsma, D., Senkowski, D., Soto-Faraco, S., & Woldorff, M. G. (2010). The multifaceted interplay between attention and multisensory integration. *Trends in Cognitive Sciences, 14*(9), 400–410.
Tanaka, J. W., & Farah, M. J. (1993). Parts and wholes in face recognition. *The Quarterly Journal of Experimental Psychology. A, Human Experimental Psychology, 46*(2), 225–245.
Turatto, M., Mazza, V., Savazzi, S., & Marzi, C. A. (2004). The role of the magnocellular and parvocellular systems in the redundant target effect. *Experimental brain research. Experimentelle Hirnforschung. Expérimentation cérébrale, 158*(2), 141–150.
Vroomen, J., Collier, R., & Mozziconacci, S. (1993). Duration and intonation in emotional speech. *Proceedings of the Third European Conference on Speech Communication and Technology, Berlin*, (pp. 577–580).
Vroomen, J., Driver, J., & de Gelder, B. (2001). Is cross-modal integration of emotional expressions independent of attentional resources? *Cognitive, Affective, & Behavioral Neuroscience, 1*(4), 382–387.
Vuilleumier, P. (2005). How brains beware: Neural mechanisms of emotional attention. *Trends in Cognitive Sciences, 9*(12), 585–594.
Vuilleumier, P., & Pourtois, G. (2007). Distributed and interactive brain mechanisms during emotion face perception: Evidence from functional neuroimaging. *Neuropsychologia, 45*(1), 174–194.
Walker, A., & Grolnick, W. (1983). Discrimination of vocal expressions by young infants. *Infant Behavior & Development, 6*, 491–498.
Walker-Andrews, A. S. (1997). Infants' Perception of expressive behaviors: Differentiation of multimodal information. *Psychological Bulletin, 121*, 437–456.
Weiskrantz, L. (1986). *Blindsight. A case study and implications*. Oxford: Oxford University Press.
Williams, C. E., & Stevens, K. N. (1972). Emotions and speech – Some acoustical correlates. *The Journal of the Acoustical Society of America, 52*(4), 1238–1250.

Chapter 11
Cross-Modal Modulation of Spatial Attention by Emotion

Tobias Brosch and Didier Grandjean

Abstract A huge amount of environmental stimulus input constantly enters the brain via the different sensory channels of the organism. Due to its limited capacity, the brain cannot process all the inputs exhaustively, and thus needs to select a subset of stimuli for further processing at the cost of others. Emotional stimuli, for example social signals such as angry faces or happy voices, are privileged in the competition for attentional processing resources: the neural representation of emotional stimuli is stronger and more robust compared to neutral stimuli; emotional stimuli are prioritized in perception, draw attention more quickly, and impede attentional disengagement longer than neutral stimuli. The representations of emotional stimuli are thus intensified at different stages of processing. This generates a vivid conscious percept allowing organisms to prepare and implement adequate responses. By modulating frontoparietal attention systems, emotional stimuli also impact on the perceptual processing of subsequent stimuli appearing at the same location as an emotional stimulus. Such neurocognitive selection mechanisms thus drastically reorganize our representation and perception of the environment by focusing on emotionally and motivationally relevant events and their immediate spatial and temporal periphery.

Until now, the effects of emotional stimuli on attentional processes have mainly been described within a sensory modality, most frequently using pictures of emotional stimuli to modulate visuospatial attention. However, in real-life situations humans typically encounter simultaneous input to several different senses, such as vision, audition, olfaction, and touch. Signals entering these different channels might originate from a common, emotionally relevant source. To receive maximal benefit

T. Brosch
New York University, Department of Psychology, 6 Washington Place, New York, NY 10003, USA
e-mail: tobias.brosch@nyu.edu

D. Grandjean (✉)
Department of Psychology, Centre Interfacultaire en Sciences Affectives (NCCR),
University of Geneva, 7 rue des Battoirs, 1205 Geneva, Switzerland
e-mail: didier.grandjean@unige.ch

P. Belin et al. (eds.), *Integrating Face and Voice in Person Perception*,
DOI 10.1007/978-1-4614-3585-3_11,

from multimodal sensory input, the brain must coordinate the input appropriately so that signals from a relevant common source are rapidly processed and integrated across the different input channels to allow for the preparation and implementation of adaptive responses.

We review the current evidence for cross-modal modulation of spatial attention by emotional information. Presenting data from behavioral and electrophysiological investigations in human subjects we illustrate, for example, the effects of emotional voices on visual attention and the effects of emotional images on haptic attention. The data converge to show that emotion modulates attentional processing across sensory modalities by boosting early sensory stages of processing, potentially implemented by a large-scale neural network centered around the amygdala, providing direct and indirect top-down signals to sensory pathways and frontoparietal pathways involved in exogenous and endogenous attentional selection processes.

This rapid cross-modal integration at multiple stages of processing may reflect a fundamental principle of human brain organization: to prioritize the processing of emotionally relevant stimuli, even if they are outside the focus of spatial attention, thus facilitating the multimodal assessment of emotionally relevant stimuli in the environment.

1 Introduction

Our environment constantly confronts us with large amounts of information. Due to capacity limits of the brain, we cannot process all the information entering our senses thoroughly, but have to select important information and prioritize its processing at the cost of other, less relevant information. This competition for neural processing capacity is driven by attentional mechanisms (Driver, 2001) which are influenced by several factors, related to the current needs and goals of the observer (*endogenous attention*) as well as to basic physical properties of the stimulus (*exogenous attention*). In addition, the emotional relevance of a stimulus constitutes an important selection criterion for prioritized processing. Efficient processing of emotional stimuli is highly adaptive, as emotion highlights the relevance of a stimulus for the well-being and survival of the organism (Scherer, 2001). Emotional stimuli should thus be noticed readily and, once detected, become the focus of attention, evaluation, and action. It has been suggested that dedicated neural circuits may underlie the prioritization of emotional stimuli (*emotional attention*, Vuilleumier, 2005). The amygdala, a limbic region critically involved in the processing of emotional information (LeDoux, 2000; Phelps, 2006; Sander, Grafman, & Zalla, 2003), is thought to play a critical role by modulating the processing of incoming sensory stimuli through direct feedback projections to sensory cortex and subsequent biasing signals to frontoparietal attention regions.

Up to now, most studies investigating the preferential role of emotional stimuli in attention and perception have examined *within-modality* effects, most frequently using pictures of emotional stimuli to modulate visual attention. However, humans typically encounter simultaneous input to several different senses, such as vision, audition, olfaction, and touch. Signals entering these different channels might originate from a common emotionally relevant source, requiring mechanisms for the integration of information conveyed by multiple sensory channels. This integration allows for a more detailed and efficient representation of the world than any single modality in isolation, as it may capitalize on the individual strengths of the different modalities. For example, audition covers a larger spatial area than vision. The rapid detection of an emotionally arousing sound may subsequently lead to an increased allocation of visual attention toward the spatial source of the sound, allowing for a more thorough analysis of the situation based on visual input.

In this chapter, we review the literature investigating cross-modal modulations of attention by emotional information. We first summarize research on the effects and mechanisms of exogenous and endogenous attention selection within and across modalities. We then highlight the special role of emotional information in attention and perception, reviewing both behavioral evidence and evidence from neuroimaging. We conclude by presenting a neurocognitive model describing the mechanisms underlying cross-modal emotional attention.

2 Mechanisms of Attentional Selection: Endogenous and Exogenous Attention

Not all incoming environmental stimulation can be processed in parallel and evaluated thoroughly due to capacity limits of the human brain (Marois & Ivanoff, 2005). To allow for a rapid and efficient analysis of behaviorally important information in the environment, dedicated attention systems therefore serve to select a subset of all incoming stimuli for more in-depth processing and preferential access to conscious awareness (Driver, 2001). Attentional prioritization leads to preferential processing via increases in sensory gain (Hillyard, Vogel, & Luck, 1998), as evidenced by perceptual enhancement such as faster stimulus detection (Posner, 1980) or increases in contrast sensitivity (Carrasco, Ling, & Read, 2004). Attentional selection can be guided by stimulus-related and by observer-dependent effects. Distinct functional subprocesses related to different selection criteria have been put forward, and their respective properties and contributions to attentional selection mechanisms have been isolated using both behavioral and brain-imaging methods. *Exogenous attention* refers to effects driven by the intrinsic physical salience of sensory inputs (Egeth & Yantis, 1997; Theeuwes, 1991; Wolfe & Horowitz, 2004). Low-level properties such as stimulus intensity, color, or size may trigger an involuntary, stimulus-driven, bottom-up attention process. Experimentally, this form of attentional selection has been demonstrated using the exogenous cueing paradigm (Posner, 1980), where participants have to indicate the location of a target that appears either at the same location as a previous exogenous cue (e.g., a bright flash) or at the opposite location. Importantly, the cue is nonpredictive of the target location, i.e., in 50% of the trials the target replaces the cue (valid trials), in 50% of the trials it appears at the opposite location (invalid trials). Faster responses to targets in valid trials indicate exogenous attention capture by the cue. This effect has been demonstrated within the visual (Posner, 1980), the auditory (Spence & Driver, 1994), and the tactile modality (Miles, Poliakoff, & Brown, 2008). Furthermore, cross-modal cueing studies have demonstrated that directing exogenous attention to a stimulus in one modality (e.g., with a nonpredictive sound) facilitates the speed of responding of spatially coincident stimuli in another modality (e.g., towards a visual or a tactile target). This cross-modal facilitation has been observed for all combinations of visual, auditory, and tactile stimuli (see Driver & Spence, 1998; Koelewijn, Bronkhorst, & Theeuwes, 2010, for reviews). Some asymmetries have been observed related to the modality of the cue: visual cues lead to a narrower focusing of the attentional field in which facilitation is achieved compared to auditory cues, an effect that may be related to the different spatial resolutions of the different sensory modalities (Spence, 2010). In contrast to the reflexive exogenous attention mechanisms, *endogenous attention* refers to a voluntary top-down process, initiated by implicit or explicit expectations for a specific object or location (Desimone & Duncan, 1995; Posner, Snyder, & Davidson, 1980). This process selects stimuli important to the current behavior and goals of the organism. This form of attentional selection has been demonstrated using the endogenous cueing task (Posner et al., 1980), in which a centrally presented arrow indicates the location where a subsequent target stimulus will probably appear, thus creating an expectation in the participants. Faster responses to validly cued targets (i.e., targets that appear at the location indicated by the arrow) reflect voluntary endogenous attention shifts. Again, this effect has been demonstrated for the visual (Posner et al., 1980), auditory (Spence & Driver, 1994), and tactile modalities (Lloyd, Bolanowski, Howard, & McGlone, 1999). Furthermore, cross-modal cueing studies have demonstrated that directing endogenous attention to one modality (e.g., creating an expectation for a sound at a specific location) facilitates the speed of responding of spatially coincident stimuli in another modality (e.g., towards a visual or a tactile target). Again, this effect has been observed for all combinations of visual, auditory and tactile stimuli (see Koelewijn et al., 2010, for a review). Besides expectations for target locations, endogenous attention can be directed toward and improve detection of other features of potential target objects such as shape, color, or direction of motion (Rossi & Paradiso, 1995) or towards complete objects (Yantis, 1992). First evidence for cross-modal object-based attention has been presented recently (Turatto, Mazza, & Umilta, 2005), demonstrating that auditory objects may affect the deployment of visual attention.

According to a recent neurocognitive model of attention, both endogenous and exogenous attention primarily implicate frontoparietal networks of cortical regions (Corbetta, Patel, & Shulman, 2008; Corbetta & Shulman, 2002; see also Peelen, Heslenfeld, & Theeuwes, 2004), with endogenous attention control being exerted by interactions of dorsal regions such as the intraparietal sulcus (IPS) and the frontal eye fields (FEF), and exogenous reorienting of the attentional focus mediated by more ventral regions in the right hemisphere such as the right ventral frontal cortex (VFC) and temporoparietal junction (TPJ). Even though most neuroimaging data investigating these two attentional networks have been collected in the visual modality, the available evidence supports a supramodal function. The ventral network is sensitive to salient events in the visual, auditory, and tactile modality, and similar ventral and dorsal frontoparietal regions are modulated by reorienting in different modalities (Corbetta et al., 2008; Eimer & Driver, 2001).

ERP studies measuring the neural effects of cross-modal endogenous and exogenous attention suggest that attentional facilitation effects are operating at early perceptual stages. Cross-modal attentional modulations affect early modality-specific ERP components (up to 200 ms after target onset), but show smaller or no effects at later components linked to post-perceptual stages (later than 200 ms, see Eimer & Driver, 2001, for a review). Studies using fMRI data and source localization models of EEG data point to the involvement of heteromodal areas such as the superior temporal sulcus (STS) as well as early modality-specific sensory areas in cross-modal attention modulation (see Koelewijn et al., 2010, for a review). McDonald, Teder-Salejarvi, Di Russo, and Hillyard (2003) measured modulations of visually evoked brain activity by nonpredictive exogenous auditory cues using ERPs and observed a first modulation in the superior temporal cortex (120–140 ms after stimulus onset), followed by a second modulation in the ventral occipital cortex of the fusiform gyrus (150–170 ms after stimulus onset). This spatiotemporal sequence suggests that enhanced visual perception produced by cross-modal exogenous attention results from feedback from multimodal superior temporal cortex to early modality-specific visual areas. Cross-modal exogenous attention may thus first facilitate processing of spatially coincident visual stimuli in the posterior parts of superior temporal gyrus and superior temporal sulcus (STG/STS), regions of multisensory convergence and integration (Hein & Knight, 2008; Kreifelts, Ethofer, Shiozawa, Grodd, & Wildgruber, 2009). Reentrant feedback from STG/STS to early visual areas may then enhance activation in early modality-specific areas by increasing sensory gain.

3 The Special Role of Emotion in Attention and Perception

In addition to endogenous and exogenous attention mechanisms, the emotional relevance of a stimulus has been shown to constitute another important feature influencing selection by attention. Behavioral findings across many different tasks and paradigms indicate that perception is facilitated and attention prioritized for emotional information. Thus, emotion processing does not only enrich our experiences with affective flavor, but can directly shape the content of our percepts and awareness. Emotional stimuli may draw attention quicker and impede attentional disengagement longer than neutral stimuli. In visual search tasks, the detection of a target among distractors is faster when the target is emotional, as opposed to neutral (e.g., Ohman, Flykt, & Esteves, 2001). Conversely, emotional distractors may prolong the search for a nonemotional target (Rinck, Reinecke, Ellwart, Heuer, & Becker, 2005). In the attentional blink task, the detection of a target word in a rapid serial visual stream (items appearing successively at fixation at ~10 Hz) is impaired when it occurs shortly after another target. However, this deficit is greatly attenuated for emotional stimuli (e.g., Anderson & Phelps, 2001). Conversely, the deficit may increase for a second neutral target following an emotional one, suggesting that the emotional meaning of items tend to grab or divert attention in situations where resources cannot be equally deployed to every stimulus (Smith, Most, Newsome, & Zald, 2006). In the visual prior-entry paradigm, two stimuli are presented simultaneously or almost simultaneously, and participants have to indicate which of the stimuli they perceived first. In this task, fearful faces are perceived earlier in time than neutral faces, reflecting accelerated perception due to attentional prioritization (West, Anderson, & Pratt, 2009). Attentional prioritization has been observed at very early cortical stages of processing, such as primary visual cortex (V1) for threatening visual stimuli (Pourtois, Grandjean, Sander, & Vuilleumier, 2004; West, Anderson, Ferber, & Pratt, 2011). Once attention has been drawn to and engaged by emotional stimuli, it may also dwell longer at their location and facilitate the processing of subsequent nonemotional target stimuli appearing at the same location. Such emotional orienting effects have been demonstrated using the dot probe task (MacLeod, Mathews, & Tata, 1986), where participants must respond to a target (a line or a dot) that replaces one of two simultaneously presented cues—one being emotionally significant (e.g., a fearful face) and the other neutral. Importantly, the cues are equated on basic physical properties such as brightness, contrast, color so that any observed preferential cueing effect is not due to exogenous attention based on low-level stimulus differences, but can be attributed to the perceived emotionality of the cues. Typical results show faster responses to targets replacing the emotional rather than the neutral cue. This effect has been demonstrated both for the visual (Brosch, Sander, & Scherer, 2007; Lipp & Derakshan, 2005) and for the auditory modality (Bertels, Kolinsky, & Morais, 2010). Emotional cueing may also increase contrast sensitivity for the subsequent target (Phelps et al., 2006). These cueing effects occur despite the fact that the cue is not predictive of target location and their emotional meaning is task-irrelevant. Modulation of attention by emotion has furthermore been observed in brain-damaged patients. The dorsal attentional network can be disrupted by stroke in the right parietal regions, resulting in neglect and/or extinction. Studies in patients with these symptoms have demonstrated that the extinction of visual and auditory stimuli can be modulated by emotional stimulus content. Pictures of spiders compared to flowers can decrease the amount of visual extinction in neglect patients (Vuilleumier & Schwartz, 2001). Similarly, emotional prosody can reduce auditory extinction in neglect patients, as demonstrated in a dichotic listening task (Grandjean, Sander, Lucas, Scherer, & Vuilleumier, 2008).

Until now, studies on the emotional modulation of spatial attention have mainly examined within-modality effects, most frequently using pictures of emotional stimuli to modulate visual attention. However, some studies have recently begun to investigate cross-modal emotional attention. In a series of studies, we adapted the emotional dot probe paradigm to investigate cross-modal bias of visual spatial attention by auditory emotion (Brosch, Grandjean, Sander, & Scherer, 2008, 2009). More specifically, we investigated whether emotional prosody (see Grandjean, Bänziger, & Scherer, 2006) influences the spatial deployment of visual attention

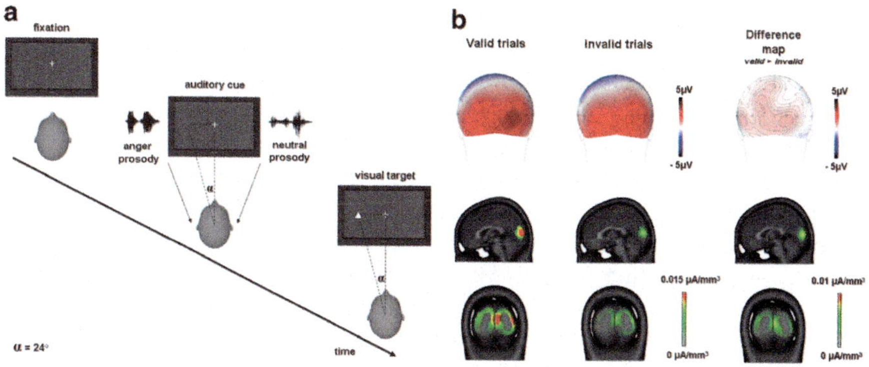

Fig. 11.1 The cross-modal emotional dot probe paradigm. (**a**) Experimental sequence of Brosch et al. (2008, 2009). Each trial started with a random time interval between 500 and 1,000 ms, after which the acoustic cue sound pair was presented. One of the sounds in the pair had emotional prosody, the other one neutral prosody. The target, a neutral geometric figure was presented with a variable cue–target stimulus onset asynchrony sound onset, on the left or right side. The angle between the target and the fixation cross was 24°, equivalent to the synthesized location of the audio stimulus pairs. In a *valid* trial, the target appeared on the side of the emotional sound, in an *invalid* trial, the target appeared on the side of the neutral sound. (**b**) Electrophysiological data confirm cross-modal effects of emotional prosody on early visual processing. *Top row*: Topographic maps for the P1 in valid and invalid trials and topographic difference map. *Middle* and *bottom rows*: Source localization revealed the intracranial generators of the P1 in striate and extrastriate visual cortex

when emotional and neutral utterances were presented simultaneously (see Fig. 11.1a). In order to give the subjective impression that the sounds originated from a specific location in space (at an angle of 24° to the left and to the right of the participants, corresponding to the locations where the visual target could appear on screen), we manipulated the interaural time difference of the sounds. We used spatially localized stimuli instead of the simpler dichotic presentation mode, as it is a closer approximation of real life contexts in which concomitant auditory and visual information can originate from a common source localized in space. We observed faster responses towards targets when they appeared at the location of the source of the emotional prosody. Importantly, this cross-modal emotional effect was not present when using synthesized control stimuli matched for the mean fundamental frequency and the amplitude envelope, two low-level acoustic parameters related to emotional prosody, of each vocal stimulus used in the experiment, ruling out the possibility that only low-level acoustic parameters trigger the cross-modal emotional effect.

Using a similar approach, Poliakoff and colleagues investigated the effect of threatening visual cues on tactile attention. In a modified cueing paradigm, visual cues were presented close to the participant's hands, which were hidden from view behind a computer screen. The cues consisted of one picture of either a threatening (snakes or spiders) or a nonthreatening stimulus (flowers or mushrooms) presented either close to the left hand or close to the right hand. Following the cue, a tactile stimulus was presented to one of the hands. Pictures of snakes led to faster responses to the tactile stimulus than nonthreatening pictures. Remarkably, this facilitation effect was enhanced in participants with high fear of snakes, showing that the cross-modal attentional facilitation is driven by the individually perceived threat value (Poliakoff, Miles, Li, & Blanchette, 2007). Following up on these results, Van Damme and colleagues compared the impact of the presentation of threatening pictures on tactile and auditory attention using the prior-entry paradigm (Van Damme, Gallace, Spence, Crombez, & Moseley, 2009). In this paradigm two target stimuli are presented simultaneously or almost simultaneously, and participants have to indicate which target they perceived first. Attentional prioritization of a target leads to accelerated perception. In some trials, participants were presented two tactile targets (vibrations with minimal stimulus onset asynchronies, between 5 and 120 ms), one to the left hand and one to the right hand. In other trials, they were presented two auditory targets emanating from two loudspeakers. In each trial, participants had to indicate which target they perceived first. Before presentation of the target pair, one of the potential target sides was cued with a picture of either a threat to the hand (such as a knife), a general threat (such as an exploding truck), or a picture with emotionally neutral content. All responses were faster when cued by threatening compared to neutral pictures, confirming cross-modal attentional bias by threat. However, in trials with tactile targets, tactile attention was modulated more strongly by pictures showing threats to the hand than by pictures showing general threat. In trials with auditory target pairs, however, attention was biased more strongly by general threat than by threat to the hand. Thus, a visual emotional stimulus indicating imminent threat to a body part leads to attentional bias toward the input from that body part, suggesting some degree of specificity in cross-modal emotional attention. In a similar vein, Schirmer and colleagues investigated to what extent being touched by a friend can modulate early stages of visual processing. Early ERP components such as the N100 and the P200 were modulated by the touch of a friend during negative and neutral pictures viewing. Furthermore, the Late Positive Component (LPC) was increased during negative picture presentations when human touch occurred compared to negative pictures without human touch (Schirmer et al., in press).

Taken together, the behavioral data reviewed here indicate that perception is facilitated and attention prioritized for emotional information. Emotional stimuli capture attention quicker and may prolong attentional disengagement relative to neutral stimuli. Depending on the task, the prioritization of emotional material can improve behavioral performance (when the target of the task is emotional), but may also lead to interference (when an emotional stimulus competes with a nonemotional target for processing resources). Longer dwelling times of attention at the location of emotional stimuli may furthermore facilitate the processing of subsequent target stimuli that appear at the same location. Whereas most studies have looked at within-modality effects of emotional attention, first studies investigating cross-modal emotional attention demonstrate that emotional attention is not restricted to one modality, but operates across modalities. Here, we reviewed evidence for the modulation of visual attention by auditory emotional information (Brosch, Grandjean et al., 2008; Brosch et al., 2009), evidence for the modulation of tactile and auditory attention by visual emotional information (Poliakoff et al., 2007; Van Damme et al., 2009), as well as evidence for the modulation of visual processing by tactile emotional information (Schirmer et al., in press).

4 Neural Mechanisms of Within-Modality Emotional Attention

Consistent with the behavioral findings reviewed above, brain imaging studies using fMRI have consistently revealed increased neural responses to many different emotional stimuli compared to emotionally neutral stimuli, both in early sensory areas like primary visual cortex, and in higher-level regions associated with object and face recognition. Enhanced responses have been observed for emotional pictures in the visual cortex (Whalen et al., 1998), emotional faces in the fusiform face area (Vuilleumier, Armony, Driver, & Dolan, 2001), and emotional body movements in the fusiform body area (Peelen, Atkinson, Andersson, & Vuilleumier, 2007). Similar results have been found in the auditory modality, in that emotional prosody increases activity in the associative auditory cortex (Ethofer, Anders, Wiethoff et al., 2006). Altogether, these findings suggest a selective modulation of brain regions involved in the processing of the specific stimulus categories by emotion. This emotional boosting of neural processing was observed even when the focus of endogenous attention was directed away from the emotional stimuli by secondary tasks, as observed both for the visual (Vuilleumier et al., 2001) and the auditory modality (Grandjean et al., 2005; Sander et al., 2005). Research using electroencephalography (EEG) has yielded similar results, revealing modulatory effects of emotion at several stages of cortical processing, including both early, sensory-related processes and later processes related to more elaborate evaluations of these stimuli, subsequent autonomic arousal, and/or memory formation (see, e.g., Eimer & Holmes, 2007; Olofsson, Nordin, Sequeira, & Polich, 2008; Vuilleumier & Pourtois, 2007, for reviews). Thus, brain imaging and electrophysiological data converge to show that emotional stimuli are represented by more robust neural signatures than neutral ones, and can consequently profit from preferential access to further cognitive processing, behavior control, and awareness.

It has been suggested that the prioritization of emotional information is driven by dedicated neural circuits (Brosch, Pourtois, Sander, & Vuilleumier, 2011; Vuilleumier, 2005; Vuilleumier & Brosch, 2009), separate from the frontoparietal networks involved in endogenous and exogenous attention allocation (Corbetta et al., 2008; Corbetta & Shulman, 2002; see also Peelen et al., 2004). In this model, the amygdala, a limbic region critically involved in the processing of emotional information (LeDoux, 2000; Phelps, 2006) is thought to play a critical role by modulating the processing of incoming sensory stimuli through direct feedback projections to visual cortex (Amaral, Behniea, & Kelly, 2003) and biasing signals to frontoparietal attention regions (Pourtois, Thut, Grave de Peralta, Michel, & Vuilleumier, 2005). Consistent with this suggestion, several PET and fMRI studies have reported that cortical increases to emotional stimuli were significantly correlated with amygdala responses, i.e., the more the amygdala was sensitive to the emotional meaning, the more the modulation observed in sensory areas.

The boosting of emotional stimuli by the amygdala not only may directly impact on sensory cortices, thus augmenting the neural representation of the emotional stimulus, but it can also recruit the frontoparietal endogenous attention network toward the location of the stimulus, so that subsequent information arising at the same location as emotional cues will benefit from enhanced processing resources. This effect has been demonstrated using the emotional dot probe task where the processing of a nonemotional target is facilitated if it appears at the same location as a previous emotional cue. A series of studies recording event-related potentials (ERPs) during the emotional dot probe task (Brosch et al., 2011; Brosch, Sander, Pourtois, & Scherer, 2008; Pourtois et al., 2004) have shown that emotional stimuli lead to a rapid gain increase in sensory cortex by means of which attended locations or stimuli receive increased perceptual processing (Hillyard et al., 1998). This gain increase is preceded by an early posterior parietal negativity, suggesting a functional coupling between activation of the frontoparietal attention network and a gain increase in early sensory cortex (Pourtois et al., 2005). Using fMRI recordings during the emotional dot probe, greater activation was observed in the intraparietal sulcus (IPS) when targets were preceded by a fearful face than a neutral face, consistent with enhanced attentional orienting and faster detection of targets on valid trials. This contrasted with strongly reduced activation on invalid trials, suggesting that IPS may become unresponsive to targets subsequent to the enhanced focusing of attention on the contralateral emotional cue task (Pourtois, Schwartz, Seghier, Lazeyras, & Vuilleumier, 2006). A recent fMRI study investigating active search for threatening stimuli reported increased connectivity between amygdala and IPS, FEF and fusiform gyrus when participants were searching for threatening compared to neutral targets (Mohanty, Egner, Monti, & Mesulam, 2009). This finding suggests that actively searching for emotional information elicits amygdalar input into the frontoparietal attention network and inferotemporal visual areas, which may facilitate the rapid detection of emotional stimuli.

Taken together, within-modality work on emotional attention has demonstrated how emotional stimuli can induce a distinctive cascade of neural events which does not only boost the processing of the stimulus itself but also influences mechanisms responsible for orienting and shifting attention in space, such that subsequent information arising at the same location as an emotional cue will also benefit from enhanced processing resources.

5 A Neurocognitive Model of Cross-Modal Emotional Attention

To receive maximal benefit from multimodal input, the brain must coordinate and integrate the input appropriately so that signals from an emotionally relevant source are prioritized across the different input channels. Thus, for example, auditory information about an emotional stimulus should lead to increased neural processing of visual information originating at the same location. This integration and cross-modal prioritization is a computational challenge, as the properties of the representation of information are highly modality-specific and differ greatly between the input channels: vision is represented retinotopically, touch somatotopically, audition first tonotopically and then head-centered (Driver & Spence, 1998). However, our attention mechanisms seem to be able to perform the necessary computations rapidly. ERP studies investigating nonemotional attention suggest that cross-modal attentional effects on early perceptual processing are based on an allocentric frame of reference reflecting common coordinates of external space (Eimer, Cockburn, Smedley, & Driver, 2001; Kennett, Eimer, Spence, & Driver, 2001). The spatial integration across modalities may be organized by convergence zones in posterior parietal areas, which have been shown to receive multimodal input and to code modality-specific coordinate frames into a common spatial representation (Andersen, Snyder, Bradley, & Xing, 1997). Additionally, single-cell recordings have confirmed the existence of heteromodal neurons with overlapping receptive fields for the different modalities, which are most sensitive to the location of an event, rather than to the modality it activates (Cerf et al., 2010; Stein & Stanford, 2008).

Most studies investigating the neural mechanisms underlying cross-modal attention have looked at the effects of nonemotional stimuli, whereas only few studies have investigated the neural correlates of cross-modal modulation of attention by emotion. Keil and colleagues used ERPs to measure resource allocation to a startle probe (a noise burst) while participants were watching emotional and neutral pictures or listening to emotional and neutral sounds. They observed a decreased amplitude of the P3 potential when startle probes were presented during emotional, as opposed to neutral, stimuli for both sound and picture foregrounds. These results indicate that emotional stimuli cross-modally attract processing resources, leading to optimized processing of the emotional stimulus and reduced processing capacity for concurrent stimuli (Keil et al., 2007). Dowman (2007) and Dowman and Ben-Avraham (2008) identified a network of brain areas involved in the detection and attentional reorienting toward the location of an unexpected painful somatosensory electrical stimulus, when endogenous attention is deployed not towards the tactile, but the visual modality. Using EEG measurements and source localization techniques, they concluded that the detection of the threatening tactile stimulus occurs in sensory cortex (somatosensory cortex and insula) during very early perceptual processing (as early as 70 ms), followed by increased activation in medial prefrontal cortex (130–300 ms), a structure sensitive to situations requiring changes in attentional control. Medial prefrontal cortex is then thought to signal to lateral prefrontal regions that endogenous attention needs to be redirected towards the threat (Bishop, Duncan, Brett, & Lawrence, 2004).

Whereas the work reviewed so far focused on the interruption of ongoing voluntary processing by emotional stimuli, another study has looked at the neural mechanisms underlying perceptual facilitation by cross-modal emotional attention. In our emotional dot probe paradigm investigating cross-modal bias of visual spatial attention by auditory emotion (Brosch, Grandjean et al., 2008; Brosch et al., 2009), we recorded ERPs to investigate at what stage of stimulus processing the deployment of visuospatial attention toward visual targets was affected by spatially congruent or incongruent emotional information conveyed in affective prosody. Faster response times to visual targets appearing at the location of the source of emotional prosody were accompanied by increased P1 amplitudes towards the target. Source localization indicated that the P1 modulation originated from generators localized in visual cortex (see Fig. 11.1b), suggesting that the cross-modal modulation of spatial attention triggered by emotional prosody affected early sensory stages of visual processing. These early effects at the level of the P1 mirror within-modality effects using the emotional dot probe paradigm (Brosch et al., 2011; Brosch, Sander et al., 2008; Pourtois et al., 2004), and imply that emotionally relevant stimuli may lead to a gain increase in early sensory cortex even when perceived in a different sensory modality. In a similar vein, the specificity of the results by Van Damme et al. (2009) presented earlier, revealing increased tactile attentional bias to a hand when a visual stimulus indicates impending threat to this hand, indirectly suggest a gain effect in primary sensory cortex S1, where somatotopic maps of the body surface have been documented (Penfield & Rasmussen, 1950).

Thus, electrophysiological studies of cross-modal emotional attention reveal that emotional information may interfere with voluntary processing across sensory modalities to boost and optimize the processing of emotional stimuli, and may furthermore amplify the early perceptual processing of multimodal information originating at the location of the emotional stimulus.

We suggest that cross-modal emotional attention may operate via two complementary pathways modulating the neural representation of emotional events across modalities (see Fig. 11.2). Previous research has shown that the amygdala plays a key role in the cross-modal integration of visual and auditory emotional information (Dolan, Morris, & de Gelder, 2001). For example, emotional prosody has been shown to lead to increased activation of the amygdala (Grandjean et al., 2005; Sander & Scheich, 2001), but also to increased activation of visual cortex (Sander et al., 2005; see also von Kriegstein, Kleinschmidt, Sterzer, & Giraud, 2005), probably reflecting a functional coupling between auditory and visual cortices. Functional connectivity analyses suggest that cross-modal effects of an emotional voice on visual processing are accompanied by increased connectivity between visual areas and the amygdala, but not directly between unimodal visual areas and auditory sensory areas (Ethofer, Anders, Erb et al., 2006). This suggests that cross-modal enhancements by emotion may not be mediated by direct coupling between modality-specific areas, but rather via supramodal relay areas. In addition to the amygdala, the superior temporal gyrus and sulcus may play an important role. Cross-modal exogenous cueing by nonemotional auditory signals has been shown to operate via reentrant feedback from STG/STS to early visual areas (McDonald et al., 2003). Posterior superior temporal sulcus acts as a convergence zone for the integration of emotional visual and auditory information and sends top-down feedback signals back to unimodal cortices (Campanella & Belin, 2007). Perceptual facilitation by cross-modal emotional attention thus may also operate by increased coupling between STG/STS and regions of unimodal cortex, potentially driven by the boosting of emotional information by the amygdala.

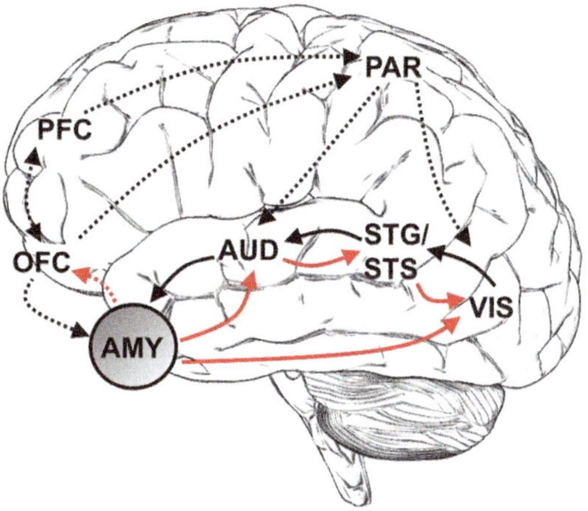

Fig. 11.2 Two neural pathways underlying cross-modal emotional attention, as illustrated here for the effects of emotional auditory information on visual perception. (**1**) Cross-modal boosting of emotional information (bold arrows): Emotional information originating in auditory cortex (AUD) is amplified by feedback signals from the amygdala (AMY). This amplification may reach visual cortex (VIS) via convergence zones such as superior temporal gyrus/sulcus (STG/STS). Additionally, the amygdala may directly mediate the functional coupling between auditory and visual unimodal cortices. (**2**) Reorienting of frontoparietal attention networks (dotted arrows): Amygdala signals may bias fronto-parietal attention regions (OFC, PFC, PAR) toward the location of emotional events to supramodally amplify information processing at this location red arrows indicate direct feedback signals originating from the amygdala

In addition to the direct enhancement of the neural representation of emotional information, the amygdala has been shown to reorient frontoparietal attention networks toward the location of an emotional stimulus (Pourtois et al., 2006; Vuilleumier & Brosch, 2009). Attentional facilitation effects of a frontoparietal reorienting have been shown to operate cross-modally for nonemotional information (Eimer & Driver, 2001). Thus, amygdala-driven recruitment of frontoparietal attention networks toward emotional stimuli will lead to benefits for subsequent information arising at the same location, independent of the modality of this information (Brosch et al., 2009). Conversely, this may lead to a reduction of processing capacities for ongoing voluntary processing in all modalities (Keil et al., 2007).

To conclude, the data reviewed here converge to show that emotion modulates attentional processing across sensory modalities by boosting early sensory stages of processing, potentially implemented by a large-scale neural network centered around the amygdala, providing direct and indirect top-down signals to sensory pathways and frontoparietal pathways involved in exogenous and endogenous attentional selection processes. This rapid cross-modal integration at multiple stages of processing may reflect a fundamental principle of human brain organization: to prioritize the processing of emotionally relevant stimuli, even if they are outside the focus of spatial attention, thus facilitating the multimodal assessment of emotionally relevant stimuli in the environment.

References

Amaral, D. G., Behniea, H., & Kelly, J. L. (2003). Topographic organization of projections from the amygdala to the visual cortex in the macaque monkey. *Neuroscience, 118*, 1099–1120.

Andersen, R. A., Snyder, L. H., Bradley, D. C., & Xing, J. (1997). Multimodal representation of space in the posterior parietal cortex and its use in planning movements. *Annual Review of Neuroscience, 20*, 303–330.

Anderson, A. K., & Phelps, E. A. (2001). Lesions of the human amygdala impair enhanced perception of emotionally salient events. *Nature, 411*, 305–309.

Bertels, J., Kolinsky, R., & Morais, J. (2010). Emotional valence of spoken words influences the spatial orienting of attention. *Acta Psychologica, 134*, 264–278.

Bishop, S. J., Duncan, J., Brett, M., & Lawrence, A. D. (2004). Prefrontal cortical function and anxiety: Controlling attention to threat-related stimuli. *Nature Neuroscience, 7*, 184–188.

Brosch, T., Grandjean, D., Sander, D., & Scherer, K. R. (2008). Behold the voice of wrath: Cross-modal modulation of visual attention by anger prosody. *Cognition, 106*, 1497–1503.

Brosch, T., Grandjean, D., Sander, D., & Scherer, K. R. (2009). Cross-modal emotional attention: Emotional voices modulate early stages of visual processing. *Journal of Cognitive Neuroscience, 21*, 1670–1679.

Brosch, T., Sander, D., Pourtois, G., & Scherer, K. R. (2008). Beyond fear: Rapid spatial orienting towards positive emotional stimuli. *Psychological Science, 19*, 362–370.

Brosch, T., Sander, D., & Scherer, K. R. (2007). That baby caught my eye… Attention capture by infant faces. *Emotion, 7*, 685–689.

Brosch, T., Pourtois, G., Sander, D., & Vuilleumier, P. (2011). Additive effects of emotional, endogenous, and exogenous attention: behavioral and electrophysiological evidence. *Neuropsychologia, 49*(7), 1779–1787.

Campanella, S., & Belin, P. (2007). Integrating face and voice in person perception. *Trends in Cognitive Sciences, 11*, 535–543.

Carrasco, M., Ling, S., & Read, S. (2004). Attention alters appearance. *Nature Neuroscience, 7*, 308–313.

Cerf, M., Thiruvengadam, N., Mormann, F., Kraskov, A., Quiroga, R. Q., Koch, C., et al. (2010). On-line, voluntary control of human temporal lobe neurons. *Nature, 467*, 1104–1108.

Corbetta, M., Patel, G., & Shulman, G. L. (2008). The reorienting system of the human brain: From environment to theory of mind. *Neuron, 58*, 306–324.

Corbetta, M., & Shulman, G. L. (2002). Control of goal-directed and stimulus-driven attention in the brain. *Nature Reviews Neuroscience, 3*, 201–215.

Desimone, R., & Duncan, J. (1995). Neural mechanisms of selective visual attention. *Annual Review of Neuroscience, 18*, 193–222.

Dolan, R. J., Morris, J. S., & de Gelder, B. (2001). Crossmodal binding of fear in voice and face. *Proceedings of the National Academy of Sciences of the United States of America, 98*, 10006–10010.

Dowman, R. (2007). Neural mechanisms of detecting and orienting attention toward unattended threatening somatosensory targets. I. Intermodal effects. *Psychophysiology, 44*, 407–419.

Dowman, R., & Ben-Avraham, D. (2008). An artificial neural network model of orienting attention toward threatening somatosensory stimuli. *Psychophysiology, 45*, 229–239.

Driver, J. (2001). A selective review of selective attention research from the past century. *British Journal of Psychology, 92*, 53–78.

Driver, J., & Spence, C. (1998). Crossmodal attention. *Current Opinion in Neurobiology, 8*, 245–253.

Egeth, H. E., & Yantis, S. (1997). Visual attention: Control, representation, and time course. *Annual Review of Psychology, 48*, 269–297.

Eimer, M., Cockburn, D., Smedley, B., & Driver, J. (2001). Cross-modal links in endogenous spatial attention are mediated by common external locations: Evidence from event-related brain potentials. *Experimental Brain Research, 139*, 398–411.

Eimer, M., & Driver, J. (2001). Crossmodal links in endogenous and exogenous spatial attention: Evidence from event-related brain potential studies. *Neuroscience and Biobehavioral Reviews, 25*, 497–511.

Eimer, M., & Holmes, A. (2007). Event-related brain potential correlates of emotional face processing. *Neuropsychologia, 45*, 15–31.

Ethofer, T., Anders, S., Erb, M., Droll, C., Royen, L., Saur, R., et al. (2006). Impact of voice on emotional judgment of faces: An event-related fMRI study. *Human Brain Mapping, 27*, 707–714.

Ethofer, T., Anders, S., Wiethoff, S., Erb, M., Herbert, C., Saur, R., et al. (2006). Effects of prosodic emotional intensity on activation of associative auditory cortex. *Neuroreport, 17*, 249–253.

Grandjean, D., Bänziger, T., & Scherer, K. R. (2006). Intonation as an interface between language and affect. *Progress in Brain Research, 156*, 235–247.

Grandjean, D., Sander, D., Lucas, N., Scherer, K. R., & Vuilleumier, P. (2008). Effects of emotional prosody on auditory extinction for voices in patients with spatial neglect. *Neuropsychologia, 46*, 487–496.

Grandjean, D., Sander, D., Pourtois, G., Schwartz, S., Seghier, M. L., Scherer, K. R., et al. (2005). The voices of wrath: Brain responses to angry prosody in meaningless speech. *Nature Neuroscience, 8*, 145–146.

Hein, G., & Knight, R. T. (2008). Superior temporal sulcus—It's my area: Or is it? *Journal of Cognitive Neuroscience, 20*, 2125–2136.

Hillyard, S. A., Vogel, E. K., & Luck, S. J. (1998). Sensory gain control (amplification) as a mechanism of selective attention: Electrophysiological and neuroimaging evidence. *Philosophical Transactions of the Royal Society B: Biological Sciences, 353*, 1257–1270.

Keil, A., Bradley, M. M., Junghofer, M., Russmann, T., Lowenthal, W., & Lang, P. J. (2007). Cross-modal attention capture by affective stimuli: Evidence from event-related potentials. *Cognitive Affective and Behavioral Neuroscience, 7*, 18–24.

Kennett, S., Eimer, M., Spence, C., & Driver, J. (2001). Tactile-visual links in exogenous spatial attention under different postures: Convergent evidence from psychophysics and ERPs. *Journal of Cognitive Neuroscience, 13*, 462–478.

Koelewijn, T., Bronkhorst, A., & Theeuwes, J. (2010). Attention and the multiple stages of multisensory integration: A review of audiovisual studies. *Acta Psychologica, 134*, 372–384.

Kreifelts, B., Ethofer, T., Shiozawa, T., Grodd, W., & Wildgruber, D. (2009). Cerebral representation of non-verbal emotional perception: fMRI reveals audiovisual integration area between voice- and face-sensitive regions in the superior temporal sulcus. *Neuropsychologia, 47*, 3059–3066.

LeDoux, J. E. (2000). Emotion circuits in the brain. *Annual Review of Neuroscience, 23*, 155–184.

Lipp, O. V., & Derakshan, N. (2005). Attentional bias to pictures of fear-relevant animals in a dot probe task. *Emotion, 5*, 365–369.

Lloyd, D. M., Bolanowski, S. J., Jr., Howard, L., & McGlone, F. (1999). Mechanisms of attention in touch. *Somatosensory and Motor Research, 16*, 3–10.

MacLeod, C., Mathews, A., & Tata, P. (1986). Attentional bias in emotional disorders. *Journal of Abnormal Psychology, 95*, 15–20.

Marois, R., & Ivanoff, J. (2005). Capacity limits of information processing in the brain. *Trends in Cognitive Sciences, 9*, 296–305.

McDonald, J. J., Teder-Salejarvi, W. A., Di Russo, F., & Hillyard, S. A. (2003). Neural substrates of perceptual enhancement by cross-modal spatial attention. *Journal of Cognitive Neuroscience, 15*, 10–19.

Miles, E., Poliakoff, E., & Brown, R. J. (2008). Investigating the time course of tactile reflexive attention using a non-spatial discrimination task. *Acta Psychologica, 128*, 210–215.

Mohanty, A., Egner, T., Monti, J. M., & Mesulam, M. M. (2009). Search for a threatening target triggers limbic guidance of spatial attention. *Journal of Neuroscience, 29*, 10563–10572.

Ohman, A., Flykt, A., & Esteves, F. (2001). Emotion drives attention: Detecting the snake in the grass. *Journal of Experimental Psychology: General, 130*, 466–478.

Olofsson, J. K., Nordin, S., Sequeira, H., & Polich, J. (2008). Affective picture processing: An integrative review of ERP findings. *Biological Psychology, 77*, 247–265.

Peelen, M., Atkinson, A., Andersson, F., & Vuilleumier, P. (2007). Emotional modulation of body-selective visual areas. *Social Cognitive and Affective Neuroscience, 2*, 274–283.
Peelen, M., Heslenfeld, D. J., & Theeuwes, J. (2004). Endogenous and exogenous attention shifts are mediated by the same large-scale neural network. *Neuroimage, 22*, 822–830.
Penfield, W., & Rasmussen, T. (1950). *The cerebral cortex of man*. New York, NY: Macmillan.
Phelps, E. A. (2006). Emotion and cognition: Insights from studies of the human amygdala. *Annual Review of Psychology, 57*, 27–53.
Phelps, E. A., Ling, S., & Carrasco, M. (2006). Emotion facilitates perception and potentiates the perceptual benefits of attention. *Psychological Science, 17*, 292–299.
Poliakoff, E., Miles, E., Li, X., & Blanchette, I. (2007). The effect of visual threat on spatial attention to touch. *Cognition, 102*, 405–414.
Posner, M. I. (1980). Orienting of attention. *Quarterly Journal of Experimental Psychology, 32*, 3–25.
Posner, M. I., Snyder, C. R., & Davidson, B. J. (1980). Attention and the detection of signals. *Journal of Experimental Psychology, 109*, 160–174.
Pourtois, G., Grandjean, D., Sander, D., & Vuilleumier, P. (2004). Electrophysiological correlates of rapid spatial orienting towards fearful faces. *Cerebral Cortex, 14*, 619–633.
Pourtois, G., Schwartz, S., Seghier, M. L., Lazeyras, F., & Vuilleumier, P. (2006). Neural systems for orienting attention to the location of threat signals: An event-related fMRI study. *Neuroimage, 31*, 920–933.
Pourtois, G., Thut, G., Grave de Peralta, R., Michel, C., & Vuilleumier, P. (2005). Two electrophysiological stages of spatial orienting towards fearful faces: Early temporo-parietal activation preceding gain control in extrastriate visual cortex. *Neuroimage, 26*, 149–163.
Rinck, M., Reinecke, A., Ellwart, T., Heuer, K., & Becker, E. S. (2005). Speeded detection and increased distraction in fear of spiders: Evidence from eye movements. *Journal of Abnormal Psychology, 114*, 235–248.
Rossi, A. F., & Paradiso, M. A. (1995). Feature-specific effects of selective visual attention. *Vision Research, 35*, 621–634.
Sander, D., Grafman, J., & Zalla, T. (2003). The human amygdala: An evolved system for relevance detection. *Reviews in the Neurosciences, 14*, 303–316.
Sander, D., Grandjean, D., Pourtois, G., Schwartz, S., Seghier, M. L., Scherer, K. R., et al. (2005). Emotion and attention interactions in social cognition: Brain regions involved in processing anger prosody. *Neuroimage, 28*, 848–858.
Sander, K., & Scheich, H. (2001). Auditory perception of laughing and crying activates human amygdala regardless of attentional state. *Cognitive Brain Research, 12*, 181–198.
Scherer, K. R. (2001). Appraisal considered as a process of multilevel sequential checking. In K. R. Scherer, A. Schorr, & T. Johnstone (Eds.), *Appraisal processes in emotion: Theory, methods, research* (pp. 92–120). New York, NY: Oxford University Press.
Schirmer, A., Teh, K. S., Wang, S., Vijayakumar, R., Ching, A., Nithianantham, D., Escoffier, N., Cheok, A.D. (2011). Squeeze me, but don't tease me: human and mechanical touch enhance visual attention and emotion discrimination. *Social Neuroscience, 6*(3), 219–230.
Smith, S. D., Most, S. B., Newsome, L. A., & Zald, D. H. (2006). An emotion-induced attentional blink elicited by aversively conditioned stimuli. *Emotion, 6*, 523–527.
Spence, C. (2010). Crossmodal spatial attention. *Annals of the New York Academy of Science, 1191*, 182–200.
Spence, C., & Driver, J. (1994). Covert spatial orienting in audition: Exogenous and endogenous mechanisms facilitate sound localization. *Journal of Experimental Psychology-Human Perception and Performance, 20*, 555–574.
Stein, B. E., & Stanford, T. R. (2008). Multisensory integration: Current issues from the perspective of the single neuron. *Nature Reviews Neuroscience, 9*, 255–266.
Theeuwes, J. (1991). Exogenous and endogenous control of attention: The effect of visual onsets and offsets. *Perception and Psychophysics, 49*, 83–90.
Turatto, M., Mazza, V., & Umilta, C. (2005). Crossmodal object-based attention: Auditory objects affect visual processing. *Cognition, 96*, B55–B64.

Van Damme, S., Gallace, A., Spence, C., Crombez, G., & Moseley, G. L. (2009). Does the sight of physical threat induce a tactile processing bias? Modality-specific attentional facilitation induced by viewing threatening pictures. *Brain Research, 1253*, 100–106.

von Kriegstein, K., Kleinschmidt, A., Sterzer, P., & Giraud, A. L. (2005). Interaction of face and voice areas during speaker recognition. *Journal of Cognitive Neuroscience, 17*, 367–376.

Vuilleumier, P. (2005). How brains beware: Neural mechanisms of emotional attention. *Trends in Cognitive Sciences, 9*, 585–594.

Vuilleumier, P., Armony, J. L., Driver, J., & Dolan, R. J. (2001). Effects of attention and emotion on face processing in the human brain: An event-related fMRI study. *Neuron, 30*, 829–841.

Vuilleumier, P., & Brosch, T. (2009). Interactions of emotion and attention. In M. S. Gazzaniga (Ed.), *The cognitive neurosciences IV* (pp. 925–934). Cambridge, MA: MIT Press.

Vuilleumier, P., & Pourtois, G. (2007). Distributed and interactive brain mechanisms during emotion face perception: Evidence from functional neuroimaging. *Neuropsychologia, 45*, 174–194.

Vuilleumier, P., & Schwartz, S. (2001). Beware and be aware: Capture of spatial attention by fear-related stimuli in neglect. *Neuroreport, 12*, 1119–1122.

West, G. L., Anderson, A. A., Ferber, S., & Pratt, J. (2011). Electrophysiological evidence for biased competition in V1 for fear expressions. *Journal of Cognitive Neuroscience, 23*(11), 3410–3418.

West, G. L., Anderson, A. A., & Pratt, J. (2009). Motivationally significant stimuli show visual prior entry: Evidence for attentional capture. *Journal of Experimental Psychology-Human Perception and Performance, 35*, 1032–1042.

Whalen, P. J., Rauch, S. L., Etcoff, N. L., McInerney, S. C., Lee, M. B., & Jenike, M. A. (1998). Masked presentations of emotional facial expressions modulate amygdala activity without explicit knowledge. *Journal of Neuroscience, 18*, 411–418.

Wolfe, J. M., & Horowitz, T. S. (2004). What attributes guide the deployment of visual attention and how do they do it? *Nature Reviews Neuroscience, 5*, 495–501.

Yantis, S. (1992). Multielement visual tracking: Attention and perceptual organization. *Cognitive Psychology, 24*, 295–340.

Chapter 12
Audiovisual Integration of Emotional Information from Voice and Face

Benjamin Kreifelts, Dirk Wildgruber, and Thomas Ethofer

Abstract When judging their social counterpart's emotional state, humans predominantly rely on nonverbal signals. In a natural environment, this nonverbal emotional communication is multimodal (i.e., facial expressions and speech melody, but also gestures, posture, or nonverbal vocalizations). Therefore, the integration of information from different sensory channels into a common percept of the current emotional state, intentions, or attitude of the social counterpart presents an elementary ability required for successful social interaction.

The first part of this chapter deals with current behavioral, neuroanatomical, electrophysiological, and neuroimaging studies on the integration of nonverbal emotional information from voice and face with special emphasis on functional magnetic resonance imaging (fMRI). The correlates of audiovisual integration of emotional information on the different levels of observation (behavioral, electrophysiological, neuroimaging) are discussed with respect to neuroanatomical data and along with methodological issues concerning current concepts of multisensory integration.

In the second part of the chapter, a methodological focus is put on the different analytical approaches (conjunction analyses, interaction analyses, correlation analyses, and connectivity analyses) used to capture and localize integration effects in the human brain as well as on the relationship between integration effects on different observational levels. We argue that none of these methods captures all facets of the integration process but that instead each of these approaches provides complementary information for the assessment of different aspects of multisensory integration of emotional signals. We demonstrate that the employment of multiple analysis techniques is necessary to dissociate effects of audiovisual emotional integration from possible confounds such as basic effects of spatiotemporal voice–face correspondence or effects of audiovisual integration of speech content.

B. Kreifelts (✉) • D. Wildgruber • T. Ethofer
Department of General Psychiatry, University of Tübingen,
Osianderstraße 24, 72076 Tübingen, Germany
e-mail: benjamin.kreifelts@med.uni-tuebingen.de; dirk.wildgruber@med.uni-tuebingen.de; tom.ethofer@gmx.de

P. Belin et al. (eds.), *Integrating Face and Voice in Person Perception*,
DOI 10.1007/978-1-4614-3585-3_12,

The third and last part of this chapter is dedicated to the alteration of audiovisual emotional integration processes in states of psychiatric disease. While processing of emotional cues in general is altered in many different psychiatric diseases, disturbance of multimodal integration occurs much less frequently. We review the yet relatively small but fast-growing number of studies in patients with schizophrenia as an exemplary psychiatric disorder with respect to alterations in behavior and neural processing of audiovisual nonverbal emotional information.

1 Part I: Introduction and Review of the Current Literature

Under naturalistic conditions, most events generate sensory stimulation via multiple channels. Multisensory integration is a process in the course of which information from the different sensory modalities is integrated by the brain into a unified, multimodal representation of the perceived event. The multimodal percept can provide additional information that is unavailable from any single sensory modality in isolation. Prerequisites for this informational gain are close spatiotemporal correspondence of sensory stimulation across different channels and semantic congruency (i.e., all information originates from the same sensory object) which can be assumed under natural conditions (Calvert, Spence, & Stein, 2004) but can be systematically altered in an experimental setting.

The first part of the chapter affords an overview of currently available data on the integration of nonverbal emotional information from voice and face. We deal with the different observational levels in turn:

1.1 Behavioral Studies

At the behavioral level, successful multisensory integration leads to shortened response latencies and heightened perceptual sensitivity (Miller, 1982; Schroger & Widmann, 1998). This behavioral integration effect is of particular importance for emotional signals as they mark events of high sociobiological relevance for the well-being and possibly survival of the individual or even for social groups as a whole. These signals can be communicated via the visual modality (e.g., facial expressions, gestures, body postures) and the auditory modality (e.g., emotional prosody, affective vocalizations, propositional content). Behavioral studies demonstrated that congruence between facial expression and prosody facilitates reactions to stimuli carrying emotional information (De Gelder & Vroomen, 2000; Dolan, Morris, & de Gelder, 2001; Massaro & Egan, 1996) (see also the chapter of Pourtois and Dhar in this book). Moreover, emotional signals perceived within one sensory channel can affect information processing in another. For instance, the perception of a facial expression can be altered by accompanying emotional prosody so that for example a facial expression is more likely being perceived as happy if accompanied by a happy (as compared to a neutral) tone of voice (De Gelder & Vroomen, 2000; Ethofer et al., 2006a; Massaro & Egan, 1996). Also, it could be demonstrated that audiovisual nonverbal emotional expressions can be classified faster and with higher accuracy when compared with unimodal auditory (emotional prosody) or visual (dynamic facial expressions) representations (Collignon et al., 2008; Kreifelts, Ethofer, Grodd, Erb, & Wildgruber, 2007). This socially highly relevant behavioral integration effect for nonverbal emotional information remains intact over a large age-span (Lambrecht et al. 2012) and was observed to be more pronounced in women than in men as women were seen to exhibit a higher degree of nonlinear probabilistic summation at the behavioral level as indicator of stronger neural integration (Collignon et al., 2010). Furthermore, as such crossmodal biases occur irrespective of the allocation of attentional resources (Collignon et al., 2008; De Gelder & Vroomen, 2000; Ethofer et al., 2006a; Vroomen, Driver & de Gelder, 2001), one may assume that the audiovisual integration of nonverbal affective information is an automatic process.

1.2 Neuroanatomical Studies and Animal Electrophysiology

A great part of our knowledge about the neuroanatomical structures subserving audiovisual integration is based on tracer and electrophysiological studies in monkeys. Here, several regions with converging projections from visual and auditory cortical areas have been located. These so-called convergence zones (Damasio, 1989) are candidate regions for the sensory integration of audiovisual information in humans as well as for the mediation of crossmodal effects (Calvert, 2001; Driver & Spence, 2000; Mesulam, 1998). These regions are located not only in cortical areas including the upper and lower banks of superior temporal sulcus (STS; Jones & Powell, 1970; Seltzer & Pandya, 1978), the orbitofrontal cortex (Chavis & Pandya, 1976; Jones & Powell, 1970), and the insula (Mesulam & Mufson, 1982), but also in subcortical regions which comprise the superior colliculus (Fries, 1984), claustrum (Pearson, Brodal, Gatter, & Powell, 1982), several nuclei within thalamus (Mufson & Mesulam, 1984) and amygdala (McDonald, 1998; Murray & Mishkin, 1985; Pitkänen, 2000).

Of these convergence zones, the most extensively studied structure is the superior colliculus (Gordon, 1973; Meredith & Stein, 1983; Peck, 1987; Wallace, Meredith, & Stein, 1993; Wallace, Wilkinson, & Stein, 1996). It plays a pivotal role with respect to orientation behavior and attention (Stein & Meredith, 1993). Following their studies of multisensory neurons in the superior colliculus Stein and Meredith (1993) phrased a set of electrophysiological "rules" for multisensory integration:

1. In multisensory neurons multimodal stimuli occurring in close proximity in space and time evoke supra-additive responses (i.e., the neuronal firing rate after a bimodal stimulus exceeds the sum of the firing rates after the respective unimodal stimulations).

2. The less effective unimodal stimuli are in generating a neuronal response the stronger are the relative effects of supra-additivity observed after bimodal stimulation. This reaction pattern in multisensory neurons was termed the rule of inverse effectiveness.
3. Spatial incongruency of the respective unimodal components of a multimodal stimulus leads to a pronounced response depression in multisensory neurons.

Similar patterns in neuronal reactivity were observed in crossmodal convergence areas in the banks of the STS (Barraclough, Xiao, Baker, Oram, & Perrett, 2005; Bruce, Desimone, & Gross, 1981; Hikosaka, Iwai, Saito, & Tanaka, 1988) and posterior insula (Fallon, Benevento, & Loe, 1978; Loe & Benevento, 1969). However, there are no direct neural connections between these cortical areas and the superior colliculus (Wallace et al., 1993). Moreover, they differ in their sensitivity to spatial factors (Stein & Wallace, 1996) which led to the assumption that they fulfill different functions in multisensory integration (perceptual judgments in the cortical areas and orientation behavior/ attention in the superior colliculus; Stein, London, Wilkinson, & Price, 1996).

A growing body of research, however, supports the idea that multisensory integration effects do not exclusively occur in brain regions previously identified as multisensory but also in brain areas thought of as unisensory, for example the auditory cortex.

In a study investigating audiovisual integration of faces and voices in macaque monkeys using local field potentials, Ghazanfar, Maier, Hoffman, and Logothetis (2005) demonstrated both, supra- and less often subadditive audiovisual integration effects occur in the core and lateral belt regions of the auditory cortex. These effects were specific for face–voice integration and obeyed the law of inverse effectiveness. These findings of audiovisual integration effects in the auditory cortex of macaque monkeys were later corroborated by Kayser, Petkov, Augath, and Logothetis (2007) who found an enhancement of cerebral responses in the core and caudal belt regions of the auditory cortex through concomitant visual stimulation, again obeying the law of inverse effectiveness, using high field fMRI.

Recently, several studies have extended these findings: In a combined local field potential and single cell electrophysiology study of the auditory cortex and the STS (Ghazanfar, Chandrasekaran, & Logothetis, 2008) it was observed that parallel face–voice stimulation increased functional interactions between the STS and the auditory cortex and these interactions were reflected in the spiking behavior of single neurons in the auditory cortex coordinated with oscillations in the STS. Similarly, a local field potential study of auditory cortex and STS using Granger causality and directed transfer functions confirmed that directed interactions from STS to auditory cortex contribute significantly to multisensory integration effects in the auditory cortex (Kayser & Logothetis, 2009). These findings speak strongly in favor of the notion that audiovisual integration effects in the auditory cortex are at least partly mediated through interactions with the multisensory STS. In the STS itself audiovisual response modulations as measured using local field potentials exhibit differences between different frequency bands with most robust audiovisual integration effects in the gamma frequency band (Chandrasekaran & Ghazanfar, 2009).

Moreover, audiovisual integration was shown to enhance the informational content of neural firing in the auditory cortex with the effect of increased correct classifications of presented naturalistic stimuli (Kayser, Logothetis, & Panzeri, 2010). Conversely, it could be shown that also neural responses to visual stimuli in the STS of the macaque monkey are modulated by simultaneous auditory stimulation with a reduction of neural information for incongruent audiovisual stimuli as opposed to congruent stimuli (Dahl, Logothetis, & Kayser, 2010).

In the domain of visual perception, Wang, Celebrini, Trotter, and Barone (2008) demonstrated that neurons in the primary visual cortex show a task-dependent reduction in their response latency to visual signals through concomitant auditory stimulation. This response time modulation of primary visual neurons occurs very early (~60 ms) and is thus unlikely to be mediated through back projections of higher order multisensory cortices.

In summary, these recent findings support the view that multimodal integration is not only subserved by higher order multimodal convergence zones, but does additionally occur at early processing stages within early sensory cortices.

1.3 Electrophysiological Studies in Humans

Previous research employing event-related potentials (ERPs, i.e., recording of electric brain responses over the human scalp) has been conducted to investigate the exact time course of crossmodal integration of emotional audiovisual signals. Based on the high temporal resolution of ERPs, it has been demonstrated that incongruent nonverbal emotional information from voice and face led to a mismatch negativity response about 180 ms after stimulation onset indicating an early modulation of auditory processing by conflicting visual information (de Gelder, Bocker, Tuomainen, Hensen, & Vroomen, 1999). In line with this finding, a subsequent study evidenced that the auditory N1 component at around 110 ms after onset of an emotional voice is enhanced by an emotionally congruent facial expression. This effect, however, is abolished through face inversion (i.e., upside down presentation; Pourtois, de Gelder, Vroomen, Rossion, & Crommelinck, 2000), a manipulation which effectively hinders recognition of emotional facial expressions (White, 1999). This result lends support to the notion that the N1 enhancement is driven by perceived facial emotion and not by low level visual features of the stimuli. Another analysis with focus on the positive deflection following the N1-P1 component about 220 ms post stimulus onset revealed a shorter latency of this deflection for emotionally congruent as compared with incongruent audiovisual stimulation (Pourtois, Debatisse, Despland, & de Gelder, 2002). These quickened ERP responses parallel behavioral facilitation with faster responses to emotionally congruent audiovisual information as compared to emotionally incongruent audiovisual information (De Gelder & Vroomen, 2000; Dolan et al., 2001; Massaro & Egan, 1996). Furthermore, an ERP study in 7-month-old infants by Grossmann, Striano, and Friederici (2006) evidenced modulations of the negative and positive ERP components by congruity of nonverbal emotional information across the auditory and visual modalities supporting the hypothesis that infants are already able to integrate and recognize audiovisual nonverbal emotional signals at early stages of their development (see also the Chap. 5 by Grossmann in this book).

Taken together, the available ERP studies provide evidence that multisensory cross talk occurs during early perceptual stages about 110–220 ms after stimulus presentation rather than during late decisional stages of information processing. This, in turn, fits in well with the observation on the behavioral level that cross-modal interaction effects occur mandatorily and irrespective of attentional resources (Collignon et al., 2008; De Gelder & Vroomen, 2000; Ethofer et al., 2006a; Vroomen et al., 2001). Thus, the ERP results point to neuronal structures which perform early steps in sensory information processing. The low spatial resolution of this technique, however, severely restricts inference on the location of the neural structures involved in the integration of audiovisual nonverbal emotional information.

A recent magnetoencephalography (MEG) study (Hagan et al., 2009) offers new insight into the spatial organization of the neural structures subserving audiovisual integration of emotional signals as MEG is an electrophysiological technique with both a high temporal and a relatively high spatial resolution. Hagan and colleagues found evidence that the right STS conforms to the supra-additivity criterion [audiovisual responses > (auditory responses + visual responses); Stein & Meredith, 1993] early after stimulus onset. This effect occurs only for congruent emotional audiovisual stimuli and not for neutral bimodal stimuli. Thus, from the perspective of electrophysiological studies, the right STS appears as the prime candidate structure for the audiovisual integration of nonverbal emotional information. It should be noted, however, that a further recent study by Chen et al. (2010) failed to detect audiovisual interaction effects on the level of unimodal auditory or visual cortices or the multisensory STS. Instead, they observed such effects in higher order association cortices in anterior frontal regions. However, no differences between emotional and neutral stimulation were reported in this study.

1.4 Neuroimaging Studies

In the attempt to transfer this knowledge of multisensory integration areas from electrophysiological studies in animals and humans using single cell recordings into the realm of human neuroimaging methods the following considerations may be helpful: In a typical human neuroimaging experiment the spatial resolution would be $3 \times 3 \times 3$ mm^3. Thus data from each voxel correspond to the averaged response of several millions of neurons (Goldman-Rakic, 1995). Further, considering the fact that even within a multisensory integration area only about 25 % of the neurons are responsive to stimuli from several modalities (Wallace, Meredith, & Stein, 1992), the neuroimaging correlates of multisensory integration can be expected to be small. Here, a restriction of the search volume informed by neuroanatomical and animal invasive electrophysiology studies may strongly improve the sensitivity to capture such neuroimaging correlates of multisensory integration by minimizing the problem of multiple comparisons (Worsley et al., 1996).

To date, there are still relatively few neuroimaging studies on audiovisual integration of nonverbal emotional information from voice and face as compared to the plethora of studies on the perception of nonverbal emotional cues from face or voice alone. Early studies (Dolan et al., 2001; Ethofer et al., 2006a; Pourtois, de Gelder, Bol, & Crommelinck, 2005) on audiovisual integration of emotion were performed using combinations of static visual stimuli (photographs of facial expressions) with dynamic acoustic stimuli (voices). To avoid confounding effects due to the temporal incongruity of such face–voice combinations, video clips expressing dynamic information in both, the auditory and visual modality have been employed in more recent experiments (Kreifelts et al., 2007; Kreifelts, Ethofer, Huberle, Grodd, & Wildgruber, 2010; Kreifelts, Ethofer, Shiozawa, Grodd, & Wildgruber, 2009; Robins, Hunyadi, & Schultz, 2009).

As yet, two groups of studies can be discerned, each with their experimental setup based on one of the three integration rules formulated by Stein and Meredith (1993). The first group of experiments capitalized on the attenuated responses of multisensory neurons to incongruent stimuli. The target contrast in such studies with two emotions combined in audiovisual stimuli is defined as

$$\left(A_{\mathrm{EMOTION1}}V_{\mathrm{EMOTION1}} + A_{\mathrm{EMOTION2}}V_{\mathrm{EMOTION2}}\right) - \left(A_{\mathrm{EMOTION1}}V_{\mathrm{EMOTION2}} + A_{\mathrm{EMOTION2}}V_{\mathrm{EMOTION1}}\right),$$

which corresponds to the interaction term in a 2×2 factorial design with A = auditory component and V = visual component.

The first experiment in this area and the very first neuroimaging study on audiovisual nonverbal emotional integration was performed by Dolan et al. (2001) who described a response enhancement in the left amygdala and the right fusiform gyrus as measured by functional magnetic resonance imaging (fMRI) and behavioral gains in response latency if the emotion expressed in voice and face (happiness or fear) matched. These findings highlight the amygdala as key structure in emotional crossmodal processing potentially mediating crossmodal perceptual biases and inducing consequent alterations in the activity of the face processing system of the fusiform gyrus through back-projections.

A second study on this topic (Ethofer et al., 2006a) extended knowledge on the crossmodal integrative function of the amygdala also using an audiovisual emotional congruency paradigm in combination with fMRI: fearful and neutral faces accompanied by a fearful voice were rated as more fearful than the same faces without concomitant auditory stimulation, and the size of this perceptual bias was related linearly to neural activity within the left amygdala. This relationship between neural and behavioral responses further supports the idea that crossmodal effects on cognitive judgments of emotional information are mediated via the amygdala. Secondly, Ethofer et al. (2006a) found enhanced blood oxygen level dependent (BOLD) responses in the right fusiform gyrus for fearful faces combined with a fearful voice as compared to the combination of a fearful face with a happy voice. This result, in turn, may indicate that activity within face processing areas in the right fusiform gyrus is modulated as result of enhanced alertness induced by the presence of additional threat-related information perceived via the auditory modality. This claim was additionally supported by a psychophysiological interaction (PPI) analysis demonstrating enhanced effective connectivity between the right fusiform gyrus and the left amygdala through fearfully spoken words as compared to happily spoken words in the presence of fearful facial expressions (Ethofer, Pourtois, & Wildgruber, 2006b). The results of these studies fit in well with existing knowledge about the function of the amygdala in emotion and especially fear processing: neuropsychological studies show that amygdala lesions may impair the recognition of fearful voices (Scott et al., 1997) and faces (Adolphs, Tranel, Damasio, & Damasio, 1994). Moreover, neuroimaging studies in healthy subjects demonstrated stronger responses of the amygdala to fear signaled via the voice (Phillips et al., 1998) and the face (Breiter et al., 1996; Morris et al., 1996).

A third and very recent fMRI experiment with a similar methodology (Müller et al. 2011) compared neural responses to congruent and incongruent pairings of emotional or neutral facial expressions with emotional or neutral nonverbal vocalizations. They replicated the previously described (Ethofer et al., 2006a) perceptual bias through fearful voices on the valence ratings of fearful or neutral faces and found audiovisual emotional incongruence effects in a cingulate-frontoparietal network which may be related to conflict monitoring and conflict resolution. However, they found no emotional incongruence effect, but evidenced generally stronger amygdala responses to emotional audiovisual combinations irrespective of congruency as compared to audiovisual pairings where either the face or the vocal element was neutral.

A second set of studies worked along a different methodological approach, namely the assumption that multisensory neurons exhibit stronger responses to multimodal stimuli than to unimodal stimulation. In this set of studies, so-called conjunction analyses and interaction analyses were used to identify audiovisual integration areas in the human brain.

We begin by reviewing those studies which employed a conjunction approach. The mathematical formulation for such a conjunction analysis is (audiovisual − auditory) ∩ (audiovisual − visual) which, applying the minimum statistic proposed by Nichols, Brett, Andersson, Wager, and Poline (2005), equals a logical AND between the two contrasts of interest.

Using a conjunction approach, Pourtois et al. (2005) found that the left middle temporal gyrus (MTG) and to a lesser degree also the left fusiform gyrus are regions with stronger cerebral activity to audiovisual pairings of happy and fearful nonverbal emotional signals than to either unimodal stimulation as measured with positron emission tomography (PET). These results marked especially the left MTG as a crossmodal integration site for nonverbal emotional information. Applying the same approach, evidence for audiovisual integration of nonverbal emotional cues in the left (Ethofer et al., 2006b; Kreifelts et al., 2007, 2010; Robins et al., 2009) and right pSTS (Kreifelts et al., 2007, 2009, 2010; Robins et al., 2009), right lower thalamus (Kreifelts et al., 2007, 2010), left hippocampus/amygdala (Kreifelts et al., 2010) and right fusiform gyrus (Kreifelts et al., 2010) has been found. Results indicating audiovisual integration of emotional signals in the posterior STS are in keeping with earlier reports demonstrating stronger responses in the posterior STS to audiovisual than to unimodal presentation of letters (e.g., van Atteveldt, Formisano, Goebel, & Blomert, 2004), speech (e.g., Calvert, Campbell, & Brammer, 2000; Stevenson & James, 2009; van Atteveldt et al., 2004; Wright, Pelphrey, Allison, McKeown, & McCarthy, 2003), objects (e.g., Beauchamp, Lee, Argall, & Martin, 2004b). This provides converging evidence implicating the posterior STS cortices in the integration of audiovisual stimuli for a broad variety of stimuli. Further research is needed to clarify whether speech, objects, and emotions share the same neural correlates for multisensory integration, or if the location of the respective integration areas can be separated using high-resolution fMRI (Fig. 12.1).

Support for a specific role of bilateral pSTS and right thalamus during processing of audiovisual nonverbal emotional cues stems from observations demonstrating a linear correlation between responses in these areas to audiovisual stimuli and the gain in behavioral accuracy for correct classification of the expressed emotions through audiovisual integration (Kreifelts et al., 2007). Moreover, a general sensitivity of these structures to a variety of emotions has been demonstrated. To further define the role of these areas during processing of affective information, we tested whether their individual BOLD integration effect in these areas estimated as $AV - \max(A,V)$ is correlated with a trait measure of emotional intelligence (self-report emotional intelligence test, SREIT; Schutte et al., 1998) (Kreifelts et al., 2010). However, of all potential integration areas for nonverbal emotional signals with significant results in the conjunction analysis $(AV - A) \cap (AV - V)$ in that study (bilateral pSTS, right thalamus, left amygdala and right fusiform gyrus) only the right pSTS exhibited a significant correlation between the BOLD integration effect, estimated as AV-max (A,V) and trait emotional intelligence. Moreover, the pSTS was the only region with a positive BOLD audiovisual integration effect showing a combined sensitivity to human voices and faces as determined by independent standard localizer experiments (Belin, Zatorre, Lafaille, Ahad, & Pike, 2000; Epstein, Harris, Stanley, & Kanwisher, 1999; Kanwisher, McDermott, & Chun, 1997). The audiovisual integration area within the pSTS can be pinpointed at the bifurcation of the STS in its two posterior ascending branches and arises exactly at a spatial overlap of the voice sensitive region in the mid portion (or posterior trunk section) of the STS and the face sensitive region in the posterior ascending branch of the STS in its posterior section (Fig. 12.2) (Kreifelts et al., 2009).

Psychophysiological interaction (PPI) analyses revealed enhanced effective connectivity of the bilateral pSTS and right thalamus with ipsilateral sensory association cortices in the fusiform gyrus and middle part of the STG during bimodal as compared to unimodal stimulation (Kreifelts et al., 2007). This increased coupling between supramodal structures and unimodal association cortices might constitute the neural mechanism for formation of the audiovisual percept of nonverbally communicated emotion. Furthermore, our findings parallel observations for visuohaptic integration evidencing an enhanced connectivity between a multimodal integration area within the parietal lobe and the respective unimodal association cortices (Macaluso, Frith, & Driver, 2000). In summary, these results speak in favor of a

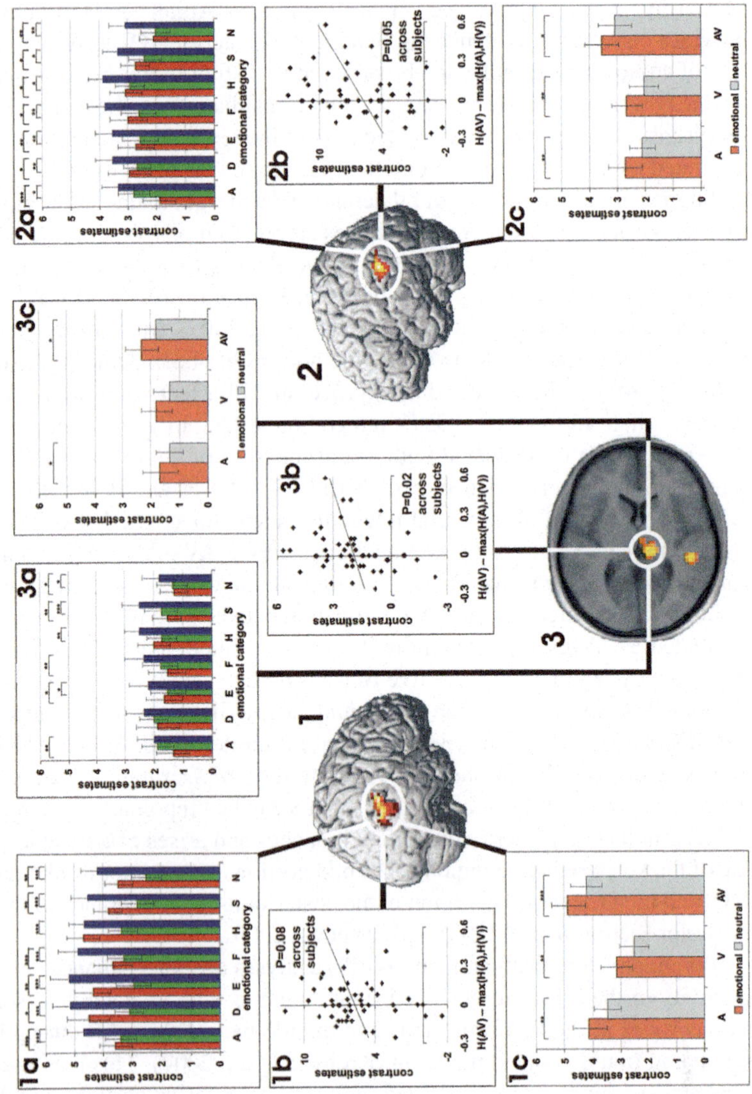

1a
1b
1c
2a
2b
2c
3a
3b
3c
1
2
3
contrast estimates
emotional category
A D E F H S N
P=0.08 across subjects
P=0.05 across subjects
P=0.02 across subjects
H(AV) – max(H(A),H(V))
emotional
neutral
A V AV

Fig. 12.1 Stronger cerebral responses during audiovisual stimulation as compared to either auditory or visual stimulation ((AV>A) ∩ (AV>V)) within right (1) and left (2) pSTG as well as right thalamus (3) ($p<0.001$, uncorrected, cluster size $k>70$, corresponding to $p<0.05$, corrected for multiple comparisons across the whole brain). Analysis of parameter estimates for auditory (*red*), visual (*green*), and audiovisual (*blue*) stimulation reveals a significant ($p \leq 0.05$) audiovisual integration effect within bilateral pSTG for all emotional expressions with the exception of happiness in right pSTG, and within right thalamus for three emotional categories and a nonsignificant tendency towards enhanced audiovisual activation in the other four categories (1–3**a**). *Asterisks* mark significant differences. All three regions exhibit stronger responses to emotional than to neutral stimuli ($p \leq 0.05$) under every experimental condition (A, V, AV) with the exception of visual stimulation in the thalamus ($p=0.06$) (1–3**b**). Positive correlation between the parameter estimates during the AV stimulation and behavioral gain, estimated as the difference between classification hit rate during the bimodal condition and the maximum of hit rates during the unimodal conditions (AV – max(A,V)), was significant ($p<0.05$) over subjects in the left pSTG and the right thalamus and exhibited a tendency versus significance ($p=0.08$) in the right pSTG (1–3**c**). Data shown from a typical subject. Figure adapted from Kreifelts et al. (2007)

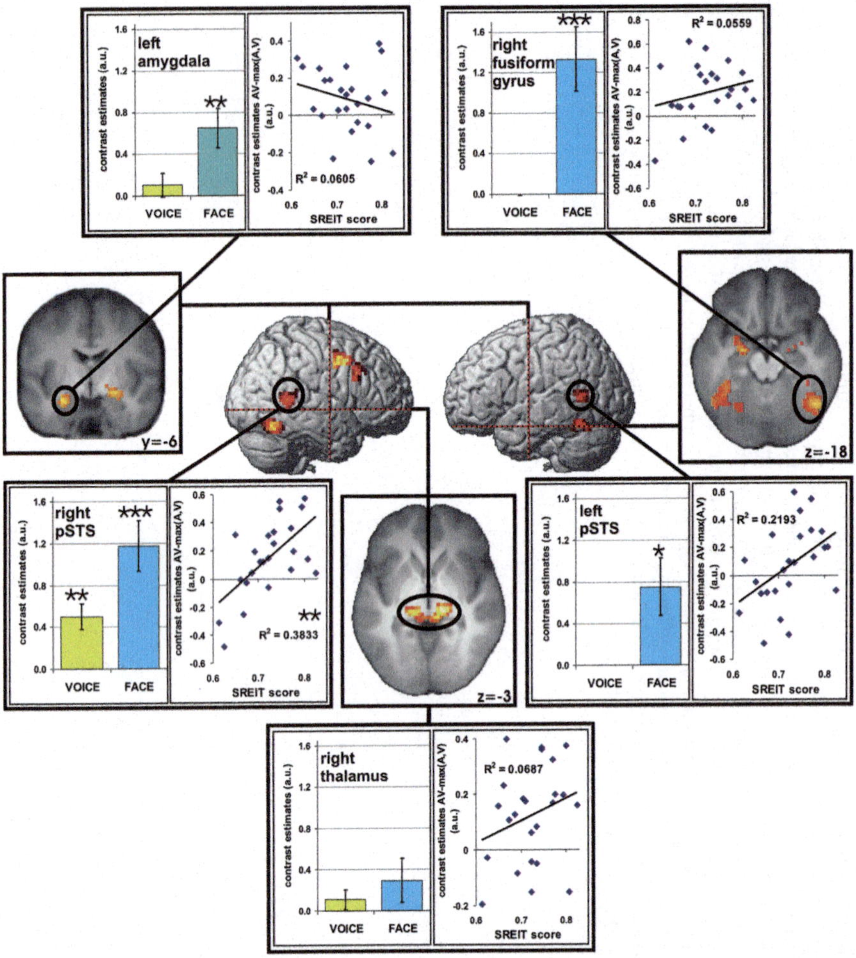

Fig. 12.2 Increased cerebral activation during audiovisual stimulation as compared to unimodal stimulation (AV>A)∩(AV>V) ($p<0.005$, uncorrected, cluster size $k>30$). *Black circles* indicate regions of interest within bilateral pSTS, right thalamus, right fusiform gyrus, and left amygdala. Activations, small volume corrected for multiple comparisons within these anatomical ROIs, are significant at $p<0.05$. Vertical-bar diagrams depict the voice sensitivity (*yellow*) and face sensitivity (*blue*) within the respective audiovisual integration areas. Results of the correlation analysis between individual BOLD integration effect, estimated as AV−max(A,V), and trait emotional intelligence (estimated by SREIT) are shown as scatter plots with regression line. Significant results are marked with *asterisks* ($^{*}p<0.05$; $^{**}p<0.01$, $^{***}p<0.001$). Error bars symbolize standard error of the mean. Figure adapted from Kreifelts et al. (2010)

mechanism with transmission of modality specific information from the respective unimodal sensory cortex to the integration area where then the bimodal percept is formed. It should be noted, however, that this model does not remain uncontested as electrophysiological studies revealed also audiovisual integration processes in modality specific cortices which cannot be simply explained by feedback loops from supramodal integration areas (Ghazanfar et al., 2005; Giard & Peronnet, 1999). Such a direct cross talk between auditory and visual cortices has also been observed in a neuroimaging study (von Kriegstein & Giraud, 2006) on face–voice associations during speaker recognition. Taken together, the coexistence of both pathways within a network comprising feed-forward as well as feedback and lateral connections between unimodal and supramodal sensory cortices as suggested by Foxe and Schroeder (review in Foxe & Schroeder, 2005) could be a potential explanation for these partially conflicting findings.

So far, the role of the thalamus during audiovisual integration remains to be clarified: Based on the findings from single cell, EEG and MEG studies, models have been proposed which implicate both cortex and thalamus in sensory integration and perceptual binding (John, 2002; Llinas & Ribary, 2001). These models assume synchronized oscillations in thalamocortical feedback loops as the neural correlate of sensory perceptual binding. Stronger activation of thalamus and pSTS during bimodal stimulation could possibly be the fMRI correlate of such synchronously oscillating thalamocortical loops during the process of audiovisual integration.

2 Part II: Methodological Considerations for Studies Targeting Audiovisual Integration of Emotional Information from Voice and Face

In fact, the definition of the most appropriate analysis of data from a multimodal neuroimaging study is less trivial than it may seem at first sight: several approaches including conjunction analyses, interaction analyses, correlation analyses with behavioral effects, and connectivity analyses have been employed so far.

The second part of this chapter details the assets and pitfalls of different analysis techniques used in neuroimaging to assess the neural correlates of audiovisual integration. Virtually all techniques applied so far are inspired by the electrophysiologically based integration rules of Stein and Meredith (1993). Interestingly, however, not all of their rules have been used so far in neuroimaging studies on audiovisual integration of emotional information.

2.1 Conjunction Analyses

The objective of conjunction analyses is to test for commonalities in brain activation patterns between two or more experimental conditions. They were introduced in neuroimaging by Price and Friston (1997). Initially they were designed to test for a global null hypothesis (H0: No effect in any of the components, H1: Significant effect in at least one of the components; Friston, Holmes, Price, Buchel, & Worsley, 1999; Friston, Penny, & Glaser, 2005). Recently, a revision has led to distinguish from these an approach which tests for a conjunction null hypothesis (H0: No effect in at least any of the components, H1: Significant effects in all of the components; Nichols et al., 2005).

It is important to note that the assumption of a logical "AND" can only be based on the rejection of the conjunction null hypothesis. This more conservative conjunction analysis was applied in most of the existing studies on audiovisual integration of nonverbal information following the conjunction approach (Ethofer et al., 2006b; Kreifelts et al., 2007, 2009, 2010).

It is an obvious property of multisensory brain areas to respond to stimuli from more than a single sensory modality. Therefore, a straightforward conjunction approach to locate such brain areas is (UNIMODAL 1 ∩ UNIMODAL 2). It has been used in studies of spatial attention to vision and touch (Macaluso et al., 2000) and audiovisual integration of motion processing (Lewis, Beauchamp, & DeYoe, 2000). A major problem with this approach is that it locates not only zones of multisensory convergence but also brain regions exhibiting unspecific activations attributable to nonsensory components of the task (e.g., working memory, response selection, motor responses). Moreover, brain areas that respond significantly exclusively to bimodal stimuli will not be located.

These problems can be overcome by computing the conjunction (BIMODAL − UNIMODAL 1) ∩ (BIMODAL − UNIMODAL 2). Here, task-related activations are removed from the contrast as they are included in each condition of the conjunction term, however, under the restriction that the task was the same for all experimental conditions included in the conjunction. This strategy found a very early application by Calvert et al. (1999) to detect brain regions involved in processing of audiovisual speech and later several times with the aim to detect integration areas for audiovisual emotional information (Ethofer et al., 2006b; Kreifelts et al., 2007, 2009, 2010; Robins et al., 2009). A particularly elaborate conjunction approach was chosen by Pourtois et al. (2005) to remove task-related brain responses assuming different attentional sets under the two unimodal conditions:

$$(\text{AV [judge A]} - \text{A [judge A]}) \cap (\text{AV [judge V]} - \text{V [judge V]}).$$

From a general perspective, it should be noted, however, that conjunction analyses based on the conjunction null hypothesis (Nichols et al., 2005) represent a very conservative strategy providing only an upper bound for the false positive rate (Friston et al., 2005). Although their conservativeness makes them remain valid even in a statistical worst-case scenario (Nichols et al., 2005), this is paid for by impaired sensitivity (Friston et al., 2005). Responding to this problem it can be very helpful to increase the analytical sensitivity through a restriction of the search volume to certain a priori defined brain regions (Worsley et al., 1996) informed by neuroanatomical or neuroimaging studies.

It should be noted, however, that conjunctions of (AV − A) ∩ (AV − V) potentially show activations in brain regions in which responses to information from auditory and visual channels simply sum up in a linear way. It has been criticized that such analyses might locate brain areas in which both neurons responsive to unimodal auditory and unimodal visual information coexist without the need of multimodal integration in these areas (Calvert, 2001; Calvert & Thesen, 2004).

2.2 *Interaction Analyses*

2.2.1 Sensory Interactions

Closer to the initial electrophysiological integration rules of Stein and Meredith (1993) and Calvert and Thesen (2004) proposed that activity in crossmodal integration areas should differ from the arithmetic sum of the respective activations to unimodal stimuli: in case the response to a bimodal stimulus exceeds the sum of the responses to either unimodal stimulation (BIMODAL > [UNIMODAL 1 + UNIMODAL 2]), this is defined as a positive interaction, while the opposite, i.e., the summed unimodal responses exceeding the bimodal response (BIMODAL < [UNIMODAL 1 + UNIMODAL 2]), is defined as negative interaction effect (Calvert, 2001; Calvert et al., 2000). Although such interactions have repeatedly been investigated and positive crossmodal interactions have been described (e.g., Calvert et al., 1999, 2000; Park et al., 2010; Stevenson, Geoghegan, & James, 2007; Werner & Noppeney, 2010b, for a review see Calvert, 2001), this analytical approach is burdened by several technical and theoretical problems when applied in neuroimaging:

The technical problem becomes obvious in the mathematical formulation of the interaction term: typically, an interaction between two factors is investigated within the framework of a 2×2 factorial design. Such a design for the investigation of interactions between two sensory modalities would consist of one bimodal and two unimodal conditions where the subject performs a stimulus-related task. It would be completed by a control condition which, apart from sensory stimulation, incorporates all components of the other conditions (e.g., working memory, response selection, motor response). Yet, it is practically impossible to implement such a control condition as it is impossible to perform a stimulus-related task without a stimulus. Consequently, this putative control condition was omitted in all hitherto imaging studies investigating crossmodal interactions with serious consequences for the interpretation of the results. For one, in brain regions where nonsensory components of the experimental procedure lead to positive brain responses in both unimodal and the bimodal condition a negative interaction effect will emerge. Secondly, brain regions with potentially unspecific deactivations to stimulus presentation under all three conditions, as may be the case in parts of the tonically active resting state network (Fox et al., 2005; Raichle et al., 2001), will produce positive interaction effects, since the single deactivation after bimodal stimulus presentation is less negative than the added deactivations of the two unimodal conditions. This vulnerability of the interaction term BIMODAL > (UNIMODAL 1 + UNIMODAL 2) to unspecific deactivations of resting state network components resulting in positive interactions and unspecific activations related to the behavioral task producing negative interactions needs to be taken into account when applying this technique. It seems advisable to carry out an inspection of the time-series and beta estimates of the general linear model as a prerequisite for the final interpretation of both, positive and negative interactions obtained from this approach. Another viable approach to circumvent this pitfall is to perform a triple conjunction analysis AV>(A+V)∩(A>REST)∩(V>REST).

The theoretical problems are related to the notion that all electrophysiological criteria for the investigation of multimodal integration can and should be applied to the BOLD effect (Calvert, 2001). Specifically, they are bound to the assumption that, according to these criteria, cells subserving crossmodal integration exhibit responses to congruent bimodal stimulation which exceeds the sum of responses of the respective unimodal stimuli, the phenomenon of supra-additivity. However, doubts have been voiced that supra-additivity on the electrophysiological level necessarily translates into supra-additive BOLD responses: there is evidence which indicates that due to a phenomenon dubbed "hemodynamic refractoriness" the BOLD response to two stimuli occurring in close temporal proximity is overpredicted by simple summation of the responses (Friston, Josephs, Rees, & Turner, 1998; Mechelli, Price, & Friston, 2001). Thus, the attenuation of the BOLD response to simultaneous bimodal stimulation might compromise the sensitivity of analysis approaches in which the responses to bimodal stimuli are expected to exceed the sum of the responses to unimodal stimulation. This problem might be accessible via the estimation of the neural responses from the BOLD response via a plausible biophysical model (Friston, Mechelli, Turner, & Price, 2000; Gitelman, Penny, Ashburner, & Friston, 2003) and a consecutive search for supra-additive neuronal responses.

However, there remains a second problem in the translation of evidence from single cell studies into investigations probing the responses of large neuronal populations. Here, evidence brought forward by the same group of researchers who initially termed the "integration rules", strongly cautions against the assumption that supra-additive crossmodal responses can be expected on the neural population level. Laurienti, Perrault, Stanford, Wallace, and Stein (2005) performed a model calculation based on pertinent data from single cell recordings and extrapolating these to a 4 mm^3 voxel in a multisensory brain area putatively investigated in a neuroimaging study. The authors demonstrate that of the 2.5 millions of neurons within such a voxel only about 7 % would exhibit supra-additive responses while the rest of the multisensory neurons (18 %) showing additive or subadditive responses. Under such circumstances, both the electrophysiological and the BOLD bimodal stimulation response on the population level can be expected to be subadditive.

Recently, Stanford and Stein (2007) once more have clarified that the optimal conditions for supra-additive multimodal responses arise when unimodal stimuli are least effective in evoking a response (rule of inverse effectiveness). Accordingly, experimental conditions capitalizing on this effect can increase the odds of observing a supra-additive bimodal response on the population level, a course of action already proposed for somewhat different reasons by Calvert (2001).

Several neuroimaging studies on crossmodal integration with BOLD responses conforming to the rule of inverse effectiveness have been performed in macaques (Kayser, Petkov, Augath, & Logothetis, 2005) and humans (Stevenson & James, 2009; Werner & Noppeney, 2010b). Stevenson and James (2009) investigated the audiovisual integration of tools and speech and demonstrated supra-additive audiovisual responses in the STS for both types of stimuli under circumstances of partial stimulus degradation according to the rule of inverse effectiveness. In Werner and Noppeney's (2010b) study on audiovisual integration of tools, supra-additive audiovisual responses in the STS, again under the condition of partial stimulus degradation were complemented with the finding that the behavioral sensory integration effect predicted the strength and nature of the audiovisual BOLD interaction effect in the STS: subjects with a behavioral gain exhibited a supra-additive crossmodal interaction, while subadditive audiovisual interactions where found in subjects without behavioral integration benefit. It still remains to be explored if and where in the brain audiovisual integration of emotional information conforms to the inverse effectiveness principle.

2.2.2 Interactions Based on Emotional Congruency/Incongruency

While the construction of a full 2×2 factorial design for the exploration of interaction effects between two sensory modalities during crossmodal integration is flawed by the lack of an appropriate control condition as laid out above, the investigation of crossmodal interactions between emotions expressed via these two modalities is not burdened with the same problem. Such a design includes solely bimodal stimulation with the bimodal stimuli being presented either under emotionally congruent (e.g., fearful voice–fearful face [$A_{FEAR}V_{FEAR}$] and happy voice–happy face [$A_{HAPPINESS}V_{HAPPINESS}$]) or incongruent conditions (e.g., fearful voice–happy face [$A_{FEAR}V_{HAPPINESS}$] and happy voice–fearful face [$A_{HAPPINESS}V_{FEAR}$]).

Such interactions of emotional information in voice and face are investigated by the term:

$$(A_{FEAR}V_{FEAR} - A_{FEAR}V_{HAPPINESS}) - (A_{HAPPINESS}V_{FEAR} - A_{HAPPINESS}V_{HAPPINESS}),$$

which is equivalent to contrasting congruent with incongruent conditions

$$\left(A_{FEAR}V_{FEAR} + A_{HAPPINESS}V_{HAPPINESS}\right) - \left(A_{FEAR}V_{HAPPINESS} + A_{HAPPINESS}V_{FEAR}\right).$$

This approach has been effectively used by several groups (Dolan et al., 2001; Ethofer et al., 2006a; Müller et al. 2011) as already reported above. As a final remark it needs to be added that the inclusion of neutral intonations/vocal expressions ($A_{NEUTRAL}$) and neutral facial expressions ($V_{NEUTRAL}$) in combination with emotional expressions ($A_{EMOTIONAL}/V_{EMOTIONAL}$) leading to the interaction

$$(A_{EMOTIONAL}V_{EMOTIONAL} - A_{EMOTIONAL}V_{NEUTRAL}) - (A_{NEUTRAL}V_{EMOTIONAL} - A_{NEUTRAL}V_{NEUTRAL}).$$

can considerably increase the interpretability of the interaction results: especially with respect to the delineation of the effects of emotional congruence from basic spatiotemporal face–voice congruence, one might expect activations following emotionally congruent bimodal stimulation to differ from activations following the neutral congruent condition which also incorporates face–voice congruence. A first step in this direction was the study by Müller et al. (2011) who used yawns as neutral vocal stimuli.

2.3 Correlation Analyses Between Brain Responses and Behavioral/Trait Measures

Enhanced stimulus classification performance and shortened response latencies under bimodal stimulation or behavioral effects of audiovisual emotional congruence/incongruence during task performance (Collignon et al., 2008; De Gelder & Vroomen, 2000; Dolan et al., 2001; Ethofer et al., 2006a; Kreifelts et al., 2007; Massaro & Egan, 1996) can be used to establish a link between behavioral and neural correlates of crossmodal integration. This type of analysis itself does not provide a metric of multisensory integration but may be a very valuable auxiliary tool demonstrating that the observed neural effects are indeed related to the investigated integration process (Ethofer et al., 2006a; Kreifelts et al., 2007; Mechelli et al., 2001; Werner & Noppeney, 2010b) and do not simply reflect low-level spatiotemporal stimulus correspondence across modalities. While Werner and Noppeney (2010b) investigating audiovisual integration of tools described a linear relationship between the BOLD integration metric and the behavioral gain during crossmodal integration in the pSTS, our own group (2007) found a similar link between crossmodal behavioral gain during an emotional classification task and BOLD activity during audiovisual stimulation in the same region. Ethofer et al. (2006a), on the other hand, demonstrated that left amygdala activity is linearly associated with the impact of fearful voices on the valence ratings of neutral faces as a strong argument for the assumption that the left amygdala constitutes one of the neural mediators of crossmodal emotional interaction effects.

In the area of analytical approaches concentrating on sensory audiovisual integration (conjunction analyses, interaction analyses), the establishment of such relationships can be especially critical under the following two aspects:

Certain brain areas, most prominently the pSTS, are implicated in the audiovisual perceptual integration of a variety of different stimuli [e.g., letters (van Atteveldt et al., 2004), speech (Calvert et al., 2000; Stevenson & James, 2009; van Atteveldt et al., 2004; Wright et al., 2003), objects (Beauchamp, Argall, Bodurka, Duyn, & Martin, 2004a; Beauchamp et al., 2004b; Stevenson & James, 2009; Werner & Noppeney, 2010b), animals (Beauchamp et al., 2004b)]: here, the link between a behavioral measure of perceptual integration of emotional signals and the neural integration metric warrants that the observed integration effect is related to emotional processing.

The second aspect pertains to the methodological limitations of neuroimaging measures of audiovisual integration as laid out above: most often audiovisual stimulation will result in a subadditive BOLD response (AV<A+V) which could theoretically be caused by parallel activation of unimodal neurons in a potential integration area without any sensory integration process necessary to explain this activation pattern. Now, an additional correlation analysis with the behavioral integration effect complements the primary audiovisual integration analysis, ascertaining that the imaging crossmodal integration metric indeed pertains to the targeted integration process.

Taken together, response-related correlation analyses between crossmodal behavioral effects and cerebral activation represent a useful approach to model systems associated with the behavioral outcome of multisensory integration.

Should, however, a direct behavioral integration measure not be accessible as during an implicit emotional perception task, it is possible to correlate the integration effect on the level of brain activation with an appropriate measure of emotional processing (e.g., measures of emotional intelligence) on the population level to evidence that the observed integration effect exceeds simple face–voice integration and is related to emotional processing. This approach is exemplified in the correlation between the BOLD audiovisual integration effect and a trait measure of emotional intelligence (self-report emotional intelligence test, SREIT; Schutte et al., 1998) revealed in the right pSTS while subjects performed a gender classification task on various auditory (emotional speech melody), visual (facial expressions), and audiovisual (emotional speech melody + facial expressions) dynamic nonverbal emotional expressions (Fig. 12.3) (Kreifelts et al., 2010).

2.4 Connectivity Analyses

Connectivity analyses are not a direct metric of multisensory integration but may help to elucidate which brain areas interact during the formation of the crossmodal percept. It can be assumed that this integration process is not only reflected in differential activation of certain brain areas but also in an increase of connectivity between those regions. While being anatomically segregated, face- and voice-processing modules need to interact during the formation of a unified percept of the emotional information expressed via different sensory channels. Analyses of connectivity have the potential to investigate the interaction between these modules and may help distinguish whether integration is achieved via supramodal relays or by direct coupling of voice- and face-sensitive cortices. Recent developments in modeling effective connectivity (Friston, Harrison, & Penny, 2003; Gitelman et al., 2003) between brain distinct areas [i.e., the influence one neural system exerts over another; Friston et al., 1997)] afford the opportunity to investigate which experimental factors result in changes in neural coupling.

To the present date, such analyses of modulations of effective connectivity with respect to the sensory aspects of audiovisual integration in fMRI have produced partially conflicting results. For one, von Kriegstein and Giraud (2006) using a psychophysiological interaction (PPI; Friston et al., 1997) analysis found evidence for an enhancement in effective connectivity directly between modality specific

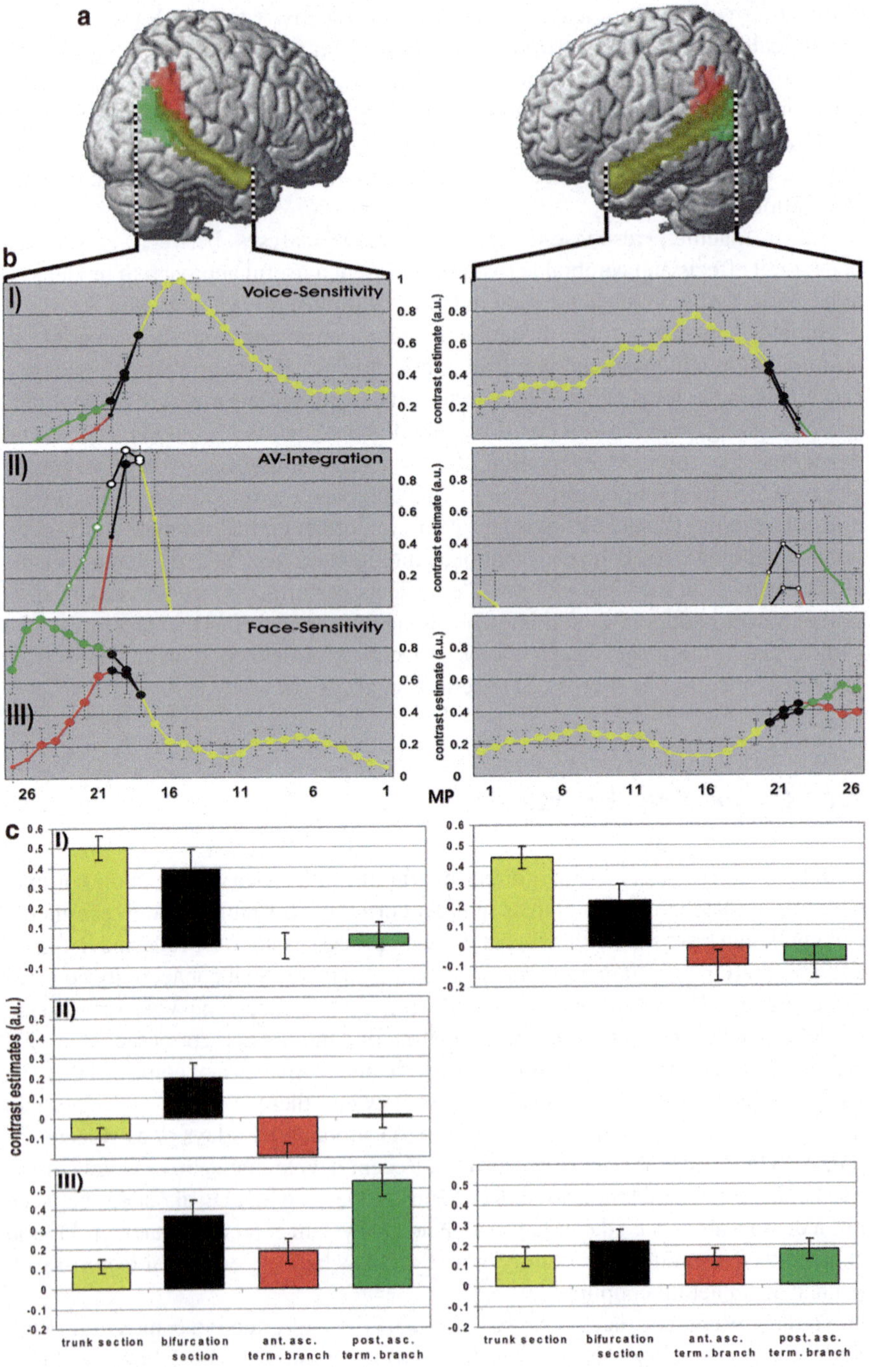
a
b
I)
Voice-Sensitivity
II)
AV-Integration
III)
Face-Sensitivity
contrast estimate (a.u.)
MP
c
contrast estimates (a.u.)
trunk section
bifurcation section
ant. asc. term. branch
post. asc. term. branch

auditory and visual cortices through learning of face–voice associations in an fMRI study on speaker recognition, while no enhancement to higher order supramodal brain areas was observed. On the other hand, we (Kreifelts et al., 2007) demonstrated enhanced connectivity through audiovisual presentation of nonverbal emotional cues as compared to either unisensory stimulation between supramodal pSTS and both auditory and visual presumably voice- and face-sensitive unimodal cortices. Werner and Noppeney (2010a) employing dynamic causal modeling, finally, observed modulations in effective connectivity both directly between auditory and visual cortices as well as between unimodal auditory and visual sensory cortices and supramodal STS during audiovisual integration of everyday actions performed with objects.

This recent work by Werner and Noppeney (2010a) might also represent a resolution of the apparent conflict between the results of the first two studies as it seems to support the coexistence of both pathways, i.e., feed-forward and feedback as well as lateral connections between unimodal and supramodal sensory cortices which has been suggested already earlier by Foxe and Schroeder (2005) also based on electrophysiological data.

3 Part III: Audiovisual Integration of Nonverbal Emotional Cues in States of Psychiatric Disease: The Example of Schizophrenia

Many psychiatric conditions (e.g., schizophrenia, autism spectrum disorders, depression, social phobia, personality disorders) are associated with alterations and impairments in the perception and correct recognition of nonverbal emotional cues. Some of these disorders are also associated with altered crossmodal integration processes with schizophrenia being the foremost to name.

Fig. 12.3 Neural representation of voice-sensitivity, audiovisual integration of nonverbal emotional information and face sensitivity in the STS. (**a**) Interindividual variability of the STS represented in form of a probability map ($n=24$). *Yellow color* marks the trunk section of the STS, *red color* marks the anterior terminal ascending branch of the STS and *green color* marks the posterior terminal ascending branch of the STS. *Stronger colors* denote higher probabilities demonstrating an increasing spatial variability along the STS from front to back. (**b**) Parameter estimates (a.u., peak-normalized) for voice sensitivity (I), audiovisual integration (II), and face sensitivity (III) for 27 measuring points along the STS. Same color code as in (**a**). Error bars represent the SEM. *Large dots* mark significant results with $p<0.05$ while *small dots* denote insignificant results. *Central white dots* represent a significant positive correlation ($p<0.05$) between the individual integration effect, AV − max (A,V) and trait emotional intelligence (as estimated by SREIT). (**c**) Averaged parameter estimates (a.u.) within separate sections of the STS (*yellow* trunk, *black* bifurcation, *red* anterior, *green* posterior terminal ascending branch) for voice-sensitivity (I), audiovisual integration (II) and face-sensitivity (III). Data from the audiovisual integration experiment in the left STS not due to lack of significant effects in any of the sections of the STS. Error bars represents the SEM. Figure adapted from Kreifelts et al. (2009)

The number of studies investigating crossmodal integration of nonverbal signals from voice and face in psychiatric diseases is still quite limited, however. While data on integration of nonverbal cues from voice and face in autism spectrum disorders are very scarce, important first steps have been taken to elucidate deficits in audiovisual perception of such cues in schizophrenia.

Several behavioral studies employing pictures, i.e., static stimulation as visual component have been performed to investigate altered sensory interactions of emotional vocal and facial cues:

Employing a 2×2 congruence/incongruence design de Jong, Hodiamont, Van den Stock, and de Gelder (2009) observed a reduced influence of visual affect on the evaluation of vocal affect and found this effect to be specific for schizophrenia when compared to two control samples one comprising healthy subjects and one patient group comprising a multitude of other forms of psychosis. Moreover, no connection of this effect to differences in vigilance or antipsychotic medication could be established. An extension of this experimental design (de Jong, Hodiamont, & de Gelder, 2010) introducing emotionally neutral visual and auditory distractors as means of modulating visual and auditory attention demonstrated that the effects of facial affect on vocal affect recognition are effectively reduced by visual distractors and less so by auditory distractors while interaction effects between visual and vocal emotion remained unaffected in schizophrenia patients. This was interpreted by the authors as evidence that regulatory effects of modality specific attention are deficient in schizophrenia. For the sake of completeness, an earlier study by the same group (de Gelder et al., 2005) with divergent results needs to be mentioned. The findings of this previous study, however, appear as less reliable as the included sample of schizophrenic patients was very small ($n = 13$), no nonschizophrenic psychosis control group was included and patient information was incomplete (de Jong et al., 2009).

No fMRI studies evaluating audiovisual integration of nonverbal emotional cues from voice and face in patients with schizophrenia have been published yet. However, first evidence from a study on audiovisual integration of speech (Szycik et al., 2009) shows altered hemodynamic activity in a mostly frontotemporal network associated with audiovisual congruence versus incongruence conditions in schizophrenia patients.

Taken together with results linking deficits in audiovisual sensory integration in schizophrenia to the symptomatology of this disease (Williams, Light, Braff, & Ramachandran, 2010) the studies reviewed above represent a promising beachhead in a fast-growing area of research exploring the psychophysical and neural underpinnings of altered multisensory perception of emotional signals in states of psychiatric disease. While one aim of such research would certainly be to deepen the understanding of disease mechanisms in relation to the complex of social cognition, another, possibly more speculative, aim could be to develop techniques to overcome disease-related difficulties in the perception of nonverbal signals and thus to alleviate the social burden of psychiatric disorders.

References

Adolphs R, Tranel D, Damasio H, Damasio A (1994) Impaired recognition of emotion in facial expressions following bilateral damage to the human amygdala. Nature 372(6507):669–672

Barraclough NE, Xiao D, Baker CI, Oram MW, Perrett DI (2005) Integration of visual and auditory information by superior temporal sulcus neurons responsive to the sight of actions. Journal of Cognitive Neuroscience 17(3):377–391

Beauchamp MS, Argall BD, Bodurka J, Duyn JH, Martin A (2004a) Unraveling multisensory integration: Patchy organization within human STS multisensory cortex. Nature Neuroscience 7(11):1190–1192

Beauchamp MS, Lee KE, Argall BD, Martin A (2004b) Integration of auditory and visual information about objects in superior temporal sulcus. Neuron 41(5):809–823

Belin P, Zatorre RJ, Lafaille P, Ahad P, Pike B (2000) Voice-selective areas in human auditory cortex. Nature 403(6767):309–312

Breiter HC, Etcoff NL, Whalen PJ, Kennedy WA, Rauch SL, Buckner RL et al (1996) Response and habituation of the human amygdala during visual processing of facial expression. Neuron 17(5):875–887

Bruce C, Desimone R, Gross CG (1981) Visual properties of neurons in a polysensory area in superior temporal sulcus of the macaque. Journal of Neurophysiology 46(2):369–384

Calvert GA (2001) Crossmodal processing in the human brain: Insights from functional neuroimaging studies. Cerebral Cortex 11(12):1110–1123

Calvert GA, Brammer MJ, Bullmore ET, Campbell R, Iversen SD, David AS (1999) Response amplification in sensory-specific cortices during crossmodal binding. NeuroReport 10(12):2619–2623

Calvert GA, Campbell R, Brammer MJ (2000) Evidence from functional magnetic resonance imaging of crossmodal binding in the human heteromodal cortex. Current Biology 10(11):649–657

Calvert GA, Spence C, Stein BE (eds) (2004) The handbook of multisensory processes. MIT Press, Cambridge, MA

Calvert GA, Thesen T (2004) Multisensory integration: Methodological approaches and emerging principles in the human brain. Journal of Physiology 98(1–3):191–205

Chandrasekaran C, Ghazanfar AA (2009) Different neural frequency bands integrate faces and voices differently in the superior temporal sulcus. Journal of Neurophysiology 101(2):773–788

Chavis DA, Pandya DN (1976) Further observations on corticofrontal connections in the rhesus monkey. Brain Research 117(3):369–386

Chen YH, Edgar JC, Holroyd T, Dammers J, Thonnessen H, Roberts TP et al (2010) Neuromagnetic oscillations to emotional faces and prosody. European Journal of Neuroscience 31(10): 1818–1827

Collignon O, Girard S, Gosselin F, Roy S, Saint-Amour D, Lassonde M et al (2008) Audio–visual integration of emotion expression. Brain Research 1242:126–135

Collignon O, Girard S, Gosselin F, Saint-Amour D, Lepore F, Lassonde M (2010) Women process multisensory emotion expressions more efficiently than men. Neuropsychologia 48(1):220–225

Dahl CD, Logothetis NK, Kayser C (2010) Modulation of visual responses in the superior temporal sulcus by audio–visual congruency. Frontiers in Integrative Neuroscience 4:10

Damasio AR (1989) Time-locked multiregional retroactivation: A systems-level proposal for the neural substrates of recall and recognition. Cognition 33(1–2):25–62

de Gelder B, Bocker KB, Tuomainen J, Hensen M, Vroomen J (1999) The combined perception of emotion from voice and face: Early interaction revealed by human electric brain responses. Neuroscience Letters 260(2):133–136

De Gelder B, Vroomen J (2000) The perception of emotions by ear and by eye. Cognition and Emotion 14(3):289–311

de Gelder B, Vroomen J, de Jong SJ, Masthoff ED, Trompenaars FJ, Hodiamont P (2005) Multisensory integration of emotional faces and voices in schizophrenics. Schizophrenia Research 72(2–3):195–203

de Jong JJ, Hodiamont PP, de Gelder B (2010) Modality-specific attention and multisensory integration of emotions in schizophrenia: Reduced regulatory effects. Schizophrenia Research 122(1–3):136–143
de Jong JJ, Hodiamont PP, Van den Stock J, de Gelder B (2009) Audiovisual emotion recognition in schizophrenia: Reduced integration of facial and vocal affect. Schizophrenia Research 107(2–3):286–293
Dolan RJ, Morris JS, de Gelder B (2001) Crossmodal binding of fear in voice and face. Proceedings of the National Academy of Sciences of the United States of America 98(17):10006–10010
Driver J, Spence C (2000) Multisensory perception: Beyond modularity and convergence. Current Biology 10(20):R731–R735
Epstein R, Harris A, Stanley D, Kanwisher N (1999) The parahippocampal place area: Recognition, navigation, or encoding? Neuron 23(1):115–125
Ethofer T, Anders S, Erb M, Droll C, Royen L, Saur R et al (2006a) Impact of voice on emotional judgment of faces: An event-related fMRI study. Human Brain Mapping 27(9):707–714
Ethofer T, Pourtois G, Wildgruber D (2006b) Investigating audiovisual integration of emotional signals in the human brain. Progress in Brain Research 156:345–361
Fallon JH, Benevento LA, Loe PR (1978) Frequency-dependent inhibition to tones in neurons of cat insular cortex (AIV). Brain Research 145(1):161–167
Fox MD, Snyder AZ, Vincent JL, Corbetta M, Van Essen DC, Raichle ME (2005) The human brain is intrinsically organized into dynamic, anticorrelated functional networks. Proceedings of the National Academy of Sciences of the United States of America 102(27):9673–9678
Foxe JJ, Schroeder CE (2005) The case for feedforward multisensory convergence during early cortical processing. NeuroReport 16(5):419–423
Fries W (1984) Cortical projections to the superior colliculus in the macaque monkey: A retrograde study using horseradish peroxidase. The Journal of Comparative Neurology 230(1):55–76
Friston KJ, Buechel C, Fink GR, Morris J, Rolls E, Dolan RJ (1997) Psychophysiological and modulatory interactions in neuroimaging. NeuroImage 6(3):218–229
Friston KJ, Harrison L, Penny W (2003) Dynamic causal modelling. NeuroImage 19(4):1273–1302
Friston KJ, Holmes AP, Price CJ, Buchel C, Worsley KJ (1999) Multisubject fMRI studies and conjunction analyses. NeuroImage 10(4):385–396
Friston KJ, Josephs O, Rees G, Turner R (1998) Nonlinear event-related responses in fMRI. Magnetic Resonance in Medicine 39(1):41–52
Friston KJ, Mechelli A, Turner R, Price CJ (2000) Nonlinear responses in fMRI: The Balloon model, Volterra kernels, and other hemodynamics. NeuroImage 12(4):466–477
Friston KJ, Penny WD, Glaser DE (2005) Conjunction revisited. NeuroImage 25(3):661–667
Ghazanfar AA, Chandrasekaran C, Logothetis NK (2008) Interactions between the superior temporal sulcus and auditory cortex mediate dynamic face/voice integration in rhesus monkeys. Journal of Neuroscience 28(17):4457–4469
Ghazanfar AA, Maier JX, Hoffman KL, Logothetis NK (2005) Multisensory integration of dynamic faces and voices in rhesus monkey auditory cortex. Journal of Neuroscience 25(20):5004–5012
Giard MH, Peronnet F (1999) Auditory–visual integration during multimodal object recognition in humans: A behavioral and electrophysiological study. Journal of Cognitive Neuroscience 11(5):473–490
Gitelman DR, Penny WD, Ashburner J, Friston KJ (2003) Modeling regional and psychophysiologic interactions in fMRI: The importance of hemodynamic deconvolution. NeuroImage 19(1):200–207
Goldman-Rakic PS (1995) Architecture of the prefrontal cortex and the central executive. Annals of the New York Academy of Sciences 769:71–83
Gordon B (1973) Receptive fields in deep layers of cat superior colliculus. Journal of Neurophysiology 36(2):157–178
Grossmann T, Striano T, Friederici AD (2006) Crossmodal integration of emotional information from face and voice in the infant brain. Developmental Science 9(3):309–315
Hagan CC, Woods W, Johnson S, Calder AJ, Green GG, Young AW (2009) MEG demonstrates a supra-additive response to facial and vocal emotion in the right superior temporal sulcus.

Proceedings of the National Academy of Sciences of the United States of America 106(47): 20010–20015
Hikosaka K, Iwai E, Saito H, Tanaka K (1988) Polysensory properties of neurons in the anterior bank of the caudal superior temporal sulcus of the macaque monkey. Journal of Neurophysiology 60(5):1615–1637
John ER (2002) The neurophysics of consciousness. Brain Research Brain Research Reviews 39(1):1–28
Jones EG, Powell TP (1970) An anatomical study of converging sensory pathways within the cerebral cortex of the monkey. Brain 93(4):793–820
Kanwisher N, McDermott J, Chun MM (1997) The fusiform face area: A module in human extrastriate cortex specialized for face perception. Journal of Neuroscience 17(11):4302–4311
Kayser C, Logothetis NK (2009) Directed interactions between auditory and superior temporal cortices and their role in sensory integration. Frontiers in Integrative Neuroscience 3:7
Kayser C, Logothetis NK, Panzeri S (2010) Visual enhancement of the information representation in auditory cortex. Current Biology 20(1):19–24
Kayser C, Petkov CI, Augath M, Logothetis NK (2005) Integration of touch and sound in auditory cortex. Neuron 48(2):373–384
Kayser C, Petkov CI, Augath M, Logothetis NK (2007) Functional imaging reveals visual modulation of specific fields in auditory cortex. Journal of Neuroscience 27(8):1824–1835
Kreifelts B, Ethofer T, Grodd W, Erb M, Wildgruber D (2007) Audiovisual integration of emotional signals in voice and face: An event-related fMRI study. NeuroImage 37(4):1445–1456
Kreifelts B, Ethofer T, Huberle E, Grodd W, Wildgruber D (2010) Association of trait emotional intelligence and individual fMRI-activation patterns during the perception of social signals from voice and face. Human Brain Mapping 31(7):979–991
Kreifelts B, Ethofer T, Shiozawa T, Grodd W, Wildgruber D (2009) Cerebral representation of non-verbal emotional perception: fMRI reveals audiovisual integration area between voice- and face-sensitive regions in the superior temporal sulcus. Neuropsychologia 47(14):3059–3066
Lambrecht L, Kreifelts B, Wildgruber D (2012) Age-related decrease in recognition of emotional facial and prosodic expressions. Emotion. Epub ahead of print.
Laurienti PJ, Perrault TJ, Stanford TR, Wallace MT, Stein BE (2005) On the use of superadditivity as a metric for characterizing multisensory integration in functional neuroimaging studies. Experimental Brain Research 166(3–4):289–297
Lewis JW, Beauchamp MS, DeYoe EA (2000) A comparison of visual and auditory motion processing in human cerebral cortex. Cerebral Cortex 10(9):873–888
Llinas R, Ribary U (2001) Consciousness and the brain. The thalamocortical dialogue in health and disease. Annals of the New York Academy of Science 929:166–175
Loe PR, Benevento LA (1969) Auditory–visual interaction in single units in the orbito-insular cortex of the cat. Electroencephalography and Clinical Neurophysiology 26(4):395–398
Macaluso E, Frith C, Driver J (2000) Selective spatial attention in vision and touch: Unimodal and multimodal mechanisms revealed by PET. Journal of Neurophysiology 83(5):3062–3075
Massaro DW, Egan PB (1996) Perceiving affect from the voice and the face. Psychonomic Bulletin and Review 3(2):215–221
McDonald AJ (1998) Cortical pathways to the mammalian amygdala. Progress in Neurobiology 55(3):257–332
Mechelli A, Price CJ, Friston KJ (2001) Nonlinear coupling between evoked rCBF and BOLD signals: A simulation study of hemodynamic responses. NeuroImage 14(4):862–872
Meredith MA, Stein BE (1983) Interactions among converging sensory inputs in the superior colliculus. Science 221(4608):389–391
Mesulam MM (1998) From sensation to cognition. Brain 121(Pt 6):1013–1052
Mesulam MM, Mufson EJ (1982) Insula of the old world monkey. III: Efferent cortical output and comments on function. The journal of Comparative Neurology 212(1):38–52
Miller J (1982) Divided attention: Evidence for coactivation with redundant signals. Cognitive Psychology 14(2):247–279

Morris JS, Frith CD, Perrett DI, Rowland D, Young AW, Calder AJ et al (1996) A differential neural response in the human amygdala to fearful and happy facial expressions. Nature 383(6603): 812–815

Mufson EJ, Mesulam MM (1984) Thalamic connections of the insula in the rhesus monkey and comments on the paralimbic connectivity of the medial pulvinar nucleus. The Journal of Comparative Neurology 227(1):109–120

Müller VI, Habel U, Derntl B, Schneider F, Zilles K, Turetsky BI et al (2011) Incongruence effects in crossmodal emotional integration. NeuroImage 54(3):2257–2266

Murray EA, Mishkin M (1985) Amygdalectomy impairs crossmodal association in monkeys. Science 228(4699):604–606

Nichols T, Brett M, Andersson J, Wager T, Poline JB (2005) Valid conjunction inference with the minimum statistic. NeuroImage 25(3):653–660

Park JY, Gu BM, Kang DH, Shin YW, Choi CH, Lee JM et al (2010) Integration of cross-modal emotional information in the human brain: An fMRI study. Cortex 46(2):161–169

Pearson RC, Brodal P, Gatter KC, Powell TP (1982) The organization of the connections between the cortex and the claustrum in the monkey. Brain Research 234(2):435–441

Peck CK (1987) Visual–auditory interactions in cat superior colliculus: Their role in the control of gaze. Brain Research 420(1):162–166

Phillips ML, Young AW, Scott SK, Calder AJ, Andrew C, Giampietro V et al (1998) Neural responses to facial and vocal expressions of fear and disgust. Proceedings of the Biological Sciences 265(1408):1809–1817

Pitkänen A (2000) Connectivity of the rat amygdaloid complex. Oxford University Press, In The Amygdala. A functional analysis. New York

Pourtois G, de Gelder B, Bol A, Crommelinck M (2005) Perception of facial expressions and voices and of their combination in the human brain. Cortex 41(1):49–59

Pourtois G, de Gelder B, Vroomen J, Rossion B, Crommelinck M (2000) The time-course of intermodal binding between seeing and hearing affective information. NeuroReport 11(6): 1329–1333

Pourtois G, Debatisse D, Despland PA, de Gelder B (2002) Facial expressions modulate the time course of long latency auditory brain potentials. Brain Research Cognitive Brain Research 14(1):99–105

Price CJ, Friston KJ (1997) Cognitive conjunction: A new approach to brain activation experiments. NeuroImage 5(4 Pt 1):261–270

Raichle ME, MacLeod AM, Snyder AZ, Powers WJ, Gusnard DA, Shulman GL (2001) A default mode of brain function. Proceedings of the National Academy of Sciences of the United States of America 98(2):676–682

Robins DL, Hunyadi E, Schultz RT (2009) Superior temporal activation in response to dynamic audio–visual emotional cues. Brain and Cognition 69(2):269–278

Schroger E, Widmann A (1998) Speeded responses to audiovisual signal changes result from bimodal integration. Psychophysiology 35(6):755–759

Schutte N, Malouff J, Hall L, Haggerty D, Cooper J, Golden C et al (1998) Development and validation of a measure of emotional intelligence. Personality and Individual Differences 25:167–177

Scott SK, Young AW, Calder AJ, Hellawell DJ, Aggleton JP, Johnson M (1997) Impaired auditory recognition of fear and anger following bilateral amygdala lesions. Nature 385(6613):254–257

Seltzer B, Pandya DN (1978) Afferent cortical connections and architectonics of the superior temporal sulcus and surrounding cortex in the rhesus monkey. Brain Research 149(1):1–24

Stanford TR, Stein BE (2007) Superadditivity in multisensory integration: Putting the computation in context. NeuroReport 18(8):787–792

Stein BE, London N, Wilkinson LK, Price DD (1996) Enhancement of perceived visual intensity by auditory stimuli: A psychophysical analysis. Journal of Cognitive Neuroscience 8:497–506

Stein BE, Meredith MA (1993) Merging of senses. MIT Press, Cambridge

Stein BE, Wallace MT (1996) Comparisons of cross-modality integration in midbrain and cortex. Progress in Brain Research 112:289–299

Stevenson RA, Geoghegan ML, James TW (2007) Superadditive BOLD activation in superior temporal sulcus with threshold non-speech objects. Experimental Brain Research 179(1):85–95

Stevenson RA, James TW (2009) Audiovisual integration in human superior temporal sulcus: Inverse effectiveness and the neural processing of speech and object recognition. NeuroImage 44(3):1210–1223

Szycik GR, Munte TF, Dillo W, Mohammadi B, Samii A, Emrich HM et al (2009) Audiovisual integration of speech is disturbed in schizophrenia: An fMRI study. Schizophrenia Research 110(1–3):111–118

van Atteveldt N, Formisano E, Goebel R, Blomert L (2004) Integration of letters and speech sounds in the human brain. Neuron 43(2):271–282

von Kriegstein K, Giraud AL (2006) Implicit multisensory associations influence voice recognition. PLoS Biology 4(10):e326

Vroomen J, Driver J, de Gelder B (2001) Is cross-modal integration of emotional expressions independent of attentional resources? Cognitive, Affective, & Behavioral Neuroscience 1(4):382–387

Wallace MT, Meredith MA, Stein BE (1992) Integration of multiple sensory modalities in cat cortex. Experimental Brain Research 91(3):484–488

Wallace MT, Meredith MA, Stein BE (1993) Converging influences from visual, auditory, and somatosensory cortices onto output neurons of the superior colliculus. Journal of Neurophysiology 69(6):1797–1809

Wallace MT, Wilkinson LK, Stein BE (1996) Representation and integration of multiple sensory inputs in primate superior colliculus. Journal of Neurophysiology 76(2):1246–1266

Wang Y, Celebrini S, Trotter Y, Barone P (2008) Visuo-auditory interactions in the primary visual cortex of the behaving monkey: Electrophysiological evidence. BMC Neuroscience 9:79

Werner S, Noppeney U (2010a) Distinct functional contributions of primary sensory and association areas to audiovisual integration in object categorization. Journal of Neuroscience 30(7):2662–2675

Werner S, Noppeney U (2010b) Superadditive responses in superior temporal sulcus predict audiovisual benefits in object categorization. Cerebral Cortex 20(8):1829–1842

White M (1999) Representation of facial expressions of emotion. The American Journal of Psychology 112(3):371–381

Williams LE, Light GA, Braff DL, Ramachandran VS (2010) Reduced multisensory integration in patients with schizophrenia on a target detection task. Neuropsychologia 48(10):3128–3136

Worsley K, Marrett S, Neelin P, Vandal AC, Friston KJ, Evans A (1996) A unified statistical approach for determining significant signals in images of cerebral activation. Human Brain Mapping 4(1):74–90

Wright TM, Pelphrey KA, Allison T, McKeown MJ, McCarthy G (2003) Polysensory interactions along lateral temporal regions evoked by audiovisual speech. Cerebral Cortex 13(10):1034–1043

Chapter 13
Emotions by Ear and by Eye

Beatrice de Gelder, Bernard M.C. Stienen, and Jan Van den Stock

Abstract Multisensory integration must stand out among the fields of research that have witnessed one of the most impressive explosions of interest this last decade, at least as measured by published papers and meetings. From a highly specialized niche occupation multisensory research has become a mainstream scientific interest in a very short time span. One of these new areas of multisensory research is emotion. Since our first exploration of this phenomenon [de Gelder Böcker, Tuomainen, Hensen, & Vroomen Neuroscience Letters 260(2):133–136, 1999], a number of studies have appeared and they have used a wide variety of behavioral, neuropsychological, and neuroimaging methods.

The goal of this chapter is fourfold. First, we review the research on audiovisual perception of emotional signals from the face and the voice. In the next section we discuss some outstanding methodological and theoretical issues followed by a report and comment on integrating emotional information provided by the voice and whole body expressions. We also include some recent work on music. Finally, we discuss findings about abnormal affective audiovisual integration in schizophrenia and in autism.

1 The Combined Perception of Facial and Vocal Expressions

In this first section we review research on audiovisual perception of emotional signals from the face and the voice. It is not our intention to review all the literature that has become available in the last decades. We only make a biased selection and linger on a few studies that have highlighted critical theoretical issues for future reference.

B. de Gelder (✉) • B.M.C. Stienen • J. Van den Stock
Cognitive and Affective Neuroscience Laboratory, Tilburg University,
Room P 511, Postbus 90153, 5000 LE Tilburg, The Netherlands
e-mail: b.degelder@uvt.nl; bstienen@gmail.com; jan.vandenstock@med.kuleuven.be

P. Belin et al. (eds.), *Integrating Face and Voice in Person Perception*,
DOI 10.1007/978-1-4614-3585-3_13, © Springer Science+Business Media New York 2013

1.1 Audiovisual Emotion Perception, the First Studies

Articles and chapters on multisensory integration inevitably open with the statement that in everyday life the perceptual system is bombarded with information that reaches the different sensory channels simultaneously. True to form, this section also starts with the remark that affective signals occurring in natural environments impinge on several sensory channels at the same time. We return to this issue later. At present we briefly review the early beginnings of this line of work in our lab and the first use of naturalistic still and video images after Massaro and Egan (1996) explored the issue with an artificial talking face.

Human emotion recognition can be based on isolated facial or vocal cues (Banse & Scherer, 1996; Scherer, Banse, Wallbott, & Goldbeck, 1991), but a combination of both modalities results in a performance increase, as shown by both increased accuracy rates and shorter response latencies (de Gelder, Böcker, Tuomainen, Hensen, & Vroomen, 1999; de Gelder & Vroomen, 2000; de Gelder, Vroomen, & Teunisse, 1995; Dolan, Morris, & de Gelder, 2001; Massaro & Egan, 1996). For detailed behavioral investigations into crossmodal influences between vocal and facial cues one needs a paradigm in which both modalities are combined to create audiovisual pairs. The manipulation ideally consists of altering both the emotional congruency between the two modalities and a task that consists of emotion categorization based on only one of both information streams. For example, de Gelder and Vroomen (2000) presented facial expressions that were morphed on a continuum between happy and sad, while at the same time a short spoken sentence was presented. This sentence had a neutral semantic meaning, but was spoken in either a happy or sad emotional tone of voice. Participants were instructed to attend to and categorize the face in a two-alternative forced-choice task, and to ignore the voice. The results showed a clear influence of the task-irrelevant auditory modality on the target visual modality. For example, sad faces were less frequently categorized as sad when they were accompanied by a happy voice. In a follow-up experiment, vocal expressions were morphed on a fear–happy continuum and presented with either a fearful or happy face, while participants were instructed to categorize the vocal expression. Again, the task-irrelevant modality (facial expressions) influenced the emotional categorization of the target modality (vocal expressions). Furthermore, this experiment was repeated under different attentional demands, but the facial expression influenced the categorization of vocal expression in every attention condition (Vroomen, Driver, & de Gelder, 2001). These findings suggest that affective multisensory integration is a mandatory and automatic process. However, based on these behavioral data, no direct claims can be made about the nature of this crossmodal bias effect. The findings could either reflect an early perceptual or later more cognitive or decisional effect.

Emotional prosody can alter facial emotion perception (Massaro & Egan, 1996) independent from conscious visual perception (de Gelder, Pourtois, & Weiskrantz, 2002), or independent from attention (Vroomen et al., 2001) and even with the explicit instruction to ignore one modality (Ethofer et al., 2006). Some basic methodological problems are associated with these paradigms and the way they are used and we comment on them as they appear in this chapter. Before doing so, one general issue arises in almost all of them and we discuss this first.

1.2 A Broader Framework: Multisensory Effects as Redundancy Reduction or as Context Effects?

As we noted above, the standard view is that multisensory stimulation represents informational redundancy and that the kind of redundancy that is created by multiple convergent stimulus inputs is beneficial for the perceiver. Indeed, affective signals often require a rapid reaction from the observer and intersensory redundancy, so it is assumed, contributes to speed by reducing uncertainty. This may seem to be particularly important for perception of emotional cues in social interactions, where convergence between facial and bodily expressions and emotional prosody facilitates rapid emotion recognition and adaptive reaction. This point is easily illustrated by looking at the details of studies that use facial expressions. The widely used stimulus set of facial expressions provided by Ekman and Friesen (1976) on average does not generate a recognition rate that is higher than 75 % with important basic emotions like fear or sadness reaching routinely lower levels of correct recognition. Even recent studies using dynamic images rather than still ones do not achieve recognition rates above 80 %. This remains a challenge for the concept of basic, hardwired emotions as seen in facial expressions, all the more so that the maximum recognition rate achieved varies with the emotion considered. But this is not a matter of concern here. What needs to be taken into account though are the implications for intersensory research because it is not clear what level of performance is best to measure the contribution of the auditory input to the visual.

In daily life facial expressions are typically accompanied by various kinds of context information like the visual scene, environmental sounds, vocal expressions, and whole body postures and movements. As of today it is still surprising that these various kinds of context have received so little attention. As a matter of fact, one wonders what changes we would need to make to mainstream models of how the brain processes facial expression and identity if indeed other visual and auditory inputs have an impact on the processing of these.

To put this debate in perspective, it may be useful to borrow from another debate: the role of context in visual perception. Phenomena of the kind seen in audiovisual emotion perception are not normally investigated under the heading of context effects.

Discussions on context influences—and their consequences for how we read and react to an emotion from the face—have a long history (Fernberger, 1928). But many of the kind of context effects that were investigated in the early days would nowadays qualify as so-called *late* effects or *postperceptual* effects, related as they are to the overall higher cognitive, conscious and deliberate (verbal) appraisal of a stimulus rather than to its online processing (Bertelson & de Gelder, 2004; de Gelder & Bertelson, 2003). And traditionally the notion of context is typically associated with late, cognitive elaborations of a percept forged at an earlier processing stage. It is integration at this early perceptual level that we have specifically targeted since our lab started to work on audiovisual perception (de Gelder et al., 1999; de Gelder & Van den Stock, 2011). And it may be useful to remind us why that work was originally viewed as controversial. As a matter of fact, traditional research on audiovisual perception used nonsense stimuli, typically consisting of short sound bursts and brief light flashes. Which is to say that for a long time the issue of semantic influences on audiovisual integration did not come to the foreground. This situation changed with the discovery by McGurk and MacDonald (1976) but did not stop the debate on the impact of semantic factors in, for example, the area of ventriloquism.

Because faces and sentence fragments were rich in content, their combination was unlikely to be rapid and automatic in the sense in which these notions are applicable to early perceptual processes. For example, it was argued initially that affective information was not a likely candidate for true multisensory perception. As a testimony to the changed intellectual climate, the notion of automatic and rapid processing of affective information is now well established in affective neuroscience. This is in part due to the increasing acknowledgement of phylogenetic continuity of the brain structures and processes involved in affective processes (de Gelder & Van den Stock, 2011). For example, the work of Panksepp (1998, 2005) has shown that so called "higher" emotions like joy are present in many species and that their neurobiological basis shows substantial continuity across different species.

1.3 Neurofunctional Basis: Initial Findings

A few aspects of human audiovisual emotion perception have already been investigated, using different neuroimaging methods. The first reports have focused on the time course of crossmodal face–voice influences and therefore made use of methods with high temporal resolution.

Studies addressing recognition and neural substrates of vocal expressions in isolation are still few (de Gelder, Vroomen, & Pourtois, 2004; George et al., 1996; Grandjean et al., 2005; Ross, 2000). However, EEG research shows that recognition of emotional prosody occurs already within the first 100–150 ms of stimulus presentation (Bostanov & Kotchoubey, 2004; Goydke, Altenmuller, Moller, & Munte, 2004) and that early integration of both modalities around 110 ms after stimulus presentation (de Gelder et al., 1999; Pourtois, de Gelder, Vroomen, Rossion, & Crommelinck, 2000), which is compatible with a perceptual stage. Supporting evidence for the automatic nature of this integration is provided by studies with blindsight patients, who are unable (due to damage in the visual cortex) to consciously perceive visual stimuli presented in a segment of the visual field. When patients are presented with auditory vocal expressions and at the same time visual facial expressions in their blind field, Dit is wat vaag. Kon niet bij de relevante papers helaas om het aan te vullen (of which they are unaware). This suggests that emotional information displayed by the unseen face is processed in these patients by alternative brain pathways influencing brain responses to the consciously perceived vocal expressions (de Gelder, Morris, & Dolan, 2005; de Gelder, Pourtois, & Weiskrantz, 2002).

Auditory and visual expressions of emotions are ecologically relevant and may therefore rely on specialized neural mechanisms as has long been recognized in animal research. Several recent studies have explored the relation between auditory and visual processing streams in nonhuman primate communication (Ghazanfar & Santos, 2004; Parr, 2004). Yet the notion of a specialized neurobiological basis is compatible with many different scenarios of multisensory perception. Three different scenarios are envisaged for multisensory emotion perception.

The first scenario focuses on convergence in cortical heteromodal areas which follows after sensory specific processes in dedicated cortices. A second possible model predicts direct corticocortical interactions between auditory and visual cortex, either unilateral or bilateral. A third alternative centers on the hypothesis of content driven processes with early extraction of the affective information by cortical and subcortical structures followed by or in parallel with cortical processing.

Extending well-established findings on integration of low-level audiovisual cues (Calvert & Thesen, 2004; Stein & Stanford, 2008), various studies have addressed the neurobiology of multisensory emotion integration. Studies in nonhuman primates revealed an ability to integrate socially relevant multimodal cues from conspecifics (Ghazanfar & Logothetis, 2003), which is characterized by responsiveness of amygdala and auditory cortex (Ghazanfar, Maier, Hoffman, & Logothetis, 2005; Remedios, Logothetis, & Kayser, 2009), superior temporal sulcus (Ghazanfar, Chandrasekaran, & Logothetis, 2008) and ventrolateral prefrontal cortex (Sugihara, Diltz, Averbeck, & Romanski, 2006). However, the main issue is whether this pattern was directly and primarily driven by emotional integration or low-level stimulus features.

The first EEG studies reported interactions of facial and vocal emotions at 110–220 ms post stimulus (de Gelder et al., 1999; Pourtois et al., 2000; Pourtois, Debatisse, Despland, & de Gelder, 2002) suggesting early convergence in primary sensory cortices. In contrast, fMRI investigations reported temporal structures as candidates for emotion integration (Ethofer et al., 2006; Kreifelts, Ethofer, Grodd, Erb, & Wildgruber, 2007; Kreifelts, Ethofer, Huberle, Grodd, & Wildgruber, 2010; Pourtois, de Gelder, Bol, & Crommelinck, 2005).

A critical question for any design to be used in multisensory studies is to decide on the critical stimulus pairs one wants to contrast or on what the relevant properties of the critical pairing conditions are. Indeed, there are many open questions on what are the constraints on multisensory integration are beyond that of spatiotemporal co-occurrence.

2 Persistent Matters of Theory and Methodology

Here we highlight some issues that we perceive as unresolved theoretical and methodological bottlenecks and provide examples and illustrations.

While the available evidence is clearly consistent with the notion that the perceptual system integrates emotional information from the face and the voice, little is known at present about possible constraints on underlying processes (de Gelder, 2000). It is for instance still unclear which spatial and temporal constraints the integration of emotion from the face and voice must obey and whether these are specific for this type of integration process. A similar question can be asked about constraints related to the information content of the inputs from the two modalities. One may speculate that the ability to combine multiple information sources in a single percept is undoubtedly advantageous for an organism, but in the absence of any limits, such an advantage would quickly be lost and as a consequence the internal processing theater would mirror the booming, buzzing confusion of the outside world. All the studies mentioned above have used face–voice stimulus pairs; they do not allow us to conclude unambiguously that it is the systemic need to exploit redundancy that drives integration. Moreover, judging a facial expression might be influenced by just about any concurrent stimulus that provides a context within which the face (or the voice) can be judged more easily. For example, we do not know if task irrelevant, redundant or secondary stimuli presented in the periphery like for example a secondary facial expression or a written emotion word would not also have an impact on rating the facial expression just like the prosody in the voice does. It is well known from studies of the Stroop task that response to a visual target is influenced by concurrently presented information that is irrelevant to the task at hand (see MacLeod & MacDonald, 2000 for review). To what extent is such an effect different from the crossmodal prosodic bias? If crossmodal bias would present the same characteristics as the Stroop effect, its advantages would be limited as it would slow down rather than speed the response and would expose the organism to vicarious influences away from the main task at hand.

One must also distinguish the advantages of intermodal interactions in normal environmental conditions from the disadvantages linked to the artificial conditions created for research purposes. Presumably, in normal situations integration serves as a powerful filter mechanism which reduces the effects of noise, spontaneous drifts, injuries or growth. When we create hyperdiscrepancies that never occur in the normal environment, we observe consequences that appear disadvantageous for the perceiver. However, they have methodological advantages and these make apparent some of the underlying mechanisms, which are best seen as the price that would be paid for an otherwise useful mechanism where unusually large discrepancies do occur. A related issue concerns the role of prosody as separate from the role of speech. The studies mentioned above do not answer the question whether the prosodic information works because it provides prosody or because it is speech.

3 Multisensory Emotion Perception Beyond the Face: Combining Whole Body Expressions with Affective Auditory Signals

We have shown previously that putting the spotlight on whole body expressions of emotion significantly widens the scope of emotion research. Bodily expressions are recognized as reliably as facial expressions and are processed under the same perceptual conditions and with the same relative independence from visual awareness or attention as facial expressions (de Gelder et al., 2010; Tamietto & de Gelder, 2010). At the same time, with facial expressions often recognized less than perfectly, bodily expressions that are emotionally congruent with the facial expression shown at the same time, improve accuracy of facial recognition while incongruent bodily information significantly hampers it (Meeren, van Heijnsbergen, & de Gelder, 2005; Van den Stock, Righart, & de Gelder, 2007). So, in some ways visual–visual combinations function perceptually in ways very similar to visual–auditory ones. We take this to be an argument in favor of viewing questions about multisensory perception in the broader framework of context effects as we discussed above.

But on the other hand, here also, like in the case of audiovisual emotion perception involving the voice and the face, a narrow focus on redundancy is misleading. Indeed, affective signals have both specificity and complementarities between them. In a nutshell, some emotions are better conveyed by the body than by the face and vice versa. For example, although one can show anger by frowning the brows, the tension in the body muscles would give away much more the strength of the angriness or the intention of the angry person while disgust is an example in point here as the facial expression is very specific while the body posture associated with disgust is less specific since it shares features with showing fear. When we add this kind of emotion specificity to the overall picture it emerges that depending on the emotion we consider, the primary sensory channel can be the face, the whole body or the voice. The point we want to make here is that simple considerations of redundancy reduction as the motor for multisensory convergence will miss some crucial facts here. Specificity is a very hard problem for the traditional approach to design audiovisual pairs that is normally based on the notion of equal contribution from each channel.

Recent studies have shown that next to facial expressions, perception of bodily expressions is also influenced by concurrent auditory information—and affective information in sounds modifies the viewers' appreciation of the affective body image. For example, recognition of dynamic whole-body expressions of emotion are influenced not only by both human and animal vocalizations (Van den Stock, Grèzes, & de Gelder, 2008), but also by instrumental music (Van den Stock, Peretz, Grèzes, & de Gelder, 2009), suggesting that the brain is efficient at extracting affective information from different sources and combining it across different sensory channels.

Many research reports have concluded that emotional information can be processed without observers being aware of it. One may ask whether bodily expressions are also automatically processed as has been shown for facial expressions. In view of the similarities between facial and bodily expressions for rapid perception and communication of emotional signals, we conjectured that perception of bodily expressions may also not necessarily require visual awareness. In a recent study of ours, participants had to detect in three separate experiments masked fearful, angry and happy bodily expressions among masked neutral bodily actions as distractors and subsequently the participants had to indicate their confidence. The onset between target and mask (Stimulus Onset Asynchrony, SOA) varied from −50 to +133 ms. Results show a lack of covariance between the objective (detection) and subjective (confidence) measurements when the participants had to detect fearful bodily expressions, yet this was not the case when participants had to detect happy or angry bodily expressions. This study provides novel evidence for the processing of fear stimuli, which apparently depends less on the visibility of the expression itself and generalizes to bodily expressions (Stienen & de Gelder, 2011).

Multisensory integration may occur independently of visual attention as previously shown with compound face–voice stimuli (Alsius, Navarra, Campbell, & Soto-Faraco, 2005; Vroomen et al., 2001). But attentional selection does not imply that one is consciously aware of the stimulus. Also, the unattended stimulus could be consciously perceived (Tamietto & de Gelder, 2010). Visual awareness of faces does not seem to be a prerequisite for audiovisual affect integration since cross-modal interactions are still observed when the face is not consciously perceived in hemianopic patients (de Gelder et al., 2002).

To address the question whether multisensory integration can occur independently of visual awareness we performed a parametric masking study in which we presented masked angry and happy bodily expressions together with congruent or incongruent human angry and happy vocalizations. The participants had to categorize the masked bodily expressions while ignoring the emotional voices and subsequently, as in the previously discussed study, they had to indicate whether they were sure of their answer or whether they were guessing. The onset between target and mask varied again from −50 to +133 ms (Stienen, Tanaka, de Gelder, 2011).

Results showed that when emotional voices and bodily postures are congruent, objective recognition of emotional bodily expressions increased regardless of SOA latency. This same effect was not seen in subjective confidence ratings where there was no facilitation effect of congruent voice information for short SOA latencies in the range of 0 to +50 ms. Conjointly, the confidence of the participants was not above zero in the SOA latency range from 0 to +33 ms while the accuracy was above chance when the emotional voice–body pairs were congruent (see Fig. 13.1). The subjective ratings can be taken as measure of the phenomenological experience of the participants' perception of the targets (Cheesman & Merikle, 1986; Esteves & Öhman, 1993). The combination of these findings shows that the emotion of the voice exerts its influence independently of the visual awareness of the target.

The lack of the interaction between congruency and SOA latency in accuracy also shows that these results do not reflect merely a decision or response bias (de Gelder & Bertelson, 2003). Such a bias would be stronger when visibility of the target is low and would thus result in an interaction of congruency and SOA latency on the categorization performance of the participants. In other words, this method shows to be a very good control to check whether such a bias is present.

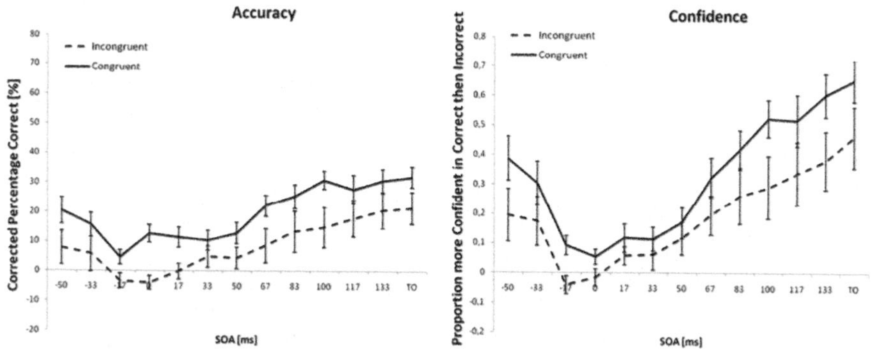

Fig. 13.1 Left: Mean categorization performance plotted as function of SOA latency corrected for chance (50 percent). Right: Mean confidence ratings plotted as function of SOA latency. Error bars represent standard error of the mean. SOA = Stimulus Onset Asynchrony, TO = Target Only

In sum these results indicate that the human voice influences the objective categorization independently of the visibility of the bodily expressions. In another experiment we addressed whether unseen bodily expressions influence our recognition of prosody in the voice. Participants had to categorize emotional spoken sentences as fearful or happy while we presented masked bodily expressions. The auditory stimuli consisted of a Dutch spoken sentence "met het vliegtuig" (which means "with the plane"), edited as to express different levels of emotion on a 7-step continuum between fearful and happy. The body postures consisted of fearful expressions and neutral actions (combing hair) and were masked with a pattern mask using an SOA latency of 33 ms. Also, a no-body condition was added to set a baseline. In these trials only the mask was presented for the duration of the target–mask combinations (Stienen, Tanaka, de Gelder, 2011).

In order to force the subjects to fixate their gaze on the screen while paying attention to the spoken sentence, we used catch trials. In 22% of the trials a fixation cross turned 45° clockwise and switched back to the original position after 133 ms. The participants were told that we were interested whether the recognition of emotion in the voice is influenced when the perceptual system is loaded with visual information. See for a schematic representation of a trial and examples of the stimuli (Fig. 13.2).

To check whether the participants had been unaware of the body stimuli we conducted an extensive semi structured exit interview and a posttest. In the posttest the participants were instructed to classify the stimuli as seen if they recollect that they had seen the bodily posture during the main experiment and as not seen when they could not recollect the bodily posture. All target stimuli were presented among a set of new bodily postures. Participants indicating having seen anything else besides the mask during the exit interview or choosing the target stimulus in the post test were excluded from analysis. This was only the case in 7 out of 31 participants.

Results showed an interaction of bodily expression and vocal expression on the categorization of the emotion of the voice revealing the influence of bodily expression presented outside of visual awareness on the perceived emotion in the voice.

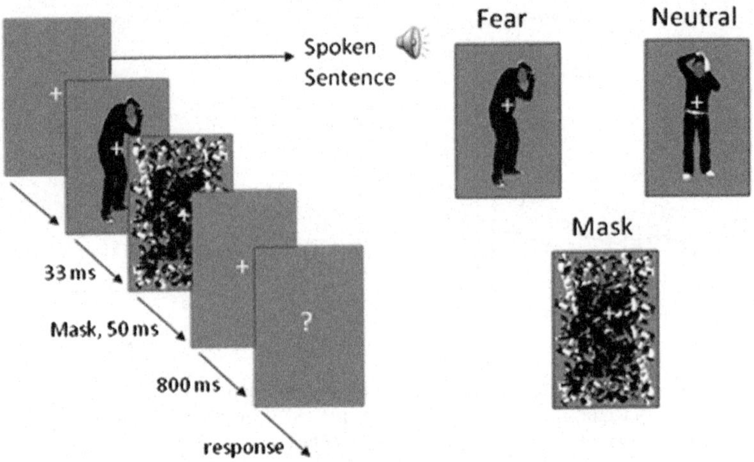

Fig. 13.2 Illustration of an example trial (left), an example of a fearful bodily expression and a neutral action (upper right) and the mask (below right). Results show that the unseen bodily expressions influence the interpretation of voice prosody

Our findings are consistent with earlier studies showing the crossmodal influence of human emotional sounds on the recognition of emotional body postures (Van den Stock et al., 2008) and the influence of emotional body postures on the interpretation of voice prosody (Van den Stock et al., 2007). We add the important notion that this crossmodal interaction is even taking place when the observer is not aware of the visual information reflecting the automaticity of the process.

4 Abnormal Integration of Multisensory Perception of Emotions in Schizophrenia and Autism

In the above sections, we have focused on audiovisual integration of emotional signals in normal subjects. Very little is known about bimodal perception of affective stimuli in clinical populations. A few isolated studies in patients with autism spectrum disorders have looked at affective multisensory integration (Magnée, de Gelder, van Engeland, & Kemner, 2007, 2008, 2011). The ability to quickly integrate multiple sources of perceptual input is important for developing adaptive social behavior. Recent data suggest that multisensory integration is impaired in individuals with autism spectrum disorder (ASD), but it remains unclear to what extent this is influenced by nonspecific stimulus or task-related factors, such as environmental noise and attention.

Recently, we measured event-related potentials in 23 high-functioning, adult ASD individuals and 24 age- and IQ-matched controls while they viewed emotionally congruent and incongruent face–voice pairs in two tasks. In the first task, the integrity of the stimuli was visually and auditory degraded by two levels of noise. In the second task multisensory integration was studied while attention was divided over the visual and audio channel, or their attention was manipulated by introducing an extra visual attention task consisting of two levels (easy and hard). ERPs were measured on typical auditory and visual processing peaks, the P2 and N170.

To control for effects of atypical unisensory processing in ASD, group differences were tested in ERP amplitudes during unisensory auditory and visual processing. No such differences between groups were found. With respect to multisensory integration, we found that ERP activity was indeed affected by manipulating environmental noise and focus of attention, and it was found to be affected differently in individuals with ASD. Results show that the amount of noise clearly influenced multisensory integration, although more vigorously in ASD individuals than in controls. Also, an important difference between the control group and ASD group was that multisensory integration was observed during divided attention and easy selective attention tasks for controls, yet for the ASD group only during the easy selective attention (Magnée et al., 2011).

The study shows that disruptions in multisensory integration of emotional stimuli are not a primary feature of ASD but are secondary to atypical processing of environmental noise and to abnormal attention mechanisms, especially those associated with divided attention. This may lead to impairments in multisensory integration under naturalistic situations, and may therefore account for several clinical characteristics of ASD.

Most studies have looked at affective multisensory integration in schizophrenia patients or patients diagnosed with a nonschizophrenic psychotic disorder (de Gelder et al., 2005b; de Jong, Hodiamont, & de Gelder, 2010; de Jong, Hodiamont, Van den Stock, & de Gelder, 2009; Van den Stock, de Jong, Hodiamont, & de Gelder, 2011). Schizophrenia is associated with deficits in affective processing (Kraepelin, 1919). Studies investigating recognition of emotions in schizophrenic patients have predominantly focussed on facial expressions (Borod, Martin, Alpert, Brozgold, & Welkowitz, 1993; Feinberg, Rifkin, Schaffer, & Walker, 1986; Heimberg, Gur, Erwin, Shtasel, & Gur, 1992; Kee, Horan, Wynn, Mintz, & Green, 2006; Kohler et al., 2003; Wolwer, Streit, Polzer, & Gaebel, 1996) and the results point to a general deficit, with an emphasis on impaired recognition of negative emotions, in particular anger and fear (see Mandal, Pandey, & Prasad, 1998 for a review). The deficits in emotion perception have been linked to the social deficits observed in schizophrenia patients (Pinkham, Hopfinger, Ruparel, & Penn, 2008).

Building on these findings and our affective audiovisual studies in normal subjects, we investigated whether schizophrenia is associated with an abnormal affective multisensory integration profile. We made use of the crossmodal bias paradigm (de Gelder & Bertelson, 2003) and presented bimodal congruent and incongruent happy and fearful face–voice combinations to patients diagnosed with schizophrenia (de Jong et al., 2009). The instruction was to categorize the vocal expression while ignoring the facial expression in a two alternative forced choice task (happy or fearful). The results showed that, compared to an age-matched control group, the schizophrenia patients were less influenced by the task irrelevant facial expression.

These findings indicate that schizophrenia is associated with abnormal affective multisensory integration. In a number of follow up experiments, we investigated whether the impact of the auditory modality under audiovisual perception conditions in schizophrenia patients is more dominant, compared to normal subjects.

We modified the design to a dual task paradigm and included an additional attention modulation (de Jong et al., 2010). Participants were presented with similar audiovisual face–voice pairs as described above (de Jong et al., 2009); however, we simultaneously presented a number on the face (“3” or “8”) in one block or a pair of tones (two low tones or one low and one high tone) in another block. After categorizing the vocal expression, participants were asked whether they saw the number “8” or whether they heard a high tone. This design provides a secondary attention manipulation and allows investigating intersensory dominance. Regarding the categorization of the vocal expression, the results showed that, compared to the original design with no secondary task, the visual secondary task affected performance (i.e. reduced crossmodal influence) in the control and nonschizophrenic psychosis group, but not in the schizophrenia group. On the other hand, a secondary auditory task reduced crossmodal influence in the control group, had no influence in the nonschizophrenic psychosis group, but increased crossmodal influence in the schizophrenia group. The combined findings are compatible with the hypothesis of an abnormal auditory dominance during audiovisual perception in schizophrenia as an explanation for altered multisensory perception. Similar findings have been reported in the domain of audiovisual speech (Ross et al., 2007).

Extending on these results, as well as on the similarities between perception of facial and bodily expressions, we presented schizophrenia patients and controls with dynamic stimuli of a person (with the facial area blurred to avoid emotion recognition from the face) engaged in common activity (picking up a glass and drinking from it). This action was performed either with a fearful or happy expression. In the bimodal blocks, the videos were simultaneously presented with either a congruent or incongruent vocal expression, which could be produced by a human or an animal. These stimuli were chosen to maximize ecological validity (de Gelder & Bertelson, 2003). The results showed that both controls and patients are influenced by the task irrelevant auditory information, when categorizing the video expression, but only when humans produce the vocal expression. This crossmodal influence was stronger in the schizophrenic group, indicating an abnormal integration of affective information across multisensory channels. When categorizing human body language, schizophrenics are more influenced by the task irrelevant auditory information, compared to the control group (Van den Stock et al., 2011).

These results are also compatible with an abnormal auditory dominance during affective multisensory perception and extend the findings to whole bodily expressions. However, we did not observe abnormal affective multisensory integration of bodily expressions paired with animal vocalizations. This finding indicates that other factors, possibly related to chance of co-occurrence in everyday life, have a mediating effect on the abnormal multisensory integration in schizophrenia patients.

At the neuroanatomical level, binding of emotional information in the face and voice has been associated with activity in the amygdala (Dolan et al., 2001). Consistent with this, abnormal amygdala activity has been reported in schizophrenia patients in response to facial expression perception (Gur et al., 2002; Michalopoulou et al., 2008; Phillips et al., 1999). Anomalous multisensory integration may partly have its roots in abnormal amygdalar activity. Further, deficient connectivity between amygdala and frontal regions has been reported in schizophrenia patients (Leitman et al., 2008). Abnormal amygdalar–frontal connectivity may either be cause or effect of dysfunctional amygdala and provides a neuroanatomical basis for the observed anomalous multisensory integration in both faces and bodies combined with vocal expressions.

References

Alsius, A., Navarra, J., Campbell, R., & Soto-Faraco, S. (2005). Audiovisual integration of speech falters under high attention demands. *Current Biology, 15*(9), 839–843.

Banse, R., & Scherer, K. R. (1996). Acoustic profiles in vocal emotion expression. *Journal of Personality and Social Psychology, 70*(3), 614–636.

Bertelson, P., & de Gelder, B. (2004). The psychology of multimodal perception. In C. Spence & J. Driver (Eds.), *Crossmodal space and crossmodal attention* (pp. 151–177). Oxford: Oxford University Press.

Borod, J. C., Martin, C. C., Alpert, M., Brozgold, A., & Welkowitz, J. (1993). Perception of facial emotion in schizophrenic and right brain-damaged patients. *The Journal of Nervous and Mental Disease, 181*(8), 494–502.

Bostanov, V., & Kotchoubey, B. (2004). Recognition of affective prosody: Continuous wavelet measures of event-related brain potentials to emotional exclamations. *Psychophysiology, 41*(2), 259–268.

Calvert, G. A., & Thesen, T. (2004). Multisensory integration: Methodological approaches and emerging principles in the human brain. *Journal of Physiology, Paris, 98*(1–3), 191–205.

Cheesman, J., & Merikle, P. M. (1986). Distinguishing conscious from unconscious perceptual processes. *Canadian Journal of Psychology, 40*(4), 343–367.

de Gelder, B. (2000). More to seeing than meets the eye. *Science, 289*(5482), 1148–1149.

de Gelder, B., & Bertelson, P. (2003). Multisensory integration, perception and ecological validity. *Trends in Cognitive Sciences, 7*(10), 460–467.

de Gelder, B., Böcker, K. B., Tuomainen, J., Hensen, M., & Vroomen, J. (1999). The combined perception of emotion from voice and face: Early interaction revealed by human electric brain responses. *Neuroscience Letters, 260*(2), 133–136.

de Gelder, B., Morris, J. S., & Dolan, R. J. (2005). Unconscious fear influences emotional awareness of faces and voices. *Proceedings of the National Academy of Sciences of the United States of America, 102*(51), 18682–18687.

de Gelder, B., Pourtois, G., & Weiskrantz, L. (2002). Fear recognition in the voice is modulated by unconsciously recognized facial expressions but not by unconsciously recognized affective pictures. *Proceedings of the National Academy of Sciences of the United States of America, 99*(6), 4121–4126.

de Gelder, B., & Van den Stock, J. (2011). Real faces, real emotions: Perceiving facial expressions in naturalistic contexts of voices, bodies and scenes. In A. J. Calder, G. Rhodes, M. H. Johnson, & J. V. Haxby (Eds.), *The Oxford handbook of face perception* (pp. 535–550). New York: Oxford University Press.

de Gelder, B., Van den Stock, J., Meeren, H. K., Sinke, C. B., Kret, M. E., & Tamietto, M. (2010). Standing up for the body. Recent progress in uncovering the networks involved in processing bodies and bodily expressions. *Neuroscience and Biobehavioral Reviews, 34*(4), 513–527.
de Gelder, B., & Vroomen, J. (2000). The perception of emotions by ear and by eye. *Cognition and Emotion, 14*(3), 289–311.
de Gelder, B., Vroomen, J., de Jong, S. J., Masthoff, E. D., Trompenaars, F. J., & Hodiamont, P. (2005). Multisensory integration of emotional faces and voices in schizophrenics. *Schizophrenia Research, 72*(2–3), 195–203.
de Gelder, B., Vroomen, J., & Pourtois, G. (2004). Multisensory perception of emotion, its time course and its neural basis. In G. Calvert, C. Spence, & B. E. Stein (Eds.), *Handbook of multisensory processes* (pp. 581–596). Cambridge, MA: MIT.
de Gelder, B., Vroomen, J., & Teunisse, J. P. (1995). Hearing smiles and seeing cries. The bimodal perception of emotion. *Bulletin of the Psychonomic Society, 29*, 309.
de Jong, J. J., Hodiamont, P. P., & de Gelder, B. (2010). Modality-specific attention and multisensory integration of emotions in schizophrenia: Reduced regulatory effects. *Schizophrenia Research, 122*(1–3), 136–143.
de Jong, J. J., Hodiamont, P. P., Van den Stock, J., & de Gelder, B. (2009). Audiovisual emotion recognition in schizophrenia: Reduced integration of facial and vocal affect. *Schizophrenia Research, 107*(2–3), 286–293.
Dolan, R. J., Morris, J. S., & de Gelder, B. (2001). Crossmodal binding of fear in voice and face. *Proceedings of the National Academy of Sciences of the United States of America, 98*(17), 10006–10010.
Ekman, P., & Friesen, W. V. (1976). *Pictures of facial affects*. Palo Alto: Consulting Psychologists Press.
Esteves, F., & Öhman, A. (1993). Masking the face: Recognition of emotional facial expressions as a function of the parameters of backward masking. *Scandinavian Journal of Psychology, 34*(1), 1–18.
Ethofer, T., Anders, S., Erb, M., Droll, C., Royen, L., Saur, R., et al. (2006). Impact of voice on emotional judgment of faces: An event-related fMRI study. *Human Brain Mapping, 27*(9), 707–714.
Feinberg, T. E., Rifkin, A., Schaffer, C., & Walker, E. (1986). Facial discrimination and emotional recognition in schizophrenia and affective disorders. *Archives of General Psychiatry, 43*(3), 276–279.
Fernberger, S. W. (1928). False suggestion and the Piderit model. *The American Journal of Psychology, 40*, 562–568.
George, M. S., Parekh, P. I., Rosinsky, N., Ketter, T. A., Kimbrell, T. A., Heilman, K. M., et al. (1996). Understanding emotional prosody activates right hemisphere regions. *Archives of Neurology, 53*(7), 665–670.
Ghazanfar, A. A., Chandrasekaran, C., & Logothetis, N. K. (2008). Interactions between the superior temporal sulcus and auditory cortex mediate dynamic face/voice integration in rhesus monkeys. *Journal of Neuroscience, 28*(17), 4457–4469.
Ghazanfar, A. A., & Logothetis, N. K. (2003). Neuroperception: Facial expressions linked to monkey calls. *Nature, 423*(6943), 937–938.
Ghazanfar, A. A., Maier, J. X., Hoffman, K. L., & Logothetis, N. K. (2005). Multisensory integration of dynamic faces and voices in rhesus monkey auditory cortex. *Journal of Neuroscience, 25*(20), 5004–5012.
Ghazanfar, A. A., & Santos, L. R. (2004). Primate brains in the wild: The sensory bases for social interactions. *Nature Reviews. Neuroscience, 5*(8), 603–616.
Goydke, K. N., Altenmuller, E., Moller, J., & Munte, T. F. (2004). Changes in emotional tone and instrumental timbre are reflected by the mismatch negativity. *Cognitive Brain Research, 21*(3), 351–359.
Grandjean, D., Sander, D., Pourtois, G., Schwartz, S., Seghier, M. L., Scherer, K. R., et al. (2005). The voices of wrath: Brain responses to angry prosody in meaningless speech. *Nature Neuroscience, 8*(2), 145–146.

Gur, R. E., McGrath, C., Chan, R. M., Schroeder, L., Turner, T., Turetsky, B. I., et al. (2002). An fMRI study of facial emotion processing in patients with schizophrenia. *The American Journal of Psychiatry, 159*(12), 1992–1999.

Heimberg, C., Gur, R. E., Erwin, R. J., Shtasel, D. L., & Gur, R. C. (1992). Facial emotion discrimination: III. Behavioral findings in schizophrenia. *Psychiatry Research, 42*(3), 253–265.

Kee, K. S., Horan, W. P., Wynn, J. K., Mintz, J., & Green, M. F. (2006). An analysis of categorical perception of facial emotion in schizophrenia. *Schizophrenia Research, 87*(1–3), 228–237.

Kohler, C. G., Turner, T. H., Bilker, W. B., Brensinger, C. M., Siegel, S. J., Kanes, S. J., et al. (2003). Facial emotion recognition in schizophrenia: Intensity effects and error pattern. *The American Journal of Psychiatry, 160*(10), 1768–1774.

Kraepelin, E. (1919). *Dementia praecox and paraphrenia.* Chicago: Medical Book Co.

Kreifelts, B., Ethofer, T., Grodd, W., Erb, M., & Wildgruber, D. (2007). Audiovisual integration of emotional signals in voice and face: An event-related fMRI study. *NeuroImage, 37*(4), 1445–1456.

Kreifelts, B., Ethofer, T., Huberle, E., Grodd, W., & Wildgruber, D. (2010). Association of trait emotional intelligence and individual fMRI-activation patterns during the perception of social signals from voice and face. *Human Brain Mapping, 31*(7), 979–991.

Leitman, D. I., Loughead, J., Wolf, D. H., Ruparel, K., Kohler, C. G., Elliott, M. A., et al. (2008). Abnormal superior temporal connectivity during fear perception in schizophrenia. *Schizophrenia Bulletin, 34*(4), 673–678.

MacLeod, C. M., & MacDonald, P. A. (2000). Interdimensional interference in the Stroop effect: Uncovering the cognitive and neural anatomy of attention. *Trends in Cognitive Science, 4*(10), 383–391.

Magnée, M. J., de Gelder, B., van Engeland, H., & Kemner, C. (2007). Facial electromyographic responses to emotional information from faces and voices in individuals with pervasive developmental disorder. *Journal of Child Psychology and Psychiatry, 48*(11), 1122–1130.

Magnée, M. J., de Gelder, B., van Engeland, H., & Kemner, C. (2008). Atypical processing of fearful face–voice pairs in pervasive developmental disorder: An ERP Study. *Clinical Neurophysiology, 119*(9), 2004–2010.

Magnée, M. J. C. M., de Gelder, B., van Engeland, H., & Kemner, C. (2011). Multisensory integration in autism spectrum disorder: Critical influence of attention and noise. *PLoS One, 6*, e24196.

Mandal, M. K., Pandey, R., & Prasad, A. B. (1998). Facial expressions of emotions and schizophrenia: A review. *Schizophrenia Bulletin, 24*(3), 399–412.

Massaro, D. W., & Egan, P. B. (1996). Perceiving affect from the voice and the face. *Psychonomic Bulletin and Review, 3*, 215–221.

McGurk, H., & MacDonald, J. (1976). Hearing lips and seeing voices. *Nature, 264*(5588), 746–748.

Meeren, H. K., van Heijnsbergen, C. C., & de Gelder, B. (2005). Rapid perceptual integration of facial expression and emotional body language. *Proceedings of the National Academy of Sciences of the United States of America, 102*(45), 16518–16523.

Michalopoulou, P. G., Surguladze, S., Morley, L. A., Giampietro, V. P., Murray, R. M., & Shergill, S. S. (2008). Facial fear processing and psychotic symptoms in schizophrenia: Functional magnetic resonance imaging study. *The British Journal of Psychiatry, 192*(3), 191–196.

Panksepp, J. (1998). *Affective neuroscience: The foundation of human and animal emotions.* New York: Oxford University Press.

Panksepp, J. (2005). Psychology. Beyond a joke: From animal laughter to human joy? *Science, 308*(5718), 62–63.

Parr, L. A. (2004). Perceptual biases for multimodal cues in chimpanzee (*Pan troglodytes*) affect recognition. *Animal Cognition, 7*(3), 171–178.

Phillips, M. L., Williams, L., Senior, C., Bullmore, E. T., Brammer, M. J., Andrew, C., et al. (1999). A differential neural response to threatening and non-threatening negative facial expressions in paranoid and non-paranoid schizophrenics. *Psychiatry Research, 92*(1), 11–31.

Pinkham, A. E., Hopfinger, J. B., Ruparel, K., & Penn, D. L. (2008). An investigation of the relationship between activation of a social cognitive neural network and social functioning. *Schizophrenia Bulletin, 34*(4), 688–697.

Pourtois, G., de Gelder, B., Bol, A., & Crommelinck, M. (2005). Perception of facial expressions and voices and of their combination in the human brain. *Cortex, 41*(1), 49–59.

Pourtois, G., de Gelder, B., Vroomen, J., Rossion, B., & Crommelinck, M. (2000). The time-course of intermodal binding between seeing and hearing affective information. *NeuroReport, 11*(6), 1329–1333.

Pourtois, G., Debatisse, D., Despland, P. A., & de Gelder, B. (2002). Facial expressions modulate the time course of long latency auditory brain potentials. *Cognitive Brain Research, 14*(1), 99–105.

Remedios, R., Logothetis, N. K., & Kayser, C. (2009). Monkey drumming reveals common networks for perceiving vocal and nonvocal communication sounds. *Proceedings of the National Academy of Sciences of the United States of America, 106*(42), 18010–18015.

Ross, E. D. (2000). Affective prosody and the aprosodias. In M. M. Mesulam (Ed.), *Principles of behavioral and cognitive neurology* (2nd ed.). London: Oxford University Press.

Ross, L. A., Saint-Amour, D., Leavitt, V. M., Molholm, S., Javitt, D. C., & Foxe, J. J. (2007). Impaired multisensory processing in schizophrenia: Deficits in the visual enhancement of speech comprehension under noisy environmental conditions. *Schizophrenia Research, 97*(1–3), 173–183.

Scherer, K. R., Banse, R., Wallbott, H. G., & Goldbeck, T. (1991). Vocal cues in emotion encoding and decoding. *Motivation and Emotion, 15*(2), 123–148.

Stein, B. E., & Stanford, T. R. (2008). Multisensory integration: Current issues from the perspective of the single neuron. *Nature Reviews. Neuroscience, 9*(4), 255–266.

Stienen, B. M. C., & de Gelder, B. (2011). Fear detection and visual awareness in perceiving bodily expressions. *Emotion, 11*(5), 1182–1189.

Stienen, B.M.C., Tanaka, A., de Gelder, B. (2011). Emotional voice and emotional body postures influence each other independently of visual awareness. PLoS ONE, *6*(10): e25517.

Sugihara, T., Diltz, M. D., Averbeck, B. B., & Romanski, L. M. (2006). Integration of auditory and visual communication information in the primate ventrolateral prefrontal cortex. *Journal of Neuroscience, 26*(43), 11138–11147.

Tamietto, M., & de Gelder, B. (2010). Neural bases of the non-conscious perception of emotional signals. *Nature Reviews. Neuroscience, 11*(10), 697–709.

Van den Stock, J., de Jong, J. J., Hodiamont, P. P. G., & de Gelder, B. (2011). Perceiving emotions from bodily expressions and multisensory integration of emotion cues in schizophrenia. *Social Neuroscience, 6*(5–6), 537–547.

Van den Stock, J., Grèzes, J., & de Gelder, B. (2008). Human and animal sounds influence recognition of body language. *Brain Research, 1242*, 185–190.

Van den Stock, J., Peretz, I., Grèzes, J., & de Gelder, B. (2009). Instrumental music influences recognition of emotional body language. *Brain Topography, 21*(3–4), 216–220.

Van den Stock, J., Righart, R., & de Gelder, B. (2007). Body expressions influence recognition of emotions in the face and voice. *Emotion, 7*(3), 487–494.

Vroomen, J., Driver, J., & de Gelder, B. (2001). Is cross-modal integration of emotional expressions independent of attentional resources? *Cognitive, Affective, & Behavioral Neuroscience, 1*(4), 382–387.

Wolwer, W., Streit, M., Polzer, U., & Gaebel, W. (1996). Facial affect recognition in the course of schizophrenia. *European Archives of Psychiatry and Clinical Neuroscience, 246*(3), 165–170.

Part IV
Impairment

Chapter 14
Crossmodal Integration of Emotional Stimuli in Alcohol Dependence

Pierre Maurage, Scott Love, and Fabien D'Hondt

Abstract Face–voice integration has been extensively explored among healthy participants during the last decades. Nevertheless, while binding alterations constitute a core feature of many psychiatric diseases, these crossmodal processing have been very little explored in these populations. This chapter presents three studies offering an integrative use of behavioural, electrophysiological and neuroimaging techniques to explore the audio–visual integration of emotional stimuli in alcohol dependence. These results constitute a preliminary step towards a multidisciplinary exploration of crossmodal processing in psychiatry, extending to other stimulations, sensorial modalities and populations. The exploration of impaired crossmodal abilities could renew the knowledge on "normal" audio–visual integration and could lead to innovative therapeutic programs.

P. Maurage (✉)
Neuroscience, Systems and Cognition (NEUROCS),
and Health and Psychological Development (CSDP) Research Units,
Institute of Psychology, Catholic University of Louvain,
Place du Cardinal Mercier, 10, B-1348 Louvain-la-Neuve, Belgium
e-mail: pierre.maurage@uclouvain.be

S. Love
Department of Psychological and Brain Sciences, Indiana University,
Bloomington, Indiana, USA
e-mail: sclove@indiana.edu

F. D'Hondt
Lille Nord de France University, Lille, France

Functional Neurosciences and Pathology Lab, UDSL, Lille, France
e-mail: fabien.dhondt@gmail.com

P. Belin et al. (eds.), *Integrating Face and Voice in Person Perception*,
DOI 10.1007/978-1-4614-3585-3_14, © Springer Science+Business Media New York 2013

1 Introduction

Attempting to obtain a comprehensive view of the audio–visual integration research field from knowledge currently available irremediably leads to a striking paradox. On the one hand, hundreds of studies have been conducted during the last two decades on crossmodal processing among healthy participants, and huge advances have undeniably been made in understanding the developmental, psychological and cerebral correlates of crossmodality (particularly of face–voice integration). The present book constitutes an up-to-date illustration of the central position that crossmodality has recently gained within the experimental psychology and neuroscience domains. It clearly shows that this blooming research domain has come to maturity, as several modelizations have been proposed to integrate the plethora of existing experimental data (e.g. Campanella & Belin, 2007). On the other hand, while the exploration of a research area is traditionally enriched by results obtained from clinical populations, very few clinical crossmodal research projects have been conducted. Face–voice integration impairments have been investigated in populations presenting perceptual impairments [e.g., visual or auditory loss (Barone, 2010; Massida et al., 2011; Zupan & Sussman, 2009)], but the exploration of crossmodal processing in neurological and psychiatric populations is still in its infancy, to say the least.

This lack of interest for impaired integration processes appears surprising, as the presence of difficulties to integrate signals coming from different sensory modalities have been suggested in a large range of pathological states. For example, schizophrenia and autism have repeatedly been described as disconnection syndromes leading to binding problems (see for example Friston & Frith, 1995 or Melillo & Leisman, 2009 for reviews). Such considerations have led to an emerging multisensory strand in the field of autism research (e.g. Kwakye, Foss-Feig, Cascio, Stone, & Wallace, 2011). Moreover, voices recently rose to promote the development of crossmodal research among clinical populations, with a double aim. First, it could lead to a better description of the impairments associated with pathological states, particularly by offering a more ecological and exhaustive evaluation of the deficits [see Campanella et al. (2010) for a thorough discussion of the usefulness of crossmodal paradigms in clinical settings]. Second, it could renew our understanding of crossmodal integration among healthy subjects: if a clinical population presents behavioural deficits in crossmodal processing, comparing the cerebral activations between this population and healthy controls will give strong insights concerning the brain regions associated with crossmodal processing. As summarized by Laurienti, Perrault, Stanford, Wallace and Stein (2005, p. 295), "the use of clinical populations can add to the battery of study designs available to the imaging scientist investigating multisensory integration". Despite the great promise of this perspective, very little research has attempted to improve our understanding of both pathological states and the mechanisms of multisensory integration in general through crossmodal research in clinical populations.

The main aim of the present chapter is thus to underline the usefulness of exploring crossmodal processing among clinical populations and to prepare the ground for the expansion of this innovative research topic. We first propose an illustration of the possibilities offered by this research field by describing our studies exploring emotional crossmodal processing in alcohol dependence. Indeed, emotional decoding impairments have been found to play a crucial role in the development and maintenance of alcohol dependence, and have been extensively investigated using visual or auditory stimulations. On the basis of these unimodal explorations, we recently conducted several studies using audio–visual bimodal paradigms in order to increase the ecological validity of the experimental designs. The complementary use of behavioural, electrophysiological and neuroimaging techniques allowed us to obtain the first insights concerning multimodal integration in alcohol dependence. In a second part, we then discuss these initial results and show how they can (together with preliminary results obtained for multimodal integration in other psychiatric states) be extended in order to develop a coherent and ambitious research program utilizing various psychiatric populations and sensory modalities. We end the chapter by underlining the various potential clinical applications and the fundamental implications that this emerging project could bring.

2 Emotions and Alcohol Dependence

Alcohol dependence is the most widespread psychiatric diagnosis and is among the more detrimental health problems in the world (Harper & Matsumoto, 2005). It affects 5–10 % of the adult population in Western countries and is directly responsible for 200,000 deaths per year in the European Union. In view of the omnipresence of this pathological state, considerable effort has been made during the last few decades to gain a better understanding of alcohol dependence's characteristics at clinical as well as theoretical levels, particularly concerning the physiological, behavioural and cerebral impairments associated with chronic excessive alcohol consumption. Alcohol dependence is known to have deleterious effects on many body systems (e.g. hepatic, cardiovascular or gastrointestinal) including the central nervous system. Indeed, it has been extensively established that alcohol dependence leads to major cerebral damage (see for example Harper, 2007; McIntosh & Chick, 2004 for reviews), particularly affecting white matter (Brooks, 2000; Oscar-Berman & Marinkovic, 2003) and also sub-cortical [e.g. amygdala (Cowen, Chen, & Lawrence, 2004; Fein et al., 2006), insula, thalamus and cerebellum (De Bellis et al., 2005; Szabo et al., 2004)] and cortical [mainly temporal and frontal lobes (Chanraud et al., 2007; Harper & Matsumoto, 2005; Kril, Halliday, Svoboda, & Cartwright, 1997)] areas. Many studies have explored the behavioural correlates of these cerebral effects and have repeatedly shown impaired performance in a large range of (neuro)psychological abilities, particularly concerning memory and executive functions (e.g. Bechara et al., 2001; Flannery et al., 2007; Oscar-Berman, Kirkley, Gansler, & Couture, 2004; Pitel et al., 2007), and also perceptual (e.g. Blusewicz, Dustman, Schenkenberg, & Beck, 1977; Kramer, Blusewicz, Robertson, & Preston, 1989; Spitzer, 1981 ; Spitzer & Ventry, 1980) and attentional (e.g. Noël et al., 2001; Smith & Oscar-Berman, 1992; Sullivan et al., 1993) abilities. In contrast with this extensive exploration of cognitive consequences, the evaluation of emotional abilities has long been neglected.

While affective states are known to have a significant influence on every aspect of our lives (e.g. memories, behaviours, choices, motivations or social interactions) and while emotional disturbances clearly appear as a central characteristic of mental diseases from a clinical point of view, the interest for experimental exploration of emotional impairments in alcohol dependence rose only during the last decade yet have led to clear results. Alcohol dependence is associated with major emotional impairments, as shown by several recent research projects which identified a reduced performance in various emotional functions among alcohol-dependent individuals, notably for alexithymia (Taieb et al., 2002; Uzun, Ats, Cansever, & Ozsahin, 2003), emotional intelligence (Cordovil de Susa Uva et al., 2010; Riley & Schutte, 2003; Szczepanska, Baran, & Mikolaszek-Boba, 2004) and empathy (Martinotti, Di Nicola, Tedeschi, Cundari, & Janiri, 2009; Maurage et al., 2011a). More centrally for the present purpose, a deficit has also been consistently observed for the decoding of the emotions expressed by faces (Clark, Oscar-Berman, Shagrin, & Pencina, 2007; Frigerio, Burt, Montagne, Murray, & Perrett, 2002; Marinkovic et al., 2009, Maurage et al., 2009a, 2011b; Oscar-Berman, Hancock, Mildworf, Hutner, & Weber, 1990) and voices (Monnot, Lovallo, Nixon, & Ross, 2002; Monnot, Nixon, Lovallo, & Ross, 2001; Uekermann, Daum, Schlebusch, & Trenckmann, 2005). Recently detoxified alcohol-dependent individuals globally overestimate the intensity of the emotions conveyed by visual and auditory stimuli, have an erroneous interpretation of emotions and are not aware of their deficit (Kornreich et al., 2001, 2002; Philippot et al., 1999). While several contradictory results have emerged, describing a preserved decoding of visual (Uekermann & Daum, 2008) or auditory (Oscar-Berman et al., 1990) stimulations, this emotional decoding deficit is now strongly established as it has been replicated in a wide variety of paradigms and stimulus sets (e.g. morphed or ambiguous faces), and among individuals with various abstinence durations (Foisy et al., 2007a; Montagne, Kessels, Wester, & de Haan, 2006; Townshend & Duka, 2003). Moreover, this impairment appears particularly present for negative emotions and is specific for emotional features because it is not observed for non-emotional complex face processing, such as gender or race identification (Foisy et al., 2007b; Maurage, Campanella, Philippot, Martin, & de Timary, 2008a).

As the development and maintenance of adapted social communication is largely based on the ability to correctly express one's own emotional states and to accurately perceive (and react to) those expressed by other individuals (Feldman, Philippot, & Custrini, 1991), these emotional deficits give rise to impaired interpersonal interactions and increase the social problems frequently observed in alcohol dependence (e.g. Maurage et al., 2009a; Uekermann, Channon, Winkel, Schlebusch, & Daum, 2007). An understanding of emotional disabilities is thus essential in clinical practice, as they play a critical role in the emergence and maintenance of alcohol dependence, notably by hampering the development of satisfactory interpersonal links, thus potentially reinforcing the excessive alcohol consumption (used as a coping strategy to face social isolation) and leading to a vicious circle (e.g. Carton,

Kessler, & Pape, 1999). To sum up, the emotion decoding impairment in alcohol dependence is now clearly identified and has a high clinical importance. However, this deficit has up to now been exclusively explored using paradigms with low ecological validity, namely using only unimodal stimuli (faces or voices). It is thus unclear whether this deficit is maintained, reduced or increased in experimental designs that are closer to real life, specifically when crossmodal stimuli are used.

3 Emotional Crossmodality in Alcohol Dependence

3.1 Rationale and Aims

As outlined above, while explored only recently, the emotional impairments associated with alcohol dependence are now strongly established, particularly concerning the deficit in the decoding of emotional faces and voices presented separately. In everyday life, as it has been repeatedly underlined in the present book, sensory events are not experienced in isolation: we are constantly immersed in a flow of multiple sensory cues carrying information from different sensory modalities. Crossmodal processing, which can globally be defined as the integration of sensory cues emanating from distinct modalities into a unified and coherent representation of the environment, is thus the rule rather than the exception and is crucial for adaptative behaviours (Driver & Spence, 2000). Crossmodal interactions are ubiquitous; the perception and production of emotional states are routinely based on several sensory aspects (e.g. emotional facial expressions and emotional prosody in crossmodal face–voice stimuli). Therefore, while constituting a valuable first exploration, the unimodal investigations of affective processing among alcohol-dependent individuals conducted up to now are insufficient to comprehend the complexity of emotion processing in this population and should be extended to more ecological crossmodal designs.

With this in mind, we now present three studies performed in our research group, which explored, for the first time, the crossmodal emotional processing of individuals with alcohol dependence. These studies combine behavioural, electrophysiological and neuroimaging techniques to determine the modification of audio–visual emotional decoding in alcohol dependence.

It should be noted that these three studies are focused on the comparison between recently detoxified alcohol-dependent participants (i.e., individuals diagnosed with alcohol dependence according to DSM-IV criteria and recruited during their third week of treatment in a detoxification centre) and healthy controls paired for age, gender and education. Moreover, alcohol-dependent participants had abstained from alcohol for at least 2 weeks before the experiment took place (in order to exclude any influence of acute alcohol intoxication) and did not present any comorbidity with other psychiatric diagnoses (thus ensuring that the emotional decoding deficits were indeed associated with alcohol consumption and not with biasing variables) nor any perceptual visual or auditory impairment.

3.2 *Behavioural Study (Maurage, Campanella, Philippot, Pham, & Joassin, 2007a)*

This first exploration of crossmodal processing of emotional stimuli in alcohol dependence was based on the elicitation of a "crossmodal facilitation effect". However, many studies exploring audio–visual integration use paradigms leading to inhibition effects (i.e., to a deteriorated performance in crossmodal conditions as compared to unimodal). Two famous examples of these crossmodal inhibition paradigms are the McGurk (McGurk & McDonald, 1976) and ventriloquist effects (e.g. Alais & Burr, 2004), in which visual stimulation alters auditory perception. But more recently, several studies have developed paradigms leading to a facilitation effect (Calvert, Hansen, Iversen, & Brammer, 2001; Frens, Van Opstal, & Van der Willigen, 1995; Latinus, VanRullen, & Taylor, 2010; Teder-Sälejärvi, McDonald, Di Russo, & Hillyard, 2002), in which congruent bimodal (audio–visual) stimulation leads to better performance (i.e., higher correct response rates and/or shorter reaction times) than unimodal. The facilitation effect has been considered as the behavioural marker of successful crossmodal integration of stimuli from different modalities (Calvert et al., 2001). Conversely, the absence of a facilitation effect in a clinical population that is observed in a paired control group would index impaired crossmodal integration in this population.

This study was thus based on a design eliciting a facilitation effect, in order to evaluate the presence of this effect among alcohol-dependent individuals. More precisely, we used an emotion–detection task in which participants were presented with emotional facial expressions and voices [i.e., audiotapes consisting in the enunciation of a semantically neutral name with an emotional prosody, taken from a validated battery (Maurage, Joassin, Philippot, & Campanella, 2007b)] depicting anger or happiness. Auditory and visual stimuli were presented separately (unimodal conditions) or simultaneously (crossmodal condition, in which faces and voices were always depicting the same emotion). As faces are classically processed more rapidly than voices (Ellis, Jones, & Mosdell, 1997; Joassin, Maurage, Bruyer, Crommelinck, & Campanella, 2004; Schweinberger, Herholz, & Sommer, 1997), we decided to manipulate the visual stimuli in order to obtain similar levels of difficulty for both vision and audition, which is necessary to enable a facilitation effect. We thus increased the perceptual difficulty of faces by means of a morphing technique in order to make them as difficult to recognize as voices (Hanley, Smith, & Hadfield, 1998; Hanley & Turner, 2000). The faces used were morphed at 40–60 % level (i.e., depicting 40 % of happiness and 60 % of anger, or conversely). Participants (20 alcohol-dependent inpatients and 20 paired controls) had to decide as quickly as possible which emotion was displayed in the stimulus (anger or happiness).

The main result of this study was that, while control participants showed a clear facilitation effect (i.e., the audio–visual condition led to significantly shorter reaction times than the unimodal auditory and visual), alcohol-dependent individuals did not present this effect, as no differences were observed in the alcohol-dependent group

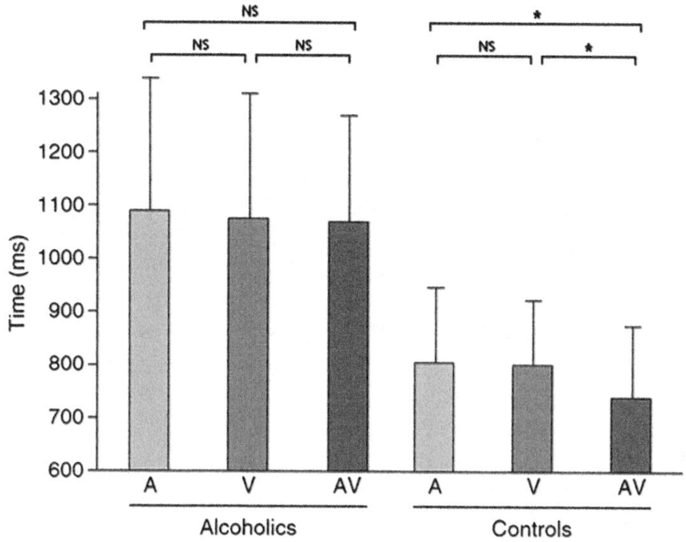

Fig. 14.1 Reaction times for alcohol-dependent (*left*) and control (*right*) participants in the emotion-detection task for visual (V), auditory (A) and audio–visual (AV) stimulations. While control participants exhibited a crossmodal facilitation effect (i.e., reduced reaction times in AV as compared to A and V), alcohol-dependent individuals did not present this effect (*NS*, non significant; $*p<0.05$). Note: Adapted from Maurage, P., Campanella, S., Philippot, P., Pham, T., & Joassin, F. (2007a). The crossmodal facilitation effect is disrupted in alcoholism: A study with emotional stimuli. *Alcohol and Alcoholism*, 42(6), p. 557. Copyright 2007 by the Medical Council on Alcohol

according to the experimental condition. In other words, alcohol dependence is associated with an absence of a crossmodal facilitation effect. As the facilitation effect is the behavioural marker of efficient crossmodal processing, these results show that alcohol dependence is associated with impaired auditory–visual integration of complex ecological stimuli. The results also showed that alcohol-dependent participants were globally slower than controls, whatever the experimental condition, which is a classical visuo-motor slowing effect associated with alcohol dependence (e.g. Fein, Bachman, Fischer, & Davenport, 1990). These results, illustrated in Fig. 14.1, constitute the first evidence of a crossmodal impairment in alcohol dependence. On their basis, two complementary studies were performed to explore the cerebral correlates of this audio–visual integration deficit.

3.3 Electrophysiological Study (Maurage et al., 2008b)

The initial study identified the existence of a specific deficit for crossmodal processing in alcohol dependence at a behavioural level but did not allow exploring the cerebral correlates of the deficit. This second study thus aimed at describing the brain alterations leading to audio–visual integration impairment, by means of event-related potentials (ERP). ERP record the brain's electrical activity during cognitive tasks with a high temporal resolution. This technique is particularly useful in attempting to identify the electrophysiological component associated with the onset of a dysfunction, and then to infer the cognitive stage related to this impairment (Campanella & Philippot, 2006). ERP have been used for decades to study alcohol-dependent participants. Most studies focused on the P3b, a long-lasting positive deflection appearing at parietal sites between 300 and 800 ms after stimulus onset, and functionally associated with the decisional stage, namely the closure of cognitive processing before starting the motor response (Polich, 2004). Alcohol dependence is associated with reduced amplitude and delayed latency of P3b (see Hansenne, 2006 for a review). However, other studies have described a deficit in earlier visual ERP components, like P100 (Ogura & Miyazato, 1991), N170 or N200 (Kathmann, Soyka, Bickel, & Engel, 1996). These deficits for P100 and more importantly for N170 (respectively linked to early visual processing and specific processing of faces) suggest that the impairment in alcohol dependence could begin before the decisional level (P3b), namely at the visuo-spatial level of cognitive processing (Maurage et al., 2007c). Therefore, ERP clearly help to identify the precise stage (e.g. perceptual, attentional or decisional) at which a deficit occurs, and hence they were used here to determine the initial cognitive stage responsible for the cross-modal integration impairment in alcohol dependence: Does the crossmodal deficit start at an early, perceptive stage or only at later processing steps? The first study described above did not differentiate the integration deficit according to the emotion depicted in the stimuli. This second study also explored the potential differential deficit observed for positive (i.e., happiness) versus negative (i.e., anger) emotions (Maurage et al., 2007c).

An emotion-detection task was performed by 15 alcohol-dependent participants and 15 paired controls, with visual (i.e., emotional facial expression) and auditory [i.e., audiotapes consisting in the enunciation of a semantically neutral name with an emotional prosody, taken from a validated battery (Maurage et al., 2007b)] stimuli presented separately or simultaneously for 700 ms. Participants had to decide as fast as possible whether the face, voice or face–voice stimulus was an angry, happy or neutral emotional expression. ERP were recorded using a 32-electrode cap (see Maurage et al., 2008b for technical details) in order to obtain, for each participant, several electrophysiological components of interest (P100, N170–N2, P3b) for each experimental condition (visual, auditory or audio–visual) and each emotion (anger, happiness, neutral).

The results confirmed the ERP deficits classically observed in alcohol dependence (e.g. Hansenne, 2006). First, alcohol-dependent individuals were slower and less accurate than control participants to identify the emotion presented in face or voice, which is in line with the repeated observation of a deficit in the emotion decoding in alcohol dependence (e.g. Maurage et al., 2009a; Philippot et al., 1999; Townshend & Duka, 2003). Second, alcohol dependence was associated with reduced amplitude and delayed latency of the N170/N2 and P3b components for visual and

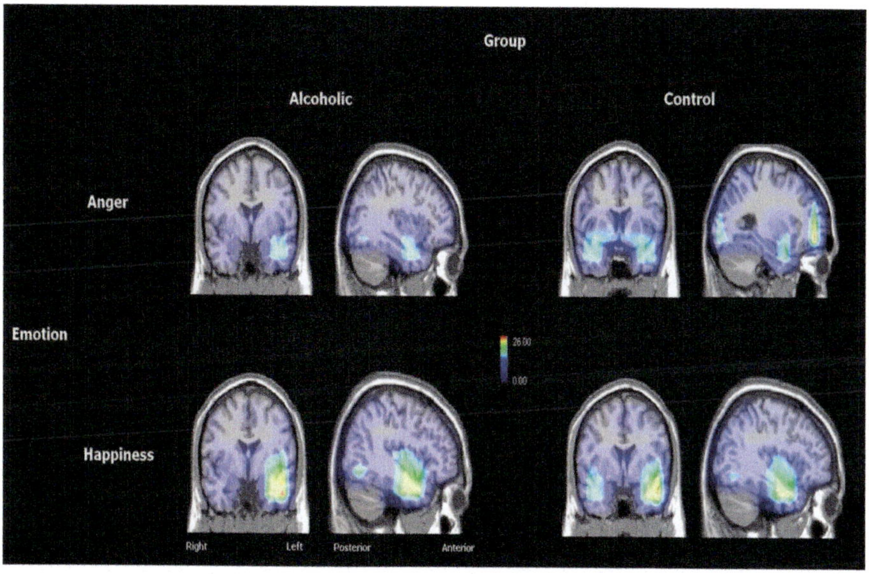

Fig. 14.2 Source reconstruction analysis of the cerebral generators among alcohol-dependent (*left*) and control (*right*) participants, for anger (*above*) and happiness (*below*) electrophysiological subtraction [AV – (A + V)] waves. Alcohol-dependent individuals exhibited a highly reduced frontal activation as compared to controls for anger stimulations. Note: Adapted from Maurage, P., Philippot, P., Joassin, F., Pauwels, L., Alonso Prieto, E., et al. (2008b). The auditory–visual integration of anger is impaired in alcoholism: An ERP Study. *Journal of Psychiatry and Neuroscience, 33*(2), p. 119. Copyright 2008 by the Canadian Medical Association

auditory stimulations, thus confirming the ERP alterations repeatedly described in this pathological state (e.g. Hansenne, 2006). But the main result of this study concerned the group differences for the cerebral activations specifically associated with crossmodal processing. Indeed, we used a subtraction technique in order to isolate the electrophysiological activities directly related to the visuo-auditory integration, as the auditory (A) and visual (V) unimodal conditions were subtracted from the auditory–visual bimodal condition (AV) using the following formula: AV – [A + V]. This method is classically used to investigate the electrophysiological correlates of crossmodal processes (e.g. Joassin et al., 2004; Teder-Sälejärvi et al., 2002). Group comparisons on these specific crossmodal activities showed that alcohol dependence leads to highly reduced brain activity during integrative processes. Moreover, this deficit is particularly present for anger stimuli, with a strong impairment starting as early as 100 ms after stimulus appearance (while the deficits for happiness and neutral stimuli appeared only after 200–300 ms and were far less marked). Finally, a source location analysis (using swLORETA method) showed that this impairment in the crossmodal processing of anger is indexed by a reduction in frontal activity, as illustrated in Fig. 14.2. These data thus complement the results obtained in the first study by showing (1) that early crossmodal processing of emotional stimulation is impaired in alcohol dependence, particularly for anger, and (2) that this deficit is associated with a reduction of the electrophysiological activations specifically linked with integrative processes, particularly in frontal areas. Nevertheless, due to their low spatial resolution, ERP are not able to precisely localize the brain areas involved in this integration deficit. Therefore, these results had to be confirmed and complemented by the use of neuroanatomical techniques, which was the central objective of the third study.

3.4 *Neuroimaging Study (Maurage et al., 2012a)*

This third study was aiming to precisely locate the cerebral regions responsible for impaired crossmodal processes in alcohol dependence, by means of functional magnetic resonance imaging (fMRI). One the one hand, alcohol dependence is known to be associated with major cerebral consequences, particularly in white matter, limbic, temporal and frontal areas. On the other hand, the emotional impairments presented by alcohol-dependent individuals are also well documented, particularly for the decoding of visual or auditory stimulation. Nevertheless, these cerebral and emotional alterations have traditionally been explored separately, and very little is known about the cerebral correlates of emotional impairments in alcohol dependence. To our knowledge, only a few studies have specifically focused on this topic, comparing the brain activations of recently detoxified alcohol-dependent participants with that of controls during the presentation of emotional scenes (Heinz et al., 2007) or emotional facial expressions (Marinkovic et al., 2009; Salloum et al., 2007). These results show that alcohol dependence is associated with a global reduction of brain activity during the processing of emotional stimuli, particularly for negative emotions, and that the most important activity reduction is observed in frontal regions, anterior cingular cortex and limbic structures (particularly the amygdala and hippocampus). A more recent study (Schulte, Müller-Oehring, Pfefferbaum, & Sullivan, 2010) also suggested that alcohol dependence is associated with white matter abnormalities, thus leading to disconnections between brain areas, and mainly between cortical and limbic structures. As the cortico–limbic connections are central for the processing and interpretation of emotional signals, this white matter deficit could play a major role in the affective disorders observed in alcohol dependence. Nevertheless, these studies were exclusively based on the presentation of visual emotional stimuli, and the brain correlates of auditory or audio–visual emotional processing remain unexplored.

On the basis of our two studies presented above and of the preliminary results suggesting that brain areas dedicated to emotional processing appear to be impaired or disconnected in alcohol dependence, we conducted an fMRI study exploring the brain correlates of crossmodal emotional processing among alcohol-dependent participants. More precisely, an emotion-detection task was administered to 12 alcohol-dependent participants and 12 paired controls while their brain activity was recorded with fMRI. The stimuli and task were identical to those presented in the first study,

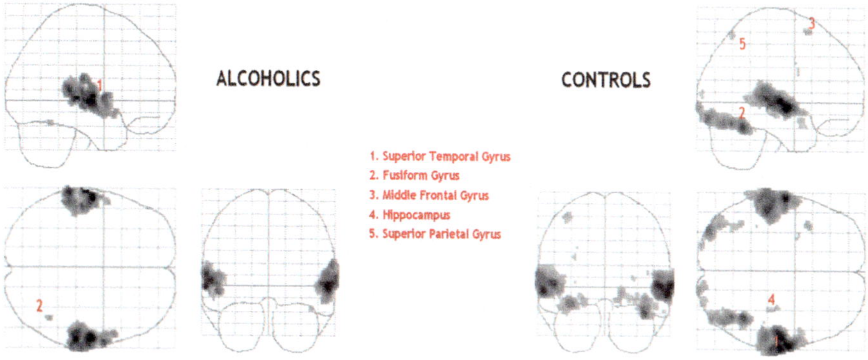

Fig. 14.3 Significant cerebral activations for AV – (A + V) subtraction, isolating the specific crossmodal activities, for alcohol-dependent (*left*) and control (*right*) participants. While controls presented a classical pattern of integrative activations in unimodal (superior temporal gyrus, fusiform gyrus) and crossmodal (middle frontal gyrus, hippocampus, superior parietal gyrus) regions, alcohol-dependent individuals did not show any significant activation in the areas specifically dedicated to crossmodal processing, but only in unimodal regions. These results are the first description of the cerebral correlates of crossmodal processing impairment in alcohol dependence

with a binary emotional decision (anger–happiness) on unimodal (morphed face or voice) or crossmodal (morphed face and voice presented simultaneously) stimulation. Brain activations during unimodal and crossmodal conditions were first computed (by subtracting from these activations those observed during a rest period without stimulation) and then the classical AV – [A + V] comparison (i.e. the subtraction between the activations observed in crossmodal condition AV and the sum of the unimodal activations A + V) was performed to isolate regions specifically involved in the integration of emotional faces and voices in both groups.

First of all, we reinforced earlier fMRI studies on visual emotional processing in alcohol dependence (Marinkovic et al., 2009; Salloum et al., 2007) by showing reduced activations in the brain areas classically associated with visual (i.e., inferior occipital gyrus, fusiform gyrus) and emotional processing (i.e., amygdala, hippocampus) among alcohol-dependent participants during emotional facial expression decoding. Moreover, we extended these previous results to emotional auditory processing, as emotional voices led, in the alcohol-dependent group, to reduced activations in the same emotional areas as visual stimuli, and also in specific auditory processing regions (i.e., superior and middle temporal gyri). These results confirm that alcohol dependence is associated with reduced brain activations during the unimodal processing of emotional stimuli. However, the central result of this study concerned crossmodal activations. As illustrated in Fig. 14.3, the subtracted activations revealed specific brain areas related to the integration of audio–visual stimulation. In the control group, this subtraction distinguished two categories of activations: on the one hand, several activations were found in unimodal regions (i.e., superior temporal gyrus for voices and fusiform gyrus for faces), showing that crossmodal stimulations provoke an enhanced activation in cerebral regions specialized in visual or auditory processing, which has been repeatedly observed among healthy participants (e.g. Calvert et al., 1999; Ghazanfar, Maier, Hoffman, & Logothetis, 2005). On the other hand, and more importantly, specific multimodal regions were revealed by the subtraction, namely middle frontal gyrus, hippocampus and superior parietal gyrus. This is in line with earlier studies (e.g. Joassin et al., 2011b; Joassin, Maurage, & Campanella, 2011a) showing that these brain regions are specifically activated in crossmodal conditions, as they receive multiple inputs from modality-specific regions and integrate them into a unitary and coherent representation of the environment (e.g. Bernstein, Auer, Wagner, & Ponton, 2008; Rämä & Courtney, 2005; see Joassin et al., 2011b for a thorough discussion on the role of these integrative regions). In the alcohol-dependent group, however, the only significant activations for crossmodal stimulations were found in the unimodal regions cited above (mainly in the auditory regions). It thus appears that alcohol dependence is associated with a large and specific crossmodal deficit, indexed here by a lack of activation in the regions normally dedicated to the integration of inputs coming from different sensory modalities: alcohol dependence is therefore associated with serious dysfunctions of the activation and connectivity between the cerebral regions involved in the multimodal perception of the social environment. These data are preliminary and will have to be confirmed in future studies using larger groups and alternative experimental designs (Goebel & van Atteveldt, 2009; Love, Pollick, & Latinus, 2011), but they reinforce our earlier behavioural and electrophysiological results showing an emotional crossmodal processing deficit in alcohol dependence, and offer the first description of the specific cerebral correlates of this impairment.

4 Clinical and Theoretical Implications for Alcohol Dependence

The three studies presented above, as they are the first to explore emotional crossmodal processing in alcohol dependence, have to be considered as preliminary and should be confirmed and extended in future work. Nevertheless, they present a coherent pattern of results, as all described similar specific audio–visual integration impairments in alcohol dependence. Moreover, the use of different experimental methods and techniques provided complementary views of this impairment from behavioural, electrophysiological and neuroimaging data. Several fundamental and clinical implications can thus already be outlined on the basis of these results, in order to lay the foundations for potential therapeutic interventions and future experimental investigation of these integrative processes.

At the experimental level, the observation that the emotional decoding deficit among alcohol-dependent individuals, widely described for unimodal stimulation (i.e. emotional facial expressions or emotional voices), is increased in crossmodal stimulation, sheds new light on these earlier results and could influence future studies in the domain. Indeed, as crossmodal situations are omnipresent, our results suggest that earlier studies based on unimodal stimulation (and often using basic stimuli) have underestimated the deficits in alcohol dependence. This crossmodal impairment could also explain the hiatus between the relatively mild deficit frequently observed when presenting unimodal stimuli in experimental situations among alcohol-dependent subjects (e.g. Beatty, Tivis, Stott, Nixon, & Parsons, 2000; Oscar-Berman et al., 1990; Uekermann et al., 2005) and the obvious impairments observed in ecological situations, and notably in clinical observations. The present results should thus lead to a re-evaluation of earlier studies using unimodal stimuli, which probably underestimated the deficit present in real-life situations. These results should also encourage future studies to use crossmodal stimulation in order to correctly evaluate various cognitive and emotional deficits in the processing of social stimuli. More generally, as emotional contexts in everyday life are most often characterized by the simultaneous perception of stimulations from different sensory modalities, our results argue for the development of more ecologically valid experimental designs, for example by means of video clips or virtual reality paradigms. Much progress has already been made in this direction for the evaluation of crossmodal processing among healthy participants (e.g., Barkhuysen, Krahmer, & Swerts, 2010; Petrini, Crabbe, Sheridan, & Pollick, 2011; Petrini, McAleer, & Pollick, 2010; Vatakis & Spence, 2006) but it has not been applied to clinical populations yet.

From a more theoretical point of view, the development of experimental work on crossmodal processing in alcohol dependence, and also in other psychiatric or neurological states, could complement the results obtained among healthy participants and help to further renew our knowledge of crossmodal integration in general. Indeed, the exploration of impaired cognitive functions among clinical populations has traditionally been used, in neurology and neuropsychology, in order to add to those observations made among healthy individuals and to give a more exhaustive description of normal functioning. For instance, studies conducted among patients with cerebral lesions provided the description of double dissociations in memory or attentional systems, thus refining the theoretical models proposed for these systems (see for example Barbeau, Pariente, Felician, & Puel, 2010; Cohn, Moscovitch, & Davidson, 2010; Listerud, Powers, Moore, Libon, & Grossman, 2009 for recent illustrations of the "double dissociation" principle). Specifically for the present topic, the results from our fMRI study, showing that the crossmodal integration impairment in alcohol dependence is associated with reduced activity in middle frontal gyrus, hippocampus and superior parietal gyrus, confirm that these regions are necessary for a correct integration between faces and voices and thus reinforce the results obtained among healthy participants. This is of course just a first step, but it underlines the fact that exploring impaired crossmodal processing can offer promising perspectives, notably a better understanding of normal integration functioning.

At the clinical level, the present results clearly confirm earlier data suggesting that emotional impairments are a critical deficit in alcohol dependence. The crossmodal paradigms used highlight that the impairments are more intense when experimental designs are closer to real-life emotional situations. It is therefore obvious that emotional perturbations are at the heart of alcohol dependence and should be considered in clinical settings. Surprisingly, the experimental studies and theoretical models proposing therapeutic programs for alcohol-dependent individuals have up to now mainly focused on cognitive and behavioural aspects (e.g. development of coping strategies and motivation to change; see DiClemente, Jordan, Marinilli, & Nidecker, 2003 for a review of the literature), and the emotional variables have been neglected. The present data, together with earlier results describing emotional alterations among alcohol-dependent individuals, should encourage the development of therapeutic approaches focused on the rehabilitation of emotional abilities. Some therapeutic programs have recently been developed to improve emotional facial expression decoding among clinical populations (e.g. FaceTales Program, Philippot & Power, 2010), but they have up to now not been applied to alcohol dependence. Future development of these therapeutic proposals should include not only visual emotional stimuli but also auditory and crossmodal stimulation, in order to develop more realistic emotion decoding rehabilitation programs. More globally, therapeutic interventions could also be improved through communication re-education programs in alcohol dependence, focusing on crossmodal processing of the expression and identification of emotions in social settings.

In line with this clinical perspective, the specific crossmodal deficit for anger stimuli described in our electrophysiological study makes particular sense at the therapeutic level. Indeed, many clinical studies (e.g. Bartek, Lindeman, & Hawks, 1999; Karno & Longabaugh, 2005) have stressed that alcohol-dependent individuals have considerable difficulties to manage their anger and to correctly react to the anger expressed by others. It has also been suggested that this anger managing inability increases interpersonal problems, which are known to be a major relapse factor after detoxification treatment. However, although some studies have suggested that alcohol dependence is associated with a relatively specific deficit in anger emotional facial expression decoding (e.g. Philippot et al., 1999), other studies have not replicated these results (Foisy, 2005; Uekermann et al., 2005) and this deficit has not been described for other stimuli (notably auditory prosody). This discrepancy between an obvious clinical deficit and contrasting experimental results could be explained by the fact that previous studies used only unimodal stimuli (mainly emotional facial expressions). These stimuli are artificial because in everyday social interactions, multimodal stimuli, and mainly auditory–visual stimuli, are much more common. Using more ecologic stimuli, our study established, at the electrophysiological level, the specific crossmodal deficit for anger in alcohol dependence that has been repeatedly observed at clinical level. The development of future therapeutic programs should thus particularly emphasize the need to take into account this particular deficit for anger expression and decoding among alcohol-dependent individuals.

5 Future Directions and Conclusion

As outlined above, our studies have to be considered as a very preliminary and exploratory step in the examination of emotional crossmodal processing among clinical populations. Indeed, we focused on a specific clinical population and used a small subset of the possible emotional stimuli and sensory modalities. Nevertheless, when combined with the few previous data sets obtained among other psychopathological populations, theses results constitute a reliable and promising basis for the development of an ambitious research program aiming at determining the behavioural and cerebral correlates of impaired crossmodal integration in psychiatry, and finally leading to strong fundamental propositions as well as clinical applications. We end this chapter by proposing three main directions that should be developed in future research, each focusing on the extension of previous results and proposing a diversification of the emotional stimuli used, sensory modalities included and psychiatric populations explored.

5.1 Using More Emotional Expressions

A main limitation of the results presented above is that they considered only a very low number of emotional states, namely happy, angry and neutral emotions. A central direction for future research will be to diversify the emotional stimulation used, in order to determine the potential differential deficits associated with different emotional states in alcohol dependence. It can indeed be hypothesized that alcohol-dependent individuals' emotional crossmodal deficit will vary according to stimulus valence. As described above, our ERP results suggested, in line with earlier results (Maurage et al. 2008c), a specific deficit for anger in alcohol dependence, as compared to happy and neutral stimuli. This specific deficit makes sense at the clinical level and could lead to the development of innovative therapeutic programs. Nevertheless, it is not clear whether this impairment is really limited to anger or is more general, as it could for example be present for every negative emotion. It is thus necessary to develop crossmodal experimental paradigms that explore a broader set of emotions, and particularly of negative ones (e.g. disgust, fear, sadness) in order to confirm our results and to separate the hypotheses of an anger-specific deficit versus a general deficit for negative emotions among alcohol-dependent individuals. This exploration of the differential crossmodal integration across emotions has already been conducted among healthy participants (e.g. Ethofer, Pourtois, & Wildgruber, 2006; Kreifelts, Ethofer, Grodd, Erb, & Wildgruber, 2007), but it has not been applied to clinical populations yet. It has also been suggested (e.g. Maurage et al. 2008c; Philippot et al., 1999) that alcohol dependence could be associated with a particular deficit for decoding and correctly reacting to emotional states which have a high interpersonal value, and particularly which are associated with a social evaluation aspect or moral judgment (e.g. anger, disgust, contempt, see Hutcherson & Gross, 2011 for a recent development on these "moral emotions"), as compared to emotions which are expressing more self-focused feelings (e.g. fear, sadness). Crossmodal paradigms, due to their high ecological validity, could be very useful in exploring these hypothetical propositions on the differential deficit between social and non-social emotions in alcohol dependence, which have still to be confirmed on the basis of sound experimental results presenting situations closer to real life.

More generally, future studies focusing on integration processes in alcohol dependence should also go beyond the exploration of classical emotion decoding. Indeed, emotional abilities are not limited to this basic emotion decoding as daily life forces us to identify and correctly react to far more various and subtle emotional signals. More complex affective abilities (e.g. empathy, emotional intelligence) are thus also needed to develop and maintain satisfactory interpersonal relations. In line with this, we recently conducted two studies exploring these complex emotional abilities among alcohol-dependent individuals. In a first study (Maurage et al., 2011b), we explored the emotion decoding abilities in a more complex task, namely the "Reading the Mind in the Eyes Test" (Baron-Cohen, Wheelwright, Hill, Raste, & Plumb, 2001). This test proposes to go further than basic emotional categories as participants have to decode subtle positive or negative mental states (e.g. interest, worry, guiltiness). We showed that alcohol dependence leads to a large deficit in the identification of these fine-grained mental states, as the impairment is even stronger than the one observed in classical emotional decoding tasks. These results, in line with recent studies exploring more complex emotional states among healthy people (e.g. Basile et al., 2011; Wagner, N'diaye, Ethofer, & Vuilleumier, 2011), underline the need to go further than conventional emotion labels and to use more subtle and ecological paradigms in order to develop a better understanding of emotional impairments in alcohol dependence. In a second study (Maurage et al., 2011a), we showed, by means of empathy questionnaires [i.e. Interpersonal Reactivity Index (Davis, 1983) and Empathy Quotient (Baron-Cohen & Wheelwright, 2004)], that alcohol-dependent individuals presented a preserved cognitive empathy but an impaired emotional one. This clearly shows that emotional abilities constitute a core impairment for alcohol-dependent individuals, and that high level emotional abilities are also impaired and should be further explored. Nevertheless, these two exploratory studies were conducted by means of unimodal visual stimuli, and thus have a low ecological validity. The use of crossmodal paradigms exploring complex emotional states and affective abilities will enrich these preliminary results by bringing experimental designs closer to daily situations and thus by offering a more valid description of alcohol-related emotional and affective disorders.

5.2 Going Beyond Auditory and Visual Modalities

As illustrated by the present book, the crossmodal literature put a strong emphasis on the integration between visual and auditory modalities. This focus is justified by the fact that vision and audition are by far the most dominant sensory modalities among human beings. Nevertheless, the near total absence of data concerning the other senses, and particularly the "chemical senses" (i.e., olfaction and taste) is surprising, as they play an underestimated but crucial role in the daily life of healthy as well as clinical populations (e.g. Schiffman, 1997). Indeed, olfactory and gustatory stimulation can also carry a strong emotional valence (e.g. Greimel, Macht, Krumhuber, & Ellgring, 2006; Shepherd, 2006; Winston, Gottfried, Kilner, & Dolan,

2005), and exploring the integration between these emotional stimulations and visual or auditory ones could constitute a promising perspective to develop and renew crossmodal integration knowledge. More specifically, olfaction has been shown to play a crucial role in the development and maintenance of alcohol dependence (e.g. Kareken et al., 2004; Little et al., 2005), but olfactory processing has up to now been studied very little in this pathology. We recently conducted a research program exploring the olfactory abilities associated with excessive alcohol consumption (Maurage, Callot, Philippot, Rombaux, & de Timary, 2011c; Maurage, Callot, Chang, Philippot, Rombaux, & de Timary, 2011d), which confirmed earlier results (e.g. Rupp et al., 2003, 2004, 2006) showing impaired processing of olfactory stimulations in alcohol dependence, and gave the first insights concerning the cerebral correlates of this deficit (by means of ERP measures). Interestingly, we showed that olfactory impairments are highly correlated with executive function deficits, and specifically with confabulation problems. These results suggest that these two abilities could rely on the same brain structures (and particularly on the orbitofrontal cortex), and that olfaction measures could be useful to shed new light on the exploration of executive and emotional impairments in alcohol dependence. This is in line with recent proposals suggesting that olfactory measures could be a reliable cognitive marker in psychiatric disorders (see Rupp, 2010 or Turetsky, Hahn, Borgmann-Winter, & Moberg, 2009 for a complete discussion on this topic). It thus appears that olfaction research is currently becoming a blooming research field among clinical populations.

But once again, all these explorations have up to now been limited to unimodal stimulation while in real-life situations, olfactory stimulations most often occur in combination with stimulation coming from other sensory modalities. This is particularly true for emotional contexts, and crossmodal explorations combining several senses (beyond audition and vision) are thus urgently needed to develop this new field of research. To our knowledge, very few studies have explored the crossmodal integration of emotional olfactory stimulation, by focusing on the influence of olfactory cues on facial expression decoding (Leppänen & Hietanen, 2003; Seubert et al., 2010a, 2010b). These preliminary results replicated the classical facilitation effect (i.e., faster reaction times and improved performance when emotional olfactory and visual stimulations are congruent), thus confirming the presence of a genuine olfactory–visual integration among healthy participants. Neuroimaging data have also suggested that, while some brain areas (e.g. middle frontal gyrus) could be activated for every crossmodal interaction, independent of the sensory modalities engaged, other structures (mostly the anterior insula) could be specialized for olfactory–visual integration (e.g. Gottfried & Dolan, 2003; Small, 2004). Finally, they showed that schizophrenic patients present an impairment of this olfactory–visual integration, particularly for negative emotional stimuli, which suggests that crossmodal impairments among psychiatric populations could be independent of the sensory modalities involved. On the basis of these innovative explorations, future studies should thus explore, among healthy as well as clinical populations, the behavioural and brain correlates of the crossmodal integration between the "chemical senses" and vision or audition. A more ambitious aim could be to go one step further towards ecological validity, by combining more than two sensorial modalities. Indeed, while our emotional experience is frequently based on the simultaneous perception of several sensory modalities, only bimodal stimulation paradigms have been proposed up to now. The recent technical advancements, and notably the growth of virtual reality, could lead to the development of experimental designs stimulating all the senses and thus open new perspectives for crossmodal processing explorations.

5.3 Applying Crossmodal Paradigms to Other Psychiatric Populations

The crossmodal studies presented in this chapter exclusively explored alcohol dependence, and more specifically recently detoxified alcohol-dependent individuals. This population is of course only a specific part of the persons presenting alcohol-related problems, and more globally of the psychiatric patients. It thus appears important to underline the potential extension of these studies to other populations, in the field of alcohol abuse and dependence, and also in other psychiatric states, with the long-term objective of developing a sound and integrative approach of crossmodal processing in clinical populations.

Concerning alcohol-related problems the literature on cerebral, cognitive and emotional impairments associated with alcohol consumption has classically been focused on installed alcohol dependence (namely on the exploration of the impairments due to chronic excessive alcohol consumption). Nevertheless, a new field of research has risen during the last decades, aiming at exploring the roots of alcohol addiction, namely the appearance and chronification of the deficits during the development of alcohol dependence. On the one hand, many studies have been conducted among populations at high risk of becoming alcohol dependent, mainly among children of alcohol-dependent individuals (see for example Lieberman, 2000; Porjesz et al., 2005 or Van der Stelt, Gunning, Snel, Zeef, & Kok, 1994 for reviews on this topic). These studies have suggested that several cerebral and cognitive impairments could be present before the development of alcohol dependence and thus be a causal factor rather than a consequence of excessive alcohol consumption. Our intention is not to go into the details of this important literature, but to underline that these explorations were once again exclusively based on unimodal stimulation. Crossmodal studies among children of alcohol-dependent individuals (notably for emotional abilities, which are still unexplored in this population) could thus give a more ecological and valid exploration of the deficits that are present before the development of alcohol dependence. On the other hand and more recently, several projects have been conducted concerning the consequences of binge drinking (i.e. the excessive but episodic alcohol consumption, typically observed among adolescents and young adults and considered to be an “entrance door” towards alcohol dependence, e.g. Enoch, 2006; McCarty et al., 2004). Recent studies have shown that binge drinking leads to cognitive effects (e.g. Giancola, 2002; Townshend & Duka, 2005; Zeigler et al., 2005), and we recently extended these results by suggesting that binge drinking habits rapidly lead to impaired processing of emotional auditory stimulation, and that this alcohol consumption pattern is particularly deleterious for brain functioning (Maurage, Pesenti, Philippot, Joassin, & Campanella, 2009b; Maurage, Joassin, Speth, Modave, Philippot, & Campanella, 2012b). Nevertheless, it is still unknown whether these deficits are modified or not when several stimulations are presented together, and crossmodal studies would thus help to extend and clarify these preliminary results. Finally, it should be noted that applying crossmodal paradigms to populations of high-risk individuals or binge drinkers would help in clarifying several points which remain unclear concerning the integration deficit observed in alcohol dependence. First, the causal link between crossmodal processing impairments and excessive alcohol consumption: exploring integration processes among at-risk individuals will indeed allow clarifying whether emotional crossmodal deficits are present before the appearance of alcohol dependence (and thus potentially playing a role in the development of this dependence, as suggested by the vicious circle described above) or are a consequence of this alcohol dependence. Second, the timing of appearance of crossmodal impairments: determining the presence of crossmodal impairments among binge drinkers, which are at the first stages of alcohol dependence, would help in understanding whether these crossmodal deficits (and notably the crossmodal brain areas dysfunctions) are appearing rapidly during the development of alcohol dependence or are only a late consequence appearing after many years of excessive alcohol consumption.

Concerning the exploration of emotional crossmodal processing in psychiatry, it is surprising to notice that very few studies have been conducted among these clinical populations. Many projects have been proposed during these last years in order to explore the visual or auditory decoding of emotional stimulations among a wide variety of psychiatric states, like depression, autism, anxiety, anorexia nervosa and drug addiction (e.g. Bhatara et al., 2010; Mejias et al., 2005; Mendlewicz, Linkowski, Bazelmans, & Philippot, 2005; Rossignol, Philippot, Douilliez, Crommelinck, & Campanella, 2005), but emotional crossmodal paradigms have been used only in a very limited number of these pathological states. Several studies (De Gelder et al., 2005; De Jong, Hodiamont, Van den Stock, & de Gelder, 2009; Pearl et al., 2009; Szycik et al., 2009) have explored the integration of emotional stimulation among schizophrenic patients and consistently described emotional crossmodal deficits in this psychiatric state, notably indexed by reduced audio–visual integration ability and by a vision–audition imbalance (i.e. a dominance of the visual stimulation on the auditory one reducing crossmodal performance). These results, together with those we have obtained in alcohol dependence, show that crossmodal processing impairments constitute a crucial aspect of psychiatric states, and should thus encourage the development of emotional crossmodal research among other psychiatric states. This is particularly true among populations which are known to present unimodal emotional decoding deficits, in order to answer the following central question: How does crossmodal integration occur when unimodal outputs are impaired? In other words, do some psychiatric populations manage to compensate their deficit in the processing of unimodal emotional stimuli by taking profit of the simultaneous presentation of two stimulations, or are all psychiatric states associated with increased processing impairments in crossmodal situations, as it has been observed in alcohol dependence and schizophrenia?

5.4 Conclusion

As it is extensively described in other chapters of the present book, the exploration of crossmodal processing among healthy controls has now become an extensive research field: Behavioural as well as cerebral correlates of the integration processes between sensory modalities have been precisely explored among animal and human populations, leading to comprehensive models on this topic. Nevertheless, this maturity of the knowledge concerning "normal" crossmodal processes appears in complete contrast with what can be observed in clinical states. Indeed, as outlined above, very little has been done up to now to attempt to understand how these crossmodal processes are impaired among neurological and psychiatric populations, and we believe that this astonishing lack of interest has had a detrimental effect on the advances that can be made in this topic.

The main aim of the present chapter was thus to underline the urgent need to explore the crossmodal integration abilities among these populations, as progressing in this direction could lead to central implications (1) for clinical aspects: using crossmodal designs among clinical populations would lead to a better understanding of the impairments presented by inpatients in real-life situations (and notably in emotional contexts). This would allow a more ecological exploration of the cognitive, cerebral and affective deficits in these populations, thus complementing and clarifying earlier results. But this could also lead to the development of new therapeutic interventions, using crossmodal clinical settings to rehabilitate impaired abilities (e.g. by means of virtual reality); (2) for fundamental research: while the data obtained among clinical populations have traditionally constituted a strong method to improve the understanding of normal abilities in neuropsychology and neuroscience (with the well-known proposition that exploring an impaired system helps to understand its healthy functioning), this approach has received very little attention in crossmodal processing research. We argue that developing the exploration of integration abilities among clinical populations could shed a new light on the several questions that are still unresolved in this research field.

By describing our research focusing on emotional crossmodal integration in alcohol dependence, we only presented here what can be considered as a modest first step towards a real and ambitious research program allowing to precisely describe the crossmodal processing abilities among psychiatric populations. We indeed believe that our work, together with the few studies conducted in schizophrenia, constitute seminal results that should be developed in the future. More specifically, studies to come should extend this exploration of crossmodal processing in three main ways, by diversifying the stimuli (i.e. using a wider range of emotional but also non-emotional stimuli), the sensory modalities (particularly by including the "chemical senses" in the crossmodal designs) and the populations explored (i.e. studying the crossmodal processes among other populations with substance abuse, and also among other psychiatric and neurologic states). These proposals for future studies are just illustrations of the many prospects offered by this largely unexplored field. In short, everything is still to be done concerning crossmodal processing in psychiatric populations, and our hope is thus that the preliminary data described in the present chapter will open the door to fresh, diverse and complementary studies.

Acknowledgments Pierre Maurage is Senior Research Assistant at the National Fund for Scientific Research (F.N.R.S., Belgium). This work was supported by grants from the French Ministry of Research.

References

Alais D, Burr D (2004) The ventriloquist effect results from near-optimal bimodal integration. Current Biology 14:257–262

Barbeau EJ, Pariente J, Felician O, Puel M (2010) Visual recognition memory: A double anatomo-functional dissociation. Hippocampus 21(9):929–934

Barkhuysen P, Krahmer E, Swerts M (2010) Crossmodal and incremental perception of audiovisual cues to emotional speech. Language and Speech 53(1):3–30

Baron-Cohen S, Wheelwright S (2004) The empathy quotient: An investigation of adults with Asperger syndrome or high functioning autism, and normal sex differences. Journal of Autism and Developmental Disorders 34(2):163–175

Baron-Cohen S, Wheelwright S, Hill J, Raste Y, Plumb I (2001) The "reading the mind in the eyes" test revised version: A study with normal adults, and adults with Asperger syndrome or high-functioning autism. Journal of Child Psychology and Psychiatry 42(241):251

Barone P (2010) Is the primary visual cortex multisensory? Comment on "Crossmodal influences on visual perception" by Prof. Ladan Shams. Physical Life Review 7(3):291–292

Bartek JK, Lindeman M, Hawks JH (1999) Clinical validation of characteristics of the alcoholic family. Nurse Diagnostic 10:158–168

Basile B, Mancini F, Macaluso E, Caltagirone C, Frackowiak RS, Bozzali M (2011) Deontological and altruistic guilt: Evidence for distinct neurobiological substrates. Human Brain Mapping 32(2):229–239

Beatty WW, Tivis R, Stott D, Nixon SJ, Parsons OA (2000) Neuropsychological deficits in sober alcoholics: Influences of chronicity and recent alcohol consumption. Alcoholism, Clinical and Experimental Research 24:149–154

Bechara A, Dolan S, Denburg N, Hindes A, Anderson SW, Nathan PE (2001) Decision making deficits, linked to a dysfunctional ventromedial prefrontal cortex, revealed in alcohol and stimulant abusers. Neuropsychologia 39:376–389

Bernstein LE, Auer ET Jr, Wagner M, Ponton CW (2008) Spatiotemporal dynamics of audiovisual speech processing. NeuroImage 39:423–435

Bhatara A, Quintin EM, Levy B, Bellugi U, Fombonne E, Levitin DJ (2010) Perception of emotion in musical performance in adolescents with autism spectrum disorders. Autism Research 3(5):214–225

Blusewicz MJ, Dustman RE, Schenkenberg T, Beck EC (1977) Neuropsychological correlates of chronic alcoholism and aging. Journal of Nervous and Mental Disorders 165:348–355

Brooks PJ (2000) Brain atrophy and neuronal loss in alcoholism: A role for DNA damage? Neurochemistry International 37:403–412

Calvert GA, Brammer MJ, Bullmore ET, Campbell R, Iversen SD, David AS (1999) Response amplification in sensory-specific cortices during crossmodal binding. NeuroReport 10:2619–2623
Calvert GA, Hansen PC, Iversen SD, Brammer MJ (2001) Detection of auditory visual integration sites in humans by application of electrophysiological criteria to the BOLD effect. NeuroImage 14:427–438
Campanella S, Belin P (2007) Integrating face and voice in person perception. Trends in Cognitive Sciences 11:535–543
Campanella S, Bruyer R, Froidbise S, Rossignol M, Joassin F, Kornreich C et al (2010) Is two better than one? A cross-modal oddball paradigm reveals greater sensitivity of the P300 to emotional face–voice associations. Clinical Neurophysiology 121:1855–1862
Campanella S, Philippot P (2006) Insights from ERPs into emotional disorders: An affective neuroscience perspective. Psychologica Belgica 46:37–53
Carton JS, Kessler EA, Pape CL (1999) Nonverbal decoding skills and relationship wellbeing in adults. Journal of Nonverbal Behavior 23:91–100
Chanraud S, Martelli C, Delain F, Kostogianni N, Douaud G, Aubin HJ et al (2007) Brain morphometry and cognitive performance in detoxified alcoholdependents with preserved psychosocial functioning. Neuropsychopharmacology 32:429–438
Clark US, Oscar-Berman M, Shagrin B, Pencina M (2007) Alcoholism and judgments of affective stimuli. Neuropsychology 21:346–362
Cohn M, Moscovitch M, Davidson PS (2010) Double dissociation between familiarity and recollection in Parkinson's disease as a function of encoding tasks. Neuropsychologia 48(14):4142–4417
Cordovil de Susa Uva M, de Timary P, Cortesi M, Mikolajczak M, Du Roy de Blicquy P, Luminet O (2010) Moderating effect of emotional intelligence on the role of negative affect in the motivation to drink in alcohol-dependent subjects. Personality and Individual Differences 48:16–21
Cowen MS, Chen F, Lawrence AJ (2004) Neuropeptides: Implications for alcoholism. Journal of Neurochemistry 89:273–285
Davis MH (1983) Measuring individual differences in empathy: Evidence from a multidimensional approach. Journal of Personality and Social Psychology 44:113–126
De Bellis MD, Narasimhan A, Thatcher DL, Keshavan MS, Soloff P, Clark DB (2005) Prefrontal cortex, thalamus, and cerebellar volumes in adolescents and young adults with adolescent-onset alcohol use disorders and comorbid mental disorders. Alcoholism, Clinical and Experimental Research 29:1590–1600
De Gelder B, Vroomen J, de Jong SJ, Masthoff ED, Trompenaars FJ, Hodiamont P (2005) Multisensory integration of emotional faces and voices in schizophrenics. Schizophrenia Research 72(2–3):195–203
De Jong JJ, Hodiamont PP, Van den Stock J, de Gelder B (2009) Audiovisual emotion recognition in schizophrenia: Reduced integration of facial and vocal affect. Schizophrenia Research 107(2–3):286–293
DiClemente CC, Jordan L, Marinilli AS, Nidecker M (2003) Psychotherapy in alcoholism treatment. In: Johnson BA, Ruiz P, Galanter M (eds) Handbook of clinical alcoholism treatment. Lippincott, Williams, & Wilkins, Baltimore, MD, pp 102–110
Driver J, Spence C (2000) Multisensory perception: Beyond modularity and convergence. Current Biology 10:731–735
Ellis HD, Jones DM, Mosdell N (1997) Intra- and inter-modal repetition priming of familiar faces and voices. British Journal of Psychology 88:143–156
Enoch MA (2006) Genetic and environmental influences on the development of alcoholism: Resilience vs risk. Annals of the New York Academy of Sciences 1094:193–201
Ethofer T, Pourtois G, Wildgruber D (2006) Investigating audiovisual integration of emotional signals in the human brain. Progress in Brain Research 156:345–361
Fein G, Bachman L, Fischer S, Davenport L (1990) Cognitive impairments in abstinent alcoholics. West Journal of Medicine 152:531–537
Fein G, Landman B, Tran H, McGillivray S, Finn P, Barakos J et al (2006) Brain atrophy in long-term abstinent alcoholics who demonstrate impairment on a simulated gambling task. NeuroImage 32:1465–1471

Feldman RS, Philippot P, Custrini RJ (1991) Social competence and non-verbal behaviour. In: Feldman RS (ed) Fundamentals of non-verbal behavior. Cambridge University Press, New York, pp 329–350

Flannery B, Fishbein D, Krupitsky E, Langevin D, Verbitskaya E, Bland C et al (2007) Gender differences in neurocognitive functioning among alcohol-dependent Russian patients. Alcoholism, Clinical and Experimental Research 31:745–754

Foisy, M.L. (2005). La reconnaissance des expressions faciales émotionnelles dans l'alcoolisme: Spécification de la nature et de l'origine des biais évaluatifs. Unpublished PhD Thesis, Catholic University of Louvain, Louvain-la-Neuve, Belgique.

Foisy ML, Kornreich C, Fobe A, D'Hondt L, Pelc I, Hanak C et al (2007a) Impaired emotional facial expression recognition in alcohol dependence: Do these deficits persist with midterm abstinence? Alcoholism, Clinical and Experimental Research 31:404–410

Foisy ML, Kornreich C, Petiau C, Parez A, Hanak C, Verbanck P et al (2007b) Impaired emotional facial expression recognition in alcoholics: Are these deficits specific to emotional cues? Psychiatry Research 150:33–41

Frens MA, Van Opstal AJ, Van der Willigen RF (1995) Spatial and temporal factors determine auditory–visual interactions in human saccadic eye movements. Perception & Psychophysics 57:802–816

Frigerio E, Burt DM, Montagne B, Murray LK, Perrett DI (2002) Facial affect perception in alcoholics. Psychiatry Research 113:161–171

Friston KJ, Frith CD (1995) Schizophrenia: A disconnection syndrome? Clinical Neuroscience 3(2):89–97

Ghazanfar AA, Maier JX, Hoffman KL, Logothetis NK (2005) Multi-sensory integration of dynamic faces and voices in rhesus monkey auditory cortex. Journal of Neuroscience 25(20):5004–5012

Giancola PR (2002) Alcohol-related aggression during the college years: Theories, risk factors and policy implications. Journal of Studies on Alcohol 14:129–139

Goebel R, van Atteveldt N (2009) Multisensory functional magnetic resonance imaging: A future perspective. Experimental Brain Research 198:153–164

Gottfried JA, Dolan RJ (2003) The nose smells what the eye sees: Crossmodal visual facilitation of human olfactory perception. Neuron 39:375–386

Greimel E, Macht M, Krumhuber E, Ellgring H (2006) Facial and affective reactions to tastes and their modulation by sadness and joy. Physiology and Behavior 89:261–269

Hanley JR, Smith ST, Hadfield J (1998) I recognise you but I can't place you: An investigation of familiar-only experiences during tests of voice and face recognition. Quarterly Journal of Experimental Psychology 51:179–195

Hanley JR, Turner JM (2000) Why are familiar-only experiences more frequent for voices than for faces? Quarterly Journal of Experimental Psychology 53:1105–1116

Hansenne M (2006) Event-related brain potentials in psychopathology: Clinical and cognitive perspectives. Psychologica Belgica 46:5–36

Harper C (2007) The neurotoxicity of alcohol. Human and Experimental Toxicology 26:251–257

Harper C, Matsumoto I (2005) Ethanol and brain damage. Current Opinion in Pharmacology 5:73–78

Heinz A, Wrase J, Kahnt T, Beck A, Bromand Z, Grüsser SM et al (2007) Brain activation elicited by affectively positive stimuli is associated with a lower risk of relapse in detoxified alcoholic subjects. Alcoholism, Clinical and Experimental Research 31:1138–1147

Hutcherson CA, Gross JJ (2011) The moral emotions: A social-functionalist account of anger, disgust, and contempt. Journal of Personality and Social Psychology 100(4):719–737

Joassin F, Maurage P, Bruyer R, Crommelinck M, Campanella S (2004) When audition alters vision: An eventrelated potential study of the cross-modal interactions between faces and voices. Neuroscience Letters 369(2):132–137

Joassin F, Maurage P, Campanella S (2011a) The neural network sustaining the crossmodal processing of human gender from faces and voices: An fMRI study. NeuroImage 54(2): 1654–1661

Joassin F, Pesenti M, Maurage P, Verreckt E, Bruyer R, Campanella S (2011b) Cross-modal interactions between human faces and voices involved in person recognition. Cortex 47(3):367–376

Kareken DA, Claus ED, Sabri M, Dzemidzic M, Kosobud AEK, Radnovich AJ et al (2004) Alcohol-related olfactory cues activate the nucleus accumbens and the ventrotemental area in high-risk drinkers: Preliminary findings. Alcoholism, Clinical and Experimental Research 28(4):550–557

Karno MP, Longabaugh R (2005) An examination of how therapist directiveness interacts with patient anger and reactance to predict alcohol use. Journal of Studies on Alcohol 66:825–832

Kathmann N, Soyka M, Bickel R, Engel RR (1996) ERP changes in alcoholics with and without alcohol psychosis. Biological Psychiatry 39:873–881

Kornreich C, Blairy S, Philippot P, Dan B, Foisy ML, Hess U et al (2001) Impaired emotional facial expression recognition in alcoholism compared with obsessive–compulsive disorder and normal controls. Psychiatry Research 102:235–248

Kornreich C, Philippot P, Foisy ML, Blairy S, Raynaud E, Dan B et al (2002) Impaired emotional facial expression recognition is associated with interpersonal problems in alcoholism. Alcohol and Alcoholism 37:394–400

Kramer JH, Blusewicz MJ, Robertson LC, Preston K (1989) Effects of chronic alcoholism on perception of hierarchical visual stimuli. Alcoholism, Clinical and Experimental Research 13:240–245

Kreifelts B, Ethofer T, Grodd W, Erb M, Wildgruber D (2007) Audiovisual integration of emotional signals in voice and face: An event-related fMRI study. NeuroImage 37(4):1445–1456

Kril JJ, Halliday GM, Svoboda MD, Cartwright H (1997) The cerebral cortex is damaged in chronic alcoholics. Neuroscience 79:983–998

Kwakye LD, Foss-Feig JH, Cascio CJ, Stone WL, Wallace MT (2011) Altered auditory and multisensory temporal processing in autism spectrum disorders. Frontiers in Integrative Neuroscience 4(129):1–11

Latinus M, VanRullen R, Taylor M (2010) Top-down and bottom-up modulation in processing bimodal face/voice stimuli. BMC Neuroscience 11:36

Laurienti PJ, Perrault TJ, Stanford TR, Wallace MT, Stein BE (2005) On the use of superadditivity as a metric for characterizing multisensory integration in functional neuroimaging studies. Experimental Brain Research 166:289–297

Leppänen JM, Hietanen JK (2003) Affect and face perception: Odors modulate the recognition advantage of happy faces. Emotion 3:315–326

Lieberman DZ (2000) Children of alcoholics: An update. Current Opinions in Pediatry 12(4):336–340

Listerud J, Powers C, Moore P, Libon DJ, Grossman M (2009) Neuropsychological patterns in magnetic resonance imaging-defined subgroups of patients with degenerative dementia. Journal of International Neuropsychological Society 15(3):459–470

Little HJ, Stephens DN, Ripley TL, Borlikova G, Duka T, Schubert M et al (2005) Alcohol withdrawal and conditioning. Alcohol, Clinical and Experimental Research 29(3):453–464

Love S, Pollick FE, Latinus M (2011) Cerebral correlates and statistical criteria of cross-modal face and voice integration. Seeing and Perceiving 24(4):351–367

Marinkovic K, Oscar-Berman M, Urban T, O'Reilly CE, Howard JA, Sawyer K et al (2009) Alcoholism and dampened temporal limbic activation to emotional faces. Alcoholism, Clinical and Experimental Research 33:1880–1892

Martinotti G, Di Nicola M, Tedeschi D, Cundari S, Janiri L (2009) Empathy ability is impaired in alcohol-dependent patients. The American Journal on Addictions 18(2):157–161

Massida Z, Belin P, James C, Rouger J, Fraysse B, Barone P, et al (2011) Voice discrimination in cochlear-implanted deaf subjects. Hear Research 275(1–2):120–129

Maurage P, Callot C, Chang B, Philippot P, Rombaux P, de Timary P (2011d) Olfactory impairment is correlated with confabulation in alcoholism: Towards a multimodal testing of orbitofrontal cortex. PLoS One 6(8):e23190

Maurage P, Callot C, Philippot P, Rombaux P, de Timary P (2011c) Chemosensory event-related potentials in alcoholism: a specific impairment for olfactory function. Biological Psychology 88(1):28–36

Maurage P, Campanella C, Philippot P, Charest I, Martin S, de Timary P (2009a) Impaired emotional facial expression decoding in alcoholism is also present for emotional prosody and body postures. Alcohol and Alcoholism 44(5):476–485

Maurage P, Campanella S, Philippot P, Martin S, de Timary P (2008a) Face processing in chronic alcoholism: A specific deficit for emotional features. Alcoholism, Clinical and Experimental Research 32(4):600–606

Maurage P, Campanella S, Philippot P, Pham T, Joassin F (2007a) The crossmodal facilitation effect is disrupted in alcoholism: A study with emotional stimuli. Alcohol and Alcoholism 42(6):552–559

Maurage P, Campanella S, Philippot P, Vermeulen N, Constant E, Luminet O et al (2008b) Electrophysiological correlates of the disrupted processing of anger in alcoholism. International Journal of Psychophysiology 70(1):50–62

Maurage P, Grynberg D, Noël X, Joassin F, Hanak C, Verbanck P et al (2011b) The "Reading the Mind in the Eyes" test as a new way to explore complex emotions decoding in alcohol dependence. Psychiatry Research 190(2–3):375–378

Maurage P, Grynberg D, Noël X, Joassin F, Philippot P, Hanak C et al (2011a) Dissociation between affective and cognitive empathy in alcoholism: a specific deficit for the emotional dimension. Alcoholism, Clinical and Experimental Research 35(9):1662–1668

Maurage P, Joassin F, Pesenti M, Grandin C, Heeren H, Philippot P et al (2012a) The neural network sustaining crossmodal integration is impaired in alcohol-dependence: an fMRI study. Cortex, in press

Maurage P, Joassin F, Philippot P, Campanella S (2007b) A validated battery of vocal emotional expressions. Neuropsychological Trends 2:63–74

Maurage P, Joassin F, Speth A, Modave J, Philippot P, Campanella S (2012b) Cerebral effects of binge drinking: Respective influences of global alcohol intake and consumption pattern. Clinical Neurophysiology 123(5):892–901

Maurage P, Pesenti M, Philippot P, Joassin F, Campanella S (2009b) Latent deleterious effects of binge drinking over a short period of time revealed by electrophysiological measures only. Journal of Psychiatry & Neuroscience 34(2):111–118

Maurage P, Philippot P, Joassin F, Pauwels L, Alonso Prieto E, Palmero Soler E et al (2008c) The auditory–visual integration of anger is impaired in alcoholism: An ERP study. Journal of Psychiatry & Neuroscience 33(2):111–122

Maurage P, Philippot P, Verbanck P, Noel X, Kornreich C, Hanak C et al (2007c) Is the P300 deficit in alcoholism associated with early visual impairments (P100, N170)? An oddball paradigm. Clinical Neurophysiology 118(3):633–644

McCarty CA, Ebel BE, Garrison MM, DiGiuseppe DL, Christakis DA, Rivara FP (2004) Continuity of binge and harmful drinking from late adolescence to early adulthood. Pediatrics 114(3): 714–719

McGurk H, McDonald J (1976) Hearing lips and seeing voices. Nature 264:746–748

McIntosh C, Chick J (2004) Alcohol and the nervous system. Journal of Neurology, Neurosurgery, and Psychiatry 75:16–21

Mejias S, Rossignol M, Debatisse D, Streel E, Servais L, Guérit JM et al (2005) Event-related potentials (ERPs) in ecstasy (MDMA) users during a visual oddball task. Biological Psychology 69(3):333–352

Melillo R, Leisman G (2009) Autistic spectrum disorders as functional disconnection syndrome. Reviews in Neuroscience 20(2):111–131

Mendlewicz L, Linkowski P, Bazelmans C, Philippot P (2005) Decoding emotional facial expressions in depressed and anorexic patients. Journal of Affective Disorders 89(1–3):195–199

Monnot M, Lovallo WR, Nixon SJ, Ross E (2002) Neurological basis of deficits in affective prosody comprehension among alcoholics and fetal alcohol-exposed adults. The Journal of Neuropsychiatry and Clinical Neurosciences 14:321–328

Monnot M, Nixon S, Lovallo W, Ross E (2001) Altered emotional perception in alcoholics: Deficits in affective prosody comprehension. Alcoholism, Clinical and Experimental Research 25: 362–369

Montagne B, Kessels RP, Wester AJ, de Haan EH (2006) Processing of emotional facial expressions in Korsakoff's syndrome. Cortex 42:705–710
Noël X, Van der Linden M, Schmidt N, Sferrazza R, Hanak C, Le Bon O et al (2001) Supervisory attentional system in nonamnesic alcoholic men. Archives of General Psychiatry 58:1152–1158
Ogura C, Miyazato Y (1991) Cognitive dysfunctions of alcohol dependence using event related potentials. Arukoru KenkyutoYakubutsu Ison 26:331–340
Oscar-Berman M, Hancock M, Mildworf B, Hutner N, Weber DA (1990) Emotional perception and memory in alcoholism and aging. Alcoholism, Clinical and Experimental Research 14:383–393
Oscar-Berman M, Kirkley SM, Gansler DA, Couture A (2004) Comparisons of Korsakoff and non-Korsakoff alcoholics on neuropsychological tests of prefrontal brain functioning. Alcoholism, Clinical and Experimental Research 28:667–675
Oscar-Berman M, Marinkovic K (2003) Alcoholism and the brain: An overview. Alcohol Research & Health 27:125–133
Pearl D, Yodashkin-Porat D, Katz N, Valevski A, Aizenberg D, Sigler M et al (2009) Differences in audiovisual integration, as measured by McGurk phenomenon, among adult and adolescent patients with schizophrenia and age-matched healthy control groups. Comprehensive Psychiatry 50(2):186–192
Petrini K, Crabbe F, Sheridan C, Pollick FE (2011) The music of your emotions: Neural substrates involved in detection of emotional correspondence between auditory and visual music actions. PLoS One 6(4)
Petrini K, McAleer P, Pollick FE (2010) Audiovisual integration of emotional signals from music improvisation does not depend on temporal correspondence. Brain Research 1323:139–148
Philippot P, Kornreich C, Blairy S, Baerts I, Den Dulk A, Le Bon O et al (1999) Alcoholics' deficits in the decoding of emotional facial expression. Alcoholism, Clinical and Experimental Research 23:1031–1038
Philippot, P., & Power, M. (2010) FaceTales. http://www.ipsp.ucl.ac.be/recherche/projets/FaceTales/en/Home.htm (last accessed date: 21st May 2012)
Pitel AL, Beaunieux H, Witkowski T, Vabret F, Guillery-Girard B, Quinette P et al (2007) Genuine episodic memory deficits and executive dysfunctions in alcoholic subjects early in abstinence. Alcoholism, Clinical and Experimental Research 31:1169–1178
Polich J (2004) Clinical application of the P300 event-related brain potential. Physical Medicine and Rehabilitation Clinics of North America 15:133–161
Porjesz B, Rangaswamy M, Kamarajan C, Jones KA, Padmanabhapillai A, Begleiter H (2005) The utility of neurophysiological markers in the study of alcoholism. Clinical Neurophysiology 116(5):993–1018
Rämä P, Courtney SM (2005) Functional topography of working memory for face or voice identity. NeuroImage 24:224–234
Riley H, Schutte NS (2003) Low emotional intelligence as a predictor of substance-use problems. Journal of Drug Education 33:391–398
Rossignol M, Philippot P, Douilliez C, Crommelinck M, Campanella S (2005) The perception of fearful and happy facial expression is modulated by anxiety: An event-related potential study. Neuroscience Letters 377(2):115–120
Rupp CI (2010) Olfactory function and schizophrenia: An update. Current Opinion in Psychiatry 23:97–102
Rupp CI, Fleischhacker W, Drexler A, Hausmann A, Hinterhuber H, Kurz M (2006) Executive function and memory in relation to olfactory deficits in alcohol-dependent patients. Alcoholism, Clinical and Experimental Research 30:1355–1362
Rupp CI, Fleischhacker W, Hausmann H, Mair D, Hinterhubern H, Kurz M (2004) Olfactory functioning in patients with alcohol dependence: Impairments in odor judgments. Alcohol and Alcoholism 39:514–519
Rupp CI, Kurz M, Kemmler G, Mair D, Hausmann A, Hinterhuber H et al (2003) Reduced olfactory sensitivity, discrimination, and identification in patients with alcohol dependence. Alcoholism, Clinical and Experimental Research 27:432–439

Salloum JB, Ramchandani VA, Bodurka J, Rawlings R, Momenan R, George D et al (2007) Blunted rostral anterior cingulate response during a simplified decoding task of negative emotional facial expressions in alcoholic patients. Alcoholism, Clinical and Experimental Research 31:1490–1504

Schiffman M (1997) Taste and smell losses in normal aging and disease. Journal of the American Medical Association 278:1357–1362

Schulte T, Müller-Oehring EM, Pfefferbaum A, Sullivan EV (2010) Neurocircuitry of emotion and cognition in alcoholism: Contributions from white matter fiber tractography. Dialogues in Clinical Neuroscience 12(4):554–560

Schweinberger SR, Herholz A, Sommer W (1997) Recognizing famous voices: Influence of stimulus duration and different types of retrieval cues. Journal of Speech, Language, and Hearing Research 40:453–463

Seubert J, Kellermann T, Loughead J, Boers F, Brensinger C, Schneider F et al (2010a) Processing of disgusted faces is facilitated by odor primes: A functional MRI study. NeuroImage 53(2):746–756

Seubert J, Loughead J, Kellermann T, Boers F, Brensinger CM, Habel U (2010b) Multisensory integration of emotionally valenced olfactory–visual information in patients with schizophrenia and healthy controls. Journal of Psychiatry & Neuroscience 35(3):185–194

Shepherd GM (2006) Smell images and the flavour system in the human brain. Nature 444: 316–321

Small DA (2004) Crossmodal integration – insights from the chemical senses. Trends in Neuroscience 27:123–124

Smith ME, Oscar-Berman M (1992) Resource-limited information processing in alcoholism. Journal of Studies on Alcohol 53:514–518

Spitzer JB (1981) Auditory effects of chronic alcoholism. Drug and Alcohol Dependence 8:317–335

Spitzer JB, Ventry IM (1980) Central auditory dysfunction among chronic alcoholics. Archives of Otolaryngology 106:224–229

Sullivan EV, Mathalon DH, Zipursky RB, Kersteen-Tucker Z, Knight RT, Pfefferbaum A (1993) Factors of the Wisconsin Card Sorting Test as measures of frontal lobe function in schizophrenia and in chronic alcoholism. Psychiatry Research 46:175–199

Szabo Z, Owonikoko T, Peyrot M, Varga J, Mathews WB, Ravert HT et al (2004) Positron emission tomography imaging of the serotonin transporter in subjects with a history of alcoholism. Biological Psychiatry 55:766–771

Szczepanska L, Baran J, Mikolaszek-Boba M (2004) Connection between personality and emotional intelligence in groups of patients after suicidal attempts and ethanol dependent persons. Przeglev Lekvi 61:287–291

Szycik GR, Münte TF, Dillo W, Mohammadi B, Samii A, Emrich HM et al (2009) Audiovisual integration of speech is disturbed in schizophrenia: An fMRI study. Schizophrenia Research 110(1–3):111–118

Taieb O, Corcos M, Loas G, Speranza M, Guilbaud O, Perez-Diaz F et al (2002) Alexithymia and alcohol dependence. Annales de Médecine Interne 153:51–60

Teder-Sälejärvi WA, McDonald JJ, Di Russo F, Hillyard SA (2002) An analysis of audiovisual crossmodal integration by means of event-related potentials (ERP) recordings. Cognitive Brain Research 14:106–114

Townshend JM, Duka T (2003) Mixed emotions: Alcoholics' impairments in the recognition of specific emotional facial expressions. Neuropsychologia 41:773–782

Townshend JM, Duka T (2005) Binge drinking, cognitive performance and mood in a population of young social drinkers. Alcoholism, Clinical and Experimental Research 29:317–325

Turetsky BI, Hahn CG, Borgmann-Winter K, Moberg PJ (2009) Scents and nonsense: Olfactory dysfunction in schizophrenia. Schizophrenia Bulletin 35:1117–1131

Uekermann J, Channon S, Winkel K, Schlebusch P, Daum I (2007) Theory of mind, humour processing and executive functioning in alcoholism. Addiction 102:232–240

Uekermann J, Daum I (2008) Social cognition in alcoholism: A link to prefrontal cortex dysfunction? Addiction 103(5):726–735

Uekermann J, Daum I, Schlebusch P, Trenckmann U (2005) Processing of affective stimuli in alcoholism. Cortex 41:189–194

Uzun O, Ats A, Cansever A, Ozsahin A (2003) Alexithymia in male alcoholics: Study in a Turkish sample. Comprehensive Psychiatry 44:349–352

Van der Stelt O, Gunning WB, Snel J, Zeef E, Kok A (1994) Children of alcoholics: Attention, information processing and event-related brain potentials. Acta Paediatrica 404(4):6

Vatakis A, Spence C (2006) Audiovisual synchrony perception for speech and music assessed using a temporal order judgment task. Neuroscience Letters 393(1):40–44

Wagner U, N'diaye K, Ethofer T, Vuilleumier P (2011) Guilt-specific processing in the prefrontal cortex. Cerebral Cortex 21(11):2461–2470

Winston JS, Gottfried JA, Kilner JM, Dolan RJ (2005) Integrated neural representations of odor intensity and affective valence in human amygdala. Journal of Neuroscience 25:8903–8907

Zeigler DW, Wang CC, Yoast RA, Dickinson BD, McCaffree MA, Robinowitz CB et al (2005) The neurocognitive effects of alcohol on adolescents and college students. Preventive Medicine 40:23–32

Zupan B, Sussman JE (2009) Auditory preferences of young children with and without hearing loss for meaningful auditory–visual compound stimuli. Journal of Communication Disorders 42(6):381–396

Chapter 15
The Role of Audition in Audiovisual Perception of Speech and Emotion in Children with Hearing Loss

Barbra Zupan

This chapter contains a video segment which can be found at the URL: http://www.springerimages.com/Belin

Abstract Integration of visual and auditory cues during perception provides us with redundant information that greatly facilitates processing. Hearing loss affects access to the acoustic cues essential to accurate perception of speech and emotion, potentially impacting audiovisual integration. This chapter explores the various factors that may impact auditory processing in persons with hearing loss, including communication environment, modality preferences, and the use of hearing aids versus cochlear implants. Audiovisual processing of both speech and emotion by children with hearing loss is also discussed.

Successful communication requires us to accurately interpret the information we receive in the speaker's message. To do this, we must make use of both the segmental and suprasegmental information in the auditory portion of the message and integrate that information with the visual cues provided from the face; a process called audiovisual integration. Audiovisual integration is both automatic and robust, leading to a super-additive effect in bimodal versus unimodal processing (Bergeson, Pisoni, & Davis, 2003; Besle, Fort, Delpuech, & Giard, 2004; Brancazio & Miller, 2005; de Gelder & Vroomen, 2000; Hietanen, Leppanen, Illi, & Surakka, 2004; Massaro & Light, 2004). Audiovisual integration is both automatic and robust, leading to a super-additive effect in bimodal versus unimodal processing (Bergeson et al., 2003 ; Besle et al., 2004; Brancazio & Miller, 2005; de Gelder & Vroomen, 2000; Hietanen et al., 2004; Massaro & Light, 2004). In other words, a person's accuracy in the processing of bimodal cues is significantly greater than would be expected from the added contributions of each unimodal modality (Hay-McCutcheon, Pisoni, & Kirk, 2005). This occurs even when both the auditory and visual signals are fully available since the information provided in each modality

B. Zupan (✉)
Department of Applied Linguistics, Brock University, 500 Glenridge Ave., St. Catharines, ON, Canada, L2S 3A1
e-mail: bzupan@brocku.ca

P. Belin et al. (eds.), *Integrating Face and Voice in Person Perception*,
DOI 10.1007/978-1-4614-3585-3_15,

is complementary, not redundant. Thus, the cues provided by the more informative modality assist in clearing up the ambiguity of the information provided by the other, leading to a robust, integrated representation of the visual and auditory cues (Massaro & Light, 2004).

Information in a bimodal signal may also become degraded if the visual or auditory information is insufficient due to a sensory impairment. Hearing loss is one such example of this. Nevertheless, persons with hearing loss have been shown to perceive audiovisual information with fairly high accuracy even though they are faced with a less than optimal auditory signal (Bergeson et al., 2003; Bergeson, Pisoni, & Davis, 2005; Grant & Seitz, 2000; Grant, Walden, & Seitz, 1998; Kaiser, Kirk, Lachs, & Pisoni, 2003; Lachs, Pisoni, & Kirk, 2001; Massaro & Light, 2004; Strelnikov, Rouger, Barone, & Deguine, 2009). Some have argued that the success of hearing impaired listeners in audiovisual processing is likely due to superior visual processing; a function of neural plasticity (Bernstein, Demorest, & Tucker, 2000; Capek et al., 2008; Champoux, Lepore, Gagne, & Theoret, 2009; Giraud, Price, Graham, Truy, & Frackowiak, 2001; Giraud & Truy, 2002; Mitchell & Maslin, 2007; Sadato et al., 2004). However, scores on audiovisual processing still exceed visual-only processing, suggesting that even when the auditory cue is less than optimal, it continues to contribute to the overall processing of a message by people with hearing loss.

When hearing status is not in question, and both the visual and auditory channels provide optimal information to the listener, the modality that is preferred in processing will more greatly influence the interpretation of the combined signal. The preferred modality for processing is developmentally determined, with infants and young children showing an auditory preference in processing and adults showing a visual preference. The developmental shift that occurs in processing is considered central to the development of language since an early focus on the auditory signal allows infants and young children the opportunity to process the transient and dynamic acoustic information (Napolitano & Sloutsky, 2004; Robinson & Sloutsky, 2004; Sloutsky & Napolitano, 2003; Zupan & Sussman, 2009). Processing preferences are seemingly automatic in nature and will certainly influence the prevalence given to one cue over another when integrating bimodal cues. However, we know very little about the impact of an anomalous hearing system on developmental processing preferences. This becomes a particularly important issue when considering the communication and treatment options available for children with hearing loss; options that range from an emphasis on visual information to an emphasis on auditory information (Zupan & Sussman, 2009).

Much of the research in the integration of bimodal cues in children has been in the area of speech perception, for both listeners with and without hearing loss. However, audiovisual processing is not unique to speech perception; visual and auditory cues of emotion are processed in much the same way. Audiovisual processing of emotion is gaining attention in the literature, but the focus has been primarily in adults without hearing loss. The chapter that follows will explore the impact of hearing loss on the perception of auditory information and how a degraded auditory signal might impact audiovisual processing of speech and emotion for children with hearing loss. Additional factors that may influence audiovisual integration in children with hearing loss will also be discussed, including the influence of developmental modality preferences and the chosen communication environment of the child.

1 The Impact of Hearing Loss on the Perception of Acoustic Cues

Studying the processing abilities of children with hearing loss allows us to further investigate the significance of acoustic cues in audiovisual processing, particularly for the speech sounds and emotion expressions in which the face does not provide distinctive information. For example, without access to auditory information, discrimination between minimal pairs such as "mom" and "mop" would be impossible based on visual cues alone. The addition of auditory information provides the listener with cues about the frequency of the sound, including the frequency transitions that occur as the tongue moves from one position in the mouth to another to articulate the consonants and vowels. Amplitude and durational cues simultaneously add information that is essential for accurate perception and discrimination of speech sounds (Chatterjee & Peng, 2008; Peng, Tomblin, & Turner, 2008). Children with hearing loss have significant difficulty processing these cues because the damaged hair cells in their cochlea significantly limit the auditory signals available. Although hearing aids and cochlear implants provide improved access to the auditory signal, acoustic cues are processed differently by these technological devices than by the human cochlea. Following is a brief description of the acoustic cues available to listeners with normal hearing for speech and emotion processing, and how these cues may be altered when being processed through a hearing aid or cochlear implant.

Fundamental frequency (F_0), or the pitch at which the sound is produced, is one of the primary acoustic cues in the perception of speech sounds and vocal expressions of emotion. The pitch of a person's voice results from the rate at which the vocal folds vibrate, an act that is biologically determined by the length and mass of the vocal folds. When our vocal folds vibrate, they create a complex sound that consists of numerous frequencies related to the F_0 of our voice; frequencies referred to as harmonics. An undamaged ear is capable of combining these harmonics in processing and encoding the fine frequency and temporal distinctions between them by tonotopically mapping them onto the cochlea when encoding them (Chatterjee & Peng, 2008). This allows listeners to hear differences between sounds that differ minimally in frequency. The frequency and temporal encoding that takes place in the cochlea provides important information about where and how in the mouth a sound was articulated, allowing us to discriminate between sounds such as /t/ and /k/ or /d/ and /n/.

Damaged hair cells in the cochlea result in poor frequency resolution because the ear is no longer capable of processing such fine frequency distinctions or the accompanying temporal information (Friesen, Shannon, Baskent, & Wang, 2001; Gantz,

Turner, Gfeller, & Lowder, 2005; Shannon, 2002; Turner, Chi, & Flock, 1999). Hearing aids will make the sound more audible for children hearing loss, but the resulting sound is still less than ideal; the sound remains distorted because damaged hair cells cannot be restored. High frequency sounds are particularly affected, negatively impacting the child's ability to perceive and identify speech sounds such as /s/ (Healy & Bacon, 2002; Stelmachowicz, Pittman, Hoover, & Lewis, 2001; Stelmachowicz, Pittman, Hoover, Lewis, & Moeller, 2004). However, listeners using hearing aids are still able to make use of temporal cues to assist with frequency perception (Turner, Souza, & Forget, 1995). Cochlear implants, on the other hand, are able to bypass the damaged hair cells and directly stimulate the nerve. However, unlike the cochlea that is designed to process very fine differences in frequency, the electrodes inserted in the cochlea during cochlear implantation are responsible for encoding frequencies that fall within a specific range. This arrangement of frequency distribution is necessary so that the limited number of electrodes can capture the overall frequency range typically encoded by the undamaged ear (Faulkner, Rosen, & Smith, 2000; Giezen, Escudero, & Baker, 2010; Kong, Stickney, & Zeng, 2005; Peng et al., 2008). However, this arrangement also results in poor frequency and temporal resolution because the frequency range assigned to each electrode does not allow for processing of the individual harmonics (Friesen et al., 2001; Gantz et al., 2005; Shannon, 2002). Additionally, because each electrode is responsible for a range of frequencies, some of the tonotopic mapping that is typical of the normal cochlea is lost, further hindering frequency resolution (Friesen et al., 2001; Geurts & Wouters, 2001; Giezen et al., 2010; Peng et al., 2008). Spectral information related to lower frequency sound is particularly affected (Chatterjee & Peng, 2008; Green, Faulkner, Rosen, & Macherey, 2005; Peng et al., 2008).

Although F_0 is biologically determined and related to the size and mass of a person's vocal folds, speech sounds are identifiable by the relevant frequency range in which they are produced. For instance, /s/ is a high frequency speech sound that is associated with frication noise at approximately 4,000 Hz while /m/ is a low frequency sound associated with nasality that occurs at approximately 250 Hz (Ling, 1989). We also regularly manipulate the pitch of our voice for linguistic and social purposes. For instance, we may increase the pitch of our voice at the end of a sentence to indicate that we are asking a question, a linguistic function often referred to as intonation. We may also produce a word or sentence at a particular pitch, or vary pitch across a word or sentence to convey a specific emotion in the message we are delivering. Fearful, for example, is associated with increased pitch as well as fluctuations in that pitch across the sentence (Zupan, Neumann, Babbage, & Willer, 2009).

A second acoustic parameter important to both speech and emotion perception is intensity. Intensity reflects the overall energy in a speech signal and is based upon amplitude changes. We are able to increase or decrease intensity by manipulating the amount of air we force through our vocal folds at a given moment in time while speaking. For instance, to increase the intensity of our voice to call out to a friend, we increase the amount of air we take into our lungs, and quickly force that air from our lungs and through our vocal folds. Sounds that are voiced, indicating that the vocal folds actually touch while vibrating, are produced with more intensity than their unvoiced counterparts. For instance, speakers produce /b/ with more intensity than /p/. Intensity then provides important information about the energy of a sound which assists listeners in discriminating between speech sounds. An undamaged ear is able to process a wide range of amplitude and intensity changes in the speaker's voice, referred to as the dynamic range. Similar to reductions in frequency, this acoustic feature is limited for persons with cochlear hearing loss, leaving them with a reduced dynamic range in processing (Fu & Shannon, 1999; Loizou, Poroy, & Dorman, 2000). However, a limited amplitude range appears to have only minimal negative effects on processing and can be maximized for cochlear implant and hearing aid users through adjustments in the processing strategies of the device (Henning & Bentler, 2008; Jenstad, Pumford, Seewald, & Cornelisse, 2000; Loizou, Poroy et al., 2000).

A typical speaker can produce sounds that range in intensity by approximately 30–60 dB (Loizou, Dorman, & Fitzke, 2000; Loizou, Poroy et al., 2000). Similar to frequency, we purposefully manipulate intensity within this range for linguistic and emotion purposes. For example, Intensity is an important indicator of the linguistic cue of stress, a cue that helps differentiate the noun "pre-SENT" from the verb "PRE-sent". Intensity can also be manipulated to convey different emotions; an increase might indicate that someone is feeling angry, while a decrease might indicate that someone is feeling sad.

The duration of an acoustic signal also contributes important information for speech and emotion perception. For speech perception, duration can provide additional cues to help listeners differentiate between consonants. For instance, the minimal pairs "bad" and "bat" are easily discriminated via durational cues: the longer duration of the /a/ vowel in "bad" primes our auditory system for the perception of the voiced consonant /d/. For emotion perception, duration relates to rate of speech, an acoustic parameter that may be further influenced by inserting pauses of various lengths. Speakers can manipulate their use of pauses and overall rate of speech in order to convey different emotion states. For instance, sad is consistently conveyed by increasing both the number and length of pauses, leading to an overall decline in speech rate.

Clearly, having sufficient access to acoustic cues is essential to successful speech and emotion perception. Speech perception is based on perception of segmental cues, cues that relay F_0, intensity and durational cues about the vowels and consonants in the speaker's message. High frequency information is particularly important for identification and discrimination of segmental cues. Conversely, emotion perception relies on perception of suprasegmental information in the auditory portion of the message, information that consists primarily of low frequency acoustic cues. As discussed above, children with hearing loss continue to receive less than optimal access to acoustic cues. Frequency parameters appear to remain the most affected even with the use of advanced technological aids such as hearing aids and cochlear implants. Hearing aids can improve the audibility of frequency information but they are unable to correct deficiencies in frequency sensitivity. Cochlear implants improve the clarity of sound and provide better access to high frequency speech sounds but they also present listeners with challenges in frequency resolution, especially for low frequency information. Thus, it seems that although the nature of audiovisual processing is similar, children with hearing loss may face different challenges in speech and emotion perception, depending in part on the technological device they are using.

1.1 Additional Factors Impacting Auditory Processing in Listeners with Hearing Loss

1.1.1 Communication Environment

In order to develop strong audiovisual perception, we need experience in processing auditory information because it is the auditory modality that more greatly contributes to the enhancement of an audiovisual speech signal (Bergeson, Houston, & Miyamoto, 2010; Bergeson et al., 2005; Massaro & Cohen, 1999). The early development of the auditory system allows auditory exposure to sound to occur even prior to birth, while development of the visual system requires up to 6 months after birth to reach the same level of functioning (Bahrick & Lickliter, 2000; Banks & Salapatek, 1983; Grimwade, Walker, Bartlett, Gordon, & Wood, 1971). Children born with hearing loss then, are at an immediate disadvantage. However, universal newborn hearing screenings have significantly decreased the average age of detection of hearing loss in children, leading to earlier intervention, both in terms of access to hearing aids and cochlear implants, as well as an opportunity for early immersion in a communication environment that provides experience in auditory processing. Improved access to acoustic cues should naturally facilitate processing of the auditory signal and lead to better integration of visual and auditory cues, but research investigating audiovisual processing in children with hearing loss has shown tremendous individual differences. Thus, it appears that audiovisual processing depends on more than access to a sufficient visual and auditory signal. Communication environment may be one of the factors that impacts audiovisual processing.

When a child is diagnosed with hearing loss, families must choose a communication option that falls along a continuum, from visually focused to orally focused language environments. American Sign Language (ASL) is a visually focused language and is comprised of manual signs, facial expressions, gestures, and postures connected via a complex grammatical system. On the opposite end of the continuum lie two primary methods of oral-focused communication options: Auditory-Verbal and Auditory-Oral. Both Auditory-Verbal and Auditory-Oral communication environments place emphasis on the acoustic signal for learning spoken language, promoting early intervention and consistent amplification. These two oral communication environments significantly differ, however, in their use of visual cues; Auditory-Verbal environments limit access to visual information while the child is acquiring language, whereas Auditory-Oral environments encourage the use of visual information and emphasize the use of lipreading, facial expressions, and gestures to enhance speech perception. In the center of the continuum lies Total Communication, a communication option that encompasses elements from both visual and oral communication environments and as such is considered a multi-modality based communication environment (Zupan & Sussman, 2009). Presumably, an emphasis on vision versus audition in the development of language should presumably have differential effects on the development of audiovisual processing for children with hearing loss. However, research comparing communication environments that span the continuum is sparse, and no one "best" approach for language development has been identified (Gravel & O'Gara, 2003).

Despite the absence of a universally accepted communication environment for children with hearing loss, recent research has explored speech perception and spoken language development in children participating in Total Communication, Auditory-Oral, and Auditory-Verbal approaches to communication. This focus likely reflects the fact that up to 96 % of children are born to hearing parents, thus families are more likely to choose a communication environment that has at least some emphasis on oral language in order to maximize the child's engagement in the family's communication exchanges (Fitzpatrick, Angus, Durieux-Smith, Graham, & Coyle, 2008; Kurtzer-White & Luterman, 2003; Mitchell & Karchmer, 2004). Furthermore, technological advances in both hearing aids and cochlear implants are providing children with hearing loss better access to the acoustic cues of speech, further encouraging oral communication environments for children with hearing loss born to hearing parents. Additionally, research suggests that there is a developmental trajectory to the cues we pay most attention to when processing audiovisual information, with auditory information being particularly important for learning language. Such processing preferences need also be considered when discussing the impact of hearing loss on the perception and integration of auditory and visual cues.

1.1.2 Modality Preferences

The early development of the auditory system likely underlies the auditory preferences in processing reported in young children, a preference also posited as an essential contributor to vocabulary, language, and literacy development (Napolitano & Sloutsky, 2004; Robinson & Sloutsky, 2004; Sloutsky & Napolitano, 2003; Zupan & Sussman, 2009). Studies in bimodal processing in young children have indicated that this preference for auditory information leads to less influence of the visual signal in processing. This is particularly evident in (McGurk and MacDonald, 1976) type tasks in which children are presented with incongruent auditory and visual signals, such as a visual /ga/ paired with an auditory /ba/. Adults are typically greatly influenced by the conflicting visual information, reporting the perception of /da/, a phoneme not presented to either channel. Children, on the other hand, are less susceptible to this effect and are more likely to report /ba/ as the perceived phoneme.

Studies that have employed a modified switch design in which participants are first habituated to a stimulus that arbitrarily pairs a visual and auditory target, and then asked to indicate changes to either the visual or auditory portion of that stimulus, have confirmed the modality preferences in processing first shown by (McGurk and MacDonald, 1976; Napolitano and Sloutsky, 2004; Robinson and Sloutsky, 2004; and Sloutsky and Napolitano 2003). For instance, Sloutsky and Napolitano (2003) habituated children and adults to paired visual landscape images with patterns of auditory tones and then presented audiovisual targets in which either the visual image or the auditory pattern changed. Results showed that children relied on the auditory portion of the audiovisual stimulus to determine change, while adults relied on the visual portion. Zupan and Sussman (2009) also lent support to developmental changes in modality preferences through a task that included audiovisual representations of familiar and unfamiliar animals. Children and adults were instructed to select the stimulus that best represented the animal. Children were more likely to select the auditory portion of the stimulus as the preferred representation while adults were more likely to select the visual stimulus.

The dominance of auditory input in processing and language development raises questions about the impact hearing loss may have on natural processing preferences and audiovisual integration. Do children with hearing loss continue to rely on auditory information in processing audiovisual language, despite their anomalous hearing systems, or do they learn to rely on their visual systems in processing? The communication environment the child is immersed in may also impact these preferences. This topic will be explored below in discussion of audiovisual processing of both speech and emotion perception in children with hearing loss.

2 Audiovisual Processing of Speech in Children with Hearing Loss

The super-additive effect that occurs with audiovisual processing seems to result from the complementary nature of the information provided by the visual and auditory channels, in that the more prominent information provided by one modality compensates for the weaker information provided by the other. For example, although it is difficult to visually discriminate between bilabial consonants, (/b/, /p/, and /m/), acoustic cues provided by the auditory channel allows the perceiver to determine if the consonant is the oral voiced bilabial /b/, oral voiceless bilabial /p/, or the nasal bilabial /m/. Thus, accuracy increases because the more salient auditory signal allows the perceiver to eliminate all but one of the numerous possibilities generated through the visual channel. Auditory and visual cues of emotion are also complementary. For instance, emotions that are poorly identified on the basis of auditory information are typically more easily identified using visual information (Burkhardt & Sendlmeier, 2000; Scherer, 2003; Zupan et al., 2009). Happy is one such example of this. Thus, when receiving an audiovisual signal, our perceptual system integrates the information in such a way that the information provided by each modality is used in the most effective way possible. If information in one modality is ambiguous or degraded in quality, the complimentary information provided by the other modality can "make up for it," allowing the perceiver to accurately interpret the message (Hay-McCutcheon et al., 2005; Kaiser et al., 2003; Massaro, 1999; Sommers, Tye-Murray, & Spehar, 2005; Tye-Murray, Sommers, & Spehar, 2007).

As discussed earlier, the auditory information in an audiovisual signal is always degraded for listeners with hearing loss, regardless of hearing technology. Acoustic cues provide essential information that allow for accurate identification and discrimination of segmental information and are thus, the more salient cues in speech processing for listeners with normal hearing (Massaro & Cohen, 1999). Without the addition of auditory information, some speech sounds are indistinguishable. However, listeners with normal hearing are able to combine the prominent acoustic cues with ambiguous visual cues for enhanced speech perception. But what happens when the auditory signal is distorted and less audible, as is the case for persons with hearing loss who rely on hearing aids to access auditory information? Erber (1972) demonstrated that even the highly degraded auditory signal provided by hearing aids can contribute to improved segmental processing when both auditory and visual information is present. Erber presented audiovisual (AV), auditory-only (AO), and visual-only (VO) consonant segments to children with normal hearing, children with severe hearing impairment, and children with profound hearing impairment. Identification accuracy for consonants presented in the VO condition was similar across all three groups of children. However, there were significant differences in the processing of AO and AV segments across these three groups. As expected, the decreased audibility of the acoustic information in children with hearing loss significantly decreased identification accuracy of AO segments for children with severe (50 %) and profound (21 %) hearing loss, as compared to children with normal hearing (99 %). Given the near perfect identification of AO segments for children with normal hearing, there was no enhancement in perception for AV consonants. However, children with severe hearing impairment were able to make substantial use of the AV signal, increasing consonant recognition from 50 to 88 %. Profoundly deaf children made only minimal gains in identification of AV segments, showing very little improvement over the VO condition (Bergeson et al., 2005; Erber, 1972).

Erber's (1972) study was one of the few studies to investigate audiovisual speech perception in hearing aid users and his work has been influential in advocating the importance of combined auditory and visual information in processing for persons with hearing loss (Power & Hyde, 1997). His results showed that even an anomalous auditory system can lead to enhanced processing, provided that the acoustic information is at least partially audible, as was the case with the listeners with severe hearing impairment. However, his results also demonstrated that listeners with profound hearing loss do not have sufficient access to the acoustic cues necessary for segmental perception through the use of hearing aids and are therefore unable to benefit from the auditory information in an AV signal. Hearing aids are simply unable to provide sound of sufficient quality and audibility to listeners with such significant hearing loss. Thus, these individuals must rely on the visual portion of an AV stimulus during perception, leading to decreased accuracy in speech perception because of the ambiguity that occurs in VO processing.

Cochlear implants have been shown to successfully resolve the speech perception challenges faced by persons with profound hearing loss by providing better access to high frequency sounds, allowing for better identification and discrimination of segmental information. Even though the signal received through the implant is lacking in frequency resolution, the auditory information provided by the electrical signal is sufficient for processing the frequency, temporal and durational cues that differentiate consonant and vowel segments. Similar to Erber's findings, better access to acoustic cues has led to improved AV processing of segmental information in cochlear implant users (Bergeson et al., 2010; Geers & Brenner, 1994; Geers, Brenner, & Davidson, 2003; Kaiser et al., 2003; Kirk et al., 2007; Lachs et al., 2001; Staller, Dowell, Beiter, & Brimacombe, 1991). The acoustic signal, even if received electrically, adds essential information to enhance audiovisual processing of words and sentences.

Numerous studies have reported AV enhancement in the processing of words and sentences for children using cochlear implants. Lachs et al. (2001) presented 27 children ranging in age from 4.2 to 8 years with a series of phrases in AO, VO, and AV conditions. Although age of implantation varied across the group, all of the children had been using their implant for 2 years. Results showed that children were most accurate in identification of the phrases when provided with combined auditory and visual information, than when given auditory or visual information in isolation. Thus, Lachs et al.'s (2001) study supports AV enhancement for listeners with cochlear implants. However, the reported AV benefit for this group of children was simply additive; in other words, the children in this study received equal benefit from the addition of auditory information to a visual signal as they did from visual information to an auditory signal. This differs from listeners with normal hearing who make greater use of the auditory than visual cues when processing segmental speech information (Bergeson et al., 2005; Massaro & Cohen, 1999).

Research in the speech and language skills of children with cochlear implants has repeatedly shown that early identification of hearing loss and subsequent early implantation leads to significantly better outcomes in auditory development and segmental perception (Bergeson et al., 2010; Giezen et al., 2010; Harrison, Gordon, & Mount, 2005; Svirsky, Robbins, Kirk, Pisoni, & Miyamoto, 2000; Wie, 2010). In fact, children who are implanted early have been reported to have similar auditory development to children with normal hearing (Robbins, Koch, Osberger, Zimmerman-Phillips, & Kishon-Rabin, 2004; Sharma, Dorman, & Spahr, 2002). They should then also be able to combine auditory information with visual information similarly to children without hearing loss. The children in the Lachs et al. (2001) study were approximately 4.5 years of age at the time they received their implants, an age considered quite late by today's standards. Thus, the delay in access to audition would have presumably made maximizing the auditory signal in audiovisual processing more challenging.

Bergeson et al. (2003) demonstrated the importance of early implantation for AV processing in a study that compared identification of words and sentences by children who had been implanted either before or after 53 months of age. Words and sentences were presented in a closed-set task under three conditions: AO, VO, and AV.

Results showed that children were most accurate in identification of the words and sentences in the AV condition, regardless of the age of implantation. However, after 2 years of implant use, children who had been implanted prior to 53 months of age showed larger improvements in AO processing, leading to more similar identification scores in the AV and AO conditions (Bergeson et al., 2003). These results show that early implantation and increased experience with the cochlear implant allow children to make adequate use of the electrical signal and integrate it with visual information to maximize speech perception in a way that is similar to persons without hearing loss (Massaro & Light, 2004).

In a follow-up study, Bergeson et al. (2005) employed an open-set sentence comprehension task with 80 children, creating an early and late implant group with the median age of 53 months. The children were presented a unique set of sentences in each presentation format: AO, VO, and AV. Similar to Bergeson et al. (2003), results indicated that children who had been implanted early and had been using their implant for a longer period of time were more accurate in identification of sentences in AO and AV conditions than sentences presented in VO conditions. Additionally, these children also received greater enhancement in processing from the addition of auditory versus visual information in the AV signal (Bergeson et al., 2005). These results are similar to what we would expect from children with normal hearing, with greater emphasis being placed on the auditory information when processing audiovisual information. However, children who were implanted later and had less auditory experience with their implants seemed to rely more on the visual information when processing, a result that is more consistent with hearing aid users with profound loss and inadequate access to the acoustic signal (Erber, 1972).

Taken together, the results of Bergeson et al.'s (2003; 2005) work suggest that even children with profound hearing loss are able to rely on auditory information to receive maximal enhancement in AV speech processing, provided they have received early access to sound and have had time to adjust to processing the electrical signal provided by the cochlear implant. Kirk et al. (2007) lent further support to this premise in a study that investigated AV, AO, and VO sentence processing in children who had received their cochlear implant prior to 24 months of age. As we would expect from children with normal hearing, children in the Kirk et al. (2007) study were most accurate when they received both auditory and visual information. Moreover, they were able to identify sentences well using only auditory information, and they relied more heavily on this auditory information when processing AV sentences.

2.1 Additional Factors in Audiovisual Processing of Speech

Early access to sound and immersion into environments that place emphasis on audition and oral language have been shown to greatly contribute to improved speech and language skills of children with hearing loss (Bergeson et al., 2003, 2005; Blamey et al., 2001; Giezen et al., 2010; Harrison et al., 2005; Houston,

Pisoni, Kirk, Ying, & Miyamoto, 2003; Meyer, Svirsky, Kirk, & Miyamoto, 1998; O'Donoghue, Nikolopoulos, & Archbold, 2000; Robbins et al., 2004; Sarant, Blamey, Dowell, Clark, & Gibson, 2001; Sharma et al., 2002; Sininger, Grimes, & Christensen, 2010; Snik, Makhdooum, Vermeulen, Brokx, & van den Broek, 1997; Svirsky et al., 2000). Hence, the communication option chosen by families will impact the auditory development of the child, and thus impact audiovisual processing of speech. Studies investigating audiovisual processing in children relying on Oral versus Total Communication environments have consistently reported that children immersed in Oral environments are better able to process the auditory signal in isolation, leading to improved audiovisual processing (Bergeson et al., 2003, 2005; Lachs et al., 2001; Meyer et al., 1998; O'Donoghue et al., 2000). Thus, it appears that being immersed in a communication environment that is more similar to children with normal hearing allows children with hearing loss to also process audiovisual information similarly. Additionally, it has been suggested that children participating in Total Communication environments may be disadvantaged in audiovisual processing because they have learned to rely equally on the visual and auditory modalities in processing, rather than focusing on extracting information from the modality providing the most salient cues (Bergeson et al., 2003, 2005).

Oral Communication environments place greater emphasis on audition and oral language than Total Communication environments. This may be an important distinction for young children who are learning language, since research in modality preferences has indicated that children focus on auditory information in processing and that this focus is essential to word learning and language development (Napolitano & Sloutsky, 2004; Robinson & Sloutsky, 2004; Sloutsky & Napolitano, 2003). The studies discussed above show that children using cochlear implants are also able to place emphasis on the auditory signal when processing audiovisual information, provided they have received access to sound early in life and are immersed in environments that focus on audition and oral language. Thus, they appear to process audiovisual information similarly to children without hearing loss. But what happens when the audiovisual signal is not complementary in nature, but is instead incongruent? Studies in modality preferences have indicated that children continue to focus on the auditory information when presented with an incongruent AV signal. But what cues do children with hearing loss use under similar conditions?

Studies investigating modality preferences in children with hearing loss are limited and inconsistent in their conclusions. In an early study investigating the relationship between hearing loss and the degree to which children rely on one sensory modality versus the other, Seewald, Ross, Giolas, and Yonovitz (1985) presented a large group of hearing aid users between 7 years, 5 months and 14 years, 8 months with AO, VO, and AV stimuli. The AV stimuli were presented under either congruent (matching visual and auditory information) or incongruent (conflicting visual and auditory information) conditions. The design was based on McGurk and MacDonald's (1976) classic study that investigated responses of children and adults to conflicting AV stimuli. However, unlike McGurk and MacDonald's study, Seewald et al. (1985) purposefully avoided presenting stimuli that would result in illusory percepts (i.e., a /da/ response to a visual /ga/ paired with an auditory /ba/). Results showed that children's modality preferences were dependent upon the degree of accessibility to the acoustic signal. Recall that all children in this study were hearing aid users; thus, children with hearing loss greater than 95 dB had very limited access to the acoustic cues necessary for perception of segmental information. Not surprisingly then, they relied primarily on visual cues when processing incongruent AV signals. However, children who had hearing loss in the 60–90 dB range and more access to acoustic cues through their hearing aids were more likely to rely on the auditory information. This was especially true for children who were immersed in oral communication environments. Seewald and colleagues concluded that modality preferences in children with hearing loss are directly related to accessibility of the acoustic information as provided by hearing aids.

The visual preference for children with profound hearing loss reported by Seewald et al. (1985) is not surprising given the very limited amount of acoustic information those children would have received through hearing aids. Children with similar losses (95 dB or greater) are now receiving cochlear implants and have better access to the auditory information necessary for segmental processing. However, Schorr, Fox, van Wassenhove, and Knudsen (2005) also found a tendency for visual preferences in speech processing in a group of cochlear implant users. They presented a McGurk task to children with and without hearing loss between the ages of 5 and 14 years. The 36 children with hearing loss had been using a cochlear implant for at least 1 year and were immersed in oral environments. Unimodal stimuli were presented in AO and VO conditions and AV stimuli were presented as congruent (i.e., visual /pa/ paired with auditory /pa/) and incongruent (i.e., visual /ka/ paired with auditory /pa/) segments. As expected, children with normal hearing performed well in AO and congruent AV trials. In the incongruent AV trials, children with normal hearing integrated the two sources of information approximately half the time, reporting hearing the illusory percept /ta/. As reported by McGurk and MacDonald (1976), a /ta/ response indicates that the visual information influenced the auditory information in processing. Children who did not integrate the auditory and visual information into an illusory percept reported perceiving the auditory portion of the stimulus (/pa/), a response that is consistent with reported modality preferences in young children. Similar to children with normal hearing, the children using cochlear implants were also accurate in their perception of AO and congruent AV trials. However, their performance on incongruent AV trials differed significantly from children with normal hearing. Results for these children showed minimal integration of the incongruent stimulus suggesting that cochlear implant users have difficulty with audiovisual processing. Additionally, children with hearing loss who did not fuse the incongruent AV stimulus into the illusory percept /ta/ were much more likely to report perceiving the visual portion of the stimulus (/ka/), a response that rarely occurred in children with normal hearing.

Overall, the results of Schorr et al. (2005) suggest that although children with hearing loss are able to integrate auditory and visual information when the cues are complementary, they do not use the auditory and visual information similarly to children with normal hearing when these cues are in conflict with one another. This was evidenced in poor fusion and increased visual responses to conflicting visual and auditory signals. However, there are a number of caveats that need to be considered when interpreting these results. First, the age of the children included in the study surpassed the upper age limit of children typically reported to have auditory preferences in processing. Such an age discrepancy may explain the tendency of the children with normal hearing in Schorr et al.'s study (Schorr et al., 2005) to be more greatly influenced by the visual information in the incongruent AV stimulus, as evidenced in their tendency to perceive /ta/. Perhaps the children showing the most prominent fusion of the conflicting signals into an illusory percept were the children in the upper range of the 5- to 14-year age-span. Similarly, the children who were not influenced by the visual information in the incongruent AV stimulus and reported perceiving only the auditory segment /pa/ may have been those children in the lower range of the included age-span. This explanation reasonably accounts for the idiosyncratic responses of children with normal hearing, yet it does not explain the lack of fusion and increased visual responses reported for children with hearing loss, results that may be explained by a second caveat to the study.

The second caveat to the interpretation of Schorr et al.'s (2005) results is the lack of information about the relationship between age of implantation, duration of implant use, communication mode, and the responses provided during the McGurk task. Studies in congruent AV processing have highlighted the importance of early implantation, increased duration of use, and immersion in oral communication environments in leading to audiovisual integration skills that approximate those seen in children without hearing loss. Although the children in Schorr et al.'s study were all reported to be using oral communication, only 1 year of cochlear implant experience was required for study inclusion. Previous research has shown that children with hearing loss were unable to maximize auditory information for improved audiovisual integration until they had been using their cochlear implant for at least 2 years (Bergeson et al., 2003, 2005; Holt, Kirk, Eisenberg, Martinez, & Campbell, 2005). Thus, the children with hearing loss in Schorr et al.'s study who reported the visual percept /ka/ in response to incongruent AV stimuli may simply not have had early enough access to sound, nor enough experience with the cochlear implant to make adequate use of the auditory information.

Zupan and Sussman (2009) also investigated modality preferences in processing in children with hearing loss using cochlear implants. The children included in the study were similar to those in Schorr et al. (2005) in that they were all immersed in oral communication environments. But they differed in several important ways. First, the children were significantly younger, ranging in age from 2 years, 6 months to 5 years, 10 months. Additionally, they had all received their cochlear implants before 4 years of age and had been using them for at least 2 years. The paradigm in Zupan and Sussman's study also differed because it was not based on incongruent AV stimuli. Instead, modality preferences were determined through examination of children's selection of preferred representations (auditory versus visual) for audiovisual presentations of animals. Results indicated that children with hearing loss using cochlear implants showed a similar auditory preference to children with normal hearing.

Clearly, research investigating the impact of hearing loss on modality preferences in the processing of segmental information needs to continue. More specifically, how do factors such as age of implantation, duration of use, and communication environment contribute to modality preferences and AV processing in children with hearing loss? Knowing more about how hearing loss impacts audiovisual processing and the cues that are most salient in processing has important implications for speech perception and language development. Will restriction of visual cues as proposed in Auditory-Verbal therapies lead children to place emphasis on the auditory modality in processing, resulting in audiovisual processing that is comparable to their peers with normal hearing? Or would such an approach disadvantage these children because they do not learn to draw additional information from the visual signal in processing, a signal that may be more salient for them because of their difficulties in accessing acoustic cues?

3 Audiovisual Processing of Emotion in Children with Hearing Loss

It is easier for us to discriminate between emotions using facial cues than it is between sounds and words (Elfenbein, Marsh, & Ambady, 2002). However, we cannot assume that isolated facial cues of emotion will lead to accurate interpretation of the speaker's message because there may be conflicting information in the voice that is essential to the meaning of the message. Sarcasm is one such example of this. Thus, access to the acoustic information is just as important for emotion perception as it is for speech perception. In order to accurately perceive emotion, children need to be able to process the segmental information that comprises the verbal content of the emotion expression as well as the suprasegmental cues, and then integrate these two sources of auditory information with available visual cues.

Children born with hearing loss do not receive the auditory exposure typical infants receive both prior to and immediately following birth. This has important implications for emotional development and social competence because it may negatively impact parent–infant interactions in these early months. Although universal newborn screenings have led to earlier identification and amplification of hearing loss in infants, it still takes at least 6 months to more than 1 year for infants to complete the necessary testing and be fit with appropriate amplification. Thus, children are without adequate access to auditory information throughout this time period and are not benefiting from the exaggerated suprasegmental information typically used by parents, a form of language called parentese (Marschark, 1993). Consequently, they are not learning how to connect these suprasegmental cues to the facial expressions they are seeing. Learning to interpret and integrate the suprasegmental cues and facial cues of emotion is essential to the development of social competence (Maxim & Nowicki, 2003; Mayer, Salovey, & Caruso, 2004). Hearing aids, and particularly the increased use of cochlear implants in recent years, have led to improved speech and language outcomes for children with hearing loss, outcomes that appear to have contributed to improved understanding of language-based concepts related to emotion (Dyck & Denver, 2003; Peters, Remmel, & Richards, 2009; Rieffe & Terwogt, 2000). But, are children who use hearing aids and cochlear implants able to adequately process the acoustic cues of emotion and integrate them with facial expressions?

Successful speech and emotion perception both require accurate integration and interpretation of the auditory and visual information in a speaker's message, yet there are reasons to believe that hearing loss may impact audiovisual perception of emotion differently than it does speech perception. First, emotion perception differs from speech perception in the modality that carries the most salient information. Many speech sounds are ambiguous and nearly impossible to differentiate without the addition of acoustic cues. Thus, even children who perform well under VO conditions will have limited success in speech perception if they are unable to integrate the visual and auditory information. Emotion, on the other hand, is readily recognizable in the face and we naturally rely more heavily on this channel during perception (Elfenbein et al., 2002). Nevertheless, we cannot ignore the contribution of the verbal content and suprasegmentals conveyed through the auditory channel because these cues will facilitate processing by either confirming our interpretation of the visual information or by indicating conditions of sarcasm or deceit when these cues are inconsistent with one another (Collignon et al., 2008; de Gelder & Vroomen, 2000; Massaro & Egan, 1996; Pell, 2005).

As described earlier, hearing aids and particularly cochlear implants are able to provide children with adequate audibility for speech perception in both AO and AV conditions. However, neither the low frequency information available to hearing aid users nor the high frequency information available to cochlear implant users have been shown to be independently sufficient for the processing of acoustic information associated with emotion expressions. Kong et al. (2005) showed that even combining the low frequency acoustic information from a hearing aid with the high frequency electrical information of a cochlear implant did not lead to improved suprasegmental perception of melody than what occurred with the use of hearing aids alone. Thus, differences in the perception of vocal expression of emotion between children with and without hearing loss are expected given the reduced audibility and limited access to frequency information available through hearing aids and cochlear implants.

Research in audiovisual perception of emotion by children with hearing aids and cochlear implants is extremely limited. Similar to investigations of emotion perception in children without hearing loss, this research also tends to focus on unimodal processing or on audiovisual processing that combines dynamic acoustic cues with static photographs. Since everyday processing of emotion expressions is generally multimodal in nature and includes dynamic facial cues, it is difficult to infer from these studies how children with hearing loss might perceive audiovisual emotion expressions in their day-to-day social interactions. Following is a brief review of the relevant studies in AO, VO, and AV emotion processing in children with hearing loss.

Oster and Risberg (1986) considered the impact of hearing loss on the ability of children to perceive emotion using only the voice. They presented 18 hearing aid users (11–13 years of age) with moderate to severe hearing loss with sentences produced in four emotional tones: happy, sad, angry, and astonished. Results showed significant difficulty in the recognition of these emotions when compared to a control group of children without hearing loss, difficulties the authors attributed to poor perception of the frequency changes in the auditory message. Although poor perception of F_0 was not indicated as the underlying cause, Hopyan-Misakyan, Gordon, Dennis, and Papsin (2009) recently reported that children with cochlear implants also have more difficulty perceiving the suprasegmentals that differentiate vocal expressions of emotion than children with normal hearing. Interestingly, the children included in the study were all oral communicators who had been implanted early and had been using their implants for an average of 7 years. It appears then that the factors that contribute to improved auditory processing of the segmental information in speech may not lead to improved suprasegmental perception. Research in unimodal facial emotion processing has shown that children with hearing loss process facial expressions of emotion similarly to children with normal hearing, suggesting that they can make similar use of the salient visual cues provided through this modality (Hopyan-Misakyan et al., 2009; Hosie, Gray, Russell, Scott, & Hunter, 1998).

What are the potential implications of these results for emotion processing? If children with hearing loss are not superior in their ability to recognize emotion using isolated facial cues, and additionally have significant difficulty processing vocal cues of emotion, they may be less likely to receive benefit from an audiovisual signal. Studies investigating audiovisual processing of emotion expressions have in fact shown deficits for these children. Most, Weisel, and Zaychik (1993) were one of the first groups of researchers to examine the impact of hearing loss on AV emotion processing. They compared the performance of 24 adolescents with severe to profound hearing loss to a group of 19 adolescents without hearing loss. All of the participants with hearing loss wore hearing aids and were reported to use oral communication. Both groups of participants were asked to identify emotion expressions portraying anger, disgust, surprise, and sadness in AO, VO, and AV presentation formats. Results indicated that the participants with hearing loss were less accurate than participants with normal hearing in identification of emotion under all three presentation formats. Additionally, although more accurate than their identification of AO expressions, there was no significant difference between the VO and AV presentations. These results suggest that the adolescents with hearing loss were relying on only visual information to interpret emotion expressions, even when additional cues were given through the auditory channel.

Most and Aviner (2009) extended Most et al.'s (1993) exploration of audiovisual emotion processing by comparing three separate groups of adolescents: one group with normal hearing, one group who wore hearing aids, and one group who used cochlear implants. A semantically neutral sentence was portrayed in various emotion expressions in AO, VO, and AV formats. Results were similar to Most et al. (1993): Adolescents with hearing loss were the only group of participants to receive benefit from the auditory signal for enhanced AV processing. These results suggest that although cochlear implants are advantageous for auditory and audiovisual processing of segmental information, they add no additional benefits to children with hearing loss than hearing aids for the perception of suprasegmentals.

Taken together, the above research indicates that unlike speech perception, the improved audibility of access to high frequency sounds does not lead to improved emotion processing in cochlear implant versus hearing aid users. These results highlight the limitations of the auditory signal provided by the cochlear implant, limitations that even early implantation, extended duration of use, and experience with oral communication cannot rectify.

3.1 Additional Factors in Audiovisual Processing of Emotion

Difficulty in the processing of vocal expressions of emotion and in poor integration of this information with facial expressions of emotion may underlie the reported social competence issues in children with hearing loss (Compton & Niemeyer, 1994; Knutson, Boyd, Reid, Mayne, & Fetrow, 1997; Marschark, 1993; Vandell & George, 1981). Schorr, Roth, and Fox (2009) recently reported a positive correlation between reported improved quality of life and the ability to identify emotional sounds, such as a crying or giggling. Still, regardless of their relative ability to identify emotional sounds, children reported significant benefits of cochlear implant use in establishing social relationships. Perhaps the suprasegmental processing and audiovisual processing of emotion expressions is less important than the ability to accurately perceive the acoustic cues that are essential to successful AO and AV speech processing. Prior to the improvements in processing capacities of hearing aids and increased use of cochlear implants, Marschark (1993) suggested that children with hearing loss were at significant risk for poor social development because parents and teachers were less likely to talk with them about emotions, and when they did, the conversations were less complex. More recent research with children with hearing loss is reporting an improved understanding of language-based concepts related to emotion, concepts such as theory of mind, suggesting that parents and teachers are beginning to talk more about emotion with children with hearing loss (Dyck & Denver, 2003; Peters et al., 2009; Rieffe & Terwogt, 2000). This suggests that improvements in AV processing of segmentals may compensate for the poor AV emotion processing in children with hearing loss by providing them access to more accurate interpretation of the verbal content of the message.

Studies investigating perception of emotion expressions that consist of conflicting suprasegmental cues and verbal content have shown that children younger than 10 years of age rely primarily on verbal content to interpret the message (Eskritt & Lee, 2003; Friend, 2000; Morton & Trehub, 2001). This suggests that children with hearing loss may interpret messages that are incongruent in suprasegmental and verbal content similarly to children with normal hearing. There are important implications to this theory in terms of social competence development. If young children,

Fig. 15.1 An example of an incongruent audiovisual emotional stimulus would be combining an angry facial expression with a sad vocal expression

regardless of hearing status, are more likely to make use of the verbal content in an audiovisual signal, then we might expect children with hearing loss to have similar types of communication failures as children with normal hearing in social interactions when the AV signal conflicts only within the auditory channel. But, children with hearing loss may still be more negatively impacted when the AV signal conflicts only in the suprasegmental information.

There is currently no published research investigating AV processing of conflicting facial and vocal cues for children with or without hearing loss. Pilot work by this author has shown similar trends in modality preferences in response to incongruent audiovisual expressions of emotion as those reported for speech perception. Figure 15.1 provides a video example of an incongruent audiovisual emotional stimulus, combining an angry facial expression with a sad vocal expression. Further exploration of audiovisual processing of congruent and incongruent emotion expression in children with hearing loss is certainly needed, including conflicting segmental and suprasegmental information within the auditory channel. Investigating processing of emotion expressions that conflict across the visual, auditory, and verbal channels may provide us more insight into the cues these children are using for emotion processing and the potential impact on social competence.

4 Conclusion

Children with hearing loss are continually challenged in their daily interactions because the auditory signal they receive through their hearing aids or cochlear implants is degraded in both quantity and quality of the acoustic cues. However, children with hearing loss are still able to process this signal accurately enough to gain perceptual benefits in audiovisual speech processing, provided they have received early and continued access to sound through appropriate technology and an oral communication environment. Less is known about the impact of hearing loss on audiovisual processing of emotion, and what little information is available suggests much less success for children with hearing loss in this area. It appears that the auditory information available to children with hearing aids and cochlear implants is simply not sufficient for the processing of the suprasegmental information in audiovisual expressions of emotion. However, as technology continues to advance, and research in this area continues, we may begin to see the same improvements in audiovisual perception of emotion as we have recently seen in audiovisual perception of speech.

References

Bahrick, L. E., & Lickliter, R. (2000). Intersensory redundancy guides attentional selectivity and perceptual learning in infancy. *Developmental Psychology, 36*(2), 190–201.

Banks, M. S., & Salapatek, P. (1983). Infant visual perception. In M. M. Haith & J. H. Campos (Eds.), *Handbook of child psychology: Infancy and developmental psychobiology* (Vol. 2). New York: Wiley.

Bergeson, T. R., Houston, D. M., & Miyamoto, R. T. (2010). Effects of congenital hearing loss and cochlear implantation on audiovisual speech perception in infants and children. *Restorative Neurology and Neuroscience, 28*, 157–165.

Bergeson, T. R., Pisoni, D. B., & Davis, R. A. O. (2003). A longitudinal study of audiovisual speech perception by children who have hearing loss who have cochlear implants. *The Volta Review, 103*(4), 347–370.

Bergeson, T. R., Pisoni, D. B., & Davis, R. A. O. (2005). Development of audiovisual comprehension skills in prelingually deaf children with cochlear implants. *Ear and Hearing, 26*, 149–164.

Bernstein, L. E., Demorest, M. E., & Tucker, P. E. (2000). Speech perception without hearing. *Perception & Psychophysics, 62*, 233–252.

Besle, J., Fort, A., Delpuech, C., & Giard, M. (2004). Bimodal speech: Early suppressive visual effects in human auditory cortex. *European Journal of Neuroscience, 20*, 2225–2234.

Blamey, P. J., Sarant, J. Z., Paatsch, L. E., Barry, J. G., Bow, C. P., Wales, R. J., et al. (2001). Relationships among perception, production, language, hearing loss, and age in children with hearing impairment. *Journal of Speech, Language, and Hearing Research, 44*, 264–285.

Brancazio, L., & Miller, J. L. (2005). Use of visual information in speech perception: Evidence for a visual rate effect both with and without a McGurk effect. *Perception & Psychophysics, 67*(5), 759–769.

Burkhardt, F., & Sendlmeier, W. F. (2000). *Verification of acoustical correlates of emotional speech using formant-synthesis*. Paper presented at the ISCA Workshop on Speech and Emotion, Northern Ireland.

Capek, C. M., MacSweeney, M., Woll, B., Waters, D., McGuire, P. K., David, A. S., et al. (2008). Cortical circuits for silent speechreading in deaf and hearing people. *Neuropsychologia, 46*, 1233–1241.

Champoux, F., Lepore, F., Gagne, J., & Theoret, H. (2009). Visual stimuli can impair auditory processing in cochlear implant users. *Neuropsychologia, 47*, 17–22.

Chatterjee, M., & Peng, S. (2008). Processing F0 with cochlear implants: Modulation frequency discrimination and speech intonation recognition. *Hearing Research, 235*, 143–156.

Collignon, O., Girard, S., Gosselin, F., Roy, S., Saint-Amour, D., Lassonde, M., et al. (2008). Audio-visual integration of emotional expression. *Brain Research, 1242*, 126–135.

Compton, M. V., & Niemeyer, J. A. (1994). Expression of affection in young children with sensory impairments: A research agenda. *Education and Treatment of Children, 17*(1), 68–85.

de Gelder, B., & Vroomen, J. (2000). The perception of emotions by ear and by eye. *Cognition and Emotion, 14*(3), 289–311.

Dyck, M. J., & Denver, E. (2003). Can the emotion recognition ability of deaf children be enhanced? A pilot study. *Journal of Deaf Studies and Deaf Education, 8*(3), 348–356.

Elfenbein, H. A., Marsh, A., & Ambady, N. (2002). Emotional intelligence and the recognition of emotion from the face. In L. F. Barrett & P. Salovey (Eds.), *The wisdom of feelings: Processes underlying emotional intelligence* (pp. 37–59). New York: Guilford.

Erber, N. P. (1972). Auditory, visual, and auditory-visual recognition of consonants by children with normal and impaired hearing. *Journal of Speech and Hearing Research, 15*, 413–422.

Eskritt, M., & Lee, K. (2003). Do actions speak louder than words? Preschool children's use of the verbal-nonverbal consistency principle during inconsistent communication. *Journal of Nonverbal Behavior, 27*(1), 25–41.

Faulkner, A., Rosen, S., & Smith, C. (2000). Effects of the salience of pitch and periodicity information on the intelligibility of four-channel vocoded speech: Implications for cochlear implants. *Journal of the Acoustical Society of America, 108*(4), 1877–1887.

Fitzpatrick, E., Angus, D., Durieux-Smith, A., Graham, I. D., & Coyle, D. (2008). Parents' needs following identification of childhood hearing loss. *American Journal of Audiology, 17*, 38–49.

Friend, M. (2000). Developmental changes in sensitivity to vocal paralanguage. *Developmental Science, 3*(2), 148–162.

Friesen, L. M., Shannon, R. V., Baskent, D., & Wang, X. (2001). Speech recognition in noise as a function of the number of spectral channels: Comparison of acoustic hearing and cochlear implants. *Journal of the Acoustical Society of America, 110*(2), 1150–1163.

Fu, Q., & Shannon, R. V. (1999). Phoneme recognition by cochlear implant users as a function of signal-to-noise ratio and nonlinear amplitude mapping. *Journal of the Acoustical Society of America, 106*(2), 18–23.

Gantz, B. J., Turner, C., Gfeller, K. E., & Lowder, M. W. (2005). Preservation of hearing in cochlear implant surgery: Advantages of combined electrical and acoustical speech processing. *The Laryngoscope, 115*, 796–802.

Geers, A., & Brenner, C. (1994). Speech perception results: Audition and lipreading enhancement. *The Volta Review, 96*, 97–108.

Geers, A., Brenner, C., & Davidson, L. (2003). Factors associated with development of speech perception skills in children implanted by age five. *Ear and Hearing, 24*(1), 24–35.

Geurts, L., & Wouters, J. (2001). Coding of the fundamental frequency in continuous interleaved sampling processors for cochlear implants. *Journal of the Acoustical Society of America, 109*(2), 713–726.

Giezen, M. R., Escudero, P., & Baker, A. (2010). Use of acoustic cues by children with cochlear implants. *Journal of Speech, Language, and Hearing Research, 53*, 1440–1457.

Giraud, A., Price, C. J., Graham, J. M., Truy, E., & Frackowiak, S. J. (2001). Cross-modal plasticity underpins language recovery after cochlear implantation. *Neuron, 30*, 657–663.

Giraud, A., & Truy, E. (2002). The contribution of visual areas to speech comprehension: A PET study in cochlear implants patients and normal-hearing subjects. *Neuropsychologia, 40*, 1562–1569.

Grant, K. W., & Seitz, P. F. (2000). The use of visible speech cues for improving auditory detection of spoken sentences. *Journal of the Acoustical Society of America, 108*(3), 1197–1208.

Grant, K. W., Walden, B. E., & Seitz, P. F. (1998). Auditory-visual speech recognition by hearing-impaired subjects: Consonant recognition, sentence recognition, and auditory-visual integration. *Journal of the Acoustical Society of America, 103*(5), 2677–2690.

Gravel, J. S., & O'Gara, J. (2003). Communication options for children with hearing loss. *Mental Retardation and Developmental Disabilities Research Reviews, 9*, 243–251.

Green, T., Faulkner, A., Rosen, S., & Macherey, O. (2005). Enhancement of temporal periodicity cues in cochlear implants: Effects on prosodic perception and vowel identification. *Journal of the Acoustical Society of America, 118*(1), 375–385.

Grimwade, J. C., Walker, D. W., Bartlett, M., Gordon, S., & Wood, C. (1971). Human fetal heart rate change and movement in response to sound and vibration. *American Journal of Obstetrics and Gynecology, 109*(1), 86–90.

Harrison, R., Gordon, K., & Mount, R. (2005). Is there a critical period of congenitally deaf children? Analyses of hearing and speech perception performance after implantation. *Developmental Psychobiology, 46*(3), 252–261.

Hay-McCutcheon, M. J., Pisoni, D. B., & Kirk, K. I. (2005). Audiovisual speech perception in elderly cochlear implant recipients. *The Laryngoscope, 115*, 1887–1894.

Healy, E. W., & Bacon, S. P. (2002). Across-frequency comparison of temporal speech information by listeners with normal and impaired hearing. *Journal of Speech, Language, and Hearing Research, 45*, 1262–1275.

Henning, R. L. W., & Bentler, R. A. (2008). The effects of hearing aid compression parameters on the short-term dynamic range of continuous speech. *Journal of Speech, Language, and Hearing Research, 51*, 471–484.

Hietanen, J., Leppanen, J., Illi, M., & Surakka, V. (2004). Evidence for the integration of audiovisual emotional information at the perceptual level of processing. *European Journal of Cognitive Psychology, 16*(6), 769–790.

Holt, R. F., Kirk, K. I., Eisenberg, L. S., Martinez, A. S., & Campbell, W. (2005). Spoken word recognition development in children with residual hearing using cochlear implants and hearing aids in opposite ears. *Ear and Hearing, 26*(4), 82–91.

Hopyan-Misakyan, T. M., Gordon, K., Dennis, M., & Papsin, B. (2009). Recognition of affective speech prosody and facial affect in deaf children with unilateral right cochlear implants. *Child Neuropsychology, 15*(2), 136–146.

Hosie, J. A., Gray, C. D., Russell, P. A., Scott, C., & Hunter, N. (1998). The matching of facial expressions by deaf and hearing children and their production and comprehension of emotion labels. *Motivation and Emotion, 22*(4), 293–313.

Houston, D. M., Pisoni, D. B., Kirk, K. I., Ying, E. A., & Miyamoto, R. T. (2003). Speech perception skills of deaf infants following cochlear implantation: A first report. *International Journal of Pediatric Otorhinolaryngology, 67*(479–495).

Jenstad, L. M., Pumford, J., Seewald, R. C., & Cornelisse, L. E. (2000). Comparison of linear gain and wide dynamic range compression hearing aid circuits II: Aided loudness measures. *Ear and Hearing, 21*(1), 32–44.

Kaiser, A. R., Kirk, K. I., Lachs, L., & Pisoni, D. B. (2003). Talker and lexical effects on audiovisual word recognition by adults with cochlear implants. *Journal of Speech, Language, and Hearing Research, 46*, 390–404.

Kirk, K. I., Hay-McCutcheon, M. J., Holt, R. F., Gao, S., Qi, R., & Gerlain, B. L. (2007). Audiovisual spoken word recognition by children with cochlear implants. *Audiological Medicine, 5*, 250–261.

Knutson, J. F., Boyd, R. C., Reid, J. B., Mayne, T., & Fetrow, R. (1997). Observational assessments of the interaction of implant recipients with family and peers: Preliminary findings. *Otolaryngology – Head and Neck Surgery, 117*(3), 196–207.

Kong, Y., Stickney, G. S., & Zeng, F. (2005). Speech and melody recognition in binaurally combined acoustic and electric hearing. *Journal of the Acoustical Society of America, 117*(3), 1351–1361.

Kurtzer-White, E., & Luterman, D. (2003). Families and children with hearing loss: Grief and coping. *Mental Retardation and Developmental Disabilities Research Reviews, 9*, 232–235.

Lachs, L., Pisoni, D. B., & Kirk, K. I. (2001). Use of audiovisual information in speech perception by prelingually deaf children with cochlear implants: A first report. *Ear and Hearing, 22*(3), 236–251.

Ling, D. (1989). *Foundations of spoken language for hearing-impaired children*. Washington, DC: Alexander Graham Bell Association for the Deaf.

Loizou, P. C., Dorman, M., & Fitzke, J. (2000). The effect of reduced dynamic range on speech understanding: Implications for patients with cochlear implants. *Ear and Hearing, 21*(1), 25–31.

Loizou, P. C., Poroy, O., & Dorman, M. (2000). The effect of parametric variations of cochlear implant processors on speech understanding. *Journal of the Acoustical Society of America, 108*(2), 790–802.

Marschark, M. (Ed.). (1993). *Psychological development of deaf children*. New York: Oxford University Press.

Massaro, D. W. (1999). Speechreading: Illusion or window into pattern recognition. *Trends in Cognitive Science, 3*(8), 310–317.

Massaro, D. W., & Cohen, M. M. (1999). Speech perception in perceivers with hearing loss: Synergy of multiple modalities. *Journal of Speech, Language, and Hearing Research, 42*, 21–41.

Massaro, D. W., & Egan, P. (1996). Perceiving affect from the voice and face. *Pscyhonomic Bulletin and Review, 3*, 215–221.

Massaro, D. W., & Light, J. (2004). Using visible speech to train perception and production of speech for individuals with hearing loss. *Journal of Speech, Language, and Hearing Research, 42*(2), 304–320.

Maxim, L. A., & Nowicki, S. J., Jr. (2003). Developmental associations between nonverbal ability and social competence. *Facta Universitatis, 2*(10), 745–758.

Mayer, J. D., Salovey, P., & Caruso, D. R. (2004). Emotional intelligence: Theory, findings, and implications. *Psychological Inquiry, 15*(3), 197–215.

McGurk, H., & MacDonald, J. (1976). Hearing lips and seeing voices. *Nature, 264*, 746–748.

Meyer, T. A., Svirsky, M. A., Kirk, K. I., & Miyamoto, R. T. (1998). Improvements in speech perception by children with profound prelingual hearing loss: Effects of device, communication mode, and chronological age. *Journal of Speech, Language, and Hearing Research, 41*, 846–858.

Mitchell, R. E., & Karchmer, M. A. (2004). Chasing the mythical ten percent: Parental hearing status of deaf and hard of hearing students in the United States. *Sign Language Studies, 4*, 138–163.

Mitchell, T. V., & Maslin, M. T. (2007). How vision matters for individuals with hearing loss. *International Journal of Audiology, 46*, 500–511.

Morton, J. B., & Trehub, S. E. (2001). Children's understanding of emotions in speech. *Child Development, 72*(3), 834–843.

Most, T., & Aviner, C. (2009). Auditory, visual, and auditory-visual perception of emotions by individuals with cochlear implants, hearing aids, and normal hearing. *Journal of Deaf Studies and Deaf Education, 14*(4), 449–464.

Most, T., Weisel, A., & Zaychik, A. (1993). Auditory, visual and auditory-visual identification of emotions by hearing and hearing-impaired adolescents. *British Journal of Audiology, 27*, 247–253.

Napolitano, A. C., & Sloutsky, V. M. (2004). Is a picture worth a thousand words? The flexible nature of modality dominance in young children. *Child Development, 75*(6), 1850–1870.

O'Donoghue, G. M., Nikolopoulos, T. P., & Archbold, S. M. (2000). Determinants of speech perception in children after cochlear implantation. *The Lancet, 356*, 466–468.

Oster, A. M., & Risberg, A. (1986). The identification of the mood of a speaker by hearing-impaired listeners. *SLT-Quarterly Progress Status Report, 4*, 79–90.

Pell, M. (2005). Nonverbal emotion priming: Evidence from the 'facial affect decision task'. *Journal of Nonverbal Behavior, 29*(1), 45–73.

Peng, S., Tomblin, J. B., & Turner, C. W. (2008). Production and perception of speech intonation in pediatric cochlear implant recipients and individuals with normal hearing. *Ear and Hearing, 29*(3), 336–351.

Peters, K., Remmel, E., & Richards, D. (2009). Language, mental state vocabulary, and false belief understanding in children with cochlear implants. *Language, Speech, and Hearing Services in Schools, 40*(3), 245–255.

Power, D. J., & Hyde, M. B. (1997). Multisensory and unisensory approaches to communicating with deaf children. *European Journal of Psychology of Education, 12*(4), 449–464.

Rieffe, C., & Terwogt, M. M. (2000). Deaf children's understanding of emotions: Desires take precedence. *Journal of Child Psychology and Psychiatry, 41*, 601–608.

Robbins, A. M., Koch, D. B., Osberger, M. J., Zimmerman-Phillips, S., & Kishon-Rabin, L. (2004). Effect of age at cochlear implantation on auditory skill development in infants and toddlers. *Archives of Otolaryngology – Head & Neck Surgery, 130*, 570–574.

Robinson, C. W., & Sloutsky, V. M. (2004). Auditory dominance and its change in the course of development. *Child Development, 75*(5), 1387–1401.

Sadato, N., Yamada, H., Okada, T., Yoshida, M., Hasegawa, T., Matsuki, K., et al. (2004). Age-dependent plasticity in the superior temporal sulcus in deaf humans: A functional fMRI study. *BMC Neuroscience, 5*(56), 1–6.

Sarant, J. Z., Blamey, P. J., Dowell, R. C., Clark, G. M., & Gibson, W. P. R. (2001). Variation in speech perception scores among children with cochlear implants. *Ear and Hearing, 22*(1), 18–28.

Scherer, K. R. (2003). Vocal communication of emotion: A review of research paradigms. *Speech Communication, 40*, 227–256.

Schorr, E. A., Fox, N. A., van Wassenhove, V., & Knudsen, E. I. (2005). Auditory-visual fusion in speech perception in children with cochlear implants. *Proceedings of the National Academy of Sciences, 102*(51), 18748–18750.

Schorr, E. A., Roth, F. P., & Fox, N. A. (2009). Quality of life for children with cochlear implants: Perceived benefits and problems and the perception of single words and emotional sounds. *Journal of Speech, Language, and Hearing Research, 52*, 141–152.

Seewald, R. C., Ross, M., Giolas, T. G., & Yonovitz, A. (1985). Primary modality for speech perception in children with normal and impaired hearing. *Journal of Speech and Hearing Research, 28*, 36–46.

Shannon, R. V. (2002). The relative importance of amplitude, temporal, and spectral cues for cochlear implant processor design. *American Journal of Audiology, 11*, 124–127.

Sharma, A., Dorman, M. F., & Spahr, A. J. (2002). A sensitive period for the development of the central auditory system in children with cochlear implants: Implications for age of implantation. *Ear and Hearing, 6*, 532–539.

Sininger, Y., Grimes, A., & Christensen, E. (2010). Auditory development in early amplified children: Factors influencing auditory-based communication outcomes in children with hearing loss. *Ear and Hearing, 31*(2), 166–185.

Sloutsky, V. M., & Napolitano, A. C. (2003). Is a picture worth a thousand words? Preference for auditory modality in young children. *Child Development, 74*(3), 822–833.

Snik, A. F. M., Makhdooum, M. J. A., Vermeulen, A. M., Brokx, J. P. L., & van den Broek, P. (1997). The relation between age at the time of cochlear implantation and long-term speech perception abilities in congenitally deaf subjects. *International Journal of Pediatric Otorhinolaryngology, 41*, 121–131.

Sommers, M. S., Tye-Murray, N., & Spehar, B. (2005). Auditory-visual speech perception and auditory-visual enhancement in normal-hearing younger and older adults. *Ear and Hearing, 26*(3), 263–275.

Staller, S. J., Dowell, R. C., Beiter, A. L., & Brimacombe, J. A. (1991). Perceptual abilities of children with the Nucleus 22-Channel cochlear implant. *Ear and Hearing, 12*(4), 34–47.

Stelmachowicz, P. G., Pittman, A. L., Hoover, B. M., & Lewis, D. E. (2001). Effect of stimulus bandwidth on the perception of /s/ in normal- and hearing-impaired children and adults. *Journal of the Acoustical Society of America, 110*(4), 2183–2190.

Stelmachowicz, P. G., Pittman, A. L., Hoover, B. M., Lewis, D. E., & Moeller, M. P. (2004). The importance of high-frequency audibility in the speech and language development of children with hearing loss. *Archives of Otolaryngology – Head & Neck Surgery, 130*, 556–562.

Strelnikov, K., Rouger, J., Barone, P., & Deguine, O. (2009). Role of speechreading in audiovisual interactions during the recovery of speech comprehension in deaf adults with cochlear implants. *Scandinavian Journal of Psychology, 50*(5), 437–444.

Svirsky, M. A., Robbins, A. M., Kirk, K. I., Pisoni, D. B., & Miyamoto, R. T. (2000). Language development in profoundly deaf children with cochlear implants. *Psychological Science, 11*(2), 153–158.

Turner, C. W., Chi, S., & Flock, S. (1999). Limiting spectral resolution in speech for listeners with sensorineural hearing loss. *Journal of Speech, Language, and Hearing Research, 42*, 773–784.

Turner, C. W., Souza, P. E., & Forget, L. N. (1995). Use of temporal envelope cues in speech recognition by normal and hearing-impaired listeners. *Journal of the Acoustical Society of America, 97*, 2568–2576.

Tye-Murray, N., Sommers, M., & Spehar, B. (2007). Auditory and visual lexical neighborhoods in audiovisual speech perception. *Trends in Amplification, 11*(4), 233–242.
Vandell, D. L., & George, L. B. (1981). Social interaction in hearing and deaf preschoolers: Successes and failures in initiations. *Child Development, 52*(2), 627–635.
Wie, O. B. (2010). Language development in children after receiving bilateral cochlear implants between 5 and 18 months. *International Journal of Pediatric Otorhinolaryngology, 74*(11), 1258–1266.
Zupan, B., Neumann, D., Babbage, D. R., & Willer, B. (2009). The importance of vocal affect to bimodal processing of emotion: Implications for individuals with traumatic brain injury. *Journal of Communication Disorders, 42*, 1–17.
Zupan, B., & Sussman, J. E. (2009). Auditory preferences of young children with and without hearing loss for meaningful auditory-visual compound stimuli. *Journal of Communication Disorders, 42*, 381–396.

Chapter 16
Searching for a Greater Sensitivity of Cognitive Event-Related Potentials Through a Crossmodal Procedure for a Better Clinical Use in Psychiatry

D. Delle-Vigne, C. Kornreich, P. Verbanck, and Salvatore Campanella

Psychiatry has never been able to satisfactorily respond to the delicate question of differential diagnosis, both in theory as well as in practice, which creates a more fundamental question: what about the discrimination between normality and pathology (e.g. Canguilhem, 1972; Duyckarts, 1964; Wakefield, 2007)? Patient's subjectivity will add some complexity to this concern, as attempts to standardize such diagnosis tools have been made. An attempt to develop a methodical procedure was partially achieved with the classification of symptoms with reference to a specific nosography, the most popular one being the Diagnostic and Statistical Manual of Mental Disorders IV-Revised (DSM) (APA, 2000). However, the stringent categories of the DSM may have diverted attention from the latent distribution of the disorders (e.g. Bender, Weisbrod, & Resch, 2007). Therefore, although this tool offers some consensus in clinical practice, the psychiatric diagnosis is still considered as a working hypothesis which will possibly evolve with respect to the therapeutic effect and the social context (Timsit-Berthier, 2003).

In the next sections, we focus on the discipline of "neuropsychiatry", which tries to bridge the gap between neurology, on the one hand, and psychiatry, on the other hand, in order to get better insight into the biological bases of psychiatric disorders (Northoff, 2008). Indeed, an increasing knowledge about anatomical structures and cellular processes underlying psychiatric disorders may help to bridge the gap

D. Delle-Vigne • C. Kornreich • P. Verbanck
Laboratory of Psychological Medicine, Free University of Brussels, Brussels, Belgium
e-mail: Dyna.Delle-Vigne@ulb.ac.be; charles.kornreich@ulb.ac.be; paul.verbank@ulb.ac.be

S. Campanella (✉)
Laboratory of Psychological Medicine, Free University of Brussels,
Brussels, Belgium

CHU Brugmann, Psychiatry Department (EEG), The Belgian Fund for Scientific Research (FNRS), 4, Place Vangehuchten, Brussels B-1020, Belgium
e-mail: salvatore.campanella@chu-brugmann.be; salvatore.campanella@ulb.ac.be

P. Belin et al. (eds.), *Integrating Face and Voice in Person Perception*,
DOI 10.1007/978-1-4614-3585-3_16, © Springer Science+Business Media New York 2013

between clinical manifestations and basic physiological processes. In this view, one essential set of tools to achieve this aim is electrophysiological assessments of psychiatric disorders, and more precisely, for our purpose, cognitive event-related potentials (ERPs).

1 Part I: Clinical Neurophysiology, Event-Related Potentials, P300, and Psychiatry

1.1 Development of a Clinical Neurophysiology

Currently, the use of DSM is under debate (e.g. Wakefield, 1992, 2007; Zimerman & Spitzer, 2005), and given the imperfection of current psychiatric diagnostic systems to capture the disorders' heterogeneity, a new ideology has emerged, which places this psychiatric nosography on neurophysiological bases (Guérit, 1998 in Timsit-Berthier, 2003).

The first objective of this cognitive neurosciences' ramification is to define physiological markers which are associated with various psychic diseases. A marker constitutes a modification of a psychobiological variable, which reflects a structural or functional disturbance, before, during, or after the disease (Campanella & Streel, 2008). If the anomaly is present during and after the morbid episode, it is qualified as a "trait marker"; if the anomaly, present during the episode, returns to normal after remission, it is qualified as a "state marker", and when a propensity to develop the pathology exists, even if not expressed, it is qualified as a "vulnerability marker" or "endophenotype" (Gottesman & Gould, 2003). Actually, an endophenotype is defined as a "heritable trait, associated with a causative pathophysiological factor in an inherited disease" (Gershon & Goldin, 1986). A vulnerability marker has to be differentiated from a risk factor, which touches any trait having a predictive validity, but not etiological meaning, for developing a psychiatric disorder (Freedman et al., 1997). Both can be used theoretically to predict psychiatric conditions, but only the vulnerability marker provides information about the disease aetiology and pathophysiology, and helps with the diagnosis and therapeutic interventions (van der Stelt & Belger, 2007). To be considered as a vulnerability marker, it is necessary that it is: (1) associated with the disease of with a given subtype, in the general population, (2) heritable, (3) state-independent, and (4) associated with the disease within pedigree (van der Stelt, 1999).

These markers, if they are found, could be used as complementary to the diagnosis, as prognostic elements, or to assist in choosing the most adequate treatment for psychiatric disorders. In fact, they can enhance our knowledge about the nature and the extent of cognitive damages, and they can offer us deeper theoretical insights into illness aetiology and pathophysiology. Taken together, it can improve the early detection of the illness, and in this way, propose more effective and targeted interventions (van der Stelt & Belger, 2007). Actually, clarifying the diagnosis will lead to the development of specific treatments (Thaker, 2008a, 2008b). The therapeutic strategies might be focused on specific pathophysiological mechanisms and cognitive dysfunctions, rather than on clinical symptoms (van der Stelt & Belger, 2007). For instance, Naismith et al. (2010) experimented with cognitive training, by using the Neuropsychological Educational Approach to Remediation (NEAR), in a population of depressed patients, which focused on improving memory. They observed that the cognitive training enhanced memory performances, and suggested that this non-pharmacological treatment could in turn act upon psychosocial functioning and reduce disability. A previous study by these same authors (2007) showed that physical disability in major depression seemed to be related to cognitive dysfunctions (such as psychomotor retardation, impaired memory retention), regardless of depression severity, while functional disability was linked to depression severity. Overall, disability was related to illness severity, but cognitive dysfunction also played a critical role, and should be targeted for cognitive interventions. It was concluded that implementation of psychosocial, cognitive, and/or vocational programs targeting psychomotor speed issues, memory or "perception" of cognitive deficits may contribute to improve cognition and also relieve symptoms, and reduce disability in major depression.

1.2 Event-Related Potentials: A Useful Method to Explore Mental Chronometry

An alternative way to explore normal and pathological cognitive processing, besides the clinical examination, is to investigate cerebral waves' differences, by means of the ERPs method. In fact, neuroelectric measures can inform us about cortical and sensory function. Briefly, this electrophysiological technique consists in recording the ongoing brain activity in "real-time", with an electrocap on the scalp, and can investigate activity's modification consecutive to a cognitive task (Duncan et al., 2009). ERPs constitute a specific part of the cerebral electric potentials, which refer to the synchronic activation of a large number of neurons, in response to the preparation or in response to a discrete event, intern or extern to the subject. A regular alternation of rhythmic and oscillatory changes over time compose this resting EEG, which can be divided into various frequency bands, associated with various behavioural states, from sleep to mental concentration (van der Stelt & Belger, 2007). These components are detected by using signal-averaging techniques in the recorded electroencephalogram (van der Stelt, 1999).

One main interest about this method when compared to other neuroimaging techniques is the high temporal resolution, at a millisecond scale (Rugg & Coles, 1995). Moreover, it allows us to infer the information processing levels damaged (Rugg & Coles, 1995) and identifies the origin and the nature of the deficits (e.g. perceptive, attentional, decisional). In fact, it can "translate" into cerebral components at the different levels of the information processing, from sensory steps to the executive ones.

Among others, a classical paradigm used to evoke these waves consists in an "oddball target detection task", where the subject has to detect as quickly as possible (typically by pressing a button) a deviant and rare stimulus (visual or auditory one) among a train of frequents stimuli, each of these having a contrasted probability of apparition. The oddball task evokes robust and reliable phenomena that have been used as markers of cognitive function (Polich & Bloom, 1999).

ERPs can be divided in two main groups: (1) exogenous potentials, associated with the physical characteristics of the stimulus, which depend on the sensory modality used, and reflects the earliest components of the information process, and (2) the cognitive potentials, or endogenous potentials, linked with internal stimulations, which reflect the latest components of the information processing. In the first group the integrity of the sensory pathways from the periphery to the cortex can be assessed, while in the second one, the experimental situation is actively involved. The cognitive potentials can be influenced by the mental state of the subject, the characteristics of the task to realize, as well as the meaning of the stimulus and the fluctuations in attention (Hansenne, 2000a, 2000b).

An ERP is composed of a series of scalp-positive and -negative voltage deflections (P or N), strictly time and phase locked to the onset of a particular stimulus event; they are followed by a number which represents their latencies. In fact, two main parameters characterize an evoked potential: (1) its latency, corresponding to the information processing speed, reflecting the activation of the first synapses. When the peak latency is reached, it represents the moment when most of the synapses are activated; in healthy individuals, a long latency is linked with a complex information treatment, while in elderly individuals such long latencies can be a sign of a degenerative illness. However, short latency is associated with higher cognitive functions (e.g. Emmerson, Dustman, Shearer, & Turner, 1989; Polich, Howard, & Starr, 1983; Polich, Ladish, & Bloom, 1990), and (2) its amplitude, corresponds to the quantity of attentional resources allocated to the task (Kramer & Strayer, 1988; Wickens, Kramer, Vanasse, & Donchin, 1983), which is proportional to the quantity of neurons involved, to the synchronization's degree and to the potential source' distance from the surface electrode.

There is a growing body of literature which identifies that specific psychiatric conditions are associated with abnormal ERP components (e.g. Katada, Sato, Ojika, & Ueda, 2004), reflecting impaired cognitive functioning. Since 1960s, one of the most studied ERP components was the P300, which is impaired in many of neuropsychological and psychiatric disorders, such that it is regarded as a marker of cognitive function in psychiatric and neurological disorders (Egerházi, Glaub, Balla, Berecz, & Degrell, 2008; Katada et al., 2004).

1.3 P300, Neuropsychology, and Psychiatry

1.3.1 The P300

A common symptom present in various psychiatric population is a deficient information processing. Castaneda, Tuulio-Henriksson, Marttunen, Suvisaari, and

Lönnqvist (2008) emphasize that cognitive alterations may constitute noteworthy factors in affecting one's ability to function socially and occupationally, in everyday life. In fact, cognitive perturbations may disturb coping abilities, which may favour relapse and may impinge on treatment compliance. For some time, it was thought that this deficit could be physiologically indexed by the deterioration of the P300 (or P3 or P3b) component (Desmedt, Debecker, & Manil, 1965; Sutton, Braren, Zubin, & John, 1965). It is a late positive wave elicited in parietal regions between 300 and 350 ms after an auditory stimulus and between 400 up to 450 ms after a visual one, when the subject detects the target (deviant) stimulus, for instance, in an oddball task. The P3 generation is determined by the psychological context of the eliciting stimulus, and is dependent on the active cognitive processing of stimulus information. Its amplitude size is elicited by task-relevant target stimuli, and its peak latency depends on the stimulus, task, and subject factors. The amplitude of the P300 is thought to index the memory processes and the allocation of attentional resources (Wickens et al., 1983), while the latency of the P300 seems to be linked with the stimulus classification speed (Duncan-Johnson, 1981; Kutas, McCarthy, & Donchin, 1977; McCarthy & Donchin, 1981), independently of behavioural response times (Ilan & Polich, 1999; Verleger, 1997), and can be considered as a motor-free measure of cognitive function. The P300 is a slow and low frequency wave, and is related to stimulus-evoked delta and theta oscillations (Başar, Başar-Eroglu, Karakaş, & Schürmann, 2001; Yordanova & Kolev, 1998). Polich and Kok (1995) which underline the fact that the P300 component is sensitive to constitutional factors (e.g. age, sex), natural (e.g. circadian rhythm, menstrual cycle) and environmental changes (e.g. exercise, caffeine, nicotine, psychotropic medication) in the individual's arousal state.

Behaviourally, the P300 is believed to reflect the "context updating" (Donchin & Coles, 1988) or the "context closure" (Desmedt, 1980; Verleger, 1988). It represents a response-related stage, specific of the decisional stage. It indexes stimulus significance and the amount of attention assigned to the evoked stimulus event, being maximal to task-relevant or attended stimuli (e.g. Picton, 1992; Polich, 1998). It corresponds to attention control and memory, both necessary for the final evaluation of a stimulus (Martín-Loeches, Muñoz, Hinojosa, Molina, & Pozo, 2001).

This P300 component can be divided in two subcomponents: the P3a and the P3b (Lembreghts, Crasson, el Ahmadi, & Timsit-Berthier, 1995; Polich, 2007). On the one hand, the P3a appears between 220 and 280 ms after the presentation of the target and refers to attentional and automatic processes, when the subject is not forced to pay attention to the discrimination task, when there is a surprise effect, i.e. a novel and salient element pops up into the environment ("novelty P300", Polich, 2007). It is elicited in frontocentral regions, and belongs to the attentional orientation complex, which represents real-time processing of involuntary attention (Iv, Zhao, Gong, Chen, & Miao, 2010). The distractor stimulus represents an orienting response, and the generators of P3a seem to be located in the dorsolateral prefrontal cortex, the anterior and posterior parts of the cingulated gyrus, and the parietal supramarginal gyrus (Andersson, Barder, Hellvin, Løvdahl, & Malt, 2008). Most of the studies do not investigate the P3a, although it maybe more sensitive to the clinical status and to the individual cognitive differences (e.g. Fein, Biggins, & MacKay, 1995; Polich, 2004; Polich & Kok, 1995; Rodríguez Holguín, Porjesz, Chorlian, Polich, & Begleiter, 1999). On the other hand, the P3b appears between 310 and 380 ms after the presentation of the target, when the subject explicitly has to assess and categorize the pertinent stimuli, and make a decision; it represents real-time processing of working memory and voluntary attention (Iv et al., 2010). It is distributed in centro-parietal regions and refers to the classical "P300" encountered in the literature, i.e. the update of working memory related to stimulus expectancy (Donchin & Coles, 1988) and the attentional allocation towards the processing of targeted events (Polich & Kok, 1995). It represents task-relevant attentional mechanisms (Polich, 2007) and its generators seem to be the ventrolateral prefrontal areas, posterior parietal and medial temporal areas, including the hippocampus (Halgren, Marinkovic, & Chauvel, 1998; Soltani & Knight, 2000).

1.3.2 Neuropsychology and P300

Some neuropsychological credit has been devoted to this P300 as an index of cognitive alteration. Patients with frontal lesions exhibited P3a deficits (e.g. Hartikainen & Knight, 2003; Knight, 1984), whereas other patients with hippocampic lesions showed a reduced P3a for novel stimuli (Knight, 1996). Some studies about individuals with epileptic foci suggested that P3b could be generated in the medial temporal lobe (e.g. Halgren et al., 1980; McCarthy, Wood, Williamson, & Spencer, 1989), whereas in other studies in individuals with a temporal lobectomy and ischemic patients, the hippocampic formation does not affect the generation of a P300 (e.g. Johnson, 1988; Polich & Squire ,1993). Tough, the integrity of the temporo-parietal lobe seems to be necessary to evoke a P300 (e.g. Knight, Scabini, Woods, & Clayworth, 1989; Verleger, Heide, Butt, & Kömpf, 1994), as well as the frontal lobe and the hippocampus for the P3a and some temporo-parietal regions for the P3b. It appears thus that the P300 seems to be due to either multiple independent generators or belongs to an integrated central system, with large connections which has an influence throughout the whole brain (Duncan, Kosmidis, & Mirsky, 2003; Nieuwenhuis, Aston-Jones, & Cohen, 2005; Pineda, Foote, & Neville, 1989; van der Stelt, 1999). Nevertheless, some cerebral structures seem to be systematically involved in the P300 generation: the hippocampus, the superior temporal sulcus, the ventrolateral prefrontal cortex, and probably the intraparietal sulcus (Halgren, Baudena, Clarke, Heit, Liégeois et al., 1995; Halgren, Baudena, Clarke, Heit, Marinkovic et al., 1995; Halgren et al., 1998; Kiss, Dashieff, & Lordeon, 1989; Smith et al., 1990).

However, the major interest of the P300 in the neuropsychological domain is probably for dementia. This interest in the P300 was strengthened notably because the differences between the P300 in healthy individuals versus demented patients reinforced the utility to use P300 in dementia's diagnosis (Braverman et al., 2006; Frodl et al., 2002; Holt et al., 1995). In dementia, its amplitude is reduced and its latency increased (e.g. Goodin, Squires, Henderson, & Starr, 1978; Goodin, Squires, & Starr, 1978; Have, Kolbeinsson, & Pétursson, 1991; Olichney and Hillert, 2004;

Polich et al., 1990), and this pattern is also found in mild dementia (e.g. Polich et al., 1990; Pfefferbaum, Ford, & Kraemer, 1990).

In addition, normal ageing implies prolongations of P300 latencies, reductions of P300 amplitudes, and a more equipotential P300 scalp distribution, so the necessity to discriminate between normal cognitive decline and demential states is crucial (Hughes & John, 1999). For example, Polich and Corey-Bloom (2005) showed that P300 amplitude in Alzheimer patients was smaller and latency longer, compared to elderly controls, across task difficulties and modalities (auditory and visual); these differences were largest for the easy visual tasks (single stimulus paradigm: respond to every stimulus occurring), offering some reliable behavioural effects to discriminate patients from controls, suggesting that the P300 can be sensitive to dementia during the early stages, and these easy discrimination tasks are much needed in clinical conditions. It seems that P300 amplitude provided more consistent differences between the groups. However, the authors do not conclude about viewing the P300 as a sensitive tool, because of measurement variability. However, it can be used in clinical routine to assess the cognitive effects of dementia. Besides, ERPs patterns are different in patients with delirium and demented patients (Jacobson, Leuchter, & Walter, 1996). Also, cholinesterase inhibitor treatment was associated with a decrease of the P300 latency in demented patients, which correlated with an amelioration of cognitive functioning in these patients (Thomas, Iacono, Bonanni, D'Andreamatteo, & Onofrj, 2001).

Another delicate point is the discrimination among various subtypes of dementia. Jiménez-Escrig et al. (2002) compared frontotemporal demented patients with Alzheimers patients and controls. There were no significant differences in P300 latency between controls and frontotemporal patients, but well between Alzheimer patients and controls and frontotemporal patients; hence, an overlap of latency values existed in the three groups. Egerházi et al. (2008) compared Alzheimer patients, mild cognitive impairments individuals, and vascular demented patients and controls. Their results revealed a longer P300 latency for both demented-patients groups, which also correlated with the severity of dementia (also see Ball, Marsh, Schubarth, Brown, & Strandburg, 1989; Goodin, Starr, Chippendale, & Squires, 1983). In the mild cognitive impairment group, the latency was significantly longer among patients with mild cerebral atrophy compared to controls Also, decreased P300 amplitude was observed in both groups of demented patients. The prolongation of P300 latency was significant among patients with both vascular and Alzheimer's dementia, and also among MCI patients with mild cerebral atrophy. It suggests that the severity of the disease is positively correlated with P300 latency; but not with the type of dementia.

1.3.3 P300 in Psychiatry

ERP studies have generally shown P300 alterations (decreased amplitude and/or delayed latency) in several psychiatric disorders (e.g. for mood disorders: Wang, Chen, & Lou, 2000; Zhu et al., 2009; for schizophrenia: Bramon et al., 2005; Ford, 1999; Mathalon et al., 2000; for chronic alcoholism: Fein & Chang, 2006; Reese & Polich, 2003). Consequently, some DSM detractors considered that P300 alterations may be a new objective tool. In this section, we will review some P300 studies in diverse pathologies, namely in mood disorders, schizophrenia, and chronic alcoholism.

Mood Disorders

Regarding the P300, some studies showed a reduced amplitude (e.g. Wang et al., 2000; Zhu et al., 2009), which varied with anhedonia, depression's severity, psychotic characteristics and suicidal antecedents (e.g. Bruder et al., 1991; Hansenne, Pitchot, Gonzalez Moreno, Zaldua, & Ansseau, 1996; Partiot et al., 1993; Pierson et al., 1991; Urcelay-Zaldua, Hansenne, & Ansseau, 1995). With respect to this link with symptoms' severity, a study by Coullaut-Valera García, Arbaiza Díaz del Rio, Coullaut-Valera García, and Ortiz (2007) exhibited a negative correlation between the P300 amplitude and the severity of the depression. Santosh, Malhotra, Raghunathan, and Mehra (1994) showed that the P300 amplitude was diminished in depressive patients with psychotic characteristics compared to depressive ones with no such traits. Hughes and John (1999) demonstrated that unipolar patients presented the exact opposite EEG pattern than schizophrenic patients, while bipolar patients presented the same EEG pattern than schizophrenics. These similarities between psychotic patients and those with psychotic depression are found again in neuropsychological measures (Castaneda et al., 2008), but are still less severe than in schizophrenia (Hill, Keshavan, Thase, & Sweeney 2004). These data invoke that psychotic depression may be resembling other psychotic disorders (Demily, Jacquet, & Marie-Cardine, 2009; Kendler et al., 1993). Thaker (2008a, 2008b) and Jabben, Arts, Krabbendam, and van Os (2009) even propose that schizophrenia and bipolar disorder may share overlapping aetiologic factors. Furthermore, the hypothesis that views these two entities as a part of a same continuum is gaining ground (e.g. Pregelj, 2009). Additionally, individuals with recurrent major depressive episodes seem to be more vulnerable to bipolar disorders and manifest more cognitive dysfunctions than depressive patients with no such features, and individuals at high risk for psychosis can develop a bipolar disorder or schizophrenia (Demily et al., 2009).

The results concerning the P300 latency are not yet consistent. Some authors do not find a prolonged latency (e.g. Gangadhar, Ancy, Janakiramaiah, & Umapathy, 1993; Gordon, Kraiuhin, Harris, Meares, & Howson, 1986; Santosh et al., 1994), unless the experimental task requires a greater amount of attention (Bruder et al., 1991); in this case, some correlation between P300 latency and depressions' severity seems to exist (Schlegel, Nieber, Herrmann, & Bakauski, 1991). Others studies do find a difference between controls and depressed individuals (e.g. Coullaut-Valera García et al., 2007; Ortiz Alonso et al., 2002). P3a latency seems to be weaker in depressed patients with blunted affects and a psychomotor retardation than in impulsive patients (Pierson et al., 1991; Partiot et al., 1993). Andersson et al. (2008) did find an increased P3a latency in female bipolar II outpatients compared to controls, but no P3b difference in latency or amplitude between the groups. Their results also indicated a general impairment in neuropsychological measures (except phonemic verbal fluency), compared to controls. The P3a was not correlated with the severity of depression, which indicates that the observed differences are not influenced by mood variation. These results involve dysfunctions associated with pre-attentive detection and automatic orientation towards stimulus change.

In a recent study, Bruder et al. (2009) showed that a novel "distractor" stimulus in a three-stimulus oddball task elicited a P3 (indistinguishable from the P3a, Simons, Graham, Miles, & Chen, 2001; Spencer, Dien, & Donchin, 1999) with shorter peak latency and more frontocentral topography than the parietal-maximum P3b to target stimuli. The novelty-P3 was appreciably reduced in depressed patients compared to controls, but the P3b was not, although a trend existed. The results of Tenke, Kayser, Stewart, and Bruder (2010) showed a decreased novelty response in depression, implying the early phase of the frontocentral novelty P3, which replicates the earlier observations of Bruder et al. (2009).

However, a large number of studies have considered the abnormal P300 amplitude as a "state marker": the amplitude reduction is related to the clinical state, and would return to normal values with remission (e.g.; Duncan et al., 1991; Gangadhar et al., 1993). Blackwood et al. (1987) showed an increment of amplitude when patients were given antidepressants during 4 weeks. Pierson, Jouvent, Quintin, Perez-Diaz, and Leboyer (2000) reported a reduced P300 amplitude and an increased P300 latency in first degree relatives of bipolar disorders I patients. Zhang, Hauser, Conty, Emrich, and Dietrich (2007) tested two groups of healthy subjects in a go/no-go task, one with no family history of depression and the other with this characteristic. The experimental group exhibited a P3b amplitude decrement, which was interpreted as a neurocognitive vulnerability marker for the development of depression.

It is apparent that the published results are very heterogeneous. This might be due to methodological or theoretical issues (see for example Hansenne, 2000a, 2000b; Polich & Kok, 1995, for a synthesis): e.g. heterogeneous samples, heterogeneous diagnoses, small samples, medication effects, differences in tests used to assess emotional processing, differences in affective stimuli used, lack of control measures, cultural differences, comorbidities, various ages, severity, and duration of symptoms. These methodological issues are encountered in both experimental and control groups. Furthermore, a more important issue should be considered in that the study populations are often built on DSM criteria, although another classification could be envisaged, built on physiopathological mechanisms, crossing over DSM categories (Guérit, 1998).

Schizophrenia

Cognitive alterations, including deficits in attention, memory, speed processing, executive functioning, may be seen as the underlying basis of schizophrenic symptoms (Matsuoka & Nakamura, 2005). It may provide phenotypic markers of the liability to illness, and data suggest the existence of a familial pattern of neurocognitive deficits (Hill, Harris, Herbener, Pavuluri, & Sweeney, 2008).

As we mentioned above, schizophrenia and bipolar disorders share some common characteristics, in ERP generation, in neuropsychological measures, in brain anatomy, and even in genetic factors (e.g. Bowie et al., 2010; Craddock & Owen, 2005; Hill et al., 2008; Hughes & John, 1999; Maier, Zobel, & Wagner, 2006; Strasser et al., 2005; Thaker, 2008a, 2008b; Zalla et al., 2004). Hill et al. (2008) and Jabben, Arts, van Os, and Krabbendam (2010) suggest that cognitive dysfunctions were more generalized, more disabling, and more severe in schizophrenia, but the level of cognitive deficits are comparable for schizophrenics and bipolar patients with a history of psychotic symptoms (Hill et al., 2008). These alterations were also present in schizophrenic relatives, but not in bipolar ones. The relation between neurocognitive impairments and psychosocial functioning was more widespread in schizophrenia. Jabben et al. (2010) proposed that cognitive perturbations constitute a stronger marker of familial vulnerability for schizophrenia than for bipolar disorder. Zalla et al. (2004) demonstrated that an increased susceptibility to interference and a reduced inhibition might be transnosographical markers for vulnerability in schizophrenia and bipolar disorders.

In ERPs studies, it has been shown that the P300 amplitude was reduced by half in schizophrenics compared to controls, particularly for auditory stimuli (e.g. Duncan, Morihisa, Fawcet, & Kirch, 1987; Ford, 1999; Roth & Cannon, 1972). A hypothesis was to link this decrement of the frontotemporal atrophy found in schizophrenics with their impairment in sustained attention (e.g. Nuechterlein, Pashler, & Subotnik, 2006). The reduction in auditory P300 amplitude has been observed in the acute phase, remission, patients under treatment, patients with no medication, and unaffected relatives (e.g. Blackwood, St Clair, Muir, & Duffy, 1991; Bramon et al., 2005; Fenton, Fenwick, Dollimore, Dunn, & Hirsch, 1980; Merrin & Floyd, 1992; Niedermeyer, 1993; Weisbrod, Hill, Niethammer, & Sauer, 1999;), making it as a potential vulnerability marker of schizophrenia (e.g. Bramon, Rabe-Hesketh, Sham, Murray, & Frangou, 2004; Jeon and Polich, 2003; Mathalon et al., 2000; Price et al., 2006). This auditory diminution was also found in individuals at ultra-high risk for psychosis (e.g. Bramon et al., 2008; Frommann et al., 2008; van der Stelt, Lieberman, & Belger, 2005). A recent study from Fisher, Labelle, and Knott (2010) exhibited a smaller P3a amplitude in hallucinating patients, compared to non-hallucinating patients and healthy controls. This P3a amplitude was correlated with auditory verbal hallucination scores, which implied that these scores were related to a dysfunctional treatment of speech. These changes have been identified at the initial as well as advanced stages of the disease (e.g. Umbricht, Bates, Lieberman, Kane, & Javitt, 2006; van der Stelt et al., 2005), and are still present in patients free of clinical symptoms and in relative remission (Ford, 1999).

Lee, Namkoong, Cho, Song, and An (2010) found in individuals at ultra-high risk for psychosis and first-episode schizophrenia, a reduction of visual P300 amplitude in a visuo-spatial oddball task, in comparison with healthy controls (e.g. Ergen, Marbach, Brand, Başar-Eroğlu, & Demiral, 2008; Strandburg et al., 1994), and it was negatively correlated with severity of negative symptoms in both groups. The alterations of amplitude and latency in auditory and visual P300 have also been reported in bipolar patients (e.g. O'Donnell, Vohs, Hetrick, Carroll, & Shekhar, 2004).

These results might indicate that the visual P300 amplitude is a neurobiological vulnerability marker, being a sign of neurophysiological alterations related to negative symptoms in schizophrenia. It suggests that the dysfunctional visuo-spatial processing begins to appear in the putative prodromal phase of the illness. It is important to underline that in Lee's study, antipsychotic medication was not correlated with P300 amplitudes, and there were no correlations between P300 latencies and symptom severity. Another study (Ozgürdal et al., 2008) has suggested that the neurophysiological changes emerged early, at the prodromal phase, and seem to have a progressive course, from prodromal to chronic state. Likewise, visual and visuo-spatial impairments were identified as potential endophenotypes of the illness (Glahn et al., 2003; Saperstein et al., 2006) and risk for psychosis (Wood et al., 2003). Therefore, auditory and visual P300 amplitude could be seen as vulnerability markers.

ERPs can potentially constitute an interesting tool to differentiate different subtypes of schizophrenia, even if some studies do not identify such distinctions. For example, Mathalon et al. (2010), using unimodal tasks (auditory and visual), failed to distinguish schizophrenic patients from schizoaffective ones, with respect to P300. Schizoaffective individuals exhibited normal P300 amplitudes.

ERP can also be combined with neuroimaging informations (Müller, Kalus, & Strik, 2001). In fact, core schizophrenia seems to be characterized by a left-temporal dysfunction, in association with verbal processing impairments; acute remitting schizophrenia-like psychosis patients exhibit signs of cerebral hyperarousal, and the drive of action of manic patients seems to rely on frontal disinhibition. Laurent et al. (1993) observed a reduction in P300 amplitude and an increase in P300 latency in a paranoid subgroup of schizophrenic patients, while the disorganized subgroup exhibited both values comparable to the controls ones. None of these measures were correlated with age, duration of illness, hospitalization, or IQ. Therefore, it would seem that different subgroups of schizophrenia might have different biological substrates.

Chronic Alcoholism

P300 amplitude presents a decrement in alcoholic patients, in relation with some genetic factors, rather than as a result of alcohol ingestion (Carlson et al., 2002). This phenomenon persists even in the remission stage. Moreover, the smaller P3 amplitude seems to predict prospectively later substance use (Berman, Whipple, Fitch, & Noble, 1993). Monteiro and Schuckit (1988) also found decreased intensity of reaction to ethanol in sons of alcoholics, compared to control subjects, smaller P300 amplitude, and differences in alpha waves. Interestingly, this reduction in amplitude was also found in individuals with family history of alcoholism (Nácher, 2000; Iacono et al., 2000; Reese and Polich, 2003), and in individuals considered at high risk to develop alcoholism (e.g. Cohen, Porjesz, Begleiter, & Wang, 1997; Porjesz et al., 1996; Ramachandran, Porjesz et al., 1996; Begleiter, & Litke, 1996). Van der Stelt, Gunning, Snel, Zeef, and Kok (1994), Hill, Shen, et al. (1999), and Hill, Yuan, and Locke (1999) observed that male children of alcoholic fathers exhibited this P300 alteration and this was also experienced in children of alcoholics, who were not yet exposed to alcohol toxicity effects (Begleiter, Porjesz, Bihari, & Kissin, 1984; Hesselbrock, Begleiter, Porjesz, O'Connor, & Bauer, 2001; Porjesz & Begleiter, 1997).

The study of Hesselbrock proposed, after genetic analysis, that the attributes of the P300 component were heritable, and a quantitative trait locus analysis found linkage to several chromosomal regions. Hill, Shen et al. (1999) and Hill, Yuan et al. (1999) also found that the P300 is transmissible in families, and that different patterns of transmission exist between families at high and low risk for alcoholism. A previous study of Hill, Steinhauer, Lowers, and Locke (1995) which investigated daughters of alcoholic mothers (biological fathers non-alcoholic) showed that these girls exhibited lower P300 than controls, suggesting that alcoholism risk transmission may be possible without a paternal alcoholism. Realmuto, Begleiter, Odencrantz, and Porjesz (1993) exhibited an auditory P3a amplitude decrement, attesting for troubles in automatic processes. Fein and Chang (2006) assessed the P300 in long-term abstinent alcoholics (mean abstinence: 6.7 years) and results showed reduced P3b amplitudes, which were not correlated with family history or alcohol use variables, and increased latencies of P3a and P3b components. They suggested that this amplitude reduction seemed to be as a result of chronic alcohol abuse.

This amplitude alteration supports the hypothesis that it reflects heritable attentional biases and information processing troubles (van der Stelt, 1999). All of this data provided support to the hypothesis that the smaller P300 amplitude is a vulnerability marker for alcoholism, and this characteristic might precede the development of the disorder (e.g. Hesselbrock et al., 2001; Hill, Shen et al., 1999; Hill, Yuan et al., 1999; Nácher, 2000; Sánchez-Turet & Serra-Grabulosa, 2002; van der Stelt et al., 1994, 1998). All these results favour the idea that genetically determined variation in neurochemical systems modulates the individual's vulnerability to alcoholism (van der Stelt, 1999).

Namkoong, Lee, Lee, Lee, and An (2004) and Bartholow, Henry, and Lust (2007) used alcohol-related cues, to show that these stimuli evoked a larger P300 amplitude in alcoholics than in controls. When they examined alcohol sensitivity, it appeared that low-sensitivity subjects exhibited larger P300 amplitude that high-sensitivity subjects, even when recent alcohol use was controlled for. The P300 elicited by alcohol-cues predicted alcohol-use at follow-up. Risk status, more than consumption history, seemed to predict cue reactivity effects. In another study by Bartholow, Lust, and Tragesser (2010), their previous findings were replicated, as well as showing that the P300 amplitudes, elicited by other targets (neutral, erotic, and adventure-related), were not different between high- and low-sensitivity individuals. These results were not correlated with impulsivity or recent alcohol consumption. They hypothesized that the P300 reactivity to alcohol cues could be considered as a new endophenotype for alcohol use disorder risk.

The P300 latency is prolonged in alcoholism, and in visual modality, P3a and P3b latencies are also longer (e.g. Biggins, MacKay, Poole, & Fein, 1995; Pfefferbaum, Rosenbloom, & Ford, 1987).

These ERP alterations testify that there are perturbations in alert processes, selective attention, and memory, and that cognitive variables are related to an increased susceptibility to develop alcoholism (Hill et al., 2004). These electrophysiological differences might have some anatomical correlates, such as suggested by differences in amygdala volume between high and low-risk adolescents (Hill, 2004). Iacono,

Carlson, Malone, and McGue (2002) showed that the reduced P3 could touch a broad field of disorder: the behavioural disinhibition spectrum, including antisocial and addictive disorders. They also proposed that the reduction of the P3 at age 17 predicted the development of substance use disorders at age 20. Somehow, the P3 amplitude may be linked with a genetic risk of disinhibiting psychiatric disorders (also see Begleiter & Porjesz, 1999; Iacono, 1998). Campanella et al. (2009) also hypothesized that attentional deficits seem to correspond to the prefrontal inhibitor deficit, observed in these patients. Cristini, Fournier, Timsit-Berthier, Bailly, and Tijus (2003) showed that electrophysiological indexes can predict relapses in abstinent patients.

The early detection of individuals at high risk represents a considerable interest: it allows us to take measures to protect the individuals from alcoholism, enlighten the aetiology of the disease, and develop new therapeutic methods and/or treatments (Eşel, 2003; Gunning, Pattiselanno, van der Stelt, & Wiers, 1994). In addition, adolescence is considered to be a critical developmental phase, which is highly vulnerable to the damaging effects of alcohol, (Guerri & Pascual, 2010) such that this approach is worthy of further development. To illustrate this, binge drinking could be considered as a risk factor to a later development of chronic alcoholism, and brief exposures to ethanol vapours can modify some electrophysiological components. In fact, repeated alcohol intake over a long period has harmful medical, as well as social consequences. For example, Ehlers et al. (2007) found in young adult Southwest California Indians that alcohol and drug exposure during adolescence associated with decreases in the latency of an early P3 component (P350). Reductions in amplitude for a later component (P450) were also found in young adults exposed to alcohol, and in those exposed to alcohol and drugs. They concluded that the results evoked some predisposing factors such as family history of alcoholism and presence of other externalizing diagnoses. The authors hypothesized that adolescent binge drinking may induce a P3 reduction in latencies and amplitudes, which might reflect a loss or delay in the development of inhibitory brain systems. Maurage, Pesenti, Philippot, Joassin, and Campanella (2009), in a study involving students displaying or not binge drinking habits, found no behavioural differences but clear electrophysiological differences (delayed latencies for P1, N2, P3b) between groups after nine months of consumption. They suggested that these latency abnormalities where similar to those evoked in long-term alcoholics, which could be interpreted as an electrophysiological marker of slowed cerebral activity due to binge drinking habits.

1.4 Part I: Conclusions

1.4.1 The P300 Alterations: Speculative Electrophysiological Markers

In the first part of this chapter, we have reviewed many studies, i.e. dementia, depression, schizophrenia, and chronic alcoholism, where changes in P300 were reported. *The results seem to suggest that P300 deficits could represent speculative markers*, since they involve a large number of psychiatric and neurological disorders. Although a lot of information has been carried out by the P300 deficits, in the following paragraphs we will discuss some methodological issues partly explaining the heterogeneous results.

In fact, P300 latency delay has been considered to be a physiological index of *dementia*, although this delay is not specific to dementia, and is also observed in various neurologic and psychiatric disorders. However, since this delay seems to be correlated with the severity of the symptoms, this could provide some prognostic indications about the course of the disease (Rodriguez et al., 1996; Soininen, Partanen, Pääkkönen, Koivisto, & Riekkinen, 1991).

The EEG beta band exhibits more power in *depressive* patients than in control (Knott, Mahoney, Kennedy, & Evans, 2001), and beta abnormalities have been linked with mental depression (Sun, Li, Zhu, Chen, & Tong, 2008). Such results support the idea that subjective symptoms can be indexed by objective bioelectrical alterations in the brain (Hinrikus et al., 2009). The observation that P300 amplitude may be a state marker is not yet consistent, although the hypothesis of the bipolar disorder pertaining to the same continuum as schizophrenia is becoming increasingly popular (e.g. Demily et al., 2009; Kendler et al., 1993).

P300 amplitude reduction in *schizophrenia* can be considered as a trait marker, because impairment occurs at every stage of the disease. Such alterations are not found in unaffected relatives and ultra-high risk of psychosis individuals. But this smaller P300 is also characteristic of other psychiatric and neurological disorders, including, as we mentioned above, bipolar affective disorders, attention-deficit hyperactivity disorder, and substance use disorders (e.g. Iacono et al., 2002; van der Stelt, van der Molen, Boudewijn Gunning, & Kok, 2001).

In *chronic alcoholism*, the P300 deficit cannot be entirely considered as a vulnerability marker, because in a similar way to schizophrenia, it is not specific to the alcohol disorder (e.g. Iacono et al., 2002). Therefore, the P300 should only be viewed as a putative pathophysiological marker, rather than a diagnostic marker (Nurnberger, 1992).

To summarize, it seems that the only "valid" information carried out by the P300 amplitudes or latency distortions is a discrimination power between health and pathology (Polich & Herbst, 2000).

1.4.2 Methodological Divergences

Some inconsistencies in above-mentioned results can be highlighted by methodological concerns. First, other factors rather than psychiatric ones can modulate the brain activity: i.e. the age of the individuals, their sex, their IQ, their personality, the ultradian cycles, fatigue, motivation, the presence of recent caffeine or food ingestion, as well as nicotine, etc. (e.g. Hansenne, 2000a, 2000b; Lembreghts et al., 1995; Polich, 2004; Polich & Herbst, 2000; Polich and Kok, 1995). These factors, easily controllable, are not often recorded.

Another important factor is the *existence of subtypes within the same pathology*, although these distinctions are not often recorded (e.g. Bruder et al., 1991; Shagass, 1981). This will generate precarious comparisons across the results, because the electrophysiological measures are not always similar. For example, Bruder et al. (1998) showed in a binaural complex tones design, a P300 right asymmetry for control participants and patients with low scores on a physical anhedonia scale, but not for patients with high anhedonia scores. Another study of Bruder et al. (1991) compared individuals having a typical major depression (melancholia or simple mood reactive depression) with individuals having an atypical depression to controls. Typical depressives showed abnormally long P300 latency for a spatial task but not a temporal task, with an abnormal lateral asymmetry (longer P300 latency for stimuli in the right hemifield). Atypical patients did not differ from controls. Correlations between longer P300 latency and ratings of insomnia was evident, while the abnormal lateral asymmetry was related to decreased right visual field advantage for syllables. Various studies of schizophrenic patients exhibited different EEG patterns, probably because different subtypes of patients were compared (e.g. John et al., 1994; Shagass & Roemer, 1991). A study of Suffin and Emory (1995) revealed that patients with analogous neurometric features responded to the same class(es) of psychopharmacologic agent(s), despite their DSM-III-R classification (patients with attention deficit disorder and affective disorders). This raises the question as to whether electrophysiological taxonomy would fit the symptoms rather better than the strict clinical one (also see Demily et al., 2009). This reality is again evident in *normal populations, who present various subclinical states*, which are not always assessed in experiments. For instance, Rossignol, Philippot, Douilliez, Crommelinck, and Campanella (2005) compared a group of low anxious students with high anxious students. Normal subjects with anxious tendencies were able to respond faster to the deviant stimuli in an emotional oddball paradigm. Moreover, the highly anxious subjects elicited earlier P3b compared to the low anxious subjects. Cavanagh and Geisler (2006) found that students with depressive tendencies showed abnormal P300 waves for happy faces. In addition, as discussed above, some differences between two separate DSM categories, such as bipolar disorder and schizophrenia, might reveal similar EEG patterns (Castaneda et al., 2008).

In addition, *the severity of the symptoms* should be taken into account, since it can elicit electrophysiological differences as well. For example, Kaustio, Partanen, Valkonen-Korhonen, Viinamäki, and Lehtonen (2002) exhibited in psychotropic drug-free depressed outpatients that affective and psychotic symptoms were associated with dissimilar types of P300 alterations. Psychotic symptoms evoked an overall reduction in P300 amplitude, marked in the left temporocentral electrode chain, which were also related to a prolonged P300 latency. Affective symptoms were linked with a relational amplitude reductions at the right temporal scalp sites. These results suggest that some different underlying neurobiological processes are involved.

Medication will also have an effect on the P300 variations: a study by Karaaslan, Gonul, Oguz, Erdinc, and Esel (2003) recorded P300 pre- and post-treatment in depressed patients with and without psychotic features. After drug treatment, delayed P300 latencies in both patient groups and decreased P300 amplitude in the patient group with psychotic features were normalized. This medication effect is also present in the control populations, notably in emotional oddballs. Kerestes et al. (2009) revealed that antidepressants may shift perceptual biases in emotional processing away from negative and towards positive stimuli. The results of Harmer et al. (2010) showed that agomelatine decreased subjective ratings of sadness, reduced recognition of sad faces, improved positive affective memory, and decreased the emotion-potentiated startle response.

Another major point is the *divergences in experimental design and acquisition methods*: no standardized outlines are respected (e.g. Duncan et al., 2009; Hansenne, 2000a, 2000b; Polich, 2004; Polich & Kok, 1995). Initially, a researcher makes partially arbitrary choices concerning the experimental design and paradigm. But it is important to remember that different choices will probably lead to different conclusions, or even conflicting results (van der Stelt & Belger, 2007). It is then crucial to match each subject within the same group and the control group with the experimental one: gender, age, education level, size of the samples, and medication should be checked before commencing the experiment (Polich, 2004). The inclusion of a control group, even if highly recommended, is sometimes missing, in order to distinguish as much as possible what is specific to the experimental group or to the control one. The subclinical tendencies of the control groups need to be assessed and short test versions exist. In the oddball tasks, presentation time, stimuli, the nature of the task, can influence the results; such bias can be eliminated by exclusively using the same testing batteries and methods for each group. For example, in dementia, it is important to obtain age-matched normative database, and to use appropriate simple P300 paradigms (Polich & Corey-Bloom, 2005). Additional cross-correlational analyses between P300, morphological and neurobiochemical data are also needed. With these elements, our knowledge about age-related cognitive changes should be enhanced.

Such standardized guidelines will contribute to an amelioration of the conclusions drawn across the results and are necessary if the P300 is to be considered as a diagnostic index (e.g. Boutros & Struve, 2002). The most important outcome is that the utilization of the P300 alone is not recommended: its clinical sensitivity has been disfavored because its alterations are diagnostically unspecific and not trustworthy for individual patients (Pogarell, Mulert, & Hegerl, 2007). For illustration, in dementia, P300 is physiologically and individually changeable, so it cannot be considered to be a fully objective diagnostic index (e.g. Chudzik, Przybyła, Kaczorowska, & Chmielewski, 2004). In depression, Bruder (1992) deplores the use of simple oddball tasks in depressed individuals that are not stimulated sufficiently to divulge cognitive dysfunctions. Bruder (1992) suggested that only some subgroups of depressed patients may have P300 deficits, but because of the heterogeneity of the samples, it is not apparent in various studies.

2 Part II: Propositions to Increase the Sensitivity of the ERP as Diagnostic Tools

Despite the fact that many interesting findings have been identified in the neuropsychiatric field, the main idea emerging from this collection of data is that *the diagnosis power of the P300 is very weak*, due to a number of factors which include individual variations, functional heterogeneity, and distributed neural generators.

However, these limitations can help us to develop tools with higher sensitivities. To illustrate this, the use of complementary techniques, such as magnetoencephalography and fMRI techniques can complete the information extracted with ERPs (e.g. Iacono et al., 2002; Van der Stelt & Belger, 2007). Behavioural methods and neuropsychological measures can also provide extra information about neurocognitive and interpersonal functioning (Andersson et al., 2008). For instance, as attention deficit may be a valid premorbid marker of memory dysfunction or dementia, Braverman et al. (2006) suggested that the results for the Test of Variables of Attention (TOVA, Greenberg, 1987) and P300, Mini-Mental Status Exam and the Weschler Memory Scale-III should be combined since it was found that the TOVA abnormalities could be considered as an indicator of the delayed P300 and attention disorder. Combining electrophysiological measures with behavioural ones may offer a more sensitive accuracy in the diagnosis and evaluation of cognitive dysfunctions.

Our own proposition to increase ERPs sensitivity, developed in this section, is to analyze various ERP components, as well as P300, through the use as of a more sophisticated oddball task, namely an emotional auditory–visual oddball. Indeed, our capacity to identify individuals constitutes the foundations of human social communication (Joassin et al., 2010). In fact, in everyday life, sensory events are not perceived separately, as it is done in the routine EEG (Maurage, Campanella, Philippot, Martin et al., 2008). Humans are constantly confronted with various stimulations, integrated into a unitary perception of the environment (Maurage, Campanella et al., 2007; Maurage, Campanella, Philippot, Martin et al., 2008). Emotions are conveyed by the five sense organs, not only by vision (e.g. Greimel, Macht, Krumhuber, & Ellgring, 2006; Shepherd, 2006; Winston, Gottfried, Kilner, & Dolan, 2005). Actually, the audio–visual integration in person recognition relies on multisensory representation of the individuals, established through experience (Campanella & Belin, 2007; Campanella et al., 2010). Binding voices and faces seems to depend on a cerebral network involving diverse facets of integration, namely sensory inputs processing, attention, and memory (Joassin et al., 2010).

2.1 *Why Are We Still Using ERPs?*

In defence of the multiple critics that can be formulated against ERP, this technique can still provide some major advantages, already outlined above, in favour of their utilization in psychiatry (e.g. Hughes, 1996; Nácher, 2000).

1. In fact, this technique is known for a long time. Its practical benefits are that it is inexpensive, non-invasive, a transportable tool, which is already implemented in the clinical psychiatric routine (e.g. Andersson et al., 2008; Fenton, 1984; Polich & Corey-Bloom, 2005). Actually, Polich and Herbst (2000) concluded that P300 information was comparable and even sometimes superior to routinely employed biomedical assay.

2. Its particularity is related to its temporal resolution, which enables the direct assessment of measures of complex cerebral processes, which are not often approachable by current clinical methods (Andersson et al., 2008; Stampfer, 1983). One additional interest is that it allows the translation of cerebral components into different levels of the information processing, from the perceptual one to the decisional one. ERPs can thus provide us with some key information for therapeutic interventions.
3. The unsolved problem about the poor spatial resolution of ERPs, probably due to the multiple underlying neural generators, can be achieved by the combined use of ERP and fMRI techniques (Matsuoka & Nakamura, 2005; Meisenzahl et al., 2004; Mulert et al., 2002; Mulert, Pogarell, & Hegerl, 2008).
4. Another major ERP's contribution is that, by investigating the real-time course of the brain activation during a cognitive event, it may be possible to distinguish different subcategories of a specific disorder, which can be based on diverse pathophysiological mechanisms, which neuropsychological tests do not always identify (Müller et al., 2001).
5. Also, a certain prediction degree to medication response can be extracted from the EEG analysis. For instance, Galderisi, Maj, Mucci, Bucci, and Kemali (1994) showed that the quantitative EEG test dose procedure could be used in the selection of the most appropriate antipsychotic medicine for schizophrenics.

Thus, it is revealed that ERPs are relatively appropriate, in comparison to others approaches, to investigate the quick changes of cerebral patterns underlying cognitive function or dysfunction (van der Stelt & Belger, 2007). Nevertheless, as highlighted by the current data, some adjustments of the technique are needed.

2.2 *Unfocused on the P300*

As P300 alterations are not exclusively found in specific psychiatric condition, there is a need to check for other biological markers (e.g. Nácher, 2000). More accurate marker might rely on the consideration of a greater number of waves, in order to detect more specific cerebral patterns per psychiatric condition. As an illustration, Foxe, Murray, and Javitt (2005) hypothesized that the P300 deficit can be a consequence of impairments that occur at early stages of the information process (e.g. perceptual or attentional stages).

2.2.1 Mood Disorders

Some former studies support this hypothesis for depression. Smith et al. (1991) showed auditory N100 and N200 problems, identifying early alterations which concerned selective attention and orientation response. Vandoolaeghe, van Hunsel, Nuyten, and Maes (1998) recorded auditory delayed P300 latency and increased

P200 amplitude in major depressed subjects without cognitive impairments compared to normal individuals; Alzheimer's dementia patients and major depressed patients with cognitive impairments had a higher P300 latency than depressed patients without cognitive alterations. Non-responders to antidepressants exhibited a higher pretreatment P300 latency and P200 amplitude than responders to treatment and normal subjects. In a study comparing subclinical anxious participants, subclinical depressive-anxious participants and a control group, Rossignol, Philippot, Crommelinck, and Campanella (2008) showed that the two anxious groups responded quicker than controls, which correlated with an earlier P3b for the anxious group; the mixed group produced higher N2b/P3a amplitudes. The authors conclude that anxious states influence later decisional stages, whereas anxious-depressive states influence early stages prior to P300 component.

Schrijvers, De Bruijn, Destoop, Hulstijn, and Sabbe (2010) have studied perfectionism and anxiety features in major depressive disorder. Cognitive control mechanisms such as action monitoring can be examined through the error-related negativity (ERN) and error positivity (Pe). Traits of perfectionism and anxiety influence ERN amplitudes in healthy subjects. Anxiety traits did not have a predictive capacity for the ERPs, while perfectionism clearly affected the ERN. Furthermore, the concern of mistake measure influenced the Pe, whereas no predictive capacity was found for anxiety traits. The impact of these variables may contribute to our understanding of the action-monitoring process and the functional significance of the Pe in depression. The divergent results could also indicate that the wide range of different affective personality styles might exert various effects on action monitoring in depression. Another study of Schrijvers et al. (2008) on depressive patients focused on psychomotor and cognitive deficits such as motor retardation and impaired executive functioning, more specifically, performance monitoring, indexed by the error negativity (Ne) or ERN components. Their results showed that severely depressed patients with retardation showed impeded action monitoring, and these two processes tended to be interdependent. Additionally, the same cerebral network was likely to be implicated in both processes.

2.2.2 Chronic Alcoholism

In chronic alcoholism, studies found increased P100/N100 latencies and reduced amplitudes, suggesting perceptive alterations (Cadaveira, Grau, Roso, & Sanchez-Turet, 1991; Maurage, Campanella, Philippot, Martin et al., 2008; Nicolás et al., 1997). In Maurage's study (2008), P100, N100, and N170 impairments were found in alcoholic patients with and without depressive tendencies, but P300 impairments were only found in depressive individuals. Some attentional deficits were also detected in other studies: N2b latency and amplitude deficits (Baguley et al., 1997; Kathmann, Soyka, Bickel, & Engel, 1996); P3a deficits (Hada, Porjesz, Begleiter, & Polich, 2000; Realmuto et al., 1993). Maurage, Joassin et al. (2007) and Maurage, Campanella, Philippot, Martin et al. (2008) observed for the first time an N170 alteration. Correlational analysis between P100/N170/P3b indicated that, for latency as well as for amplitude, P300 perturbations were directly proportional to those present at prior stages. The extent of the decisional perturbations was directly dependent on the extent of perceptual perturbations. Miyazato and Ogura (1993) showed reduced amplitudes of N100, N200, and P300 in alcoholics and increased P300 latency in alcoholics compared to controls.

Some deficits in the MMN wave are also found in chronic alcoholics, after acute alcohol ingestion, and even during the remission period (e.g. Jääskeläinen, Schröger, & Näätänen, 1999; Kathmann, Wagner, Rendtorff, & Engel, 1995; Realmuto et al., 1993). These MMN variations (decreased amplitude and increased latency) can reflect alterations in the pre-attentional ability to passively detect auditory changes. However, such alterations were not found with magnetoencephalography (Pekkonen et al., 1998). Interestingly, van der Stelt, Gunning, Snel, and Kok (1997) did not find MMN differences in children of alcoholics, suggesting that MMN alterations could be considered as a state marker rather than a vulnerability one, meaning that attentive dysfunctions (reflected by P300 alterations in children of alcoholics), rather than automatic ones, might be linked with alcoholism vulnerability. Zhang, Cohen, Porjesz, and Begleiter (2001) exhibited larger MMN responses in high-risk individuals for alcoholism, compared to low-risk individuals, which suggested that a deficit of inhibition characterizes high-risk individuals. But it is important to emphasize that MMN as a marker has clinical limitations as well: concerning P300, the MMN differences are replicated between groups (Escera & Grau, 1996; Pekkonen, Rinne, & Näätänen, 1995), but not individually, which imply that this wave cannot be considered as a marker between healthy MMN and pathological MMN (Sánchez-Turet & Serra-Grabulosa, 2002). Contradictory results (e.g. Fein, Whitlow, & Finn, 2004) are also partly due to methodological differences between studies (Sánchez-Turet & Serra-Grabulosa, 2002). De Cesarei, Codispoti, Schupp, and Stegagno (2006) emphasize that alcohol effects on cognitive, emotional, and behavioural processes are linked to an impairment of attention. In an ERP-categorization task, they exhibited that alcohol intoxication had deleterious effects at the perceptual level of processing as well as the post-perceptual processes. Ahveninen, Escera, Polo, Grau, and Jääskeläinen (2000) support the use of MMN in chronic alcoholism. In fact, even a low dose of acute alcohol significantly impairs automatic change detection and involuntary attention shifting. In turn, auditory sensory traces decay slightly faster and are more vulnerable to the distracting effect of backward masking in alcoholics than in healthy subjects. Furthermore, chronic alcohol abuse might accelerate the age-related alteration of automatic change detection. Also, MMN changes might predict altered performance in behavioural memory and attention tasks in alcoholics. It appears thus that MMN could constitute an objective non-invasive tool for exploring the neurophysiological functional deficits related to both acute alcohol intoxication and chronic alcoholism.

2.2.3 Schizophrenia

In schizophrenia, data are also in favour of this hypothesis. Lebedeva, Kaleda, Abramova, Barkhatova, and Omel'chenko (2008) exhibited decreased N100 and P300 amplitude. MMN amplitude and latency are also altered in schizophrenics,

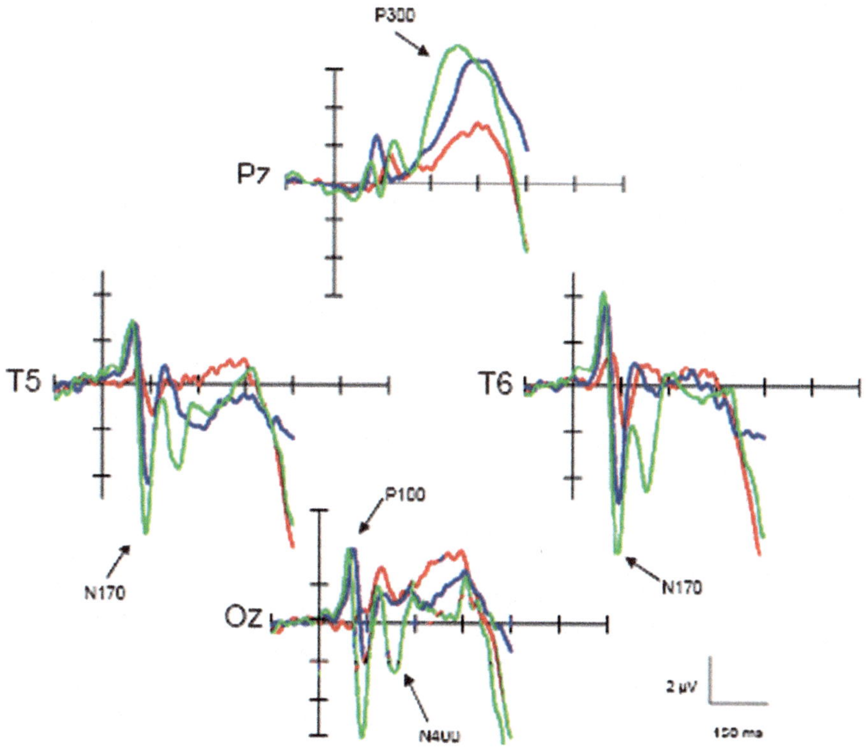

Fig. 16.1 Illustration of the P100 (Oz), the N170 (T5, T6), the P300 (Pz), and the N400 (T5, Oz, T6) recorded in response to emotional deviant faces for control subjects (*green waves*), low- (*blue*) and high- (*red*) schizophrenic patients (from Campanella et al., 2006)

compared to controls (e.g. Kawakubo, Rogers, & Kasai, 2006; Umbricht et al., 2006), and generally associated with higher cognitive deficits and global alterations in social and everyday functioning (Javitt, Doneshka, Grochowski, & Ritter, 1995; Kawakubo et al., 2006; Light & Braff, 2005). Campanella, Montedoro, Streel, Verbanck, and Rosier (2006) showed a reduction in amplitude of the N170, for emotional as well as identity faces, which was positively correlated to positive symptoms of schizophrenia. The amplitude of the P100 was also decreased. Probably in schizophrenia, there is an involvement of early visual processing which might outline the decrement of amplitude and the increment of latencies of later P300 and N400 components (Fig. 16.1).

van der Stelt and Belger (2007) emphasized that in schizophrenia, MMN alterations seem to be sensitive to premorbid cognitive status and family history risk status, and is also more dominant in chronically ill patients. MMN abnormalities could thus reflect premorbid or trait features of the illness, and post-onset progressive disease pathology in cerebral regions mediating auditory perception and language processing. Korostenskaja and Kähkönen (2009) postulated that P300 and MMN responses could be viewed as potential candidates for monitoring the cognitive changes caused by neurochemical variations during antipsychotic treatment in patients, since neurotransmitters play important roles in the generation of these components.

2.2.4 Summary

Thus, including prior waves analyses to P300 data may enable the refinement of cerebral patterns exhibited in various psychiatric disorders (e.g. Maurage, Campanella, Philippot, Martin et al., 2008; Rossignol et al., 2005, 2008). As a consequence, it can facilitate the discrimination between diverse psychiatric populations, and even within the same pathology, increase the sensitivity of ERPs. For instance, schizophrenics evoked smaller error-negativity amplitudes compared to controls, while depression and anxiety patients showed an increment of amplitude of the ERN component (Balogh & Czobor, 2010). Zhu et al. (2009) explored ERP between anxious and/or depressed patients: anxious patients exhibited longer P3a and P3b latencies, and lower N2-P3b amplitudes, whereas depressed patients presented lower N2-P3b amplitudes; patients with both anxiety and depression showed longer P3a latencies and lower N2-P3b amplitudes. P3a-b latencies were longer in anxious and anxious-depressed patients, compared to depressed ones, and N2 latencies were longer in anxious depressed patients compared to anxious or depressed ones.

2.3 More Sensitive Designs Are Required: An Example: The Bimodal Oddball

We have reviewed some P300 studies, to show that its isolated use is uninformative, but when combined to the analysis of other waves, may be helpful to discriminate healthy subjects from psychiatric ones, and even subgroups belonging to the same psychiatric disorder. We have also identified the fact that maybe other ERP components can be candidates to constitute biological markers, as it appears that the current markers tend to be "putative markers" rather than "real markers".

All these data inform us that ERPs possess their own particularities compared to other neuroimaging techniques, which enables the discovery of more accurate information about the functions and dysfunctions of the cognitive processing in psychiatric populations.

2.3.1 Affective Disorders Lead to the Development of the Affective Neurosciences

In addition to having an efficient cognitive functioning, creating adequate social interactions with other human beings is crucial to cooperating for survival (Darwin, 1872; Feldman, Philippot, & Custrini, 1991), and maintaining healthy psychological functioning (Carton, Kessler, & Pape, 1999). Possessing an appropriate decoding system for emotions may thus constitute the basis of successful interpersonal relationships, in that it may facilitate correct interpretations of a partner's intentions (Patterson, 1999) and offer a satisfactory subjective sense of well-being (Seiferth et al., 2008).

In contrast, currently it is admitted that deficits in social cognition produce misunderstandings during communication, which leads to the disturbed interpersonal functioning present in most psychiatric populations (Kornreich & Philippot, 2006; Montag et al., 2010; Patterson, 1999). Nowadays, it is commonly accepted that emotional processing disorder is the common symptom in various psychiatric states (e.g. Brüne, 2005; Frigerio, Burt, Montagne, Murray, & Perrett, 2002; Surguladze et al., 2004; Uekermann, Daum, Schlebusch, & Trenckmann, 2005).

It is not so surprising that a huge number of studies investigating emotions have been carried out. Among these, many empirical studies used emotional facial expressions (EFE). De facto, the face is one of the most important channels of communication (Buck, 1984; Hess, Kappas, & Scherer, 1988) and can convey information, such as person's mental state, intention, or disposition (Chang, Xu, Shi, Zhang, & Zhao, 2010). Research has shown that individuals who are less skilled in facial decoding also possess less social competences, and are less liked by their peers (Edwards, Manstead, & Macdonald, 1984; Feldman et al., 1991; Philippot & Feldman, 1990). Relationships between the ability to decode EFE and social skills in general have been extensively documented (Edwards et al., 1984; Feldman et al., 1991; Patterson, 1999; Philippot & Feldman, 1990).

Consequently, during the past few years, a growing number of contributions to the emotion literature from the cognitively oriented tradition have given rise to a new area of research, commonly known as "Affective Neurosciences" (Davidson, Pizzagalli, Nitschke, & Putnam, 2002). This new discipline explores the neural bases of mood and emotion, in order to obtain a better understanding of the brain circuitry underlying emotional processing. One main goal of this discipline is to establish "a scientific psychopathology", by describing normal interactions between emotional and cognitive processes, in order to understand how the impairment of these processes could lead to different clinical conditions where pathology is associated with emotional disorders (Andreasen, 1997).

2.3.2 Synchronized Emotional Stimuli: The Emotional Bimodal Oddball

In an "affective neurosciences" way, combining the ERPs technique and a need to develop more ecological stimuli, we have been working on the elaboration of a more refined design, using bimodal emotional stimulations, rather than unimodal bips and flashes, commonly used in the psychiatry routine (e.g. Brefczynski-Lewis, Lowitszch, Parsons, Lemieux, & Puce, 2009).

Clearly, recognition of emotions in one modality is biased towards the emotion expressed in a simultaneously presented but task irrelevant modality (Van den Stock, Peretz, Grèzes, & de Gelder, 2009). Crossmodal actions imply complex integrative processes, different from the unimodal ones, and in that way, these multimodal procedures might be specifically and independently impaired in psychiatric disorders (Campanella et al., 2010). For instance, de Jong, Hodiamont, Van den Stock, and de Gelder (2009) showed crossmodal face–voice alterations in schizophrenic patients, but not unimodal ones. At a behavioural level, crossmodal effects are mainly indexed by faster reaction times (Maurage, Campanella, Philippot, de Timary et al., 2008). Schweinberger, Robertson, and Kaufmann (2007) demonstrated that voice recognition was facilitated when simultaneously presented with a congruent face, and impaired when presented with a face not sharing the same identity. Thus, subjects cannot ignore a face when it is presented in time synchrony with a voice.

Our Proposition

In the light of this reality, we suggest that a more complex design would ameliorate the existing P300 differences between the healthy and clinical clusters (Campanella et al., 2010). In this perspective, in order to enhance electrophysiological measures' sensitivity, our laboratory has run ERPs studies using a crossmodal emotional oddball, which allows us to compare this condition to unimodal ones, usually used separately in the fields of psychology and psychiatry. In fact, using a task requiring emotion perception is nowadays considered as a robust multisensory situation (Collignon et al., 2008). The integration of the face and voice of the interlocutor may optimize event identification.

Therefore, our adapted emotion detection task consists of simultaneously presenting emotional and neutral faces from Ekman and Friesen (1976) with a congruent neutral word ("paper") pronounced either neutrally or with an emotional tone, in an emotion detection task (Joassin, Maurage, Bruyer, Crommelinck, & Campanella, 2004; Maurage, Campanella et al., 2007). Synchronized stimulations are preferred because these are more realistic with everyday life situations, than incongruent ones (Campanella et al., 2010). Maurage, Campanella et al. (2007) showed that the auditory–visual integration presents a better perceptual sensitivity, and displays a "crossmodal facilitation effect" in an oddball paradigm: the detection of the bimodal stimuli is faster and more accurate than the detection of the unimodal ones, both visual and auditory (de Gelder & Vroomen, 1995, 2000; Giard & Peronnet, 1999; Molholm et al., 2002).

ERPs constitute an extremely accurate tool to detect the most subtle cognitive perturbations (Maurage, Campanella et al., 2009; Rugg & Coles, 1995). By using more elaborate tools this will enable us to detect these minor changes between experimental clusters, even at a subclinical level. Actually, some authors hypothesized that the crossmodal condition might amplify the differences between groups, thus allowing us to discriminate more easily between normal and pathology, and between the different subcategories of a pathology (Campanella & Belin, 2007; Campanella et al., 2010; Campanella & Philippot, 2006). This may lead to the creation of more homogeneous patient clusters for research studies, with respect to their cognitive and neurophysiological abilities, which represents a revision in patient care (Campanella & Guérit, 2009).

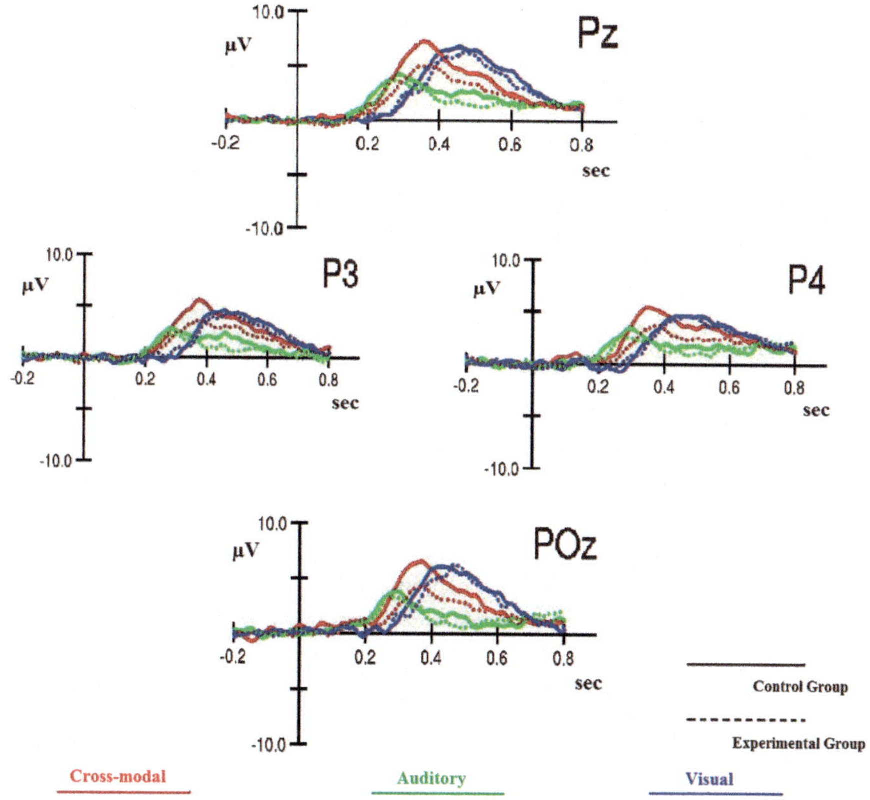

Fig. 16.2 The P300 component recorded on four parietal electrodes (P3, Pz, POz, P4) for each modality in the control and subclinical groups. The two groups did not differ on V and A modalites, but well on AV (bimodal) one (from Campanella et al., 2010)

Already Some Original Data

Some interesting outcomes have been obtained in psychopathology with these integrated stimuli. For instance, Campanella et al. (2010) found that students with anxious-depressive tendencies when compared to controls exhibited significant decreased P3b amplitude, but only in the crossmodal detection condition, and not in the unimodal ones. Thus, the neuropsychological deficits regarding their depression and anxiety scores, which distinguished the two healthy groups in the neuropsychological tests, but not in their reaction times, were electrophysiologically differently indexed, but only by the bimodal condition. Therefore, when the procedure uses more complex, sophisticated and ecological stimuli, the P300 is sensitive enough to discriminate different groups, even among the healthy ones with subclinical symptoms (Fig. 16.2).

In chronic alcoholism, (Maurage, Philippot et al., 2007) obtained, at a behavioural level, no crossmodal facilitation effect, compared to controls. These data may suggest that the crossmodal impairment in alcoholism could partly clarify the disparity between experimental data describing mild emotional facial expressions' impairments in recognition tasks (e.g. Oscar-Berman, Hancock, Mildworf, Hutner, & Weber, 1990) and the many clinical data identifying massive problems. In an ERP study from the same author (Maurage, Campanella, Philippot, Vermeulen et al., 2008), alcoholics, compared to healthy controls, were impaired in the bimodal condition, especially for the angry stimuli but not for processing of happy and neutral audio–visual stimuli. The specific deficit in alcoholics, meaning in processing anger stimuli, extensively described in clinical situations but not clearly identified in previous studies using unimodal stimuli, is revealed by the crossmodal condition.

2.4 Part II: Conclusions

2.4.1 A New Oddball Design

Starting from the principle that psychiatric populations share a common symptom, namely troubles in social cognition, we proposed here a more sophisticated oddball paradigm, intending to be as much as possible closely related to real social situations.

To obtain more precise information about the cerebral activity of these populations, we propose that a crossmodal emotional oddball may be more adequate. Even though, this paradigm could be extended to various conditions: dynamic stimuli (e.g. Atkinson, Dittrich, Gemmell, & Young, 2004; Brefczynski-Lewis et al., 2009; Gray et al., 2006; Rich et al., 2008; Summers, Papadopoulou, Bruno, Cipolotti, & Ron, 2006; Venn et al., 2004), or incongruent ones (since the ability to correctly judge the emotional content in ambiguous situations is an important communicational skill too). Actually, this could play a role in the interpersonal difficulties encountered in psychiatric populations (Frühholz, Fehr, & Herrmann, 2009; Nixon, Tivis, & Parsons, 1992; Zhu, Zhang, Wu, Luo, & Luo, 2010). Also, stimulations including the whole body's information, and not just the face can be used (e.g. Bannerman, Milders, de Gelder, & Sahraie, 2009; Stekelenburg & de Gelder, 2004; Van den Stock et al., 2007), or movement-related potentials, (e.g. Bender et al., 2007); odorant primes, (Seubert et al., 2010); musical excerpts, (Naranjo et al., 2010).

The revision of the classical oddball currently used in clinical psychiatric routine could be the first step of the legitimate rehabilitation of the utilization of ERPs in psychiatry. By employing a more complex and a more ecological paradigm, the retrieved information concerning the cognitive processing of emotional stimulations should be more useful in a therapeutic perspective.

2.4.2 Implications and Perspectives

Having reviewed some P300 literature, which showed its limitations in terms of sensitivity and specificity, we have proposed some key arguments to continue to use ERP in psychiatric routine. By rethinking some technical parameters to fit more accurately with the social reality of patients, we suggest that the tools have to evolve by "fitting" to our current knowledge of the diseases.

So, what have we learned?

1. ERPs cannot be ignored in terms of creating preventive tools as well as adequate cognitive and medicinal treatments. The finality to possess sensitive tools is to enhance patients' social skills, well-being, self-confidence, and autonomy. (Barch, 2009; Sablier, Stip, & Franck, 2009).As we mentioned earlier, the detection of a patient-specific pattern of deficits, through ERPs, is shown to be decisive, as it will quite likely lead to a customized cognitive remediation program, and consequently, could positively act upon their motivation (Sablier et al., 2009). It could therefore promote the development of effective therapeutic strategies, focused on particular pathophysiological mechanisms and cognitive dysfunctions, rather than only on the clinical symptoms (van der Stelt & Belger, 2007). Actually, the exact psychopathological characterization of patients seems to be more appropriate than the use of an artificial classification system (Bender et al., 2007; Berrettini, 2005). Understanding every detail of information processing dysfunctions in patients could provide us with some new treatment methods, including specific neuropsychological rehabilitation procedures. (Campanella & Philippot, 2006; Delle-Vigne et al., 2011). For instance, cognitive impairments, rather than the positive or negative symptoms of schizophrenia, predict poor performance in basic activities of daily living (Raffard, Gely-Nargeot, Capdevielle, Bayard, & Boulenger, 2009; Sablier et al., 2009). Although it is possible to reduce psychotic symptoms and to prevent relapses with antipsychotic medication, this is not yet applicable for cognitive deficits.
2. Furthermore, ERPs, combined with other research tools, may participate in the development of our comprehension of pathology, and a better understanding of the cognitive and cerebral functions engaged in psychiatric disorders as well as the drug-induced changes on the neural substrates of information processing. For instance, detecting the specific information stage damaged during the real-time course of the cerebral activity is one of the main interests of using ERPs in the psychiatric routine. Bender et al. (2007) also insists on the fact that ERPs can show characteristic patterns of the underlying neuronal mechanisms behind the behavioural symptoms, and complete clinical instruments of psychiatric disorder detection.

2.5 General Summary

The ability to decode other's emotion, in order to develop successful social interactions, is a fundamental competence for human life (Collignon et al., 2008). Our ability to integrate face and voice of our partner in a unique percept is a key determinant for prosperous relationships.

With this in mind, our main argument is that the current use of ERPs has to evolve towards more ecological and more elaborate stimulations using this emotional information, to capture the complexity of the integrative processes. Actually, we have suggested that these upper levels of information processes can be specifically altered in some psychiatric disorders, whereas the lower levels can be spared. A more refined paradigm will lead to enhanced refined results, which in turns can help to set up the most adequate therapeutic interventions. By acting on the cognitive dysfunctions this may spread to psychological well-being, and could constitute the first step to the road of recovery.

References

Ahveninen, J., Escera, C., Polo, M. D., Grau, C., & Jääskeläinen, I. P. (2000). Acute and chronic effects of alcohol on preattentive auditory processing as reflected by mismatch negativity. *Audiology & Neuro-Otology, 5*(6), 303–311.

American Psychiatric Association. (2000). Diagnostic and statistical manual of mental disorders IV-TR, Edittions Masson.

Andersson, S., Barder, H. E., Hellvin, T., Løvdahl, H., & Malt, U. F. (2008). Neuropsychological and electrophysiological indices of neurocognitive dysfunction in bipolar II disorder. *Bipolar Disorders, 10*(8), 888–899.

Andreasen, N. C. (1997). Linking mind and brain in the study of mental illnesses: A project for a scientific psychopathology. *Science, 275*(5306), 1586–1593.

Atkinson, A. P., Dittrich, W. H., Gemmell, A. J., & Young, A. W. (2004). Emotion perception from dynamic and static body expressions in point-light and full-light displays. *Perception, 33*(6), 717–746.

Baguley, I. J., Felmingham, K. L., Lahz, S., Gordan, E., Lazzaro, I., & Schotte, D. E. (1997). Alcohol abuse and traumatic brain injury: Effect on event-related potentials. *Archives of Physical Medicine and Rehabilitation, 78*(11), 1248–1253.

Ball, S. S., Marsh, J. T., Schubarth, G., Brown, W. S., & Strandburg, R. (1989). Longitudinal P300 latency changes in Alzheimer's disease. *Journal of Gerontology, 44*(6), M195–M200.

Balogh, L., & Czobor, P. (2010). Event-related EEG potentials associated with error detection in psychiatric disorder: Literature review. *Psychiatria Hungarica: A Magyar Pszichiátriai Társaság Tudományos Folyóirata, 25*(2), 121–132.

Bannerman, R. L., Milders, M., de Gelder, B., & Sahraie, A. (2009). Orienting to threat: Faster localization of fearful facial expressions and body postures revealed by saccadic eye movements. *Proceedings of the The Royal Society B: Biological Sciences/, 276*(1662), 1635–1641.

Barch, D. M. (2009). Neuropsychological abnormalities in schizophrenia and major mood disorders: Similarities and differences. *Current Psychiatry Reports, 11*(4), 313–319.

Bartholow, B. D., Henry, E. A., & Lust, S. A. (2007). Effects of alcohol sensitivity on P3 event-related potential reactivity to alcohol cues. *Psychology of Addictive Behaviors: Journal of the Society of Psychologists in Addictive Behaviors, 21*(4), 555–563.

Bartholow, B. D., Lust, S. A., & Tragesser, S. L. (2010). Specificity of P3 event-related potential reactivity to alcohol cues in individuals low in alcohol sensitivity. *Psychology of Addictive Behaviors: Journal of the Society of Psychologists in Addictive Behaviors, 24*(2), 220–228.

Başar, E., Başar-Eroglu, C., Karakaş, S., & Schürmann, M. (2001). Gamma, alpha, delta, and theta oscillations govern cognitive processes. *International Journal of Psychophysiology: Official Journal of the International Organization of Psychophysiology, 39*(2–3), 241–248.

Begleiter, H., & Porjesz, B. (1999). What is inherited in the predisposition toward alcoholism? A proposed model. *Alcoholism, Clinical and Experimental Research, 23*(7), 1125–1135.

Begleiter, H., Porjesz, B., Bihari, B., & Kissin, B. (1984). Event-related brain potentials in boys at risk for alcoholism. *Science, 225*(4669), 1493–1496.

Bender, S., Weisbrod, M., & Resch, F. (2007). Which perspectives can endophenotypes and biological markers offer in the early recognition of schizophrenia? *Journal of Neural Transmission, 114*(9), 1199–1215.

Berman, S. M., Whipple, S. C., Fitch, R. J., & Noble, E. P. (1993). P3 in young boys as a predictor of adolescent substance use. *Alcohol, 10*(1), 69–76.

Berrettini, W. H. (2005). Genetic bases for endophenotypes in psychiatric disorders. *Dialogues in Clinical Neuroscience, 7*(2), 95–101.

Biggins, C. A., MacKay, S., Poole, N., & Fein, G. (1995). Delayed P3A in abstinent elderly male chronic alcoholics. *Alcoholism, Clinical and Experimental Research, 19*(4), 1032–1042.

Blackwood, D. H., St Clair, D. M., Muir, W. J., & Duffy, J. C. (1991). Auditory P300 and eye tracking dysfunction in schizophrenic pedigrees. *Archives of General Psychiatry, 48*(10), 899–909.

Blackwood, D. H., Whalley, L. J., Christie, J. E., Blackburn, I. M., St Clair, D. M., & McInnes, A. (1987). Changes in auditory P3 event-related potential in schizophrenia and depression. *The British Journal of Psychiatry: The Journal of Mental Science, 150*, 154–160.

Boutros, N. N., & Struve, F. (2002). Electrophysiological assessment of neuropsychiatric disorders. *Seminars in Clinical Neuropsychiatry, 7*(1), 30–41.

Bowie, C. R., Depp, C., McGrath, J. A., Wolyniec, P., Mausbach, B. T., Thornquist, M. H., et al. (2010). Prediction of real-world functional disability in chronic mental disorders: A comparison of schizophrenia and bipolar disorder. *The American Journal of Psychiatry, 167*(9), 1116–1124.

Bramon, E., McDonald, C., Croft, R. J., Landau, S., Filbey, F., Gruzelier, J. H., et al. (2005). Is the P300 wave an endophenotype for schizophrenia? A meta-analysis and a family study. *NeuroImage, 27*(4), 960–968.

Bramon, E., Rabe-Hesketh, S., Sham, P., Murray, R. M., & Frangou, S. (2004). Meta-analysis of the P300 and P50 waveforms in schizophrenia. *Schizophrenia Research, 70*(2–3), 315–329.

Bramon, E., Shaikh, M., Broome, M., Lappin, J., Bergé, D., Day, F., et al. (2008). Abnormal P300 in people with high risk of developing psychosis. *NeuroImage, 41*(2), 553–560.

Braverman, E. R., Chen, T. J. H., Schoolfield, J., Martinez-Pons, M., Arcuri, V., Varshavskiy, M., et al. (2006). Delayed P300 latency correlates with abnormal Test of Variables of Attention (TOVA) in adults and predicts early cognitive decline in a clinical setting. *Advances in Therapy, 23*(4), 582–600.

Brefczynski-Lewis, J., Lowitszch, S., Parsons, M., Lemieux, S., & Puce, A. (2009). Audiovisual non-verbal dynamic faces elicit converging fMRI and ERP responses. *Brain Topography, 21*(3–4), 193–206.

Bruder, G. E. (1992). P300 findings for depressive and anxiety disorders. *Annals of the New York Academy of Sciences, 658*, 205–222.

Bruder, G. E., Kroppmann, C. J., Kayser, J., Stewart, J. W., McGrath, P. J., & Tenke, C. E. (2009). Reduced brain responses to novel sounds in depression: P3 findings in a novelty oddball task. *Psychiatry Research, 170*(2–3), 218–223.

Bruder, G. E., Tenke, C. E., Towey, J. P., Leite, P., Fong, R., Stewart, J. E., et al. (1998). Brain ERPs of depressed patients to complex tones in an oddball task: relation of reduced P3 asymmetry to physical anhedonia. *Psychophysiology, 35*(1), 54–63.

Bruder, G. E., Towey, J. P., Stewart, J. W., Friedman, D., Tenke, C., & Quitkin, F. M. (1991). Event-related potentials in depression: Influence of task, stimulus hemifield and clinical features on P3 latency. *Biological Psychiatry, 30*(3), 233–246.

Brüne, M. (2005). Emotion recognition, 'theory of mind', and social behavior in schizophrenia. *Psychiatry Research, 133*(2–3), 135–147.

Buck, R. (1984). *The communication of emotion*. New York: Guilford.

Cadaveira, F., Grau, C., Roso, M., & Sanchez-Turet, M. (1991). Multimodality exploration of event-related potentials in chronic alcoholics. *Alcoholism, Clinical and Experimental Research, 15*(4), 607–611.

Campanella, S., & Belin, P. (2007). Integrating face and voice in person perception. *Trends in Cognitive Sciences, 11*(12), 535–543.

Campanella, S., Bruyer, R., Froidbise, S., Rossignol, M., Joassin, F., Kornreich, C., et al. (2010). Is two better than one? A cross-modal oddball paradigm reveals greater sensitivity of the P300 to emotional face-voice associations. *Clinical Neurophysiology: Official Journal of the International Federation of Clinical Neurophysiology, 121*(11), 1855–1862.
Campanella, S., & Guérit, J. (2009). How clinical neurophysiology may contribute to the understanding of a psychiatric disease such as schizophrenia. *Neurophysiologie Clinique/ Clinical Neurophysiology, 39*(1), 31–39.
Campanella, S., Montedoro, C., Streel, E., Verbanck, P., & Rosier, V. (2006). Early visual components (P100, N170) are disrupted in chronic schizophrenic patients: An event-related potentials study. *Neurophysiologie Clinique/Clinical Neurophysiology, 36*(2), 71–78.
Campanella, S., Petit, G., Maurage, P., Kornreich, C., Verbanck, P., & Noël, X. (2009). Chronic alcoholism: Insights from neurophysiology. *Neurophysiologie Clinique/Clinical Neurophysiology, 39*(4–5), 191–207.
Campanella, S., & Philippot, P. (2006). Insights from ERPs into emotional disorders: An affective neuroscience perspective. *Psychologica Belgica, 46*(1–2), 37–53.
Campanella, S., & Streel, E. (Eds.). (2008). *Psychopathologie et neurosciences: Questions actuelles de neurosciences cognitives et affectives*. Bruxelles: De Boeck.
Canguilhem, G. (1972). *Le normal et le pathologique*. Paris: Presses Universitaires de France.
Carlson, S. R., Iacono, W. G., & McGue, M. (2002). P300 amplitude in adolescent twins discordant and concordant for alcohol use disorders. *Biological Psychology, 61*(1–2), 203–227.
Carton, J. S., Kessler, E. A., & Pape, C. L. (1999). Nonverbal decoding skills and relationship well-being in adults. *Journal of Nonverbal Behavior, 23*(1), 91–100.
Castaneda, A. E., Tuulio-Henriksson, A., Marttunen, M., Suvisaari, J., & Lönnqvist, J. (2008). A review on cognitive impairments in depressive and anxiety disorders with a focus on young adults. *Journal of Affective Disorders, 106*(1–2), 1–27.
Cavanagh, J., & Geisler, M. W. (2006). Mood effects on the ERP processing of emotional intensity in faces: A P3 investigation with depressed students. *International Journal of Psychophysiology: Official Journal of the International Organization of Psychophysiology, 60*(1), 27–33.
Chang, Y., Xu, J., Shi, N., Zhang, B., & Zhao, L. (2010). Dysfunction of processing task-irrelevant emotional faces in major depressive disorder patients revealed by expression-related visual MMN. *Neuroscience Letters, 472*(1), 33–37.
Chudzik, W., Przybyła, M., Kaczorowska, B., & Chmielewski, H. (2004). Potential P300 in diagnostics of cognitive functions. *Wiadomości Lekarskie (Warsaw, Poland: 1960), 57*(7–8), 356–359.
Cohen, H. L., Porjesz, B., Begleiter, H., & Wang, W. (1997). Neurophysiological correlates of response production and inhibition in alcoholics. *Alcoholism, Clinical and Experimental Research, 21*(8), 1398–1406.
Collignon, O., Girard, S., Gosselin, F., Roy, S., Saint-Amour, D., Lassonde, M., et al. (2008). Audio-visual integration of emotion expression. *Brain Research, 124*, 126–35.
Coullaut-Valera García, J., Díaz, A., del Rio, I., Coullaut-Valera García, R., & Ortiz, T. (2007). Alterations of P300 wave in occipital lobe in depressive patients. *Actas Españolas De Psiquiatría, 35*(4), 243–248.
Craddock, N., & Owen, M. J. (2005). The beginning of the end for the Kraepelinian dichotomy. *The British Journal of Psychiatry: The Journal of Mental Science, 186*, 364–366.
Cristini, P., Fournier, C., Timsit-Berthier, M., Bailly, M., & Tijus, C. (2003). ERPs (N200, P300 and CNV) in alcoholics: Relapse risk assessment. *Neurophysiologie Clinique/Clinical Neurophysiology, 33*(3), 103–119.
Darwin, C. (1872). *The expression of emotions in man and animals*. London: John Murray.
Davidson, R. J., Pizzagalli, D., Nitschke, J. B., & Putnam, K. (2002). Depression: Perspectives from affective neuroscience. *Annual Review of Psychology, 53*, 545–574.
Delle-Vigne, D., Campanella, S., Kajosch, H., Verbanck, P., & Kornreich, C. (2011) [Increasing P300 sensitivity trough and emotional bimodal oddball]. *Acta Psychiatica Belgica, 111*(1), 29–44.
De Cesarei, A., Codispoti, M., Schupp, H. T., & Stegagno, L. (2006). Selectively attending to natural scenes after alcohol consumption: An ERP analysis. *Biological Psychology, 72*(1), 35–45.

De Gelder, B., & Vroomen, J. (1995). Hearing smiles and seeing cries: The bimodal perception of emotions. *Bulletin of the Psychonomic Society, 30*, 15.

De Gelder, B., & Vroomen, J. (2000). The perception of emotions by ear and by eye. *Cognition and Emotion, 14*, 289–311.

de Jong, J. J., Hodiamont, P. P. G., Van den Stock, J., & de Gelder, B. (2009). Audiovisual emotion recognition in schizophrenia: Reduced integration of facial and vocal affect. *Schizophrenia Research, 107*(2–3), 286–293.

Demily, C., Jacquet, P., & Marie-Cardine, M. (2009). How to differentiate schizophrenia from bipolar disorder using cognitive assessment? *L'Encéphale, 35*(2), 139–145.

Desmedt, J. E. (1980). P300 in serial tasks: An essential post-decision closure mechanism. *Progress in Brain Research, 54*, 682–686.

Desmedt, J. E., Debecker, J., & Manil, J. (1965). Demonstration of a cerebral electric sign associated with the detection by the subject of a tactile sensorial stimulus. The analysis of cerebral evoked potentials derived from the scalp with the aid of numerical ordinates. *Bulletin De l'Académie Royale De Médecine De Belgique, 5*(11), 887–936.

Donchin, E., & Coles, M. (1988). Is the P300 component a manifestation of context updating? *Behavioral and Brain Sciences, 11*, 357–374.

Duncan, C. C., Barry, R. J., Connolly, J. F., Fischer, C., Michie, P. T., Näätänen, R., et al. (2009). Event-related potentials in clinical research: Guidelines for eliciting, recording, and quantifying mismatch negativity, P300, and N400. *Clinical Neurophysiology: Official Journal of the International Federation of Clinical Neurophysiology, 120*(11), 1883–1908.

Duncan, C. C., Kosmidis, M. H., & Mirsky, A. F. (2003). Event-related potential assessment of information processing after closed head injury. *Psychophysiology, 40*(1), 45–59.

Duncan, C. C., Mirsky, A. F., Deldin, P. J., Skwerer, R. G., Jacobsen, F. M., & Rosenthal, N. E. (1991). Brain potentials index treatment response in seasonal affective disorder. In M. Ansseau, R. von Frenckell, & G. Franck (Eds.), *Biological markers of depression: State of the art* (pp. 117–120). Amsterdam: Elsevier Science Publishers B.V.

Duncan, C. C., Morihisa, J. M., Fawcet, R., & Kirch, D. (1987). P300 in schizophrenia: State or trait marker? *Psychopharmacology Bulletin, 23*, 497–501.

Duncan-Johnson, C. C. (1981). Young Psychophysiologist Award address, 1980. P300 latency: A new metric of information processing. *Psychophysiology, 18*(3), 207–215.

Duyckarts, F. (1964). *La notion du normal en psychologie clinique*. Paris: Vrin.

Edwards, R., Manstead, A. S. R., & Macdonald, C. J. (1984). The relationship between children's sociometric status and ability to recognize facial expressions of emotion. *European Journal of Social Psychology, 14*(2), 235–238.

Egerházi, A., Glaub, T., Balla, P., Berecz, R., & Degrell, I. (2008). P300 in mild cognitive impairment and in dementia. *Psychiatria Hungarica: A Magyar Pszichiátriai Társaság Tudományos Folyóirata, 23*(5), 349–357.

Ehlers, C. L., Phillips, E., Finnerman, G., Gilder, D., Lau, P., & Criado, J. (2007). P3 components and adolescent binge drinking in Southwest California Indians. *Neurotoxicology and Teratology, 29*(1), 153–163.

Ekman, P., & Friesen, W. (1976). *Pictures of facial affect*. Palo Alto, CA: Consulting Psychologists Press.

Emmerson, R. Y., Dustman, R. E., Shearer, D. E., & Turner, C. W. (1989). P3 latency and symbol digit performance correlations in aging. *Experimental Aging Research, 15*(3–4), 151–159.

Ergen, M., Marbach, S., Brand, A., Başar-Eroğlu, C., & Demiralp, T. (2008). P3 and delta band responses in visual oddball paradigm in schizophrenia. *Neuroscience Letters, 440*(3), 304–308.

Escera, C., & Grau, C. (1996). Short-term replicability of the mismatch negativity. *Electroencephalography and Clinical Neurophysiology, 100*(6), 549–554.

Eşel, E. (2003). Biological trait markers of alcohol dependence. *Türk Psikiyatri Dergisi/Turkish Journal of Psychiatry, 14*(1), 60–71.

Fein, G., Biggins, C. A., & MacKay, S. (1995). Alcohol abuse and HIV infection have additive effects on frontal cortex function as measured by auditory evoked potential P3A latency. *Biological Psychiatry, 37*(3), 183–195.

Fein, G., & Chang, M. (2006). Visual P300s in long-term abstinent chronic alcoholics. *Alcoholism, Clinical and Experimental Research, 30*(12), 2000–2007.

Fein, G., Whitlow, B., & Finn, P. (2004). Mismatch negativity: No difference between controls and abstinent alcoholics. *Alcoholism, Clinical and Experimental Research, 28*(1), 137–142.

Feldman, R., Philippot, P., & Custrini, R. (1991). Social competence and non-verbal behaviour. In R. Feldman (Ed.), *Fundamentals of nonverbal behaviour* (pp. 329–350). New York: Cambridge University Press.

Fenton, G. W. (1984). The electroencephalogram in psychiatry: Clinical and research applications. *Psychiatric Developments, 2*(1), 53–75.

Fenton, G. W., Fenwick, P. B., Dollimore, J., Dunn, T. L., & Hirsch, S. R. (1980). EEG spectral analysis in schizophrenia. *The British Journal of Psychiatry: The Journal of Mental Science, 136*, 445–455.

Fisher, D. J., Labelle, A., & Knott, V. J. (2010). Auditory hallucinations and the P3a: Attention-switching to speech in schizophrenia. *Biological Psychology, 85*(3), 417–423.

Ford, J. M. (1999). Schizophrenia: The broken P300 and beyond. *Psychophysiology, 36*(6), 667–682.

Foxe, J. J., Murray, M. M., & Javitt, D. C. (2005). Filling-in in schizophrenia: A high-density electrical mapping and source-analysis investigation of illusory contour processing. *Cerebral Cortex, 15*(12), 1914–1927.

Freedman, R., Coon, H., Myles-Worsley, M., Orr-Urtreger, A., Olincy, A., Davis, A., et al. (1997). Linkage of a neurophysiological deficit in schizophrenia to a chromosome 15 locus. *Proceedings of the National Academy of Sciences of the United States of America, 94*(2), 587–592.

Frigerio, E., Burt, D. M., Montagne, B., Murray, L. K., & Perrett, D. I. (2002). Facial affect perception in alcoholics. *Psychiatry Research, 113*(1–2), 161–171.

Frodl, T., Hampel, H., Juckel, G., Bürger, K., Padberg, F., Engel, R. R., et al. (2002). Value of event-related P300 subcomponents in the clinical diagnosis of mild cognitive impairment and Alzheimer's Disease. *Psychophysiology, 39*(2), 175–181.

Frommann, I., Brinkmeyer, J., Ruhrmann, S., Hack, E., Brockhaus-Dumke, A., Bechdolf, A., et al. (2008). Auditory P300 in individuals clinically at risk for psychosis. *International Journal of Psychophysiology: Official Journal of the International Organization of Psychophysiology, 70*(3), 192–205.

Frühholz, S., Fehr, T., & Herrmann, M. (2009). Early and late temporo-spatial effects of contextual interference during perception of facial affect. *International Journal of Psychophysiology: Official Journal of the International Organization of Psychophysiology, 74*(1), 1–13.

Galderisi, S., Maj, M., Mucci, A., Bucci, P., & Kemali, D. (1994). QEEG alpha 1 changes after a single dose of high-potency neuroleptics as a predictor of short-term response to treatment in schizophrenic patients. *Biological Psychiatry, 35*(6), 367–374.

Gangadhar, B. N., Ancy, J., Janakiramaiah, N., & Umapathy, C. (1993). P300 amplitude in non-bipolar, melancholic depression. *Journal of Affective Disorders, 28*(1), 57–60.

Gershon, E. S., & Goldin, L. R. (1986). Clinical methods in psychiatric genetics. I. Robustness of genetic marker investigative strategies. *Acta Psychiatrica Scandinavica, 74*(2), 113–118.

Giard, M. H., & Peronnet, F. (1999). Auditory-visual integration during multimodal object recognition in humans: A behavioral and electrophysiological study. *Journal of Cognitive Neuroscience, 11*(5), 473–490.

Glahn, D. C., Therman, S., Manninen, M., Huttunen, M., Kaprio, J., Lönnqvist, J., et al. (2003). Spatial working memory as an endophenotype for schizophrenia. *Biological Psychiatry, 53*(7), 624–626.

Goodin, D. S., Squires, K. C., Henderson, B. H., & Starr, A. (1978). Age-related variations in evoked potentials to auditory stimuli in normal human subjects. *Electroencephalography and Clinical Neurophysiology, 44*(4), 447–458.

Goodin, D. S., Squires, K. C., & Starr, A. (1978). Long latency event-related components of the auditory evoked potential in dementia. *Brain: A Journal of Neurology, 101*(4), 635–648.

Goodin, D. S., Starr, A., Chippendale, T., & Squires, K. C. (1983). Sequential changes in the P3 component of the auditory evoked potential in confusional states and dementing illnesses. *Neurology, 33*(9), 1215–1218.

Gordon, E., Kraiuhin, C., Harris, A., Meares, R., & Howson, A. (1986). The differential diagnosis of dementia using P300 latency. *Biological Psychiatry, 21*(12), 1123–1132.

Gottesman, I. I., & Gould, T. D. (2003). The endophenotype concept in psychiatry: Etymology and strategic intentions. *The American Journal of Psychiatry, 160*(4), 636–645.

Gray, J., Venn, H., Montagne, B., Murray, L., Burt, M., Frigerio, E., et al. (2006). Bipolar patients show mood-congruent biases in sensitivity to facial expressions of emotion when exhibiting depressed symptoms, but not when exhibiting manic symptoms. *Cognitive Neuropsychiatry, 11*(6), 505–520.

Greenberg, L. M. (1987). An objective measure of methylphenidate response: Clinical use of the MCA. *Psychopharmacology Bulletin, 23*, 279–282.

Greimel, E., Macht, M., Krumhuber, E., & Ellgring, H. (2006). Facial and affective reactions to tastes and their modulation by sadness and joy. *Physiology & Behavior, 89*(2), 261–269.

Guérit, J. M. (1998). Revue de la littérature. *Electroencephalography and Clinical Neurophysiology, 28*, 92–93.

Guerri, C., & Pascual, M. (2010). Mechanisms involved in the neurotoxic, cognitive, and neurobehavioral effects of alcohol consumption during adolescence. *Alcohol, 44*(1), 15–26.

Gunning, W. B., Pattiselanno, S. E., van der Stelt, O., & Wiers, R. W. (1994). Children of alcoholics. Predictors for psychopathology and addiction. *Acta Paediatrica Supplement, 404*, 7–8.

Hada, M., Porjesz, B., Begleiter, H., & Polich, J. (2000). Auditory P3a assessment of male alcoholics. *Biological Psychiatry, 48*(4), 276–286.

Halgren, E., Baudena, P., Clarke, J. M., Heit, G., Liégeois, C., Chauvel, P., et al. (1995). Intracerebral potentials to rare target and distractor auditory and visual stimuli. I. Superior temporal plane and parietal lobe. *Electroencephalography and Clinical Neurophysiology, 94*(3), 191–220.

Halgren, E., Baudena, P., Clarke, J. M., Heit, G., Marinkovic, K., Devaux, B., et al. (1995). Intracerebral potentials to rare target and distractor auditory and visual stimuli II Medial, lateral and posterior temporal lobe. *Electroencephalography and Clinical Neurophysiology, 94*(4), 229–250.

Halgren, E., Marinkovic, K., & Chauvel, P. (1998). Generators of the late cognitive potentials in auditory and visual oddball tasks. *Electroencephalography and Clinical Neurophysiology, 106*(2), 156–164.

Halgren, E., Squires, N. K., Wilson, C. L., Rohrbaugh, J. W., Babb, T. L., & Crandall, P. H. (1980). Endogenous potentials generated in the human hippocampal formation and amygdala by infrequent events. *Science, 210*(4471), 803–805.

Hansenne, M. (2000a). The P300 cognitive event-related potential. I. Theoretical and psychobiologic perspectives. *Neurophysiologie Clinique/Clinical Neurophysiology, 30*(4), 191–210.

Hansenne, M. (2000b). The P300 cognitive event-related potential. II. Individual variability and clinical application in psychopathology. *Neurophysiologie Clinique/Clinical Neurophysiology, 30*(4), 211–231.

Hansenne, M., Pitchot, W., Gonzalez Moreno, A., Zaldua, I. U., & Ansseau, M. (1996). Suicidal behavior in depressive disorder: An event-related potential study. *Biological Psychiatry, 40*(2), 116–122.

Harmer, C. J., de Bodinat, C., Dawson, G. R., Dourish, C. T., Waldenmaier, L., Adams, S., et al. (2010). Agomelatine facilitates positive versus negative affective processing in healthy volunteer models. *Journal of Psychopharmacology, 25*, 1159–1167.

Hartikainen, K., & Knight, R. T. (2003). Lateral and orbital prefrontal cortex contributions to attention. In J. Polich (Ed.), *Detection of change: Event-related potential and fMRI findings* (pp. 99–116). Norwell, MA: Kluwer Academic Press.

Have, G., Kolbeinsson, H., & Pétursson, H. (1991). Dementia and depression in old age: Psychophysiological aspects. *Acta Psychiatrica Scandinavica, 83*(5), 329–333.

Hess, U., Kappas, A., & Scherer, K. (1988). Mutlichannel communication of emotion: Synthetic signal production. In K. Sherer (Ed.), *Facets of emotion: Recent research* (pp. 161–182). Hillsdale, NJ: Lawrence Erbaulm Associates.

Hesselbrock, V., Begleiter, H., Porjesz, B., O'Connor, S., & Bauer, L. (2001). P300 event-related potential amplitude as an endophenotype of alcoholism—Evidence from the collaborative study on the genetics of alcoholism. *Journal of Biomedical Science, 8*(1), 77–82.

Hill, S.K., Keshavan, M.S., Thase, M.E., & Sweeney, J.A. (2004). Neuropsychological dysfunction in antipsychotic-naive first-episode unipolar psychotic depression. The *American Journal of Psychiatry, 161*(6), 996–1003.

Hill, S. K., Harris, M. S. H., Herbener, E. S., Pavuluri, M., & Sweeney, J. A. (2008). Neurocognitive allied phenotypes for schizophrenia and bipolar disorder. *Schizophrenia Bulletin, 34*(4), 743–759.

Hill, S. Y., Shen, S., Locke, J., Steinhauer, S. R., Konicky, C., Lowers, L., et al. (1999). Developmental delay in P300 production in children at high risk for developing alcohol-related disorders. *Biological Psychiatry, 46*(7), 970–981.

Hill, S. Y., Steinhauer, S., Lowers, L., & Locke, J. (1995). Eight-year longitudinal follow-up of P300 and clinical outcome in children from high-risk for alcoholism families. *Biological Psychiatry, 37*(11), 823–827.

Hill, S. Y., Yuan, H., & Locke, J. (1999). Path analysis of P300 amplitude of individuals from families at high and low risk for developing alcoholism. *Biological Psychiatry, 45*(3), 346–359.

Hinrikus, H., Suhhova, A., Bachmann, M., Aadamsoo, K., Võhma, U., Lass, J., et al. (2009). Electroencephalographic spectral asymmetry index for detection of depression. *Medical & Biological Engineering & Computing, 47*(12), 1291–1299.

Holt, L. E., Raine, A., Pa, G., Schneider, L. S., Henderson, V. W., & Pollock, V. E. (1995). P300 topography in Alzheimer's disease. *Psychophysiology, 32*(3), 257–265.

Hughes, J. R. (1996). A review of the usefulness of the standard EEG in psychiatry. *Clinical EEG (Electroencephalography), 27*(1), 35–39.

Hughes, J. R., & John, E. R. (1999). Conventional and quantitative electroencephalography in psychiatry. *The Journal of Neuropsychiatry and Clinical Neurosciences, 11*(2), 190–208.

Iacono, W. G. (1998). Identifying psychophysiological risk for psychopathology: examples from substance abuse and schizophrenia research. *Psychophysiology, 35*(6), 621–637.

Iacono, W. G., Carlson, S. R., & Malone, S. M. (2000). Identifying a multivariate endophenotype for substance use disorders using psychophysiological measures. *International Journal of Psychophysiology, 38*(1), 81–96.

Iacono, W. G., Carlson, S. R., Malone, S. M., & McGue, M. (2002). P3 event-related potential amplitude and the risk for disinhibitory disorders in adolescent boys. *Archives of General Psychiatry, 59*(8), 750–757.

Ilan, A. B., & Polich, J. (1999). P300 and response time from a manual Stroop task. *Clinical Neurophysiology: Official Journal of the International Federation of Clinical Neurophysiology, 110*(2), 367–373.

Iv, J., Zhao, L., Gong, J., Chen, C., & Miao, D. (2010). Event-related potential based evidence of cognitive dysfunction in patients during the first episode of depression using a novelty oddball task. *Psychiatry Research, 182*(1), 58–66.

Jääskeläinen, I. P., Schröger, E., & Näätänen, R. (1999). Electrophysiological indices of acute effects of ethanol on involuntary attention shifting. *Psychopharmacology, 141*(1), 16–21.

Jabben, N., Arts, B., Krabbendam, L., & van Os, J. (2009). Investigating the association between neurocognition and psychosis in bipolar disorder: Further evidence for the overlap with schizophrenia. *Bipolar Disorders, 11*(2), 166–177.

Jabben, N., Arts, B., van Os, J., & Krabbendam, L. (2010). Neurocognitive functioning as intermediary phenotype and predictor of psychosocial functioning across the psychosis continuum: Studies in schizophrenia and bipolar disorder. *The Journal of Clinical Psychiatry, 71*(6), 764–774.

Jacobson, S., Leuchter, A. F., & Walter, D. (1996). Conventional and quantitative EEG in the diagnosis of delirium among the elderly. *Journal of Neurology, Neurosurgery, and Psychiatry, 56*(Suppl. 2), 153–158.

Javitt, D. C., Doneshka, P., Grochowski, S., & Ritter, W. (1995). Impaired mismatch negativity generation reflects widespread dysfunction of working memory in schizophrenia. *Archives of General Psychiatry, 52*(7), 550–558.

Jeon, Y., & Polich, J. (2003). Meta-analysis of P300 and schizophrenia: Patients, paradigms, and practical implications. *Psychophysiology, 40*(5), 684–701.

Jiménez-Escrig, A., Fernandez-Lorente, J., Herrero, A., Baron, M., Lousa, M., de Blas, G., et al. (2002). Event-related evoked potential P300 in frontotemporal dementia. *Dementia and Geriatric Cognitive Disorders, 13*(1), 27–32.

Joassin, F., Maurage, P., Bruyer, R., Crommelinck, M., & Campanella, S. (2004). When audition alters vision: An event-related potential study of the cross-modal interactions between faces and voices. *Neuroscience Letters, 369*(2), 132–137.

Joassin, F., Pesenti, M., Maurage, P., Verreckt, E., Bruyer, R., & Campanella, S. (2010). Cross-modal interactions between human faces and voices involved in person recognition. *Cortex; A Journal Devoted to the Study of the Nervous System and Behavior, 47*, 367–376.

John, E. R., Prichep, L. S., Alper, K. R., Mas, F. G., Cancro, R., Easton, P., et al. (1994). Quantitative electrophysiological characteristics and subtyping of schizophrenia. *Biological Psychiatry, 36*(12), 801–826.

Johnson, R. (1988). Scalp-recorded P300 activity in patients following unilateral temporal lobectomy. *Brain: A Journal of Neurology, 111*(Pt 6), 1517–1529.

Karaaslan, F., Gonul, A. S., Oguz, A., Erdinc, E., & Esel, E. (2003). P300 changes in major depressive disorders with and without psychotic features. *Journal of Affective Disorders, 73*(3), 283–287.

Katada, E., Sato, K., Ojika, K., & Ueda, R. (2004). Cognitive event-related potentials: Useful clinical information in Alzheimer's disease. *Current Alzheimer Research, 1*(1), 63–69.

Kathmann, N., Soyka, M., Bickel, R., & Engel, R. R. (1996). ERP changes in alcoholics with and without alcohol psychosis. *Biological Psychiatry, 39*(10), 873–881.

Kathmann, N., Wagner, M., Rendtorff, N., & Engel, R. R. (1995). Delayed peak latency of the mismatch negativity in schizophrenics and alcoholics. *Biological Psychiatry, 37*(10), 754–757.

Kaustio, O., Partanen, J., Valkonen-Korhonen, M., Viinamäki, H., & Lehtonen, J. (2002). Affective and psychotic symptoms relate to different types of P300 alteration in depressive disorder. *Journal of Affective Disorders, 71*(1–3), 43–50.

Kawakubo, Y., Rogers, M. A., & Kasai, K. (2006). Procedural memory predicts social skills in persons with schizophrenia. *The Journal of Nervous and Mental Disease, 194*(8), 625–627.

Kendler, K. S., McGuire, M., Gruenberg, A. M., Spellman, M., O'Hare, A., & Walsh, D. (1993). The Roscommon Family Study. II. The risk of nonschizophrenic nonaffective psychoses in relatives. *Archives of General Psychiatry, 50*(8), 645–652.

Kerestes, R., Labuschagne, I., Croft, R. J., O'Neill, B. V., Bhagwagar, Z., Phan, K. L., et al. (2009). Evidence for modulation of facial emotional processing bias during emotional expression decoding by serotonergic and noradrenergic antidepressants: An event-related potential (ERP) study. *Psychopharmacology, 202*(4), 621–634.

Kiss, I., Dashieff, R. M., & Lordeon, P. (1989). A parieto-occipital generator for P300: Evidence from human intracranial recordings. *The International Journal of Neuroscience, 49*(1–2), 133–139.

Knight, R. T. (1984). Decreased response to novel stimuli after prefrontal lesions in man. *Electroencephalography and Clinical Neurophysiology, 59*(1), 9–20.

Knight, R. T. (1996). Contribution of human hippocampal region to novelty detection. *Nature, 383*(6597), 256–259.

Knight, R. T., Scabini, D., Woods, D. L., & Clayworth, C. C. (1989). Contributions of temporal-parietal junction to the human auditory P3. *Brain Research, 502*(1), 109–116.

Knott, V., Mahoney, C., Kennedy, S., & Evans, K. (2001). EEG power, frequency, asymmetry and coherence in male depression. *Psychiatry Research, 106*(2), 123–140.

Kornreich, C., & Philippot, P. (2006). Dysfunctions of facial emotion recognition in adult neuropsychiatric disorders: Influence on interpersonal difficulties. *Psychologica Belgica, 46*(1/2), 79–98.

Korostenskaja, M., & Kähkönen, S. (2009). What do ERPs and ERFs reveal about the effect of antipsychotic treatment on cognition in schizophrenia? *Current Pharmaceutical Design, 15*(22), 2573–2593.

Kramer, A. F., & Strayer, D. L. (1988). Assessing the development of automatic processing: An application of dual-task and event-related brain potential methodologies. *Biological Psychology, 26*(1–3), 231–267.

Kutas, M., McCarthy, G., & Donchin, E. (1977). Augmenting mental chronometry: The P300 as a measure of stimulus evaluation time. *Science, 197*(4305), 792–795.

Laurent, A., Garcia-Larrea, L., Dalery, J., Terra, J. L., D'Amato, T., Marie-Cardine, M., et al. (1993). The P 300 potential in schizophrenia. *L'Encéphale, 19*(3), 221–227.

Lebedeva, I. S., Kaleda, V. G., Abramova, L. I., Barkhatova, A. N., & Omel'chenko, M. A. (2008). Neurophysiological abnormalities in the P300 paradigm as endophenotypes of schizophrenia. *Zhurnal Nevrologii I Psikhiatrii Imeni S.S. Korsakova/Ministerstvo Zdravookhraneniia I Meditsinskoĭ Promyshlennosti Rossiĭskoĭ Federatsii, Vserossiĭskoe Obshchestvo Nevrologov [i] Vserossiĭskoe Obshchestvo Psikhiatrov, 108*(1), 61–70.

Lee, S. Y., Namkoong, K., Cho, H. H., Song, D., & An, S. K. (2010). Reduced visual P300 amplitudes in individuals at ultra-high risk for psychosis and first-episode schizophrenia. *Neuroscience Letters, 486*(3), 156–160.

Lembreghts, M., Crasson, M., el Ahmadi, A., & Timsit-Berthier, M. (1995). Interindividual variability of exogenous and endogenous auditory evoked potentials in a condition of voluntary attention. *Neurophysiologie Clinique/Clinical Neurophysiology, 25*(4), 203–223.

Light, G. A., & Braff, D. L. (2005). Mismatch negativity deficits are associated with poor functioning in schizophrenia patients. *Archives of General Psychiatry, 62*(2), 127–136.

Maier, W., Zobel, A., & Wagner, M. (2006). Schizophrenia and bipolar disorder: Differences and overlaps. *Current Opinion in Psychiatry, 19*(2), 165–170.

Martín-Loeches, M., Muñoz, F., Hinojosa, J. A., Molina, V., & Pozo, M. A. (2001). The P300 component of evoked potentials in the evaluation of schizophrenia: New evidence and future visions. *Revista De Neurologia, 32*(3), 250–258.

Mathalon, D. H., Ford, J. M., & Pfefferbaum, A. (2000). Trait and state aspects of P300 amplitude reduction in schizophrenia: A retrospective longitudinal study. *Biological Psychiatry, 47*(5), 434–449.

Mathalon, D. H., Hoffman, R. E., Watson, T. D., Miller, R. M., Roach, B. J., & Ford, J. M. (2010). Neurophysiological Distinction between Schizophrenia and Schizoaffective Disorder. *Frontiers in Human Neuroscience, 29,* 3–70.

Matsuoka, H., & Nakamura, M. (2005). Cognitive dysfunction and electroencephalogram in schizophrenia. *Seishin Shinkeigaku Zasshi/Psychiatria Et Neurologia Japonica, 107*(4), 307–322.

Maurage, P., Campanella, S., Philippot, P., Charest, I., Martin, S., & de Timary, P. (2009). Impaired emotional facial expression decoding in alcoholism is also present for emotional prosody and body postures. *Alcohol and Alcoholism, 44*(5), 476–485.

Maurage, P., Campanella, S., Philippot, P., de Timary, P., Constant, E., Gauthier, S., et al. (2008). Alcoholism leads to early perceptive alterations, independently of comorbid depressed state: An ERP study. *Neurophysiologie Clinique/Clinical Neurophysiology, 38*(2), 83–97.

Maurage, P., Campanella, S., Philippot, P., Martin, S., & de Timary, P. (2008). Face processing in chronic alcoholism: A specific deficit for emotional features. *Alcoholism, Clinical and Experimental Research, 32*(4), 600–606.

Maurage, P., Campanella, S., Philippot, P., Pham, T. H., & Joassin, F. (2007). The crossmodal facilitation effect is disrupted in alcoholism: A study with emotional stimuli. *Alcohol and Alcoholism, 42*(6), 552–559.

Maurage, P., Campanella, S., Philippot, P., Vermeulen, N., Constant, E., Luminet, O., et al. (2008). Electrophysiological correlates of the disrupted processing of anger in alcoholism. *International Journal of Psychophysiology: Official Journal of the International Organization of Psychophysiology, 70*(1), 50–62.

Maurage, P., Joassin, F., Philippot, P., & Campanella, S. (2007). A validated battery of vocal emotional expressions. *Neuropsychological Trends, 2*, 63–74.

Maurage, P., Pesenti, M., Philippot, P., Joassin, F., & Campanella, S. (2009). Latent deleterious effects of binge drinking over a short period of time revealed only by electrophysiological measures. *Journal of Psychiatry & Neuroscience: JPN, 34*(2), 111–118.

Maurage, P., Philippot, P., Joassin, F., Pauwels, L., Pham, T., Prieto, E. A., et al. (2007). The auditory-visual integration of anger is impaired in alcoholism: An event-related potentials study. *Journal of Psychiatry & Neuroscience: JPN, 33*(2), 111–122.

Maurage, P., Philippot, P., Verbanck, P., Noel, X., Kornreich, C., Hanak, C., et al. (2007). Is the P300 deficit in alcoholism associated with early visual impairments (P100, N170)? An oddball paradigm. *Clinical Neurophysiology: Official Journal of the International Federation of Clinical Neurophysiology, 118*(3), 633–644.

McCarthy, G., & Donchin, E. (1981). A metric for thought: A comparison of P300 latency and reaction time. *Science, 211*(4477), 77–80.

McCarthy, G., Wood, C. C., Williamson, P. D., & Spencer, D. D. (1989). Task-dependent field potentials in human hippocampal formation. *The Journal of Neuroscience: The Official Journal of the Society for Neuroscience, 9*(12), 4253–4268.

Meisenzahl, E. M., Frodl, T., Müller, D., Schmitt, G., Gallinat, J., Zetzsche, T., et al. (2004). Superior temporal gyrus and P300 in schizophrenia: A combined ERP/structural magnetic resonance imaging investigation. *Journal of Psychiatric Research, 38*(2), 153–162.

Merrin, E. L., & Floyd, T. C. (1992). Negative symptoms and EEG alpha activity in schizophrenic patients. *Schizophrenia Research, 8*(1), 11–20.

Miyazato, Y., & Ogura, C. (1993). Abnormalities in event-related potentials: N100, N200 and P300 topography in alcoholics. *The Japanese Journal of Psychiatry and Neurology, 47*(4), 853–862.

Molholm, S., Ritter, W., Murray, M. M., Javitt, D. C., Schroeder, C. E., & Foxe, J. J. (2002). Multisensory auditory-visual interactions during early sensory processing in humans: A high-density electrical mapping study. *Brain Research. Cognitive Brain Research, 14*(1), 115–128.

Montag, C., Ehrlich, A., Neuhaus, K., Dziobek, I., Heekeren, H. R., Heinz, A., et al. (2010). Theory of mind impairments in euthymic bipolar patients. *Journal of Affective Disorders, 123*(1–3), 264–269.

Monteiro, M. G., & Schuckit, M. A. (1988). Populations at high alcoholism risk: Recent findings. *The Journal of Clinical Psychiatry, 49*(Suppl), 3–7.

Mulert, C., Jäger, L., Pogarell, O., Bussfeld, P., Schmitt, R., Juckel, G., et al. (2002). Simultaneous ERP and event-related fMRI: Focus on the time course of brain activity in target detection. *Methods and Findings in Experimental and Clinical Pharmacology, 24*(Suppl. D), 17–20.

Mulert, C., Pogarell, O., & Hegerl, U. (2008). Simultaneous EEG-fMRI: Perspectives in psychiatry. *Clinical EEG and Neuroscience: Official Journal of the EEG and Clinical Neuroscience Society (ENCS), 39*(2), 61–64.

Müller, T. J., Kalus, P., & Strik, W. K. (2001). The neurophysiological meaning of auditory P300 in subtypes of schizophrenia. *The World Journal of Biological Psychiatry: The Official Journal of the World Federation of Societies of Biological Psychiatry, 2*(1), 9–17.

Nácher, V. (2000). Genetic association between the reduced amplitude of the P300 and the allele A1 of the gene which codifies the D2 dopamine receptor (DRD2) as possible biological markers for alcoholism. *Revista De Neurologia, 30*(8), 756–763.

Naismith S. L., Redoblado-Hodge M. A., Lewis S. J., Scott E. M., Hickie I. B. (2010). Cognitive training in affective disorders improves memory: a preliminary study using the NEAR approach. *Journal of Affective Disorders, 121*(3), 258–262.

Namkoong, K., Lee, E., Lee, C. H., Lee, B. O., & An, S. K. (2004). Increased P3 amplitudes induced by alcohol-related pictures in patients with alcohol dependence. *Alcoholism, Clinical and Experimental Research, 28*(9), 1317–1323.

Naranjo, C., Kornreich, C., Campanella, S., Noël, X., Vandriette, Y., Gillain, B., et al. (2010). Major depression is associated with impaired processing of emotion in music as well as in facial and vocal stimuli. *Journal of Affective Disorders, 128*(3), 243–251.

Nicolás, J. M., Estruch, R., Salamero, M., Orteu, N., Fernandez-Solà, J., Sacanella, E., et al. (1997). Brain impairment in well-nourished chronic alcoholics is related to ethanol intake. *Annals of Neurology, 41*(5), 590–598.

Niedermeyer, E. (1993). Principles of neurometric analysis of EEG and evoked potentials. In E. Niedermeyer & F. Lopes da Silva (Eds.), *EEG: Basic principles, clinical applications and related fields* (pp. 97–117). Baltimore, MD: Williams & Wilkins.

Nieuwenhuis, S., Aston-Jones, G., & Cohen, J. D. (2005). Decision making, the P3, and the locus coeruleus-norepinephrine system. *Psychological Bulletin, 131*(4), 510–532.

Nixon, S. J., Tivis, R., & Parsons, O. A. (1992). Interpersonal problem-solving in male and female alcoholics. *Alcoholism, Clinical and Experimental Research, 16*(4), 684–687.
Northoff, G. (2008). Neuropsychiatry. An old discipline in a new gestalt bridging biological psychiatry, neuropsychology, and cognitive neurology. *European Archives of Psychiatry and Clinical Neuroscience, 258*(4), 226–238.
Nuechterlein, K. H., Pashler, H. E., & Subotnik, K. L. (2006). Translating basic attentional paradigms to schizophrenia research: Reconsidering the nature of the deficits. *Development and Psychopathology, 18*(3), 831–851.
Nurnberger, J. I. (1992). Should be biologic marker be sensitive and specific? *Acta Psychiatrica Scandinavica, 86*(1), 1–4.
O'Donnell, B. F., Vohs, J. L., Hetrick, W. P., Carroll, C. A., & Shekhar, A. (2004). Auditory event-related potential abnormalities in bipolar disorder and schizophrenia. *International Journal of Psychophysiology: Official Journal of the International Organization of Psychophysiology, 53*(1), 45–55.
Olichney, J. M., & Hillert, D. G. (2004). Clinical applications of cognitive event-related potentials in Alzheimer's disease. *Physical Medicine and Rehabilitation Clinics of North America, 15*(1), 205–233.
Ortiz Alonso, T., Pérez-Serrano, J. M., Zaglul Zaiter, C., Coullaut García, J., Coullaut García, R., & Criado Rodríguez, J. (2002). P300 clinical utility in major depression. *Actas Españolas De Psiquiatría, 30*(1), 1–6.
Oscar-Berman, M., Hancock, M., Mildworf, B., Hutner, N., & Weber, D. A. (1990). Emotional perception and memory in alcoholism and aging. *Alcoholism, Clinical and Experimental Research, 14*(3), 383–393.
Ozgürdal, S., Gudlowski, Y., Witthaus, H., Kawohl, W., Uhl, I., Hauser, M., et al. (2008). Reduction of auditory event-related P300 amplitude in subjects with at-risk mental state for schizophrenia. *Schizophrenia Research, 105*(1–3), 272–278.
Partiot, A., Pierson, A., Le Houezec, J., Dodin, V., Renault, B., & Jouvent, R. (1993). Loss of automatic processes and blunted-affect in depression: A P3 study. *European Psychiatry, 8*(6), 309–318.
Patterson, M. (1999). The evolution of a parallel process model of non-verbal communication. In P. Philippot, R. Feldman, & E. Coats (Eds.), *The social context of non-verbal behavior* (pp. 317–347). New York: Cambridge University Press.
Pekkonen, E., Ahveninen, J., Jääskeläinen, I. P., Seppä, K., Näätänen, R., & Sillanaukee, P. (1998). Selective acceleration of auditory processing in chronic alcoholics during abstinence. *Alcoholism, Clinical and Experimental Research, 22*(3), 605–609.
Pekkonen, E., Rinne, T., & Näätänen, R. (1995). Variability and replicability of the mismatch negativity. *Electroencephalography and Clinical Neurophysiology, 96*(6), 546–554.
Pfefferbaum, A., Ford, J. M., & Kraemer, H. C. (1990). Clinical utility of long latency "cognitive" event-related potentials (P3): The cons. *Electroencephalography and Clinical Neurophysiology, 76*(1), 6–12. discussion 1.
Pfefferbaum, A., Rosenbloom, M., & Ford, J. M. (1987). Late event-related potential changes in alcoholics. *Alcohol, 4*(4), 275–281.
Philippot, P., & Feldman, R. S. (1990). Age and social competence in preschoolers' decoding of facial expression. *The British Journal of Social Psychology / the British Psychological Society, 29*(Pt 1), 43–54.
Picton, T. W. (1992). The P300 wave of the human event-related potential. *Journal of Clinical Neurophysiology: Official Publication of the American Electroencephalographic Society, 9*(4), 456–479.
Pierson, A., Jouvent, R., Quintin, P., Perez-Diaz, F., & Leboyer, M. (2000). Information processing deficits in relatives of manic depressive patients. *Psychological Medicine, 30*(3), 545–555.
Pierson, A., Partiot, A., Ammar, S., Dodin, V., Loas, G., Jouvent, R., et al. (1991). ERP differences between anxious-impulsive and blunted-affect depressive inpatients. In M. Ansseau, R. von Frenckell, & G. Franck (Eds.), *Biological markers of depression: State of the art* (pp. 121–129). Amsterdam: Elsevier Science Publishers B.V.

Pineda, J. A., Foote, S. L., & Neville, H. J. (1989). Effects of locus coeruleus lesions on auditory, long-latency, event-related potentials in monkey. *The Journal of Neuroscience: The Official Journal of the Society for Neuroscience, 9*(1), 81–93.

Pogarell, O., Mulert, C., & Hegerl, U. (2007). Event-related potentials in psychiatry. *Clinical EEG and Neuroscience: Official Journal of the EEG and Clinical Neuroscience Society (ENCS), 38*(1), 25–34.

Polich, J. (1998). P300 clinical utility and control of variability. *Journal of Clinical Neurophysiology: Official Publication of the American Electroencephalographic Society, 15*(1), 14–33.

Polich, J. (2004). Clinical application of the P300 event-related brain potential. *Physical Medicine and Rehabilitation Clinics of North America, 15*(1), 133–161.

Polich, J. (2007). Updating P300: An integrative theory of P3a and P3b. *Clinical Neurophysiology: Official Journal of the International Federation of Clinical Neurophysiology, 118*(10), 2128–2148.

Polich, J., & Bloom, F. E. (1999). P300, alcoholism heritability, and stimulus modality. *Alcohol, 17*(2), 149–156.

Polich, J., & Corey-Bloom, J. (2005). Alzheimer's disease and P300: Review and evaluation of task and modality. *Current Alzheimer Research, 2*(5), 515–525.

Polich, J., & Herbst, K. L. (2000). P300 as a clinical assay: Rationale, evaluation, and findings. *International Journal of Psychophysiology: Official Journal of the International Organization of Psychophysiology, 38*(1), 3–19.

Polich, J., Howard, L., & Starr, A. (1983). P300 latency correlates with digit span. *Psychophysiology, 20*(6), 665–669.

Polich, J., & Kok, A. (1995). Cognitive and biological determinants of P300: An integrative review. *Biological Psychology, 41*(2), 103–146.

Polich, J., Ladish, C., & Bloom, F. E. (1990). P300 assessment of early Alzheimer's disease. *Electroencephalography and Clinical Neurophysiology, 77*(3), 179–189.

Polich, J., & Squire, L. R. (1993). P300 from amnesic patients with bilateral hippocampal lesions. *Electroencephalography and Clinical Neurophysiology, 86*(6), 408–417.

Porjesz, B., & Begleiter, H. (1997). Event-related potentials in COA's. *Alcohol Health and Research World, 21*(3), 236–240.

Porjesz, B., Begleiter, H., Litke, A., Bauer, L. O., Kuperman, S., O'Connor, S. J., et al. (1996). Visual P3 as a potential phenotypic marker for alcoholism: Evidence from the COGA national project. In: Ogura, C., Koga, Y., Shimokochi, M. (eds.), *Recent Advances in Event-Related Brain Potential Research.* Elsevier Science, Holland, 539–549.

Pregelj, P. (2009). Psychosis and depression: A neurobiological view. *Psychiatria Danubina, 21*(Suppl. 1), 102–105.

Price, G. W., Michie, P. T., Johnston, J., Innes-Brown, H., Kent, A., Clissa, P., et al. (2006). A multivariate electrophysiological endophenotype, from a unitary cohort, shows greater research utility than any single feature in the Western Australian family study of schizophrenia. *Biological Psychiatry, 60*(1), 1–10.

Raffard, S., Gely-Nargeot, M., Capdevielle, D., Bayard, S., & Boulenger, J. (2009). Learning potential and cognitive remediation in schizophrenia. *L'Encéphale, 35*(4), 353–360.

Ramachandran, G., Porjesz, B., Begleiter, H., & Litke, A. (1996). A simple auditory oddball task in young adult males at high risk for alcoholism. *Alcoholism, Clinical and Experimental Research, 20*(1), 9–15.

Realmuto, G., Begleiter, H., Odencrantz, J., & Porjesz, B. (1993). Event-related potential evidence of dysfunction in automatic processing in abstinent alcoholics. *Biological Psychiatry, 33*(8–9), 594–601.

Reese, C., & Polich, J. (2003). Alcoholism risk and the P300 event-related brain potential: Modality, task, and gender effects. *Brain and Cognition, 53*(1), 46–57.

Rich, B. A., Grimley, M. E., Schmajuk, M., Blair, K. S., Blair, R. J. R., & Leibenluft, E. (2008). Face emotion labeling deficits in children with bipolar disorder and severe mood dysregulation. *Development and Psychopathology, 20*(2), 529–546.

Rodríguez Holguín, S., Porjesz, B., Chorlian, D. B., Polich, J., & Begleiter, H. (1999). Visual P3a in male subjects at high risk for alcoholism. *Biological Psychiatry, 46*(2), 281–291.

Rodriguez, G., Nobili, F., Arrigo, A., Priano, F., De Carli, F., Francione, S., et al. (1996). Prognostic significance of quantitative electroencephalography in Alzheimer patients: Preliminary observations. *Electroencephalography and Clinical Neurophysiology, 99*(2), 123–128.

Rossignol, M., Philippot, P., Crommelinck, M., & Campanella, S. (2008). Visual processing of emotional expressions in mixed anxious-depressed subclinical state: An event-related potential study on a female sample. *Neurophysiologie Clinique/Clinical Neurophysiology, 38*(5), 267–275.

Rossignol, M., Philippot, P., Douilliez, C., Crommelinck, M., & Campanella, S. (2005). The perception of fearful and happy facial expression is modulated by anxiety: An event-related potential study. *Neuroscience Letters, 377*(2), 115–120.

Roth, W. T., & Cannon, E. H. (1972). Some features of the auditory evoked response in schizophrenics. *Archives of General Psychiatry, 27*(4), 466–471.

Rugg, M., & Coles, M. (Eds.). (1995). *Electrophysiology of mind: Event-related brain potentials and cognition*. Oxford: Oxford University Press.

Sablier, J., Stip, E., & Franck, N. (2009). Cognitive remediation and cognitive assistive technologies in schizophrenia. *L'Encéphale, 35*(2), 160–167.

Sánchez-Turet, M., & Serra-Grabulosa, J. M. (2002). Auditory evoked potentials and alcohol: Characteristics of the mismatch negativity component in alcoholism. *Revista De Neurologia, 35*(11), 1049–1055.

Santosh, P. J., Malhotra, S., Raghunathan, M., & Mehra, Y. N. (1994). A study of P300 in melancholic depression–correlation with psychotic features. *Biological Psychiatry, 35*(7), 474–479.

Saperstein, A. M., Fuller, R. L., Avila, M. T., Adami, H., McMahon, R. P., Thaker, G. K., et al. (2006). Spatial working memory as a cognitive endophenotype of schizophrenia: Assessing risk for pathophysiological dysfunction. *Schizophrenia Bulletin, 32*(3), 498–506.

Schlegel, S., Nieber, D., Herrmann, C., & Bakauski, E. (1991). Latencies of the P300 component of the auditory event-related potential in depression are related to the Bech-Rafaelsen Melancholia Scale but not to the Hamilton Rating Scale for Depression. *Acta Psychiatrica Scandinavica, 83*(6), 438–440.

Schrijvers, D. L., De Bruijn, E. R. A., Destoop, M., Hulstijn, W., & Sabbe, B. G. C. (2010). The impact of perfectionism and anxiety traits on action monitoring in major depressive disorder. *Journal of Neural Transmission, 117*(7), 869–880.

Schrijvers, D., de Bruijn, E. R. A., Maas, Y., De Grave, C., Sabbe, B. G. C., & Hulstijn, W. (2008). Action monitoring in major depressive disorder with psychomotor retardation. *Cortex; A Journal Devoted to the Study of the Nervous System and Behavior, 44*(5), 569–579.

Schweinberger, S. R., Robertson, D., & Kaufmann, J. M. (2007). Hearing facial identities. *Quarterly Journal of Experimental Psychology, 60*(10), 1446–1456.

Seiferth, N. Y., Pauly, K., Habel, U., Kellermann, T., Shah, N. J., Ruhrmann, S., et al. (2008). Increased neural response related to neutral faces in individuals at risk for psychosis. *NeuroImage, 40*(1), 289–297.

Seubert, J., Loughead, J., Kellermann, T., Boers, F., Brensinger, C. M., & Habel, U. (2010). Multisensory integration of emotionally valenced olfactory-visual information in patients with schizophrenia and healthy controls. *Journal of Psychiatry & Neuroscience: JPN, 35*(3), 185–194.

Shagass, C. (1981). Neurophysiological evidence for different types of depression. *Journal of Behavior Therapy and Experimental Psychiatry, 12*(2), 99–111.

Shagass, C., & Roemer, R. (1991). Evoked potential topography in unmedicated and medicated schizophrenics. *International Journal of Psychophysiology: Official Journal of the International Organization of Psychophysiology, 10*(3), 213–224.

Shepherd, G. M. (2006). Smell images and the flavour system in the human brain. *Nature, 444*(7117), 316–321.

Simons, R. F., Graham, F. K., Miles, M. A., & Chen, X. (2001). On the relationship of P3a and the Novelty-P3. *Biological Psychology, 56*(3), 207–218.

Smith, M. E., Banquet, J., El Massioui, F., & Widlocher, D. (1991). Measuring cognitive deficits in depressives through ERPs. In M. Ansseau, R. von Frenckell, & G. Franck (Eds.), *Biological Markers of depression: State of the art* (pp. 131–144). Amsterdam: Elsevier Science Publishers B.V.

Smith, M. E., Halgren, E., Sokolik, M., Baudena, P., Musolino, A., Liegeois-Chauvel, C., et al. (1990). The intracranial topography of the P3 event-related potential elicited during auditory oddball. *Electroencephalography and Clinical Neurophysiology, 76*(3), 235–248.
Soininen, H., Partanen, J., Pääkkönen, A., Koivisto, E., & Riekkinen, P. J. (1991). Changes in absolute power values of EEG spectra in the follow-up of Alzheimer's disease. *Acta Neurologica Scandinavica, 83*(2), 133–136.
Soltani, M., & Knight, R. T. (2000). Neural origins of the P300. *Critical Reviews in Neurobiology, 14*(3–4), 199–224.
Spencer, K. M., Dien, J., & Donchin, E. (1999). A componential analysis of the ERP elicited by novel events using a dense electrode array. *Psychophysiology, 36*(3), 409–414.
Stampfer, H. G. (1983). Event-related potentials in psychiatry: Approaches to research and clinical applications. *The Australian and New Zealand Journal of Psychiatry, 17*(4), 307–318.
Stekelenburg, J. J., & de Gelder, B. (2004). The neural correlates of perceiving human bodies: An ERP study on the body-inversion effect. *NeuroReport, 15*(5), 777–780.
Strandburg, R. J., Marsh, J. T., Brown, W. S., Asarnow, R. F., Guthrie, D., Higa, J., et al. (1994). Reduced attention-related negative potentials in schizophrenic adults. *Psychophysiology, 31*(3), 272–281.
Strasser, H. C., Lilyestrom, J., Ashby, E. R., Honeycutt, N. A., Schretlen, D. J., Pulver, A. E., et al. (2005). Hippocampal and ventricular volumes in psychotic and nonpsychotic bipolar patients compared with schizophrenia patients and community control subjects: A pilot study. *Biological Psychiatry, 57*(6), 633–639.
Suffin, S. C., & Emory, W. H. (1995). Neurometric subgroups in attentional and affective disorders and their association with pharmacotherapeutic outcome. *Clinical EEG (Electroencephalography), 26*(2), 76–83.
Summers, M., Papadopoulou, K., Bruno, S., Cipolotti, L., & Ron, M. A. (2006). Bipolar I and bipolar II disorder: Cognition and emotion processing. *Psychological Medicine, 36*(12), 1799–1809.
Sun, Y., Li, Y., Zhu, Y., Chen, X., & Tong, S. (2008). Electroencephalographic differences between depressed and control subjects: An aspect of interdependence analysis. *Brain Research Bulletin, 76*(6), 559–564.
Surguladze, S. A., Young, A. W., Senior, C., Brébion, G., Travis, M. J., & Phillips, M. L. (2004). Recognition accuracy and response bias to happy and sad facial expressions in patients with major depression. *Neuropsychology, 18*(2), 212–218.
Sutton, S., Braren, M., Zubin, J., & John, E. R. (1965). Evoked-potential correlates of stimulus uncertainty. *Science, 150*(700), 1187–1188.
Tenke, C. E., Kayser, J., Stewart, J. W., & Bruder, G. E. (2010). Novelty P3 reductions in depression: Characterization using principal components analysis (PCA) of current source density (CSD) waveforms. *Psychophysiology, 47*(1), 133–146.
Thaker, G. (2008a). Psychosis endophenotypes in schizophrenia and bipolar disorder. *Schizophrenia Bulletin, 34*(4), 720–721.
Thaker, G. K. (2008b). Neurophysiological endophenotypes across bipolar and schizophrenia psychosis. *Schizophrenia Bulletin, 34*(4), 760–773.
Thomas, A., Iacono, D., Bonanni, L., D'Andreamatteo, G., & Onofrj, M. (2001). Donepezil, rivastigmine, and vitamin E in Alzheimer disease: A combined P300 event-related potentials/neuropsychologic evaluation over 6 months. *Clinical Neuropharmacology, 24*(1), 31–42.
Timsit-Berthier, M. (2003). Interest of neurophysiological exploration in clinical psychiatry. *Neurophysiologie Clinique/Clinical Neurophysiology, 33*(2), 67–77.
Uekermann, J., Daum, I., Schlebusch, P., & Trenckmann, U. (2005). Processing of affective stimuli in alcoholism. *Cortex; A Journal Devoted to the Study of the Nervous System and Behavior, 41*(2), 189–194.
Umbricht, D. S. G., Bates, J. A., Lieberman, J. A., Kane, J. M., & Javitt, D. C. (2006). Electrophysiological indices of automatic and controlled auditory information processing in first-episode, recent-onset and chronic schizophrenia. *Biological Psychiatry, 59*(8), 762–772.
Urcelay-Zaldua, I., Hansenne, M., & Ansseau, M. (1995). The influence of suicide risk and despondency on the amplitude of P300 in major depression. *Neurophysiologie Clinique/Clinical Neurophysiology, 25*(5), 291–296.

Van den Stock, J., Peretz, I., Grèzes, J., & de Gelder, B. (2009). Instrumental music influences recognition of emotional body language. *Brain Topography, 21*(3–4), 216–220.

Van den Stock, J., Righart, R., & de Gelder, B. (2007). Body expressions influence recognition of emotions in the face and voice. *Emotion, 7*(3), 487–494.

Van Der Stelt, O. (1999). ESBRA-Nordmann 1998 Award Lecture: Visual P3 as a potential vulnerability marker of alcoholism: Evidence from the Amsterdam study of children of alcoholics. European Society for Biomedical Research on Alcoholism. *Alcohol and Alcoholism, 34*(3), 267–282.

van der Stelt, O., & Belger, A. (2007). Application of electroencephalography to the study of cognitive and brain functions in schizophrenia. *Schizophrenia Bulletin, 33*(4), 955–970.

van der Stelt, O., Gunning, W. B., Snel, J., & Kok, A. (1997). No electrocortical evidence of automatic mismatch dysfunction in children of alcoholics. *Alcoholism, Clinical and Experimental Research, 21*(4), 569–575.

van der Stelt, O., Gunning, W. B., Snel, J., Zeef, E., & Kok, A. (1994). Children of alcoholics: Attention, information processing and event-related brain potentials. *Acta Paediatrica Supplement, 404*, 4–6.

van der Stelt, O., Kok, A., Smulders, F. T., Snel, J., & Boudewijn Gunning, W. (1998). Cerebral event-related potentials associated with selective attention to color: Developmental changes from childhood to adulthood. *Psychophysiology, 35*(3), 227–239.

van der Stelt, O., Lieberman, J. A., & Belger, A. (2005). Auditory P300 in high-risk, recent-onset and chronic schizophrenia. *Schizophrenia Research, 77*(2–3), 309–320.

van der Stelt, O., van der Molen, M., Boudewijn Gunning, W., & Kok, A. (2001). Neuroelectrical signs of selective attention to color in boys with attention-deficit hyperactivity disorder. *Brain Research. Cognitive Brain Research, 12*(2), 245–264.

Vandoolaeghe, E., van Hunsel, F., Nuyten, D., & Maes, M. (1998). Auditory event related potentials in major depression: Prolonged P300 latency and increased P200 amplitude. *Journal of Affective Disorders, 48*(2–3), 105–113.

Venn, H. R., Gray, J. M., Montagne, B., Murray, L. K., Michael Burt, D., Frigerio, E., et al. (2004). Perception of facial expressions of emotion in bipolar disorder. *Bipolar Disorders, 6*(4), 286–293.

Verleger, R. (1988). Event-related potentials and cognition: A critique of the context updating hypothesis and an alternative interpretation of P3. *Behavioral and Brain Sciences, 11*, 343–356.

Verleger, R. (1997). On the utility of P3 latency as an index of mental chronometry. *Psychophysiology, 34*(2), 131–156.

Verleger, R., Heide, W., Butt, C., & Kömpf, D. (1994). Reduction of P3b in patients with temporoparietal lesions. *Brain Research. Cognitive Brain Research, 2*(2), 103–116.

Wakefield, J. C. (1992). Disorder as harmful dysfunction: A conceptual critique of DSM-III-R's definition of mental disorder. *Psychological Review, 99*(2), 232–247.

Wakefield, J. C. (2007). The concept of mental disorder: Diagnostic implications of the harmful dysfunction analysis. *World Psychiatry: Official Journal of the World Psychiatric Association (WPA), 6*(3), 149–156.

Wang, J., Chen, X., & Lou, F. (2000). Event-related potentials and suicide behavior in patients with affective disorder. *Zhonghua Yi Xue Za Zhi, 80*(4), 275–277.

Weisbrod, M., Hill, H., Niethammer, R., & Sauer, H. (1999). Genetic influence on auditory information processing in schizophrenia: P300 in monozygotic twins. *Biological Psychiatry, 46*(5), 721–725.

Wickens, C., Kramer, A., Vanasse, L., & Donchin, E. (1983). Performance of concurrent tasks: A psychophysiological analysis of the reciprocity of information-processing resources. *Science, 221*(4615), 1080–1082.

Winston, J. S., Gottfried, J. A., Kilner, J. M., & Dolan, R. J. (2005). Integrated neural representations of odor intensity and affective valence in human amygdala. *The Journal of Neuroscience: The Official Journal of the Society for Neuroscience, 25*(39), 8903–8907.

Wood, S. J., Pantelis, C., Proffitt, T., Phillips, L. J., Stuart, G. W., Buchanan, J. A., et al. (2003). Spatial working memory ability is a marker of risk-for-psychosis. *Psychological Medicine, 33*(7), 1239–1247.

Yordanova, J., & Kolev, V. (1998). Single-sweep analysis of the theta frequency band during an auditory oddball task. *Psychophysiology, 35*(1), 116–126.

Zalla, T., Joyce, C., Szöke, A., Schürhoff, F., Pillon, B., Komano, O., et al. (2004). Executive dysfunctions as potential markers of familial vulnerability to bipolar disorder and schizophrenia. *Psychiatry Research, 121*(3), 207–217.

Zhang, X. L., Cohen, H. L., Porjesz, B., & Begleiter, H. (2001). Mismatch negativity in subjects at high risk for alcoholism. *Alcoholism, Clinical and Experimental Research, 25*(3), 330–337.

Zhang, Y., Hauser, U., Conty, C., Emrich, H. M., & Dietrich, D. E. (2007). Familial risk for depression and p3b component as a possible neurocognitive vulnerability marker. *Neuropsychobiology, 55*(1), 14–20.

Zhu, X., Zhang, H., Wu, T., Luo, W., & Luo, Y. (2010). Emotional conflict occurs at an early stage: Evidence from the emotional face-word Stroop task. *Neuroscience Letters, 478*(1), 1–4.

Zhu, C., Zheng, Z., Qiu, C., Zou, K., Nie, X., Feng, Y., et al. (2009). Brain evoked potentials in patients with depression or anxiety. *Sichuan Da Xue Xue Bao Yi Xue Ban/Journal of Sichuan University. Medical Science Edition, 40*(4), 708–711.

Zimerman, M., & Spitzer, R. (2005). Psychiatric classification. In V. Sadock (Ed.), *Kaplan & Sadock's comprehensive textbook of psychiatry* (8th ed., pp. 1003–1034). Baltimore, MD: Lippincott Williams and Wilkins.

Index

A
Addiction
  alcohol, 288
  drug, 289
Alcohol dependence
  auditory and visual modalities
    "chemical senses", 286
    ecological validity, 287–288
    olfactory and gustatory, 286, 287
    schizophrenic patients, 287
  behavioural study
    facial expressions and voices, 276
    facilitation effect, 276
    reaction times, 277
    ventriloquist effects, 276
  clinical and theoretical implications
    anger stimuli, 284
    ecologic stimuli, 284
    emotional decoding, 282
    neurological states/psychiatric, 283
    psychopathological populations, 284–285
    sensory modalities, 283
    therapeutic programs, 283–284
  clinical aspects, 290
  electrophysiological study
    cerebral activations, 279
    ERP, 277–278
    face–voice stimulus, 278
    neuroanatomical techniques, 280
    source location analysis, 279
  emotions
    cerebral damage, 273
    crossmodal experimental paradigms, 285
    experimental designs, 275
    exploratory studies, 286
    integration processes, 286
    mental diseases, 274
    mental states, 286
    negative, 285
    psychiatric diagnosis, 273
    social communication, 274
  fundamental research, 290
  healthy controls, 290
  neuroimaging study, 280–282
  preliminary data, 291
  psychiatric populations (*see* Psychiatry)
  rationale and aims
    detoxified alcohol, 275
    multiple sensory, 275
  sensory modalities, 272
  unimodal explorations, 273
  visual–audio integration, 272
Amodal information, 101
  infants' recognition, faces and voices, 80
  and modality-specific
    description, 74
    multimodal stimulation, 75
    predictions, intersensory redundancy hypothesis, 75, 76
AMY. *See* Amygdala (AMY)
Amygdala (AMY), 231, 232, 242
  fronto-parietal attention, 219
  large-scale neural network, 219
Arbitrary face-voice relations
  amodal properties, 83
  habituation infants, 81
  infants examination, 82
  intersensory redundancy hypothesis, 81
  modality-specific properties, 81
  pairings, 82–83
  voice and visual appearance, 81

P. Belin et al. (eds.), *Integrating Face and Voice in Person Perception*,
DOI 10.1007/978-1-4614-3585-3, © Springer Science+Business Media New York 2013

ASD. *See* Autism spectrum disorder (ASD)
Attention, 75–77, 142–146, 193–194, 262–264, 327–330, 332–334, 337–339, 341–344
  amodal property, 75, 81
  audiovisual condition
    face, 140, 144, 145
    uncontrolled, 142–143
    voice, 140, 143–144
  experiment late processing results, 87, 88
  facilitation *vs.* attenuation, 76
  fronto-parietal networks, 210
  intersensory redundancy hypothesis, 83
  multimodal stimulation, 77, 81
  neurophysiological process, 86
  and neurophysiological synchrony, 72
  observer-dependent effects, 209
  rapid cross-modal integration, 219
  redundant multimodal stimulation, 75
  sense modality, 75
  synchronous bimodal speech, 86
  top-down approach, 72
AUD. *See* Auditory cortex (AUD)
Audiovisual (AV). *See* Hearing loss, children
Audiovisual condition, 165, 169–171, 197, 276
  attention to face, 140, 144, 145
  attention to voice
    curves differentiation, 140, 143–144
    non-significant effect, face information, 144
  audio only, 140
  no direction to modality, 140
  uncontrolled attention
    2D plot, categorisation response, 143
    3D plot, categorisation response, 142, 143
    individual categorization strategy, 143
  video only, 140–141
Audiovisual integration (AVI), 135–147, 186–188, 193, 195, 201, 299–300. *See also* Audiovisual integration (AVI), voice and face; Face-voice gender, AVI
  ability, human social functioning, 119
  auditory speech comprehension, 121
  communication situations, 120
  crossmodal adaptation (*see* Crossmodal adaptation)
  description, 119–120
  face-voice integration
    corresponding and noncorresponding identities, 123–129
    speaker identification, 121–122
  IAC model and PIN, 120
  neural mechanisms
    auditory stimulation, 122
    description, 122
    face-voice integration, 122
    functional connectivity analysis, 122–123
    speaker identification, 122
  neurophysiological recordings, 121
  neuropsychological evidence, 120
  temporal correspondence, face and voice, 121
  unisensory areas, 120
Audiovisual integration (AVI), voice and face
  audiovisual nonverbal emotional expressions, 227
  conjunction analyses
    brain activation, 237
    global null hypothesis, 237–238
    unimodal conditions, 238–239
  connectivity analyses
    crossmodal percept, 243
    psychophysiological interaction, 243, 245
    unimodal and supramodal sensory cortices, 245
  correlation analyses
    bimodal stimulation, 242
    BOLD, 243
    neural representation, 243–245
    pSTS, 242
  electrophysiological studies, humans
    ERP, 229, 230
    MEG, 230
    nonverbal emotional information, 229
  emotional congruency/incongruency, 241–242
  emotional signals, 226
  facial expression and prosody facilitates reactions, 226
  neuroanatomical studies and animal electrophysiology
    convergence zones, 227
    potential and single cell electrophysiology, 228
    rules, multisensory integration, 227–228
    spatial factors, 228
    STS, 228, 229
    visual cortex, 229
  neuroimaging studies
    amygdala and fusiform gyrus, 231
    BOLD, 231, 233
    cerebral response, 233–235
    conjunction analyses and interaction analyses, 232, 233
    fMRI experiment, 232

multisensory integration, 230
PPI analysis, 232
STS and face sensitive region, 233, 236
temporal incongruity, 231
thalamus, 237
psychiatric disease, 245–246
sensory interactions
BOLD, 240, 241
putative control condition, 239
single cell studies, 240
supra-additive multimodal, 240, 241
triple conjunction analysis, 240
unimodal and bimodal stimulus, 239
sensory stimulation, 226
Audition, audiovisual perception
bimodal *vs.* unimodal processing, 299
cochlear implants, 317, 318
hearing loss (*see* Hearing loss, children)
neural plasticity, 300
optimal information, 300
signal, 300
Auditory cortex (AUD), 218, 219
Auditory-visual representations, 152, 306
animals
cross-modal interference, 33–34
cross-modal matching-to-sample, 34–36
expectancy violation, 33
preferential looking, 31, 33
humans
sound-source identification, 36–37
vocal types, 38–39
Autism spectrum disorder (ASD), 262–263
AVI. *See* Audiovisual integration (AVI)

B

Bimodal, 7, 77, 84, 157, 192, 238, 239, 300, 305, 346–350
Bimodal oddball, ERP
affective disorders
decoding system, 347
EFE, 347
emotional processing disorder, 347
neurosciences, 347
design, 350
patient-specific pattern, 351
psychiatric disorders, 351
synchronized emotional stimuli
accurate and elaborate tool, 348
chronic alcoholism, 349–350
crossmodal action, 347–348
crossmodal facilitation effect, 348
P300 component recorded, four parietal electrodes, 349
Blood oxygen level dependent (BOLD), 231, 233, 240, 241, 243
BOLD. *See* Blood oxygen level dependent (BOLD)
Brain-imaging methods, 209

C

Catechol-*O*-methyltransferase (COMT)
dopamine (DA) elimination, 106
ERPs, 107, 109
genotype, emotional stimuli processing, 106
infants' recovery, distress, 109
infant temperament analysis, 109
negative emotions, infants, 108
prefrontal brain process, 106, 107
Children. *See* Hearing loss, children
Chronic alcoholism, P300
ERB, unfocused
correlational analysis, 343–344
increased P100/N100 latencies and reduced amplitudes, 343
MMN, 344
N2b latency, amplitude, and P3a deficits, 343
psychiatry
adolescence, 337
alcohol toxicity effects, 335
broad field, disorder, 337
description, 335
electrophysiological differences and amplitude alteration supports, 336
genetic analysis and girls exhibited lower, 336
P350 and P450, 337
sensitivity and risk status, 336
COMT. *See* Catechol-*O*-methyltransferase (COMT)
Cross-modal, 97, 100, 121, 122, 231, 254, 258
adaptation, 129–130
bias effect, 192
effects, 185, 186, 188, 227
influence, 264
integration, 239, 241–243, 245
multimodal/heteromodal brain regions, 188
multisensory perception effects, 189
nonverbal emotional information, 232
ventriloquist illusion, 191
Crossmodal adaptation
ERP topography, 130
hypothesis, 130
scalp voltage maps, 128, 130

Crossmodal adaptation (*cont.*)
silent videos, female/male speakers, 129
stimulus attributes, 129
systematic contrastive aftereffects, 129
Cross-modal integration, 218, 219, 278, 287
auditory-visual integration process, 150
description, 150
developmental and pervasive disorders, 158
electrophysiological studies and data, 150
explicit/controlled *vs.* implicit/automatic aspects, 158
face and voice categorization task, 158
faces and voices
gender processing (*see* Gender processing, cross-modal integration)
identity processing (*see* Identity processing, cross-modal integration)
hierarchical network, cerebral areas, 150
neural mechanisms, 151
neuro-functional impairments, 158–159
Cross-modal interference
auditory go/no-go task, Guinea baboons, 34
description, 33
primary and secondary caretaker's voice, 34
violation paradigms, 34
visual matching-to-sample task, squirrel monkeys, 34
Crossmodality. *See* Alcohol dependence
Cross-modal matching-to-sample
arbitrary relationship, 35
auditory association task, 34
bonobo and domestic dogs, 35
congruent movie, chimpanzee, 36
description, 34
expectancy violation paradigm, 36
individual recognition task, 35
intensive training, 35
objects, sounds, 35
paradigms, 35
Cross-modal modulation, spatial attention
amygdala, 208
attentional selection mechanisms, 209
auditory cortex (AUD), 218, 219
auditory cues, 210
electrophysiological data, 213, 218
emotion
attentional blink task, 211
brain-damaged patients, 212
cueing effects, 212
dot probe paradigm, 213
LPC, 214
prioritization, 212
tactile targets, 214
visual cues, 213
within-modality effects, 212, 214
endogenous and exogenous
crossmodal cueing studies, 210
description, 209
ERP studies, 211
fronto-parietal networks, 210
IPS, 210
neural mechanisms, within-modality
fronto-parietal attention, 216
visual cortex, 215
neural processing capacity, 208
neurocognitive model
amygdala, 219
auditory information, 216
electrophysiological studies, 218
emotional prosody, 218
exogenous cueing, 218
functional connectivity analyses, 218
resource allocation, startle probe, 217
source localization, 217–218
observer-dependent effects, 209
within-modality effects, 208
Cross-modal representation
auditory-visual representations, animals
behavioral studies, 31, 32
cross-modal interference, 33–34
cross-modal matching-to-sample, 34–36
expectancy violation, 33
preferential looking, 31, 33
description, 29
evolutional specializations, 30
human auditory-visual representation
sound-source identification, 36–37
vocal types, 38–39
neural mechanisms, 30
sound source identity and vocal types, 30
stimuli, visual-tactile matching-to-sample task, 29–30
Cueing effects, 212

D

Diagnostic and statistical manual of mental disorders (DSM), 325
DLPFC. *See* Dorsolateral prefrontal cortex (DLPFC)
Dorsolateral prefrontal cortex (DLPFC)
auditory localization, 52
visuospatial processing, 53

E

EEG. *See* Electroencephalogram (EEG)
EFE. *See* Emotional facial expressions (EFE)
Electroencephalogram (EEG)
  bimodal and unimodal stimulation, 84
  multisensory *vs.* unisensory processing, 168
  neurophysiological base, 83
  oscillatory response, 174
  peak amplitudes and latencies, 169
  postsynaptic potential activity, 171
Electrophysiology
  animal, 227–229
  humans, 229–230
Emotional facial expressions (EFE), 192, 195, 284, 347
Emotions
  brain-damaged patients, 212
  description, 95
  directions
    definition, human mirror neuron system, 110
    electrophysiological work, 110
    neural source identification, 111
    NIRS systems, 112
  electrophysiological studies, 97
  ERP findings
    advantages, auditory sensory modality, 103
    audiovisual/auditory changes, 103
    developmental sequence, 103–104
    electrophysiological correlation, 105
    fMRI studies, 102
    intermodal invariants, multimodal contexts, 104
    live interactions *vs.* experimental studies, 104–105
    maternal smiles, 105
    mother-infant studies, 104
    multimodal information availability, 104
    sensory-specific and sensory-unspecific effects, 102–103
  exogenous attention, 208
  face-like and non-face-like stimuli, 96
  facial expressions, 96
  genetic factors
    clinical disorders, 108
    complex phenotypes, 110
    COMT and SLC6A4/5-HTTLPR genotypes, 106, 108
    description, 105
    endophenotypes, 109
    ERP studies, 106–107
    fMRI study, adults, 106
    hypersensitivity, adults, 108–109
    mirroring/simulation mechanisms, 109
    neuroimaging studies, adults, 108
    neurotransmitter systems, 106
    occipital sites, ERP data, 109
    Pc and Nc components, 107
    polymorphisms, 107
    positive *vs.* negative emotions, 108
    structural analysis, 106
  infants' perception, ERPs (*see* Event-related potentials (ERPs))
  interpersonal interactions, 96
  postnatal maturation, sensory systems, 96
  sensitivity, auditory information, 96
  social stimuli, infants, 96
  top-down feedback, 218
Emotions by ear and eye
  audiovisual perception
    human, 254
    perceptual system, 254
    prosody, 254–255
  multisensory effects
    debate, 255
    dynamic images, 255
    faces and sentence fragments, 256
    facial expressions, 255
    informational redundancy, 255
    late effects/postperceptual effects, 255–256
  neurofunctional basis
    auditory and visual expressions, 257
    cortical heteromodal areas, 257
    crossmodal face–voice influences, 256
    EEG, 256, 257
  schizophrenia and autism, 262–265
  theory and methodology
    environmental conditions, 258
    spatial and temporal constraints, 258
    Stroop effect, 258
  whole body expressions, auditory signals
    crossmodal interaction, 262
    description, 259
    fearful, angry and happy, 260
    human voice influences, 261
    multisensory integration, 260
    SOA, 260, 261
    trial, stimuli, 261, 262
ERN. *See* Error-related negativity (ERN)
ERPs. *See* Event-related potentials (ERPs)
Error-related negativity (ERN), 343
Event-related potentials (ERPs)

Event-related potentials (ERPs) (*cont.*)
advantages, 341–342
anticipatory visual motion, 127
audiovisual speaker identification paradigm, 126
behavioral methods, 97
bimodal and unimodal stimulation, 84
bimodal oddball
affective disorders, 346–347
synchronized emotional stimuli, 347–350
cerebral activity, left angular gyrus, 152
components, 341
diagnosis power, P300, 340
emotional information, 97
explore mental chronometry
cerebral electric potentials, 327
cortical and sensory function, 327
exogenous and cognitive/endogenous potentials, 328
latency and amplitude, 328
neuroimaging and electrophysiological techniques, 327
oddball target detection task, 327–328
scalp-positive and-negative voltage deflections, 328
face and voice processing
behavioral findings, 101
congruent and incongruent words, 100
electrophysiological evidence, 100–101
multimodal audiovisual events, 101
Nc and Pc components, 101
social interactions, 102
unimodal emotional information, 100
face identity, 150
face processing
adult-like electrophysiological response, 98
behavioral looking methods, 99
differences, emotions, 97–98
experiences, 98
happy and angry expressions, 97
locomotor and prelocomotor infants, 98
neurocognitive process, 99
topography, brain systems, 98
visual-paired comparison task, 99
face-sensitive N250r component, 128
face-voice identity correspondence, 128
face-voice integration, 127
FFA and TVA, 129
infants' visual acuity, 97
magnetoencephalography and fMRI techniques, 341
psychiatric disorders, 352
results, infants and adults, 85
signal-to-noise ratio, 127
topography, audiovisual correspondence, 130
TOVA abnormalities, 341
unfocused P300, 342–346
unimodal auditory and visual response, 84
voice processing
fMRI evidence, 100
infants', emotional speech, 99
negative shifts, infant and mother, 99
positive slow waves, 99–100
Expectancy violation, 33

F

Face. *See* Face and voice integration, emotion perception
Face and voice integration, emotion perception
facial expressions, 183
multisensory integration and stimulus redundancy, 183
multisensory perception
emotion, 190–194
emotion *vs.* emotion stimulus redundancy (*see* Multisensory)
object-based (*see* Object-based multisensory perception)
multisensory stimulation, 182
prosodic features, 183–184
psychoacoustical parameters, 183
social/cognitive sciences, 182
Face processing, NHPs, 103, 107, 150, 231
behavioral response
habituation–dishabituation paradigm, 47
interactions, social cues, 46
match-to-sample paradigm, 46
occurrence relationship, 46
primary configural information, 46
neuronal response, temporal lobe
anterior IT cortical neurons, 47
awake behaving monkeys, 47
description, 47
face view and gaze direction, 48
regions, temporal cortex, 48
PFC
definition, 48
delay activity, 48–49
face response, 49, 50
face–vocalization cells location, 50, 61
memory and decision-making tasks, fMRI, 49
neural recordings, human brain, 49

neurons clusters, 50
OFC and VLPFC, rhesus monkey, 49
"patches", 51
visual response, 48
Face-voice gender, AVI
ANOVAs, 146
articulatory movements, 136
audiovisual condition
attention to face, 144
attention to voice, 143–144
uncontrolled attention, 142–143
audiovisual, design and procedure
attend to face, 140
attend to voice, 140
audio only, 140
no direction to modality, 140
video only, 140–141
crossmodal interactions, 137
dynamic *vs.* static, 141
end-point face and voice morphs, unimodal stimuli, 145
facial movement effect, 145
fundamental frequencies, 147
gender discriminators, 145
Matlab 2007b and PTB3 extensions, 139
neuroimaging technique, 135–136
objectives, 144
paralinguistic information, 135
participants, 146
person's biological characteristics, 135
real-life conditions, 136
state-of-the-art facial and vocal morphing techniques, 136
stimuli methods
audiovisual video production, 139
auditory morphing, 138–139
face morphing, 138
video recording and editing, 137–138
subjects method, 137
and unimodal information, 136–137
unimodal *vs.* audiovisual, 141–142
Face–voice integration
corresponding and noncorresponding identities
asynchrony, AVI, 125
dynamic audiovisual trial, VO timings, 123–124
ERPs, 126–127
experimental evidence, 123
face-sensitive N170 component, 128
FFA and TVA, 129
independent, speaker correspondence, 127, 128
mean RTs, 126, 127
person recognition and speech perception, AVI, 126
reaction time differences, 124–125
response categories, audiovisual asynchrony, 125, 126
scalp voltage maps, 128–129
synchronized and static face comparison, 125
speaker identification
audiovisual speech facility, 122
long-term repetition priming effect, 122
postperceptual processing stage, 121–122
voice recognition, 122
Facial expression, 10–13, 46–48, 50–51, 191–196, 254–255, 257–259
in adolescent twins, 107
amodal information, infants, 101
COMT and SLC6A4/5-HTTLPR genotypes, 106, 107
congruent/incongruent voice, 100
differential responsiveness, 103
ERPs, 97, 110
fMRI study, adults, 98
genetic variation, 105
infants' visual acuity, 96, 97
limbic and prefrontal brain regions, 106
locomotion/affective experience function, 98
Nc and Pb components, 107
newborns discrimination, 96
visual-paired comparison task, 99
FEF. *See* Frontal eye fields (FEF)
FFA. *See* Fusiform face area (FFA)
FLMP. *See* Fuzzy logical model of perception (FLMP)
fMRI. *See* Functional magnetic resonance imaging (fMRI)
Frontal eye fields (FEF), 210
Functional magnetic resonance imaging (fMRI)
amygdala activity, faces, 108
angry *vs.* facial expressions, 98
brain activity, 152
comparison, infant NIRS data, 111
limbic and prefrontal brain region activity, 106
multisensory interaction effects, 169
multisensory processing, 173
noninvasive EEG, 168
noninvasive human studies, 172
sensory responses, 102
source modeling purpose, 175
voice-face associations, 152, 153
voice-face integration, 155

Fusiform face area (FFA), 122, 123, 129, 150, 151
Fuzzy logical model of perception (FLMP), 188–189

G

Gender perception. *See* Face-voice gender, AVI
Gender processing, cross-modal integration
  androgynous faces, genders, 154
  audiovisual interactions, 154–155
  auditory and visual information interaction, 153–154
  brain sections, contrast [FV-(V + F)], 156
  dopaminergic channels, 157
  experimental paradigm, 155
  face-voice associations, 156, 157
  fMRI design, 155
  integrative activations, left superior parietal gyrus, 156–157
  left parietal gyrus activation, 157, 158
  PPI analysis, 157
  super-and sub-additive neurons activity, 155
  unimodal visual and auditory regions, 157

H

Hearing loss, children
  acoustic parameter, 302
  audiovisual processing, speech
    acoustic information, 307
    age limit, 312
    ambiguous/degrade, 307
    bilabial consonants, 306
    cochlear implants, 308, 312
    communication mode, 312
    development and segmental perception, 308
    factors, 313
    idiosyncratic responses, 312
    oral communication environments, 310
    sensory modality, 310
    signal, 307
    sound and immersion, environments, 309
    speech processing, 311
    unimodal stimuli, 311
    words and sentences, identification, 309
  bimodal processing, 305
  changes, visual/auditory, 306
  cochlear implants, 301, 302
  communication environment
    auditory system, 304
    auditory-verbal and auditory-oral communication, 304, 305
    newborn hearing screenings, 304
    vision *vs.* audition, 305
  damaged hair cells, 301–302
  electrode, 302
  emotion, audiovisual processing
    acoustic information, 314
    expressions, 317
    facial cues, 313
    incongruent, 317
    neutral sentence, 315
    sounds, 316
    speech perception, 314
    suprasegmental processing, 316
    tones, 315
    uniersal newborn screenings, 313
    unimodal and multimodal processing, 314
  frequency parameters, 303
  fundamental frequency, 301, 302
  intensity, 303
  speech sounds and emotion expressions, 301
  undamaged ear, 303
  vocabulary, language, and literacy development, 305
Human faces and vocalizations, brains
  concurrent, neural signatures
    audiovisual stimulus, 164
    averaged ERPs, 165
    description, 164
    harmonic-to-noise ratios, 164
    low-level auditory stimulus, 165
    multisensory incongruity response, 167
    N140, congruous audiovisual stimulus pairings, 165, 166
    N170, unisensory paradigm, 168
    scalp topographic voltage maps, 165, 166
    types, grayscale visual stimuli, 164
    vertex N140s and P400s, 165, 166
    visual stimulus category, 167
  dynamic, information integration
    audio and visual stimulus, 169
    auditory and visual, N140, 169
    cognitive processing, 170
    EEG rhythms, 171
    electrophysiological response, 171
    ERP components, 169
    ERP findings, face-voice audiovisual study, 169, 170
    frame-by-frame basis, 168
    multisensory experiment data, 169

multisensory incongruity P400, 170
N170 amplitude difference, sensory stimulation, 168
non-verbal vocalizations, 168
post-stimulus period, 171
temporal dynamics and neuroanatomical loci, 168

I

Identity, corresponding and noncorresponding. *See* Face–voice integration
Identity processing, cross-modal integration
brain activity, fMRI design, 152, 153
brain sections, contrast [VF-(V + F)], 152, 154
cognitive models, face identification, 153
comparisons, bimodal and unimodal condition, 151–152
ERPs, 152
functional connectivity, 152
human social interactions, 151
neural network, supramodal convergence regions, 151
unimodal regions, sensory stimuli, 151
voice recognition, 151
Infancy. *See also* Intersensory perception, infants
amodal and modality-specific information processing, 101
COMT and 5-HTTLPR, 109, 110
emotion perception development, 96
facial expressions, 107
gene effect development, 106
genetic polymorphisms, 107
hypersensitivity, 108–109
Interactive activation and competition (IAC) model, 120
Intersensory perception, infants
amodal and modality-specific information, 74–76
amodal properties, 80
arbitrary face–voice relations (*see* Arbitrary face-voice relations)description, 71–73
faces and voices, 79–80
intermodal perceptual binding, 73
intersensory redundancy hypothesis
affective expressions, 77–79
rhythm and tempo, 76–77
tempo and rhythm, 76
nature, information, 74
neurophysiological foundations
EEG and fMRI evidence, 83
EEG/ERP response, 84
ERP results, infants and adults, 84, 85
experiment 2 late processing results, 87, 88
experiment 1 results, 86, 87
face–voice synchrony, 85–86
multimodal and unimodal stimulation, 83
multimodal stimulation, 84
onset/offset events, 87
temporal synchrony, 85, 86
Intersensory redundancy
description, 74
infants' discrimination
affective expressions, 77–79
rhythm and tempo, 76–77
tempo and rhythm, 76
Intraparietal sulcus (IPS), 210
IPS. *See* Intraparietal sulcus (IPS)

L

Late positive component (LPC), 214
Local field potential (LFP) response
audiovisual vocalizations, 22
concurrent recordings, 18
facial and vocal signals, monkeys, 16
grunt vocalizations, 17
lateral belt auditory cortex, 17
unimodal auditory condition, 16
LPC. *See* Late positive component (LPC)

M

Magnetoencephaloraphy (MEG) study, 173–175
Mate choice, monkeys
description, 7
Geospiza fortis, morphological variants, 7
integration ability, 8
song properties, females, 8
territorial males, 7–8
visual and acoustic signals, 8
MMN, 344–345
Mood disorders, P300
ERB, unfocused
cognitive control mechanisms, 343
ERN and anxious groups, 343
error positivity (Pe) and error negativity (Ne), 343
P200 amplitude, 342–343
psychiatry
negative correlation, 332
novel "distractor" stimulus and DSM, 333
P3a latency, 332–333

Mood disorders, P300 (*cont.*)
P3b amplitude decrement and state marker, 333
schizophrenia and bipolar, 332
unipolar and bipolar patients, 332
Morphing
auditory, 138–139
face, 138
Multisensory, 59–62, 167–175, 226–233, 260, 264, 348
emotion *vs.* emotion stimulus redundancy
affective pairings and crossmodal bias, 195
ANOVA, 197
attentional bottleneck, 200
audiovisual integration, emotion/visual redundancy, 195
audio-visual pairings, 200
bimodal face-voice parings, 196
early/preattentive stage, 195
face-voice combinations, 200
mean RTs, 198, 199
methods, 196–197
post hoc comparisons, 198
putative limitation, 200
sad and angry faces, 198
statistical analysis, 197
object-based perception (*see* Object-based multisensory perception)
perception of emotion
attention in, 193–194
behavioral evidence, 191–192
functions, 190–191
Multisensory recognition, vertebrates
acoustic and visual signals, 4–5
aggression and territorial defense
accuracy, 5
animal's recognition ability, 5
dart-poison frogs (*Epipedobates femoralis*), 6, 7
pied currawong (*Strepera graculina*), 5, 6
resources, 5
temporal integration experiment, 7
territorial Bornean ranid frog (*Staurois guttatus*), 5
vocalizations, geckos (*Ptenopus garrulus garrulus*), 5–6
audiovisual processing, NHPs (*see* Nonhuman primates (NHPs))
communication modality, 4
face-voice signals (*see* Neocortical processing, face–voice signals in monkeys)
mate choice
description, 7
Geospiza fortis, morphological variants, 7
integration ability, 8
song properties, females, 8
territorial males, 7–8
visual and acoustic signals, 8
predator and prey interactions
acoustic-visual integration, 9
body posture change, 9
description, 8
robot squirrel alarm behavior, real squirrels, 6, 9

N

NEAR. *See* Neuropsychological educational approach to remediation (NEAR)
Near-infrared spectroscopy (NIRS), 111
Neocortical processing, face–voice signals in monkeys
description, 16
eye movements and auditory cortex
average fixation, eye *vs.* mouth region, 20, 21
complex sequence, sensory events, 21
LFP-derived current-source density activity, 20
monkeys saccade, mouth region, 20, 21
proprioceptive signal, 21
face-sensitive input, auditory cortex
auditory cortex and STS, 18
connectivity and response properties, 20
faces and biological motion, 18
population phase concentration, 18, 19
reciprocal connections, 18
salient audiovisual events, 20
spike-field cross-spectrogram illustration, 18, 19
time-frequency plots illustration, 18, 19
functions, association areas, 16
multisensory behavior, neurophysiology
example, audiovisual integrative response, 22
intertrial phase coherence, 22–23
neural activity, 22
phase-resetting hypothesis, 23
spiking activity and LFPs, 22
multisensory LFP response, 17
multisensory vocal perception, 16
single neuron examples, multisensory integration, 16, 17
species-typical vocalizations, 16

specificity, face-voice integrative response, 17
unimodal and bimodal versions, 16
visual sensations, 16
Neurocognitive model, 216–219
Neuroimaging, 51, 62, 108, 136, 230–237, 327
audio–visual stimulation, 281
brain activations, 280, 281
cortico–limbic connections, 280
fMRI, 280
multimodal regions, 282
unimodal regions, 281, 282
Neurophysiological correlation
audiovisual stimuli, animals, 163–164
datas, individual, 163
human faces and vocalizations, brains
concurrent, neural signatures, 164–168
dynamic, information integration, 168–171
non-verbal behaviors, 163
relevance, current findings
audiovisual response, ERP data, 172
beta frequency range, 174
cortical function and putative hierarchical processing, 172
EEG activity, 174–175
epilepsy surgery patients, 173
ERP findings, face-voice audiovisual integration study, 170, 171
evoked and induced activity, 174
functional connectivity, 173
inverse effectiveness principle, 172
magnetoencephalography (MEG) study, 173
multimodal concurrent EEG and fMRI, 175
multisensory stimulation, 170–171
neurophysiological response, 173
noninvasive human studies, fMRI, 172
oscillatory response, 174
transcranial magnetic stimulation, 174
Neuropsychiatry
description, 325–326
DSM, 325
ERPs (*see* Event-related potentials (ERPs))
methodological divergences
binaural complex tones design, 339
brain activity, 338
correlations, 339
DSM-III-R classification, 339
existence, subtypes, 338
experimental design and acquisition, 340
medication, 339–340
severity, symptoms, 339
NEAR, 327
normality and pathology, 325
P300, 328–337
therapeutic strategies, 326–327
trait, state, and vulnerability marker, 326
Neuropsychological educational approach to remediation (NEAR), 327
N170, face-sensitive component, 107, 128, 150, 166–169, 171, 263, 278, 343, 345
NHPs. *See* Nonhuman primates (NHPs)
NIRS. *See* Near-infrared spectroscopy (NIRS)
Nonhuman primates (NHPs). *See also* Face processing, NHPs
audiovisual communication, 9–10
description, 9
face and voice integration
behavioral datas, 14
free-response paradigm task structure, 14, 15
mean RTs, 14, 15
vocal components, 14
waveform, coo vocalizations, 14, 15
facial to vocal expressions
exemplars, Rhesus monkeys, 10–13
human speech, 10
unique lip configurations and mandibular positions, 10
human *vs.* nonhuman multisensory ability, 9
Nonverbal emotional information, 226, 227, 229, 231, 245–246

O

Object-based multisensory perception
audiovisual pairings, 184–185
behavioral effects and cognitive models
bimodal/audiovisual stimulus, 187
cumulative probability, RTs, 187–188
FLMP, 188
neuroimaging/neurophysiology, 188
RSE and RTE, 187
speech perception, 189
crossmodal paradigm, 189
description, 184
domains, human cognition, 185
natural and arbitrary pairs, 185
sets of constraints, 186
space perception, 186
temporal and spatial coincidence, 186
ventriloquist effect/illusion, 185
Observer-dependent effects, 209
OFC. *See* Orbitofrontal cortex (OFC)

Orbitofrontal cortex (OFC), 219
face-responsive neurons, 49
monkey and human face categories, 50–51
neurophysiological recordings, 51
recordings, 50

P

P300
amplitude and latency, 329
cognitive perturbations, 329
context updating and context closure, 329
ERB, unfocused
chronic alcoholism, 343–344
mood disorders, 342–343
schizophrenia, 344–346
neuropsychology
Alzheimer patients and cholinesterase inhibitor treatment, 331
cerebral structures, 330
dementia's diagnosis, 330–331
frontal lobe and hippocampus, 330
novel stimuli and epileptic foci, 330
prolongations, latencies, 331
sensitive tool and MCI group, 331
P3a, 329–330
P3b, 330
P3 generation, 329
psychiatry
chronic alcoholism, 335–337
mood disorders, 332–333
schizophrenia, 333–335
slow and low frequency wave, 329
speculative electrophysiological markers
deficits, 337–338
EEG, 338
physiological index, dementia, 338
putative pathophysiological marker, 338
Person identity node (PIN)
definition, 120
postperceptual processing, 123
speaker recognition, 120–121
Person recognition, 147, 151, 152, 341
AVI, speech perception, 126
face and voice, 119
unimodal stimuli, 120
PIN. *See* Person identity node (PIN)
Posterior STS (pSTS), 232–233, 242
PPI. *See* Psychophysiological interaction (PPI)
Preferential looking, 10, 11, 13
auditory-visual representation, monkeys, 31
cross-modal representation, 31
studies, preverbal infants, 31
visual stimulus, 31
vocal individuality, mangabeys, 33
Prefrontal cortex (PFC)
face and voice integration
dynamic movie clips, monkeys, 59
face–vocalization cells location, 59, 61
multisensory interactions, 59, 60
polymodal regions, STS, 59, 61
social communication, 61
studies, NHPs, 60
VLPFC, 59–60
primate ventral PFC (*see* Primate ventral PFC)
Primate ventral PFC
description, 45
face and voice integration, (*see* Prefrontal cortex (PFC))
face processing, NHPs
behavioral response, 46–47
neuronal response, temporal lobe, 47–48
PFC, 48–51
vocalization processing
auditory projections, 52–54
auditory responsive domain, VLPFC, 54–55
description, 51
PFC and auditory, NHPs, 51–52
representation, VLPFC, 55–59
Prioritization
emotional attention, 208
neural circuits, 215
prior-entry paradigm, 214
Prosody, 226–227
pSTS. *See* Posterior STS (pSTS)
Psychiatric disease, 105
description, 245
employing pictures, 246
fMRI, 246
nonverbal signals, 246
schizophrenia, 246
Psychiatry
cerebral, cognitive and emotional impairments, 288
crossmodal studies, 288, 289
emotional auditory stimulation, 289
exploring integration processes, 289
stages, alcohol dependence, 289
unimodal emotional stimuli, 289–290
vision–audition imbalance, 289
Psychophysiological interaction (PPI), 157, 232, 233, 243, 245

R

Redundancy, 74–83, 85, 88, 89, 195–201
Redundant signal effect (RSE), 187
Redundant target effect (RTE), 187
Response latencies (RT)
  congruent bimodal stimulus pairs, 191, 192
  distribution, 188
  emotional facial expression, 198
RSE. *See* Redundant signal effect (RSE)
RT. *See* Response latencies (RT)
RTE. *See* Redundant target effect (RTE)

S

Schizophrenia and autism
  amygdala, 265
  ASD, 262, 263
  audiovisual studies, 263, 264
  emotional signals, 262
  facial and bodily expressions, 264
  nonschizophrenic psychotic disorder, 263
  visual and audio channel, 263
Schizophrenia, P300
  ERB, unfocused
    amplitude reduction, N170, 345
    MMN, 345–346
    N100, 344
  psychiatry
    auditory diminution and ERPs, 334
    dysfunctional visuo-spatial process, 335
    frontotemporal atrophy, 334
    and neurocognitive impairments, 334
    P3a amplitude and ultra-high risk, 334
    phenotypic markers, 333
    unimodal tasks and neuroimaging, 335
    visual and visuo-spatial impairments, 335
Serotonintransporter (SLC6A4/5-HTTLPR)
  distress recovery, 109
  emotional sensitivity, 109
  ERP analysis, 109
  genetic variation, 110
  genotype, emotional stimuli processing, 106
  happy facial expressions, 107
  occipital electrodes, brain processing, 107
SOA. *See* Stimulus onset asynchrony (SOA)
Sound-source identification
  auditory-visual individual recognition, 37
  behavioral studies, nonhuman animals, 32, 36
  cross-modal identity representation, 37
  preferential looking paradigm, rhesus monkeys, 37
  sex differences, chimpanzees, 37
  visual memory, 37
Speaker identification. *See* Audiovisual integration (AVI)
Speech perception, 300, 303, 305, 309, 314
Stimulus onset asynchrony (SOA), 174, 260, 261
STS. *See* Superior temporal sulcus (STS)
Superior temporal gyrus (STG), 218
Superior temporal sulcus (STS), 227, 228, 233, 241, 244, 245

T

Temporal synchronization
  amodal properties, 83
  audio–visual integration, 86
  auditory and visual stimulation, 78
  human infants neurophysiological response, 84
  infants' sensitivity, 85
  intersensory redundancy hypothesis, 86
  stimulus/subject gender, 82
Temporal voice area (TVA), 122–123, 129, 130
Test of variables of attention (TOVA) abnormalities, 341
TVA. *See* Temporal voice area (TVA)

U

Unimodal *vs.* audiovisual integration
  face categorization
    female, green, 141, 142
    male, beige, 141, 142
  voice categorization
    comparison, 142
    female, green, 142
    male, beige, 142

V

Ventrolateral prefrontal cortex (VLPFC)
  auditory responsive domain, 54–55
  face and vocal information, 61
  face-responsive neurons, 49
  face-responsive "patches", 51
  human neuroimaging studies, 62
  representation, VLPFC, 55–59
  ventral auditory stream identification, 53

Visual cortex, 152, 172, 174, 212, 215, 218, 229, 257
VLPFC. *See* Ventrolateral prefrontal cortex (VLPFC)
Vocal expression, 10–13, 20, 241, 253–257
  emotional expressions, 110
  ERP studies and experimental investigations, 104
  infants' ability, 104
  positive and negative, infants, 103
Vocalization processing, 23
  auditory projections
    analysis, anatomical connections, 53, 54
    lateral belt auditory areas, 53
    multisensory area TPO and TAa, 53
    rostrocaudal topography, 52
    temporoprefrontal connections, 52
  auditory responsive domain, VLPFC, 54–55
  description, 51
  PFC and auditory, NHPs
    auditory discrimination tasks, 51
    DLPFC neurons, 51
    neurophysiological recordings, 51
  representation, VLPFC, 55–59
Vocal type representations
  auditory-visual events, 38
  cross-modal representation, 39
  human phonemes, phonological phenomena, 39
  movie clips, 38
  in nonhuman animals, 38–39
  preferential looking paradigm, 38
  synchronization, sounds and movies, 38
  temporal synchrony and phonetic correspondence, 38
Voice. *See* Face and voice integration, emotion perception

W

Within-modality effects
  emotional modulation, 212
  neural mechanisms, 215–216
  P1 mirror, 218

Printed by Printforce, the Netherlands